Extrapolation and Rational Approximation

Claude Brezinski • Michela Redivo-Zaglia

Extrapolation and Rational Approximation

The Works of the Main Contributors

 Springer

Claude Brezinski
Laboratoire Paul Painlevé
University of Lille
Villeneuve d'Ascq, France

Michela Redivo-Zaglia
Department of Mathematics
"Tullio Levi-Civita"
University of Padua
Padua, Italy

ISBN 978-3-030-58420-7   ISBN 978-3-030-58418-4   (eBook)
https://doi.org/10.1007/978-3-030-58418-4

Mathematics Subject Classification: 65B05, 40A15, 41A20, 65F10, 65H10

This Springer imprint is published by the registered company Springer Nature Switzerland AG
The registered company address is: Gewerbestrasse 11, 6330 Cham, Switzerland

Contents

not belong to its kernel, the transformation produces a limit that depends on the first term used in the interpolation process. Thus, it is only an approximation of the limit of the initial sequence. Therefore, increasing the index of the first term used in the interpolation process leads to a different extrapolated limit, and it gives rise, step by step, to a transformed sequence, each term of which is only an approximation of the limit of the original sequence. When a sequence is not convergent but is related in a certain way to a number (or a vector, a matrix, a tensor,...) called its *antilimit* (a word introduced by Shanks in [567]; see Sect. 3.2.4), the same process can be used to transform it into a convergent sequence. This is, in particular, the case for divergent series. More detailed explanations about extrapolation methods are given in Sect. 2.2.

Approximations of the limit, or antilimit, of a formal power series can be obtained by methods different from, but related, to the transformation of the sequence of its partial sums. Rational approximations of functions defined by their formal series expansion can be obtained by *Padé approximants* and *Padé-type approximants*, which are rational functions whose power series expansions in ascending powers of the variable agree with that of the given power series as far as possible. Series can also be transformed into *continued fractions* of various types by different methods. The successive *convergents* of a continued fraction are also rational functions, and some of them can coincide with Padé approximants. These two topics are strongly related to the theory of *formal orthogonal polynomials* (FOPs), a generalization of the usual ones. Other types of series such as Fourier series and series in Chebyshev polynomials can also be approximated by rational functions generalizing Padé approximants. Best rational approximation of functions is not considered in this book.

Continued fractions have a very long history, almost as long as the history of mathematics itself. For example, they are implied in Euclid's method for the greatest common divisor and in the proofs of the transcendental character of the numbers π and e. They were also instrumental in the development of orthogonal polynomials and in the spectral theory of operators. Padé approximants are more recent, but they go back several centuries [102]. They are much used in number theory and in the computation of the special functions encountered in mathematical physics.

The convergence and acceleration properties of sequence transformations have to be theoretically studied, and the degree of approximation and the convergence of these rational approximations also. Since they need to be implemented on a computer, it is necessary to design effective (usually recursive) algorithms for that purpose, and to code them in a programming language. Their numerical stability has to be studied, and the propagation of rounding errors as well. Finally, useful applications have to be found.

As can be seen in the figure below,[1] sequence transformations, Padé approximants, and continued fractions have many interconnections with other important matters in applied mathematics and numerical analysis, some of which are developed herein.

---

[1] Adapted from [108, p. 63].

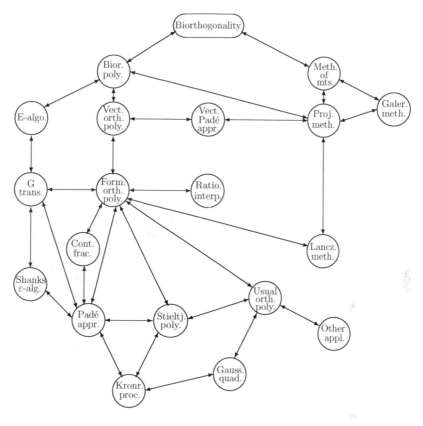

The interconnections between topics.

This book paints a fresco of the field of extrapolation and rational approximation over the last several centuries to the present through the works of their contributors. Simultaneously, it can serve as an introduction to the topics covered, which are extrapolation methods, Padé approximation, orthogonal polynomials, continued fractions, Lanczos-type methods, etc., and it describes the many links between them. In order that the reader may fully appreciate the works of the contributors, the book contains a quite precise description of the mathematical landscape on these subjects from the nineteenth century to the first part of the twentieth. Then, after an analysis of the works produced after that period (in particular those of Richardson, Aitken, Shanks, Wynn, and others), we give a review of the most recent developments of the domains concerned. Research is carried out by both women and men. This is why we also provide testimonies by several researchers who worked (and are still working) on these themes. They show the human side of research, and they furnish complementary views, and additional explanations and results. For the same reason, we sometimes indicate some personal facts about the people encountered. Biographies of various actors mentioned during our journey end the book.

After a mathematical introduction to the topics treated in this book, we begin our guided tour with Lewis Fry Richardson (1881–1953),[2] an original character, who, as early as 1910, was the first to openly speak of extrapolation methods. He also worked on the numerical solution of ordinary and partial differential equations; he proposed an iterative method for solving systems of linear equations; he was a pioneer in the study of fractals, a meteorologist who attempted to formulate a mathematical model for weather prediction and its mathematical solution; he was interested in turbulence theory; and he studied statistics and collected a vast amount of data on the causes and the dynamics of wars. He was a free thinker, an independent spirit, an unorthodox scientist, never in the main stream, but his works are still studied, used, and pursued.

Alexander Craig Aitken (1895–1967)[3] was one of New Zealand's most eminent mathematicians. He was a specialist in linear algebra, and he introduced the concept of generalized least squares used in the linear regression model. He proposed a quite efficient extrapolation method known as the Aitken Δ^2, process which was the starting point of almost all modern nonlinear sequence transformations for accelerating the convergence of sequences.

Daniel Shanks (1917–1996)[4] was an American physicist and mathematician. He was interested in numerical analysis, and later became a recognized expert in number theory. In 1949, he invented the sequence transformation that bears his name [567]. Due to its efficiency, its potentialities, and its numerous connections with other topics (continued fractions, Padé approximants, orthogonal polynomials,...), this transformation had a great influence on the development of extrapolation methods, and of iterative procedures for solving systems of nonlinear equations.

Among the main contributors, we must not forget Henri Eugène Padé (1863–1953).[5] Although he was not the discoverer of the rational approximants bearing his name, his contributions played an important role in their dissemination. He also promoted the theory of continued fractions.

Peter Wynn (1931–2017)[6] was a British mathematician, numerical analyst, and computer scientist who worked in several European and American countries. He had, and continues to have, a great influence on the development of nonlinear extrapolation methods, Padé approximation, continued fractions, formal orthogonal polynomials, and related topics. We learned of Wynn's death during the final stage of the preparation of a special issue of the journal *Numerical Algorithms* devoted to the Shanks transformation and its extensions, and it was decided to dedicate it to him [447].

At the beginning of our work, we (the authors of this book) had in mind to write a review paper on Wynn's works. Since they consist of about 100 publications, our

---

[2] See Sect. 8.7 for his biography.
[3] See Sect. 8.1 for his biography.
[4] See Sect. 8.12 for his biography.
[5] See Sect. 8.6 for his biography.
[6] See Chap. 4 for his biography.

project grew rapidly. Then, obviously, an introduction to the topics treated by Wynn had to be added for readers who were not too familiar with them. And the project expanded even further. But we also decided to extend our project and to include an analysis of the works of the other researchers who worked in the numerous areas connected with these topics.

It was also interesting to present the state of the art before and around 1950 at time when numerical analysis was rapidly expanding, Richardson, Aitken, and Shanks were designing their transformations; and Wynn was beginning his career; still another extension of the project, which then became too big for a paper and evolved into this book. Then, after the analysis of the works of these pioneers, it was significant to discuss the further developments of the topics considered. Research is a human adventure carried out by individuals, each of them bringing his own cornerstone, often with a different point of view, and his own results. This is why we asked several researchers involved in the domains for their personal testimonies in order to enlarge the scope of the book and to emphasize their personal contributions. For the same reason, we also added short biographies of some of the actors encountered during our trip. Thus, the genesis of this book explains to readers why an important part of it is devoted to Wynn's works (some of them mentioned here for the first time).

During the last stage of preparation of this book, we received a message from Francis Alexander (Sandy) Norman (b. 1952), professor at the University of Texas at San Antonio, informing us that Wynn had left several boxes of documents at the house of his friends Maria Antonietta and Manuel Philip Berriozábal (b. 1931). He proposed to us that he digs into those documents and that he would send them to us. Of course, we gladly accepted, and we plan to analyze carefully this unpublished material in a future publication [122].

When writing this book, we also realized that in fact, we had been much more influenced than we thought in our own research by the works of Wynn, even if we had not yet read all his papers and reports. Like Wynn, we worked on convergence acceleration, formal orthogonal polynomials, Padé approximation, and continued fractions. However, we stayed more on the purely numerical analysis side of these domains, and we never ventured too much into the complex analysis or the purely mathematical algebraic aspects of these topics. Although Wynn often provided some numerical examples in his papers and reports, we gave many more applications of extrapolation methods and Padé approximants to the numerical solution of various problems in numerical analysis and applied mathematics: convergence acceleration, treatment of the Gibbs phenomenon, solution of integral equations, numerical linear algebra, estimation of the errors in the solution of linear systems, Tikhonov regularization, the PageRank and the multi PageRank problems, tensor eigenvalues, treatment of breakdown and near-breakdown in Lanczos-type methods, model reduction, solution of nonlinear equations, computation of matrix functions, and, like Wynn himself, algorithms and production of the related numerical software. Thus, thanks to the large community of themes treated by Wynn and us, we were forced to quote (too) many of our own publications. However, we tried to limit them as much as we could. Our complete lists can be found on our respective homepages.

Let us now describe the contents of this book. Chapter 2 is an introduction to the main topics concerned: extrapolation methods, Padé approximation, formal orthogonal polynomials, and continued fractions (for a bibliography until 1990, see [103]). In Chap. 3, we give an account of the mathematical knowledge in these fields to the mid-twentieth century, that is, until the time Richardson, Aitken, Shanks, and Wynn produced their extrapolation methods. The life of Peter Wynn is reported in Chap. 4, and Chap. 5 is devoted to the analyses of his works, together with some comments and additional mathematical details when necessary. Chapter 5 also contains a short account of the discovery of Wynn's scientific legacy, and of its contents that will later be separately analyzed. Detailed commentaries on the works described in the previous chapters, and descriptions of their further developments are presented in Chap. 6. As said above, several researchers whose works are related to those discussed herein sent us their personal testimonies. They are gathered in Chap. 7. In Chap. 8, we give a brief biography of the main actors met during our journey. Our personal feelings are given as a conclusion in the last chapter.

The first time an author is mentioned, we give, when known, the dates of her/his birth and death. They are also given in the index. Quotations of parts of original publications are in small roman characters, and our inside comments are in italics surrounded by square brackets. Quotations by authors which are inside our text are in italics, and, if necessary, comments are in roman. In the bibliographies (those of the personal testimonies of Chap. 7 and those at the end of the book), we indicated, in italics, the page(s) where each reference is cited. In the index, keywords are in roman characters, while authors' names are in italics.

Chapter 2
Mathematical Background

In this chapter, we give an introductory account of the main topics treated in their publications by the authors we will encounter in this book. Explanations of the other topics they worked on will be given later, when necessary. More details and historical comments can be easily found in the available literature, in particular, for extrapolation methods in [90, 92, 94, 123, 196, 587, 692, 699], for Padé approximation in [26, 31, 94, 271, 284], for continued fractions in [359, 360, 418, 493, 683], for formal orthogonal polynomials in [94, 206, 577]. Since these topics are inter–related, some references concern several of them. An interesting discussion on convergence acceleration by Dirk Laurie (1946–2019) can be found in [74, pp. 227–261]. The corresponding Matlab and Octave programs of many of the acceleration methods can be obtained from [397]. Fortran subroutines for several of these methods are described in [123], and can be downloaded from [127]. Algol subroutines for certain convergence acceleration methods are given in [260]. Matlab procedures can be obtained from [133] (see [132] for the corresponding explanations about this package). For the works and a biography of Padé, see [483] (in French). A history of continued fractions and Padé approximants is provided in [102].

2.1 Mathematical Material

We begin by introducing some mathematical material that is useful for our purpose, namely, determinantal identities, Schur's complement and determinantal formula, and designants. No prerequisite is needed.

2.1.1 Determinantal Identities

Determinantal identities due James Joseph Sylvester (1814–1897) [630] and Franz Ferdinand Schweins (1780–1856) [561, pp. 317–431] are instrumental in the deriva-

© Springer Nature Switzerland AG 2020
C. Brezinski, M. Redivo-Zaglia, *Extrapolation and Rational Approximation*,
https://doi.org/10.1007/978-3-030-58418-4_2

tion of several recursive algorithms for the implementation of the sequence transformations presented below. Let us give them in their full generality. In particular, they were used by Wynn [713] for obtaining his scalar ε-algorithm. On these identities, consult [459, 509], and [460, pp. 159–175] for an analysis of the work of Schweins.

Let $b_1, \ldots, b_k$ be elements of a vector space E, and let the a_{ij} be scalars. Then Sylvester's identity is

$$
\begin{vmatrix}
b_1 & b_2 & \cdots & b_{k-1} & b_k \\
a_{11} & a_{12} & \cdots & a_{1,k-1} & a_{1k} \\
\vdots & \vdots & & \vdots & \vdots \\
a_{k-2,1} & a_{k-2,2} & \cdots & a_{k-2,k-1} & a_{k-2,k} \\
a_{k-1,1} & a_{k-1,2} & \cdots & a_{k-1,k-1} & a_{k-1,k}
\end{vmatrix}
\begin{vmatrix}
a_{12} & \cdots & a_{1,k-1} \\
\vdots & & \vdots \\
a_{k-2,2} & \cdots & a_{k-2,k-1}
\end{vmatrix}
$$

$$
=
\begin{vmatrix}
b_1 & b_2 & \cdots & b_{k-1} \\
a_{11} & a_{12} & \cdots & a_{1,k-1} \\
\vdots & \vdots & & \vdots \\
a_{k-2,1} & a_{k-2,2} & \cdots & a_{k-2,k-1}
\end{vmatrix}
\begin{vmatrix}
a_{12} & \cdots & a_{1,k-1} & a_{1k} \\
\vdots & & \vdots & \vdots \\
a_{k-2,2} & \cdots & a_{k-2,k-1} & a_{k-2,k} \\
a_{k-1,2} & \cdots & a_{k-1,k-1} & a_{k-1,k}
\end{vmatrix}
\tag{2.1}
$$

$$
-
\begin{vmatrix}
b_2 & \cdots & b_{k-1} & b_k \\
a_{12} & \cdots & a_{1,k-1} & a_{1k} \\
\vdots & & \vdots & \vdots \\
a_{k-2,2} & \cdots & a_{k-2,k-1} & a_{k-2,k}
\end{vmatrix}
\begin{vmatrix}
a_{11} & a_{12} & \cdots & a_{1,k-1} \\
\vdots & \vdots & & \vdots \\
a_{k-2,1} & a_{k-2,2} & \cdots & a_{k-2,k-1} \\
a_{k-1,1} & a_{k-1,2} & \cdots & a_{k-1,k-1}
\end{vmatrix} .
$$

If we denote the largest determinant by D, by C its center, that is, the determinant of order $k-2$ obtained by deleting the first and the last rows and columns of D, by NW the determinant of order $k-1$ in the upper left corner of D, by NE the one in the upper right corner, by SW the determinant in the lower left corner, and by SE the determinant in the lower right corner of D, Sylvester's identity is

$$
D \cdot C = NW \cdot SE - NE \cdot SW.
$$

Let $c_1, \ldots, c_k$ be scalars. Schweins's identity is

$$
\begin{vmatrix}
b_1 & b_2 & \cdots & b_{k-1} & b_k \\
a_{11} & a_{12} & \cdots & a_{1,k-1} & a_{1k} \\
\vdots & \vdots & & \vdots & \vdots \\
a_{k-2,1} & a_{k-2,2} & \cdots & a_{k-2,k-1} & a_{k-2,k} \\
a_{k-1,1} & a_{k-1,2} & \cdots & a_{k-1,k-1} & a_{k-1,k}
\end{vmatrix}
\begin{vmatrix}
c_1 & c_2 & \cdots & c_{k-1} \\
a_{11} & a_{12} & \cdots & a_{1,k-1} \\
\vdots & \vdots & & \vdots \\
a_{k-2,1} & a_{k-2,2} & \cdots & a_{k-2,k-1}
\end{vmatrix}
$$

$$-\begin{vmatrix} b_1 & b_2 & \cdots & b_{k-1} \\ a_{11} & a_{12} & \cdots & a_{1,k-1} \\ \vdots & \vdots & & \vdots \\ a_{k-2,1} & a_{k-2,2} & \cdots & a_{k-2,k-1} \end{vmatrix} \begin{vmatrix} c_1 & c_2 & \cdots & c_{k-1} & c_k \\ a_{11} & a_{12} & \cdots & a_{1,k-1} & a_{1k} \\ \vdots & \vdots & & \vdots & \vdots \\ a_{k-2,1} & a_{k-2,2} & \cdots & a_{k-2,k-1} & a_{k-2,k} \\ a_{k-1,1} & a_{k-1,2} & \cdots & a_{k-1,k-1} & a_{k-1,k} \end{vmatrix}$$

$$=\begin{vmatrix} b_1 & b_2 & \cdots & b_{k-1} & b_k \\ c_1 & c_2 & \cdots & c_{k-1} & c_k \\ a_{11} & a_{12} & \cdots & a_{1,k-1} & a_{1k} \\ \vdots & \vdots & & \vdots & \vdots \\ a_{k-2,1} & a_{k-2,2} & \cdots & a_{k-2,k-1} & a_{k-2,k} \end{vmatrix} \begin{vmatrix} a_{11} & a_{12} & \cdots & a_{1,k-1} \\ \vdots & \vdots & & \vdots \\ a_{k-2,1} & a_{k-2,2} & \cdots & a_{k-2,k-1} \\ a_{k-1,1} & a_{k-1,2} & \cdots & a_{k-1,k-1} \end{vmatrix}.$$

If $c_1 = 1$ and $c_2 = \cdots = c_k = 0$, Schweins's identity reduces to Sylvester's.

Sylvester's and Schweins's identities also hold if $b_1, \ldots, b_k$ are scalars and $c_1, \ldots, c_k$ elements of a vector space E, and vice versa. In these identities, a determinant with a row of vectors of E designates the vector of E obtained by expanding it with respect to this row of vectors using the classical rules for expanding a determinant. For example, the determinant

$$\begin{vmatrix} b_1 & b_2 \\ a_{11} & a_{12} \end{vmatrix},$$

where b_1 and b_2 are vectors of E and a_{11} and a_{12} scalars, is the vector $a_{12}b_1 - a_{11}b_2$. For higher dimensions, the procedure for expanding such a determinant is the same. Obviously, such a definition can be extended to the case in which the b_i are elements of a general vector space. In the particular case $E = \mathbb{C}^n$, such a determinant represents a vector of $\mathbb{C}^n$, and each of its components is the usual determinant obtained by considering only the corresponding component of each vector in the vector row.

2.1.2 Schur's Complement and Identity

Other important notions that will be used below are that of *Schur's complement* and *Schur's determinantal identity*; see [706]. Let us consider the partitioned matrix

$$M = \begin{pmatrix} A & B \\ C & D \end{pmatrix},$$

where the submatrix A is assumed to be square and nonsingular. The *Schur's complement* of A in M, denoted by (M/A), is the matrix

$$(M/A) = D - CA^{-1}B.$$

When M is square, the notion of Schur's complement is connected to block Gaussian elimination

$$M = \begin{pmatrix} A & B \\ C & D \end{pmatrix} = \begin{pmatrix} I & 0 \\ CA^{-1} & I \end{pmatrix} \begin{pmatrix} A & B \\ 0 & (M/A) \end{pmatrix}.$$

Moreover, we have the following *Schur's determinantal identity*,

$$\det M = \det A \cdot \det(M/A),$$

an identity first proved by Issai Schur (1875–1941) [560]. Partitioning the matrix M such that the upper left-hand corner is nonsingular is a matter of convention. Thus, we can similarly define the following Schur's complements, where the matrix that is inverted is assumed to be square and nonsingular

$$(M/B) = C - DB^{-1}A,$$
$$(M/C) = B - AC^{-1}D,$$
$$(M/D) = A - BD^{-1}C.$$

We have

$$M = \begin{pmatrix} I & 0 \\ DB^{-1} & I \end{pmatrix} \begin{pmatrix} A & B \\ (M/B) & 0 \end{pmatrix}, \qquad \det M = -\det B \cdot \det(M/B),$$

$$M = \begin{pmatrix} I & AC^{-1} \\ 0 & I \end{pmatrix} \begin{pmatrix} 0 & (M/C) \\ C & D \end{pmatrix}, \qquad \det M = -\det C \cdot \det(M/C),$$

$$M = \begin{pmatrix} I & BD^{-1} \\ 0 & I \end{pmatrix} \begin{pmatrix} (M/D) & 0 \\ C & D \end{pmatrix}, \qquad \det M = \det D \cdot \det(M/D). \qquad (2.2)$$

If, for example, A belongs to a vector space E, $B = [b_1, \dots, b_k]$ with $b_i \in E$, C is a vector of dimension k, and D a square and invertible $k \times k$ matrix, the last relation is still valid. In this case $\det M$ is the element of E obtained by expanding M with respect to its first row of elements of E by the classical rules, and $(M/D) \in E$.

2.1.3 Designants

Let us now recall the definition of designants and give some of their properties. They were introduced by Arend Heyting (1898–1980) [328] as a generalization of determinants in a noncommutative algebra.

Let $\mathcal{M}_q$ be the algebra of $q \times q$ matrices with complex entries, and $\mathcal{M}_{p,q}$ the set of $p \times q$ matrices with complex entries. We consider the system of linear equations

$$x_1 a_{11} + x_2 a_{12} = b_1,$$
$$x_1 a_{21} + x_2 a_{22} = b_2,$$

where $a_{ij} \in \mathcal{M}_q$, $b_i \in \mathcal{M}_{p,q}$, and the unknown matrices x_i belong to $\mathcal{M}_{p,q}$. Let us remark that in this system, the unknowns are multiplied on the right by the coefficients.

Assume that the matrix a_{11} is regular. For eliminating the first unknown x_1 from the first equation of this system, let us proceed as in Gaussian elimination, that is, let us multiply it by a_{11}^{-1} on the right, then by a_{21}, and finally subtract it from the second equation. We obtain

$$x_2(a_{22} - a_{12}a_{11}^{-1}a_{21}) = b_2 - b_1 a_{11}^{-1} a_{21}.$$

The expression $a_{22} - a_{12}a_{11}^{-1}a_{21}$ is called the *designant* (more precisely, the right-designant) of the system, and it is denoted by

$$\begin{vmatrix} a_{11} & a_{12} \\ a_{21} & a_{22} \end{vmatrix}_r, \tag{2.3}$$

where the subscript r stands for "right." Left designants can be defined similarly by considering a system of equations with coefficients on the left

$$a_{11}x_1 + a_{12}x_2 = b_1,$$
$$a_{21}x_1 + a_{22}x_2 = b_2.$$

Assuming that $a_{11} \neq 0$ and eliminating x_1 in the second equation leads to

$$(a_{22} - a_{21}a_{11}^{-1}a_{12})x_2 = 0,$$

and the left designant is

$$\begin{vmatrix} a_{11} & a_{12} \\ a_{21} & a_{22} \end{vmatrix}_l = a_{22} - a_{21}a_{11}^{-1}a_{12}. \tag{2.4}$$

If the right designant is regular, then

$$x_2 = (b_2 - b_1 a_{11}^{-1}a_{21})(a_{22} - a_{12}a_{11}^{-1}a_{21})^{-1},$$

that is, using the notation introduced above for designants,

$$x_2 = \begin{vmatrix} a_{11} & b_1 \\ a_{21} & b_2 \end{vmatrix}_r \begin{vmatrix} a_{11} & a_{12} \\ a_{21} & a_{22} \end{vmatrix}_r^{-1}.$$

Similarly, the other unknown matrix is given by

$$x_1 = \begin{vmatrix} b_1 & a_{12} \\ b_2 & a_{22} \end{vmatrix}_r \begin{vmatrix} a_{11} & a_{12} \\ a_{21} & a_{22} \end{vmatrix}_r^{-1}.$$

The notion of designant is related to that of Schur's complement. Indeed, the Schur's complement (M^T/a_{11}) of a_{11} in the matrix

$$M = \begin{pmatrix} a_{11} & a_{12} \\ a_{21} & a_{22} \end{pmatrix}$$

is the right designant (2.3), while the left designant (2.4) is the Schur's complement (M/a_{11}) of a_{11} in M.

The notion of designant can be generalized to systems in a noncommutative algebra with an arbitrary number of equations and unknowns. We consider the system

$$x_1 a_{11} + \cdots + x_n a_{1n} = b_1$$
$$\vdots \qquad\qquad \vdots \qquad \vdots$$
$$x_1 a_{n1} + \cdots + x_n a_{nn} = b_n.$$

We have

$$x_n = \begin{vmatrix} a_{11} & \cdots & a_{1,n-1} & b_1 \\ \vdots & & \vdots & \vdots \\ a_{n1} & \cdots & a_{n,n-1} & b_n \end{vmatrix}_r \begin{vmatrix} a_{11} & \cdots & a_{1n} \\ \vdots & & \vdots \\ a_{n1} & \cdots & a_{nn} \end{vmatrix}_r^{-1}.$$

Let us set

$$\Delta_n = \begin{vmatrix} a_{11} & \cdots & a_{1n} \\ \vdots & & \vdots \\ a_{n1} & \cdots & a_{nn} \end{vmatrix}_r$$

and let $A_{r,s}^{(n-2)}$ be the designant of order $n-1$ obtained by keeping the rows $1, 2, \ldots, n-2, r$ and the columns $1, 2, \ldots, n-2, s$ of Δ_n. Then Δ_n can be written as a designant and computed from $A_{n-1,n-1}^{(n-2)}, A_{n-1,n}^{(n-2)}, A_{n,n-1}^{(n-2)}$ and $A_{n,n}^{(n-2)}$ using Sylvester's identity

$$\Delta_n = \begin{vmatrix} A_{n-1,n-1}^{(n-2)} & A_{n-1,n}^{(n-2)} \\ A_{n,n-1}^{(n-2)} & A_{n,n}^{(n-2)} \end{vmatrix}_r = A_{n,n}^{(n-2)} - A_{n-1,n}^{(n-2)} \left[A_{n-1,n-1}^{(n-2)} \right]^{-1} A_{n,n-1}^{(n-2)}.$$

Left designants can be defined in a similar way for systems of equations in which the unknowns x_j are multiplied on the left by the coefficients a_{ij}. In this case

$$\Delta_n = \begin{vmatrix} A_{n-1,n-1}^{(n-2)} & A_{n-1,n}^{(n-2)} \\ A_{n,n-1}^{(n-2)} & A_{n,n}^{(n-2)} \end{vmatrix}_l = A_{n,n}^{(n-2)} - A_{n,n-1}^{(n-2)} \left[A_{n-1,n-1}^{(n-2)} \right]^{-1} A_{n-1,n}^{(n-2)},$$

and the Schur's complement is recovered.

More information about designants can be found in [328, 546].

2.2 Extrapolation Methods

When a scalar sequence (S_n) is slowly converging to its limit S, it is sometimes possible to modify the process that produces it for obtaining a method with a faster convergence. This is the case, for example, when the Gauss–Seidel method is modified to produce the successive overrelaxation method (SOR). But in other cases, the sequence is produced by a black box without any access to it. Then one can *transform* this sequence into a new sequence (T_n) which, under some assumptions, *converges faster* to the same limit, that is,

$$\lim_{n \to \infty} \frac{T_n - S}{S_n - S} = 0.$$

The procedure $T : (S_n) \longmapsto (T_n)$ is called a *sequence transformation*, and in this case, we say that T *accelerates the convergence* of (S_n). A sequence transformation is also called a *convergence acceleration method* or an *extrapolation method*.

The idea behind such transformations is to assume that the sequence (S_n) to be transformed behaves like a model sequence (u_n) whose limit can be exactly computed by an algebraic process. In other words, applying the transformation T to (u_n) produces a constant sequence all terms of which are equal to its limit or antilimit. The set $\mathcal{K}_T$ of these model sequences is called the *kernel* of the transformation, and it is characteristic of it. Starting from a given index n, a certain number of consecutive terms of the sequence to be transformed are interpolated by the corresponding terms of a model sequence belonging to the kernel. Then, the limit of this model sequence is taken as an estimate of the limit or the antilimit S of (S_n). This is an extrapolation process. Since this estimate depends on the starting index n, the sequence (S_n) has been transformed into a new sequence (T_n). If (S_n) belongs to the kernel of the transformation T, then for all n, $T_n = S$. Usually, a sequence transformation is built starting from its kernel, and it is implemented by a recursive algorithm. However, in some cases, a new sequence transformation can be directly obtained by modifying the algorithm corresponding to another transformation, a procedure that makes difficult, or even impossible, the characterization of its kernel.

These concepts, and how a sequence transformation is defined, will become clear with the examples that follow. In this chapter, we consider only the case of scalar sequences. Transformations for other types of sequences will be discussed in the next chapters. We begin with Richardson's extrapolation method [520, 523], an important well–known linear sequence transformation. It allows one to easily understand the derivation and the mechanism behind extrapolation methods. Its historical development will be described in Sect. 3.2.1.

In the sequel, the forward difference operator Δ is defined by $\Delta u_n = u_{n+1} - u_n$, and its powers are given by $\Delta^{k+1}(u_n) = \Delta(\Delta^k(u_n)) = \Delta^k(u_{n+1}) - \Delta^k(u_n)$, where Δ^0 is the identity.

2.2.1 Richardson's Extrapolation Method

It is assumed that the sequence (S_n) depends on a known auxiliary sequence (x_n) tending to zero. The kernel of Richardson's extrapolation method is the set of sequences such that for all n, S_n is a polynomial in x_n. The process consists in interpolating the terms $S_n, S_{n+1}, \ldots, S_{n+k}$ by a polynomial of degree k in a variable x at the points $x_n, x_{n+1}, \ldots, x_{n+k}$, and then to extrapolate it at $x = 0$. Thus, since its value depends on n and also on the degree k, it is denoted by $T_k^{(n)}$. The mathematical formula for the interpolation polynomial is well–known. Writing it at $x = 0$ gives

$$
T_k^{(n)} = \begin{vmatrix} S_n & \cdots & S_{n+k} \\ x_n & \cdots & x_{n+k} \\ \vdots & & \vdots \\ x_n^k & \cdots & x_{n+k}^k \end{vmatrix} \bigg/ \begin{vmatrix} 1 & \cdots & 1 \\ x_n & \cdots & x_{n+k} \\ \vdots & & \vdots \\ x_n^k & \cdots & x_{n+k}^k \end{vmatrix}
$$

$$
= \begin{vmatrix} S_n & \Delta S_n & \cdots & \Delta S_{n+k-1} \\ x_n & \Delta x_n & \cdots & \Delta x_{n+k-1} \\ \vdots & \vdots & & \vdots \\ x_n^k & \Delta x_n^k & \cdots & \Delta x_{n+k-1}^k \end{vmatrix} \bigg/ \begin{vmatrix} \Delta x_n & \cdots & \Delta x_{n+k-1} \\ \vdots & & \vdots \\ \Delta x_n^k & \cdots & \Delta x_{n+k-1}^k \end{vmatrix},
$$

where the operator Δ acts on the lower indices.

By Schur's determinantal formula (2.2), the second formula shows that $T_k^{(n)}$ can be written as a Schur's complement [706],

$$
T_k^{(n)} = S_n - (\Delta S_n, \ldots, \Delta S_{n+k-1}) \begin{pmatrix} \Delta x_n & \cdots & \Delta x_{n+k-1} \\ \vdots & & \vdots \\ \Delta x_n^k & \cdots & \Delta x_{n+k-1}^k \end{pmatrix}^{-1} \begin{pmatrix} x_n \\ \vdots \\ x_n^k \end{pmatrix}.
$$

Interpolation polynomials at one point can be recursively computed by the Neville–Aitken scheme, which at $x = 0$ gives Richardson's extrapolation method [520, 523]

$$
T_{k+1}^{(n)} = T_k^{(n)} - \frac{T_k^{(n+1)} - T_k^{(n)}}{x_{n+k+1} - x_n} x_n, \quad k, n = 0, 1, \ldots, \tag{2.5}
$$

with $T_0^{(n)} = S_n$ for $n = 0, 1, \ldots$.

It can be proved that this recurrence relation holds by applying Sylvester's determinantal identity separately to the numerator and to the denominator of the formula for $T_k^{(n)}$.

Richardson's method is a linear transformation, the quantities $T_k^{(n)}$ are usually displayed in a double entry table, and the preceding formula relates three of them located at the vertices of a triangle, as shown in Fig. 2.1. The lower index k represents a column, and the upper index n a descending diagonal.

Convergence results on the set of sequences $\{(T_k^{(n)})\}$ were obtained by Pierre-Jean Laurent (b. 1937) [395].[1]

Theorem 2.1 *A necessary and sufficient condition that*

$$
\forall k, \quad \lim_{n \to \infty} T_k^{(n)} = S,
$$

$$
\forall n, \quad \lim_{k \to \infty} T_k^{(n)} = S,
$$

---

[1] See his testimony in Sect. 7.7.

for all sequences (S_n) that converge to S is that $\exists \, \alpha$ and β, with $\alpha < 1 < \beta$, such that $\forall n$, $x_{n+1}/x_n \notin [\alpha, \beta]$.

We also have the following result

Theorem 2.2 *Assume that the conditions of Theorem 2.1 are satisfied. Then a necessary and sufficient condition that*

$$\lim_{n \to \infty} \frac{T_{k+1}^{(n)} - S}{T_k^{(n)} - S} = 0$$

is that

$$\lim_{n \to \infty} \frac{T_k^{(n+1)} - S}{T_k^{(n)} - S} = \lim_{n \to \infty} \frac{x_{n+k+1}}{x_n}.$$

Following this result, each column in the array of Fig. 2.1 converges faster than the preceding one if the conditions of both theorems are satisfied.

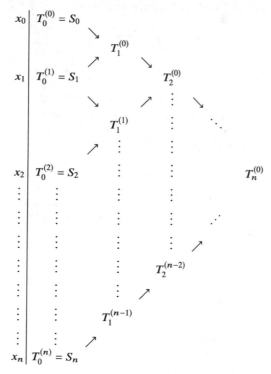

Fig. 2.1 Richardson's process

If S_n is the result obtained by the trapezoidal rule for computing a definite integral with the step size $h_n = h_0/2^n$, and if we set $x_n = h_n^2$ in Richardson's method (2.5), then Romberg's method [535] is recovered. It is well known that this method, although it is a linear one, is able to accelerate the convergence of (S_n) under mild conditions on the differentiability of the integrand.

2.2.2 The Aitken Process

In this section,[2] we begin by introducing the Aitken Δ^2 process [5] by different approaches. Our ultimate goal is to build a sequence transformation $T : (S_n) \longmapsto (T_n)$ able to accelerate *all* sequences such that

$$\exists S, \exists \lambda \neq 0, 1, \quad \lim_{n \to \infty} (S_{n+1} - S)/(S_n - S) = \lambda.$$

We begin by considering a sequence (S_n) of the form

$$S_{n+1} - S = \lambda(S_n - S), \quad n = 0, 1, \ldots, \tag{2.6}$$

where λ is *known*. Obviously, the sequence transformation T defined by

$$T_n = \frac{S_{n+1} - \lambda S_n}{1 - \lambda}, \quad n = 0, 1, \ldots, \tag{2.7}$$

is such that $\forall n, T_n = S$.

The *kernel* $\mathcal{K}_T$ of this transformation is the set of sequences that satisfy a relation of the form (2.6). A necessary and sufficient condition that $\forall n, T_n = S$ is that $(S_n) \in \mathcal{K}_T$. The relation (2.6) is called the *implicit form* of the kernel. Let us remark that any sequence of $\mathcal{K}_T$ can be written

$$S_n = S + \alpha \lambda^n, \quad n = 0, 1, \ldots, \tag{2.8}$$

with $\alpha = S_0 - S$ unknown. This is the *explicit form* of the kernel.

Remark 2.1 The preceding results are relevant with the only assumption that $\lambda \neq 1$. If $|\lambda| < 1$, then (S_n) converges to its limit S. If $|\lambda| > 1$, (S_n) does not converge and S is called its *antilimit*.

Let us now look at the convergence and acceleration properties of this transformation T. First, we see that for every convergent sequence (S_n), the sequence (T_n) converges, and, moreover, that it converges to the same limit.

Then what is the set of sequences that are accelerated by T? We have

$$\frac{T_n - S}{S_n - S} = \frac{\dfrac{S_{n+1} - S}{S_n - S} - \lambda}{1 - \lambda}.$$

Thus if $\lambda \neq 1$, this formula shows that (T_n) converges faster than (S_n) if and only if

---

[2] This section is an adapted extract, with permission, of [135].

$$\lim_{n\to\infty} \frac{S_{n+1} - S}{S_n - S} = \lambda.$$

Obviously, since the value of λ has to be known, this class of sequences is quite restricted.

It has to be noticed that this sequence transformation is *linear*. As we already pointed out, most of the linear sequence transformations are only able to accelerate the convergence of restricted classes of sequences.

This is the reason why we will now consider a sequence of the form (2.6), but where λ is an *unknown* parameter.

• Applying the operator Δ to a sequence of the form (2.6) gives, for all n,

$$\lambda = \frac{\Delta S_{n+1}}{\Delta S_n}.$$

If (S_n) does not satisfy (2.6), this ratio depends on n. However, putting this value in (2.7), we obtain the transformation

$$
\begin{aligned}
T_n &= \frac{S_{n+1} - \dfrac{\Delta S_{n+1}}{\Delta S_n} S_n}{1 - \dfrac{\Delta S_{n+1}}{\Delta S_n}} \\
&= \frac{S_n \Delta S_{n+1} - S_{n+1} \Delta S_n}{\Delta^2 S_n} \\
&= \frac{S_n S_{n+2} - S_{n+1}^2}{S_{n+2} - 2S_{n+1} + S_n}, \quad n = 0, 1, \dots.
\end{aligned}
$$

This transformation is the so-called *Aitken Δ^2 process*. Notice that in contrast to the sequence transformation defined by (2.7), the Aitken process is *nonlinear*. Its kernel is the set of sequences satisfying (2.6), or equivalently, (2.8), but where now $\lambda \neq 1$ and $\alpha \neq 0$ are any unknown constants.

Let us immediately advise that from the numerical point of view, the preceding expressions are unstable, and that one of the following equivalent ones has to be used:

$$T_n = S_n - \frac{(\Delta S_n)^2}{\Delta^2 S_n} \tag{2.9}$$

$$= S_{n+1} - \frac{\Delta S_n \Delta S_{n+1}}{\Delta^2 S_n}$$

$$= S_{n+2} - \frac{(\Delta S_{n+1})^2}{\Delta^2 S_n}.$$

It has to be mentioned that applying the Aitken process to a convergent sequence does not always produce a sequence that converges. Counterexamples in which

(T_n) has several accumulation points were given by Samuel Lubkin (1906–1972) [423] and Imanuel Marx (1918–1964) [438] (see [645] on this paper). However, if (S_n) and (T_n) converge, they have the same limit as proved by Richard Ray Tucker (1926–1994) [644, 646] (see also [647]).

The formulas giving the Aitken process can be recovered in different ways. Let us explain how, since they will be of interest below.

• It is equivalent to assume that the following three interpolation conditions defining the *explicit form* of the kernel hold,

$$S_n = S + \alpha\lambda^n,$$
$$S_{n+1} = S + \alpha\lambda^{n+1},$$
$$S_{n+2} = S + \alpha\lambda^{n+2},$$

and then to solve the system for the three unknowns α, λ, and S (S will be denoted by T_n, since these unknowns depend on the index n if (S_n) does not belong to the explicit kernel of the transformation). We get

$$\Delta S_n = \alpha(\lambda - 1)\lambda^n,$$
$$\Delta S_{n+1} = \alpha(\lambda - 1)\lambda^{n+1}.$$

Thus

$$\lambda = \frac{\Delta S_{n+1}}{\Delta S_n},$$
$$\alpha\lambda^n = \frac{\Delta S_n}{\lambda - 1},$$

and replacing these two quantities by their expressions in (2.8) leads to (2.9).

If $|\lambda| < 1$, every sequence of the form (2.8) converges to S. Thus the Aitken process corresponds to an interpolation by a sequence of this form, followed by its extrapolation at infinity. This is a common feature of many convergence acceleration methods, and this is why they are often referred to as *extrapolation methods*.

• We want to construct a sequence transformation such that if $\forall n$,

$$a_0(S_n - S) + a_1(S_{n+1} - S) = 0, \tag{2.10}$$

with $a_0 a_1 \neq 0$ and $a_0 + a_1 \neq 0$, then $\forall n$, $T_n = S$. This is the *implicit form* of the kernel of the transformation, another way of writing (2.6) that is recovered with $\lambda = -a_0/a_1 \neq 1$, a condition equivalent to $a_0 + a_1 \neq 0$. Consider the system

$$a_0(S_n - S) + a_1(S_{n+1} - S) = 0,$$
$$a_0(S_{n+1} - S) + a_1(S_{n+2} - S) = 0.$$

This is a homogeneous system of equations with a nonzero solution. Thus its determinant must vanish, that is,

$$\begin{vmatrix} S_n - S & S_{n+1} - S \\ S_{n+1} - S & S_{n+2} - S \end{vmatrix} = 0.$$

Replacing the second row by its difference with the first one, we obtain

$$\begin{vmatrix} S_n & S_{n+1} \\ \Delta S_n & \Delta S_{n+1} \end{vmatrix} - S \begin{vmatrix} 1 & 1 \\ \Delta S_n & \Delta S_{n+1} \end{vmatrix} = 0,$$

and it follows that

$$S = \begin{vmatrix} S_n & S_{n+1} \\ \Delta S_n & \Delta S_{n+1} \end{vmatrix} \Big/ \begin{vmatrix} 1 & 1 \\ \Delta S_n & \Delta S_{n+1} \end{vmatrix}. \tag{2.11}$$

Denoting by T_n this ratio of determinants (since it depends on n when (S_n) does not belong to the kernel), the Aitken process is recovered.

Thus, using any of these three approaches, the sequence (S_n) is transformed into the sequence (T_n) given by the Aitken Δ^2 process.

It is easy to prove the following important acceleration result.

Theorem 2.3 *Let (S_n) be a sequence converging to S. If there exists $\lambda \neq 1$ such that*

$$\lim_{n \to \infty} (S_{n+1} - S)/(S_n - S) = \lambda,$$

then

$$\lim_{n \to \infty} (T_n - S)/(S_{n+1} - S) = 0.$$

Moreover, if $\lambda \neq 0$, then

$$\lim_{n \to \infty} (T_n - S)/(S_{n+2} - S) = 0.$$

A result proved by Jean-Paul Delahaye (b. 1952) [195] is useful for checking this condition.

Theorem 2.4 *Assume that $|\lambda| \neq 1$. A necessary and sufficient condition that*

$$\lim_{n \to \infty} (S_{n+1} - S)/(S_n - S) = \lambda$$

is that

$$\lim_{n \to \infty} \Delta S_{n+1}/\Delta S_n = \lambda.$$

2.2.3 The Shanks Transformation

For explaining the Shanks transformation, we proceed as in the third way used for the Aitken process.[3] We want to construct a sequence transformation whose *implicit form* of the kernel is the set of sequences satisfying, for all n, the linear homogeneous difference equation of order k

$$a_0(S_n - S) + a_1(S_{n+1} - S) + \cdots + a_k(S_{n+k} - S) = 0, \qquad (2.12)$$

where $a_1, \ldots, a_k$ are arbitrary unknown constants with $a_0 a_k \neq 0$ (otherwise k could be replaced by a lower index), and $a_0 + \cdots + a_k \neq 0$ (otherwise S is not uniquely defined). Obviously, this formula generalizes (2.10). The problem is to find a formula for the unknown S. Writing the relation (2.12) for the indices $n, \ldots, n + k$, we obtain a homogeneous linear system with a nonzero solution. Thus, its determinant must be equal to zero:

$$
\begin{vmatrix}
S_n - S & S_{n+1} - S & \cdots & S_{n+k} - S \\
S_{n+1} - S & S_{n+2} - S & \cdots & S_{n+k+1} - S \\
\vdots & \vdots & & \vdots \\
S_{n+k} - S & S_{n+k+1} - S & \cdots & S_{n+2k} - S
\end{vmatrix} = 0.
$$

Replacing each row by its difference with the previous one, and separating it into two determinants, we obtain

$$
\begin{vmatrix}
S_n & S_{n+1} & \cdots & S_{n+k} \\
\Delta S_n & \Delta S_{n+1} & \cdots & \Delta S_{n+k} \\
\vdots & \vdots & & \vdots \\
\Delta S_{n+k-1} & \Delta S_{n+k} & \cdots & \Delta S_{n+2k-1}
\end{vmatrix}
- S
\begin{vmatrix}
1 & 1 & \cdots & 1 \\
\Delta S_n & \Delta S_{n+1} & \cdots & \Delta S_{n+k} \\
\vdots & \vdots & & \vdots \\
\Delta S_{n+k-1} & \Delta S_{n+k} & \cdots & \Delta S_{n+2k-1}
\end{vmatrix} = 0,
$$

which gives S as a ratio of determinants.

If (S_n) does not satisfy (2.12), these determinants can still be computed, but their ratio is no longer equal to S for all n, but to a number depending on k and n, and denoted by $e_k(S_n)$. Thus, the sequence (S_n) has been transformed into the set of sequences $\{(e_k(S_n))\}$. This is the definition of the *Shanks transformation* [567, 568]. It is given by the first expression below. In the numerator, replacing each row by its sum with the preceding one, and in the denominator, each column by its difference with the preceding one, leads to the second expression in (2.13):

---

[3] Parts of this section are adapted extracts, with permission, from [135].

$$
e_k(S_n) = \frac{\begin{vmatrix} S_n & S_{n+1} & \cdots & S_{n+k} \\ \Delta S_n & \Delta S_{n+1} & \cdots & \Delta S_{n+k} \\ \vdots & \vdots & & \vdots \\ \Delta S_{n+k-1} & \Delta S_{n+k} & \cdots & \Delta S_{n+2k-1} \\ 1 & 1 & \cdots & 1 \\ \Delta S_n & \Delta S_{n+1} & \cdots & \Delta S_{n+k} \\ \vdots & \vdots & & \vdots \\ \Delta S_{n+k-1} & \Delta S_{n+k} & \cdots & \Delta S_{n+2k-1} \end{vmatrix}}{} = \frac{H_{k+1}(S_n)}{H_k(\Delta^2 S_n)}, \quad k, n = 0, 1, \ldots, \quad (2.13)
$$

where $H_k(u_n)$ denotes the *Hankel determinant* (introduced by Hermann Hankel (1839–1873) in his dissertation [305])

$$
H_k(u_n) = \begin{vmatrix} u_n & u_{n+1} & \cdots & u_{n+k-1} \\ u_{n+1} & u_{n+2} & \cdots & u_{n+k} \\ \vdots & \vdots & & \vdots \\ u_{n+k-1} & u_{n+k} & \cdots & u_{n+2k-2} \end{vmatrix},
$$

with $H_0(u_n) = 1$. Sometimes, this determinant will simply be denoted by $H_k^{(n)}$ when there is no confusion about the sequence to which it applies. The formula (2.13) is the definition of the Shanks transformation [567, 568]. Notice that the Aitken Δ^2 process is recovered for $k = 1$.

Formula (2.13) shows, after replacing each column in the numerator and in the denominator by its difference with the preceding one, that $e_k(S_n)$ is a Schur's complement, that is,

$$
e_k(S_n) = S_n - (\Delta S_n, \ldots, \Delta S_{n+k-1}) \begin{pmatrix} \Delta^2 S_n & \cdots & \Delta^2 S_{n+k-1} \\ \vdots & & \vdots \\ \Delta^2 S_{n+k-1} & \cdots & \Delta^2 S_{n+2k-2} \end{pmatrix}^{-1} \begin{pmatrix} \Delta S_n \\ \vdots \\ \Delta S_{n+k-1} \end{pmatrix}.
$$

Let us notice that the kernel of the Shanks transformation as given by (2.12) can also be written

$$
b_0(S_n - S) + b_1 \Delta S_n + \cdots + b_k \Delta^k S_n = 0, \quad (2.14)
$$

with $b_0 b_k \neq 0$, and thus the Shanks transformation is also given by the ratio

$$
e_k(S_n) = \frac{\begin{vmatrix} S_n & \Delta S_n & \cdots & \Delta^k S_n \\ \Delta S_n & \Delta^2 S_n & \cdots & \Delta^{k+1} S_n \\ \vdots & \vdots & & \vdots \\ \Delta^k S_n & \Delta^{k+1} S_n & \cdots & \Delta^{2k} S_n \end{vmatrix}}{\begin{vmatrix} \Delta^2 S_n & \cdots & \Delta^{k+1} S_n \\ \vdots & & \vdots \\ \Delta^{k+1} S_n & \cdots & \Delta^{2k} S_n \end{vmatrix}}. \quad (2.15)
$$

We have the following theorem.

Theorem 2.5 *A sufficient condition that $e_k(S_n) = S$, for all n, is that the sequence (S_n) satisfy (2.12) with $a_0 a_k \neq 0$ and $a_0 + \cdots + a_k \neq 0$.*
If $\forall n$, $H_k(\Delta S_n) \neq 0$, the condition is also necessary.

As we will see in the next section, the condition $\forall n$, $H_k(\Delta S_n) \neq 0$ is no longer required for proving the necessary condition of this theorem if the Shanks transformation is implemented via the ε-algorithm.

The difference equation (2.12) can be solved by writing its characteristic polynomial and expressing its zeros. It leads to a closed-form expression for the sequences satisfying it [118]. It is the *explicit form* of the kernel of the Shanks transformation given by the following result.

Theorem 2.6 *A necessary and sufficient condition that $e_k(S_n) = S$, for all n, is that the sequence (S_n) have the form*

$$S_n = S + \sum_{i=1}^{p} A_i(n) r_i^n + \sum_{i=p+1}^{q} [B_i(n)\cos(\beta_i n) + C_i(n)\sin(\beta_i n)]e^{\omega_i n} + \sum_{i=0}^{m} c_i \delta_{in},$$

where A_i, B_i, and C_i are polynomials in n such that if d_i is the degree of A_i plus 1 for $i = 1, \ldots, p$, and the maximum of the degrees of B_i and C_i plus 1 for $i = p+1, \ldots, q$, one has

$$m + 1 + \sum_{i=1}^{p} d_i + 2 \sum_{i=p+1}^{q} d_i = k,$$

with the conventions that the second sum vanishes if there are no complex zeros, and $m = -1$ if there is no term in δ_{in} (Kronecker's symbol).

See [135] for detailed explanations of the Shanks transformation and its genesis.

2.2.4 The ε-Algorithm

In practice, the determinants in (2.13) and (2.15) are not easy to compute, and a recursive algorithm for computing the $e_k(S_n)$ is needed. This is the role of the ε-algorithm due to Peter Wynn [713], and whose rule is

$$\varepsilon_{k+1}^{(n)} = \varepsilon_{k-1}^{(n+1)} + \frac{1}{\varepsilon_k^{(n+1)} - \varepsilon_k^{(n)}}, \quad k, n = 0, 1, \ldots,$$

with $\forall n$, $\varepsilon_{-1}^{(n)} = 0$ and $\varepsilon_0^{(n)} = S_n$.

Using Sylvester's and Schweins's determinatal identities, he proved the following equalities.

Theorem 2.7

$$\varepsilon_{2k}^{(n)} = e_k(S_n) = \frac{H_{k+1}(S_n)}{H_k(\Delta^2 S_n)}, \quad \varepsilon_{2k+1}^{(n)} = \frac{1}{e_k(\Delta S_n)} = \frac{H_k(\Delta^3 S_n)}{H_{k+1}(\Delta^2 S_n)}.$$

Thus the quantities $\varepsilon_{2k+1}^{(n)}$ are only intermediate computations. However, they are useful in the proofs of some theoretical results.

Let us now come back to the proof of the necessary condition of Theorem 2.5 and show that we have a more complete result if the Shanks transformation is implemented by the ε-algorithm [135]:

Theorem 2.8 *A necessary and sufficient condition that* $\varepsilon_{2k}^{(n)} = S$, *for all n, is that the sequence* (S_n) *satisfy (2.12) with* $a_0 a_k \neq 0$ *and* $a_0 + \cdots + a_k \neq 0$.

The difference between Theorem 2.5 and this new result is that Theorem 2.5 concerns the Shanks transformation while Theorem 2.8 is related to the recursive algorithm used for its implementation. Since the Shanks transformation can be implemented by other algorithms, the condition $\forall n, H_k(\Delta S_n) \neq 0$ may have to be reintroduced for them, and perhaps Theorem 2.8 is valid only for the ε-algorithm. This is an open question.

The quantities $\varepsilon_k^{(n)}$ are usually displayed in a two-dimensional array (the ε-*array*) where the lower index k remains the same in a column of the table, and the upper index n is the same in a descending diagonal. The rule of the ε-algorithm relates quantities located at the four vertices of a lozenge (also called a rhombus) in three different columns and two descending diagonals, as shown in Fig. 2.2.

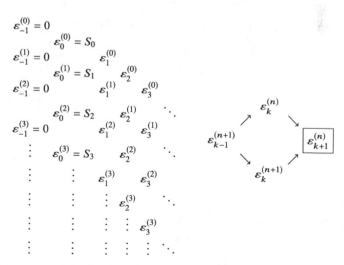

Fig. 2.2 The ε-array and the rhombus rule

This algorithm will be discussed later in more detail.

2.2.5 Confluent Extrapolation

Let f be a function such that $\lim_{t\to\infty} f(t) = S$. If the convergence is slow, the function f can be transformed, by a *function transformation*, into a new function F converging faster to the same limit under certain assumptions, that is, such that for any appropriate norm,

$$\lim_{t\to\infty} \|F(t) - S\| / \|f(t) - S\| = 0.$$

Following an idea similar to that of the Shanks transformation for scalar sequences, a function transformation $f \longmapsto e_k$ such that for all t, $e_k(f(t)) = S$ if (and only if) for all t, f satisfies the linear differential equation of order k

$$b_0(f(t) - S) + b_1 f'(t) + \cdots + b_k f^{(k)}(t) = 0, \tag{2.16}$$

with $b_0 b_k \neq 0$. Clearly, this equation is the continuous analogue of the difference equation (2.14) defining the kernel of the scalar Shanks transformation. If the coefficients b_i can be computed, then a transformation possessing the desired property can be constructed. For that, let us differentiate (2.16) k times. We thus obtain a homogeneous linear system with a nonzero solution. Thus, its determinant must be equal to zero, that is,

$$\begin{vmatrix} f(t) - S & f'(t) & \cdots & f^{(k)}(t) \\ f'(t) & f''(t) & \cdots & f^{(k+1)}(t) \\ \vdots & \vdots & & \vdots \\ f^{(k)}(t) & f^{(k+1)}(t) & \cdots & f^{(2k)}(t) \end{vmatrix} = 0.$$

Therefore, by construction, the transformation defined by

$$e_k(f(t)) = \frac{\begin{vmatrix} f(t) & f'(t) & \cdots & f^{(k)}(t) \\ f'(t) & f''(t) & \cdots & f^{(k+1)}(t) \\ \vdots & \vdots & & \vdots \\ f^{(k)}(t) & f^{(k+1)}(t) & \cdots & f^{(2k)}(t) \end{vmatrix}}{\begin{vmatrix} f''(t) & \cdots & f^{(k+1)}(t) \\ \vdots & & \vdots \\ f^{(k+1)}(t) & \cdots & f^{(2k)}(t) \end{vmatrix}}$$

will be such that for all t, $e_k(f(t)) = S$ if and only if (2.16) (the *implicit kernel* of the transformation) holds for all t. This formula is clearly the continuous analogue of (2.15). It is called its *confluent form*. Defining the functional Hankel determinant $H_k^{(n)}(t)$ by

$$H_k^{(n)}(t) = \begin{vmatrix} f^{(n)}(t) & \cdots & f^{(n+k-1)}(t) \\ \vdots & & \vdots \\ f^{(n+k-1)}(t) & \cdots & f^{(n+2k-2)}(t) \end{vmatrix},$$

we have $e_k(f(t)) = H_{k+1}^{(0)}(t)/H_k^{(2)}(t)$, a formula similar to the second ratio in (2.13). The *explicit kernel* is given by a theorem similar to Theorem 2.6.

Theorem 2.9 *A necessary and sufficient condition that $e_k(f(t)) = S$, for all t, is that the function f have the form*

$$f(t) = S + \sum_{i=1}^{p} A_i(t)e^{r_i t} + \sum_{i=p+1}^{q} [B_i(t)\cos(\beta_i t) + C_i(t)\sin(\beta_i t)]e^{\omega_i t},$$

where A_i, B_i, and C_i are polynomials in t such that if d_i is the degree of A_i plus 1 for $i = 1, \ldots, p$, and the maximum of the degrees of B_i and C_i plus 1 for $i = p+1, \ldots, q$, one has

$$\sum_{i=1}^{p} d_i + 2\sum_{i=p+1}^{q} d_i = k.$$

An algorithm able to compute recursively the e_k is needed. It is the *confluent ε-algorithm*, due to Wynn [721]. Its rule is

$$\varepsilon_{k+1}(t) = \varepsilon_{k-1}(t) + 1/\varepsilon_k'(t),$$

with $\varepsilon_{-1}(t) = 0$ and $\varepsilon_0(t) = f(t)$, and we have

$$\varepsilon_{2k}(t) = e_k(f(t)) = \frac{H_{k+1}^{(0)}(t)}{H_k^{(2)}(t)}, \quad \varepsilon_{2k+1}(t) = \frac{1}{e_k(f'(t))} = \frac{H_k^{(3)}(t)}{H_{k+1}^{(1)}(t)}.$$

2.3 Padé Approximation

In this section, we will give the definition of Padé approximants and review their most interesting properties for our purpose. On Padé approximation, see [31, 94].

2.3.1 Definition and Algebraic Theory

Let f be a formal power series (the Taylor expansion of a function around zero)

$$f(t) = \sum_{i=0}^{\infty} c_i t^i,$$

where the coefficients c_i and the variable t can be complex. A Padé approximant of f is a rational function with a numerator of degree p at most and a denominator of degree q at most such that its power series expansion (obtained by the Euclidean division of the numerator by the denominator in ascending powers of the variable) agrees with f as far as possible, that is, up to the degree $p + q$ inclusively. Such an approximant is denoted by $[p/q]_f$ and, by construction, one has

$$[p/q]_f(t) - f(t) = O(t^{p+q+1}) \quad (t \to 0).$$

These conditions are sometimes known as the *accuracy-through-order conditions*.

Let us set $[p/q]_f(t) = N_p(t)/D_q(t)$ with

$$N_p(t) = a_0 + a_1 t + \cdots + a_p t^p \quad \text{and} \quad D_q(t) = b_0 + b_1 t + \cdots + b_q t^q.$$

Then writing the conditions of the definition as $f(t)D_q(t) - N_p(t) = O(t^{p+q+1})$ leads to

$$a_0 = c_0 b_0,$$
$$a_1 = c_1 b_0 + c_0 b_1,$$
$$\vdots$$
$$a_p = c_p b_0 + c_{p-1} b_1 + \cdots + c_{p-q} b_q,$$
$$0 = c_{p+1} b_0 + c_p b_1 + \cdots + c_{p-q+1} b_q,$$
$$\vdots$$
$$0 = c_{p+q} b_0 + c_{p+q-1} b_1 + \cdots + c_p b_q,$$

with the convention that $c_i = 0$ for $i < 0$, which allows one to treat simultaneously the cases $p \le q$ and $p \ge q$.

The last q equations contain $q + 1$ unknowns $b_0, \ldots, b_q$, and thus this system has a nontrivial solution. Usually, b_0 is set to 1, which allows one to solve the system formed by the last q preceding equations for the coefficients $b_1, \ldots, b_q$. Knowing the b_i, the first $p + 1$ equations directly provide the a_i. This direct approach was first given by Johann Heinrich Lambert (1728–1777) in 1758 [383]. A different one, via continued fractions, was obtained by Joseph Louis Lagrange (1736–1813) in 1776 [382] (see [102] for the complete history).

Let us mention that Padé approximants at infinity for the function $f(t) = \sum_{i=0}^{\infty} c_i t^{(-i+1)}$ can be defined for $p = q - 1$ and the condition $f(t)D_q(t) - N_{q-1}(t) = O(t^{-q-1})$.

Relying on the works of Augustin Louis Cauchy (1789–1857) [164, pp. 527–529] and Carl Gustav Jacobi (1804–1851) [348] on rational interpolation, Ferdinand Georg Frobenius (1849–1917) proved in 1881 that [251]

$$[p/q]_f(t) = \begin{vmatrix} t^q f_{p-q}(t) & t^{q-1} f_{p-q+1}(t) & \cdots & f_p(t) \\ c_{p-q+1} & c_{p-q+2} & \cdots & c_{p+1} \\ \vdots & \vdots & & \vdots \\ c_p & c_{p+1} & \cdots & c_{p+q} \end{vmatrix} \bigg/ \begin{vmatrix} t^q & t^{q-1} & \cdots & 1 \\ c_{p-q+1} & c_{p-q+2} & \cdots & c_{p+1} \\ \vdots & \vdots & & \vdots \\ c_p & c_{p+1} & \cdots & c_{p+q} \end{vmatrix},$$

where f_m is the partial sum of f up to the term of degree m inclusively. Let us mention that in [305], Hankel also implicitly gave this formula in 1861.

Comparing this formula with (2.13), it is easy to see, after some simple manipulations of the rows and the columns, that if the Shanks transformation (or the ε-algorithm) is applied to the partial sums of the formal power series f, then $\forall k, n$,
$$e_k(f_n) = \varepsilon_{2k}^{(n)} = [n + k/k]_f(t).$$

Usually Padé approximants are arranged in the double entry table shown in Fig. 2.3, called the *Padé table*:

$$\begin{array}{cccc} [0/0] & [0/1] & [0/2] & [0/3] & \cdots \\ [1/0] & [1/1] & [1/2] & [1/3] & \cdots \\ [2/0] & [2/1] & [2/2] & [2/3] & \cdots \\ [3/0] & [3/1] & [3/2] & [3/3] & \cdots \\ \vdots & \vdots & \vdots & \vdots & \ddots \end{array}$$

Fig. 2.3 The Padé table

Let g be the reciprocal series of f, which is formally defined by $g(t)f(t) = 1$. Writing $g(t) = d_0 + d_1 t + d_2 t^2 + \cdots$, this condition gives, by identification of the coefficients of the successive powers of t, $c_0 d_0 = 1$ and $c_0 d_i + c_1 d_{i-1} + \cdots + c_i d_0 = 0$ for $i \geq 1$. The series g exists if and only if $c_0 \neq 0$, and it can be proved that $[q/p]_g(t)[p/q]_f(t) = 1$. Thus, applying the ε-algorithm to the partial sums g_n of g leads to $e_k(g_n) = \varepsilon_{2k}^{(n)} = [n + k/k]_g(t) = 1/[k/n + k]_f(t)$, which shows that the ε-algorithm allows one to compute both halves of the Padé table. The first column of this table contains the partial sums of the series f, and the first row those of g.

An important property of Padé approximants is the *consistency property*. Namely, if f is an irreducible rational function with a numerator of degree p and a denominator of degree q, then for all $i, j \geq 0$, $[p + i/q + j]_f \equiv f$.

Padé approximants are uniquely defined. Indeed, two solutions of the problem lead to the same rational function, since $N_1(t) - f(t)D_1(t) = O(t^{p+q+1})$ and $N_2(t) - f(t)D_2(t) = O(t^{p+q+1})$ implies $N_1(t)D_2(t) - D_1(t)N_2(t) = O(t^{p+q+1})$. But the degree of $N_1 D_2 - D_1 N_2$ is at most $p + q$, and thus $N_1(t)D_2(t)$ is identical to $D_1(t)N_2(t)$.

Consider a rational function $R(t) = N(t)/D(t)$. The polynomials N and D can have a common factor. In particular, if t^k is a factor of D, it is also a factor of N, as can be seen from the previous system and thus $N(t)/D(t)$ cannot have a pole at the origin. Dividing by the highest power k in t contained in D gives a solution with $D(0) \neq 0$ and degree $N \leq p - k$, degree $D \leq q - k$, $N(t) - f(t)D(t) = O(t^{p+q+1-k})$, and we have the following result.

Theorem 2.10 *Let $R(t) = N(t)/D(t)$ be an irreducible rational function where N has degree $p - k$, D has degree $q - k$, with $k \geq 0$, and such that*

$$N(t) - f(t)D(t) = O(t^{p+q+1-k}) \quad (t \to 0).$$

Then, for $i, j = 0, \ldots, k$,

$$[p - k + i/q - k + j]_f(t) \equiv R(t)$$

and no other Padé approximant is identical to R if k is maximal.

This result, obtained by Padé in his doctoral thesis [481], follows from the definition of the approximants and their uniqueness. This identity between Padé approximants can hold for all i and j, and in that case, f is a rational function with a numerator of degree $p - k$ and a denominator of degree $q - k$, or there may exist a maximal value of k for which it holds, and in that case, no other Padé approximant is identical to R. The preceding theorem says that identical Padé approximants can be located only in square blocks of the table, a property known as the *block structure* of the Padé table. A Padé table with no blocks is called *normal*.

To end this presentation let us say a few words about Padé–Hermite approximants, since they had an important historical impact. Let $f_1, \ldots, f_k$ be given formal power series. The Padé–Hermite approximation problem consists in finding the polynomials $P_1, \ldots, P_k$ of respective degrees $n_1, \ldots, n_k$ such that

$$P_1(x)f_1(x) + \cdots + P_k(x)f_k(x) = O(x^{n_1 + \cdots + n_k + k - 1}).$$

This problem was studied by Charles Hermite (1822–1901) in 1893 [324], and Henri Padé in 1894 [482]. It was using this type of simultaneous approximations, but for several numbers instead of series, that Hermite proved the transcendence of the number e in 1873 [323]. His fundamental idea was quite close to a work of Jacobi [349], only found seventeen years after his death, on the simultaneous rational approximation of two real numbers. It consists in finding the polynomials $p_1, \ldots, p_m$ of respective degrees $N - n_1, \ldots, N - n_m$ with $N = n_1 + \cdots + n_m$ such that, for $i, j = 1, \ldots, m$, one has $e^{k_i x} p_j(x) - e^{k_j x} p_i(x) = O(x^{N+1})$, where $k_1, \ldots, k_m$ are arbitrary positive integers. Nine years later, walking in Hermite's footsteps, Carl Louis Ferdinand von Lindemann (1852–1939) proved the transcendence of π [410], thus closing by a negative result the problem of the quadrature of the circle, an open problem for more than 2000 years (see [430, Chap. IX] for simple proofs). A new proof that π is transcendental was given in 1996 by Vladimir Nikolaevich Sorokin [606]. It is based on a simultaneous Padé approximation problem involving certain multiple polylogarithms. On this type of approximants, let us also mention the work of Hendrik Jager (b. 1933) [350], a student of Jan Popken (1905–1970).

2.3.2 Convergence

By convergence of Padé approximants we mean the convergence of an infinite sequence of approximants in the Padé table as at least one of the degrees goes to infinity. This sequence can be formed by the approximants lying on a diagonal or a paradiagonal, or in a column of the table, or in a row, or on any ray sequence. Convergence can be understood as pointwise convergence, or uniform convergence on compact subsets of the complex plane, or convergence in measure or even in capacity. It can concern a particular function, or a class of functions, or functions with branch points, or entire functions, or meromorphic functions. In any case, it is a difficult problem of complex analysis. For precise definitions and explanations about the various sequences and concepts considered, consult [26, 31, 142, 143].

To see the difficulties about convergence, let us give a simple example taken from [51]. We consider the following series in the complex variable z:

$$f(z) = \frac{10 + z}{1 - z^2} = \sum_{i=0}^{\infty} c_i z^i,$$

with $c_{2i} = 10$ and $c_{2i+1} = 1$. It converges for $|z| < 1$. We have

$$[k/1] = \sum_{i=0}^{k-1} c_i z^i + \frac{c_k z^k}{1 - c_{k+1} z / c_k}.$$

When k is odd, $[k/1]$ has a simple pole at $z = 1/10$, while f has no pole. Thus the sequence $([k/1])$ cannot converge to f in $|z| < 1$. The poles of the Padé approximants can prevent convergence, and a sequence of them can be nonconvergent in a domain where the series converges. In order to prove the convergence of a sequence of approximants in a certain domain, it must be proved that the spurious (or dubious) poles (that is, those that do not approximate poles of f) move out of the domain of convergence as the degree(s) tend to infinity. But more surprisingly, the zeros of the approximants can also prevent convergence. Let $g(z) = (1 - z^2)/(10 + z)$ be the reciprocal series of f. It converges in $|z| < 10$. We have $[k/1]_f(z)[1/k]_g(z) = 1$. Since $[1/2k + 1]_g(0.1) = 0$ and $g(0.1) \neq 0$, the sequence $([1/k]_g)$ cannot converge in $|z| < 10$, where the series g does. These examples show the crucial importance of the study of the location of the zeros and the poles of Padé approximants. This study is based on the asymptotic behavior of the related families of formal orthogonal polynomials.

The existence of Froissart doublets, that is, nearby pole–zero pairs, is also problematic, and they have to be eliminated [272]. As mentioned in [37, pp. 296–298], these doublets were first observed around 1970 by Marcel Froissart (1934–2014), a French theoretical nuclear physicist, professor at *Collège de France*, who proved in 1961 that total scattering cross-sections among particles (quantities related to their interaction probabilities) do not increase faster than the square of the logarithm of the energy when this energy increases [252]. This was an important result both for theorists and experimentalists.

Let us now give one optimistic result followed by one that seriously limits our ambitions. The first result concerns the exponential function.

Theorem 2.11 *Let $f(z) = e^z$. For any integer sequence (m_i, n_i), $i \geq 1$, such that $\lim_{i \to \infty}(m_i + n_i) = \infty$, the poles of the Padé approximants $[m_i/n_i]_f$ tend to infinity and $\lim_{i \to \infty}[m_i/n_i]_f(z) = e^z$ uniformly on any compact subset of $\mathbb{C}$.*

The next result is due to Hans Wallin (b. 1936) [684].

Theorem 2.12 *There exists an entire function f such that the sequence of diagonal Padé approximants $([k/k]_f)$ is unbounded at every point of the complex plane except at zero, and so no convergence result can be expected on any open set of the complex plane.*

Let us now give a quite general theorem due to William Branham Jones (b. 1931)[4] and Wolfgang Joseph Thron (1918–2001) [359].

Theorem 2.13 *Let (m_k) and (n_k) be two sequences of nonnegative integers such that $\lim_{k \to \infty} \max(m_k, n_k) = \infty$, let $R_k(z) = [m_k/n_k]_f(z)$, and let D be a domain of the complex plane containing the origin. Then*

- *(R_k) converges uniformly on any compact subset of D if and only if $\{R_k(z)\}$ is uniformly bounded on any compact subset of D.*
- *If (R_k) converges uniformly on any compact subset of D, then $f(z) = \lim_{k \to \infty} R_k(z)$ is holomorphic in D and the series is the Taylor expansion of the function f about the origin.*

The most famous theorem in this area is due to Robert Fernand Bernard, Viscount de Montessus de Ballore (1870–1937) [456].[5] It concerns one and only one column of the Padé table.

Theorem 2.14 *Let f be analytic at $z = 0$ and meromorphic with exactly n poles $\alpha_1, \ldots, \alpha_n$ counted with their multiplicity in the disk $D_R = \{z, |z| < R\}$. Let D be the domain $D_R - \{\alpha_1, \ldots, \alpha_n\}$. Then, the sequence $([m/n]_f)$ converges to f uniformly on any compact subset of D as m goes to infinity, and the poles of $[m/n]_f$ converge to those of f.*

This theorem gives a result only for the nth column of the Padé table. If f has fewer poles, only some of the poles of $[m/n]_f$ are attracted by those of f, and the other ones may go anywhere, thus destroying the convergence. On the genesis and the circulation of this theorem, see [400].

---

[4] See his testimony in Sect. 7.6.

[5] See http://leferrand.perso.math.cnrs.fr/RMB.html for a picture of Montessus and additional information (in French).

An important class of series is that of Stieltjes series. It is a series of the form $S(z) = \sum_{i=0}^{\infty}(-1)^i c_i z^i$ with $c_i = \int_0^{\infty} x^i \, d\varphi(x)$, where φ is a positive, bounded, and nondecreasing measure. The function f associated with this series is $f(z) = \int_0^{\infty} d\varphi(x)/(1 + xz)$. The series S is the expansion of f into a formal power series that may not converge except at $z = 0$, while the function is analytic in the cut plane $\mathbb{C} - (-\infty, 0)$. In the case of a Stieltjes series of radius $R \neq 0$, f is analytic in $\mathbb{C} - (-\infty, -R)$, and for all $j \geq 0$, we have uniform convergence to f of the sequence $([n + j/n]_f)$ in a certain domain of the complex plane as n goes to infinity. We will return to these series when analyzing the corresponding works of Wynn.

The last result we will consider is that of the so-called Pólya frequency series [21].

Theorem 2.15 *Consider the series expansion $\sum_{i=0}^{\infty} c_i z^i$ of the function*

$$f(z) = a_0 e^{\gamma z} \prod_{i \geq 0} \left(\frac{1 + \alpha_i z}{1 - \beta_i z} \right),$$

with $a_0 > 0, \gamma \geq 0, \alpha_i \geq 0, \beta_i \geq 0, \sum_i(\alpha_i + \beta_i) < \infty$. Then

- *The Padé table is normal, and $(-1)^{m(m-1)/2} H_n(c_{n-m+1}) > 0$.*

- *Let (m_k, n_k) be a ray sequence such that $\lim_{k \to \infty} m_k = \infty$ and $\lim_{k \to \infty} m_k/n_k = \omega$ with $0 \leq \omega \leq \infty$, and let $[m_k/n_k]_f = N_{m_k,n_k}/D_{m_k,n_k}$. Then*

$$\lim_{k \to \infty} N_{m_k,n_k} = a_0 e^{\omega \gamma z/(1-\omega)} \prod_{i \geq 0}(1 + \alpha_i z),$$

$$\lim_{k \to \infty} D_{m_k,n_k} = a_0 e^{-\gamma z/(1+\omega)} \prod_{i \geq 0}(1 - \beta_i z).$$

The convergence of both sequences is uniform on any compact subset of the complex plane.

2.4 Formal Orthogonal Polynomials

The theory of classical orthogonal polynomials emerges from that of continued fractions [102]. As stated in the preface of [612]:

> The theory of orthogonal polynomials can be divided into two loosely related parts. One of them is the formal, algebraic aspect of the theory, which has close connections with special functions, combinatorics, and algebra, and it is mainly devoted to concrete orthogonal systems or hierarchies of systems such as the Jacobi, Hahn, Askey–Wilson ... polynomials.

> The investigation of more general orthogonal polynomials with methods of mathematical analysis belongs to the other part of the theory. Here the central questions are the asymptotic behavior of the polynomials and their zeros, recovering the measure of orthogonality, and so forth. This part has applications to approximation processes such as polynomial and rational

interpolation, Padé approximation, and best rational approximation, to Fourier expansions, quadrature processes, eigenvalue problems, and so forth.

The algebraic part of the theory of orthogonal polynomials is related to the definition, construction, and algebraic properties of Padé approximants, while the analytic part plays a fundamental role in their convergence.

After defining what formal orthogonal polynomials are, we explain their link with Padé approximants. Then some of their important algebraic properties are reviewed.

2.4.1 Definition and Link with Padé Approximants

Let (c_i) be a sequence of real or complex numbers. We consider the linear functional $c^{(n)}$ on the vector space of polynomials defined by $c^{(n)}(x^i) = c_{n+i}$, and let $P_k^{(n)}(x) = a_0 + a_1 x + \cdots + a_k x^k$ be the polynomial of exact degree k satisfying the conditions

$$c^{(n)}(x^i P_k^{(n)}(x)) = 0 \quad \text{for} \quad i = 0, \ldots, k-1,$$

that is,

$$a_0 c_{n+i} + a_1 c_{n+i+1} + \cdots + a_k c_{n+i+k} = 0, \quad i = 0, \ldots, k-1.$$

Such a polynomial exists if the Hankel determinant $H_k(c_n)$ is nonzero. If this condition is satisfied for all k and a fixed value of n, the set $\{P_k^{(n)}\}$ is the *family of formal orthogonal polynomials* (denoted by FOP) with respect to $c^{(n)}$, which is said to be *definite*. For different values of n, they are called *adjacent* families of FOPs.

Let $Q_k^{(n)}$ be the polynomial of degree $k - 1$ *associated* to $P_k^{(n)}$, and defined by

$$Q_k^{(n)}(t) = c^{(n)} \left(\frac{P_k^{(n)}(x) - P_k^{(n)}(t)}{x - t} \right),$$

where $c^{(n)}$ acts on x and t is a parameter. It is a polynomial of degree $k - 1$.

It is easy to see from the equations giving the coefficients of the numerators and the denominators of Padé approximants that we have

$$[n + k/k]_f(t) = \sum_{i=0}^{n} c_i t^i + t^{n+1} \widetilde{Q}_k^{(n+1)}(t) / \widetilde{P}_k^{(n+1)}(t),$$

with $\widetilde{Q}_k^{(n+1)}(t) = t^{k-1} Q_k^{(n+1)}(t^{-1})$ and $\widetilde{P}_k^{(n+1)}(t) = t^k P_k^{(n+1)}(t^{-1})$ (which corresponds to a reversal of the numbering of their coefficients).

There exists a bunch of recurrence relations between these families of polynomials and their associated polynomials. They allow one to compute recursively any sequence of Padé approximants in the Padé table. Some of them were obtained one by one and independently by various researchers without any link between them or any guiding idea as to how they were derived (see, for example, [25]). The theory of formal orthogonal polynomials showed that they had, in fact, a common origin. Moreover, new relations for their recursive computation were obtained; see [94].

It could of interest to explain how formal orthogonal polynomials get into the theory of Padé approximation [114]. Let again $f(t) = \sum_{i=0}^{\infty} c_i t^i$ be a formal power series, and let c be the linear functional on the vector space of polynomials defined by (it is identical to $c^{(0)}$ as previously defined)

$$c(x^i) = c_i, \quad i = 0, 1, \ldots.$$

We have, with c acting on x and t a parameter,

$$
\begin{aligned}
c\left(\frac{1}{1 - xt}\right) &= c(1 + xt + x^2 t^2 + \cdots) \\
&= c(x^0) + c(x^1)t + c(x^2)t^2 + \cdots \\
&= c_0 + c_1 t + c_2 t^2 + \cdots \\
&= f(t).
\end{aligned}
$$

Similarly to what is done for building an interpolatory quadrature method, replacing $1/(1 - xt)$ by its interpolation polynomial R_{k-1} of degree $k - 1$ at k points leads to an approximation $c(R_{k-1})$ of $f(t)$. Let v_k be a polynomial of degree exactly k, and let $x_1, \ldots, x_n$ be its distinct zeros, each one with its multiplicity k_i for $i = 1, \ldots, n$, and with $k_1 + \cdots + k_n = k$. The Hermite interpolation polynomial R_{k-1} of $1/(1 - xt)$ at the zeros of v_k is given by the formula

$$R_{k-1}(x) = \frac{1}{1 - xt}\left(1 - \frac{v_k(x)}{v_k(t^{-1})}\right).$$

Indeed, it can be seen that R_{k-1} is a polynomial of degree $k - 1$ in x (and also in t) and that it satisfies

$$R_{k-1}^{(j)}(x_i) = \left.\frac{d^j}{dx^j}\left(\frac{1}{1 - xt}\right)\right|_{x=x_i},$$

for $j = 1, \ldots, k_i - 1$ and $i = 1, \ldots, n$. For a generalization of this formula, see [345, 347] and [346, pp. 59–61].

An approximation of $f(t) = c(1/(1 - xt))$ is obtained by

$$c(R_{k-1}(x)) = \frac{1}{t^k v_k(t^{-1})} t^{k-1} c\left(\frac{v_k(t^{-1}) - v_k(x)}{t^{-1} - x}\right).$$

Let us define the polynomial w_k of degree $k - 1$ associated to v_k by

$$w_k(t) = c\left(\frac{v_k(x) - v_k(t)}{x - t}\right).$$

Setting $\widetilde{v}_k(t) = t^k v_k(t^{-1})$ and $\widetilde{w}_k(t) = t^{k-1} w_k(t^{-1})$ (which corresponds to a reversing of the numbering of their coefficients), we have $c(R_{k-1}(x)) = \widetilde{w}_k(t)/\widetilde{v}_k(t)$. Replacing

$R_{k-1}(x)$ by its expression, yields

$$c(R_{k-1}(x)) = c\left(\frac{1}{1-xt}\right) - \frac{t^k}{\widetilde{v}_k(t)} c\left(\frac{v_k(x)}{1-xt}\right) = f(t) + O(t^k).$$

The rational approximant $c(R_{k-1}(x))$ has a numerator of degree $k - 1$ and a denominator of degree k; it is called a *Padé-type approximant* of f, and it is denoted by $(k - 1/k)_f(t)$ [93]. Thus, this approximation result is similar to the property of an interpolatory quadrature method of being exact on the vector space of polynomials of degree at most $k - 1$ [94]. The polynomial v_k is called its *generating polynomial*.

For computing an integral, it is well known that the degree of approximation can be improved by taking the interpolation points as the zeros of the corresponding orthogonal polynomial, thus leading to a *Gaussian quadrature method*. We will now play a similar game with $c(R_{k-1}(x))$. Looking back to the formula given above, we have

$$c(R_{k-1}(x)) - f(t) = -\frac{t^k}{\widetilde{v}_k(t)} c\left(\frac{v_k(x)}{1-xt}\right)$$

$$= -\frac{t^k}{\widetilde{v}_k(t)} c(v_k(x)(1 + xt + \cdots + x^{k-1}t^{k-1} + x^k t^k/(1 - xt))).$$

If we impose that the polynomial v_k satisfies $c(v_k(x)) = 0$, then the first term in this expansion of the error disappears. If we also impose that $c(xv_k(x)) = 0$, then the second term of the error also disappears, and so on. Since v_k is a polynomial of degree k, and since its $k + 1$ coefficients are defined up to a multiplying factor, we can impose on it only k conditions, namely

$$c(x^i v_k(x)) = 0, \quad i = 0, \ldots, k - 1.$$

Thus, the polynomial v_k is the polynomial of degree k belonging to the family of *formal orthogonal polynomials* with respect to the linear functional c. It will now be denoted by P_k instead of v_k to use a more conventional notation. It corresponds to the polynomial $P_k^{(0)}$ of our general definition, and it is given by

$$P_k(x) = (D_k)^{-1} \begin{vmatrix} 1 & x & \cdots & x^k \\ c_0 & c_1 & \cdots & c_k \\ c_1 & c_2 & \cdots & c_{k+1} \\ \vdots & \vdots & & \vdots \\ c_{k-1} & c_k & \cdots & c_{2k-1} \end{vmatrix}, \tag{2.17}$$

where $(D_k)^{-1} \neq 0$ is a normalization coefficient. The preceding expression of the error becomes

$$c(R_{k-1}(x)) - f(t) = -\frac{t^{2k}}{\widetilde{P}_k(t)} c\left(\frac{x^k P_k(x)}{1-xt}\right).$$

The rational approximant $c(R_{k-1}(x))$ is now the Padé approximant $[k - 1/k]_f(t)$ as previously defined. By writing $f(t) = \sum_{i=0}^{n-1} c_i t^i + t^n c^{(n)}(x^n/(1 - xt))$ with $c^{(n)}$ as defined above, the other Padé approximants can be recovered with the adjacent families of formal orthogonal polynomials corresponding to $c^{(n)}$ [89]. Thus, Padé approximants can be viewed as formal Gaussian quadrature methods [114]. The case of the Stieltjes series was already treated in [13].

Since the error of these quadrature methods can be estimated via a procedure due to Alexandr Semenovich Kronrod (1921–1986) [379], it was extended to Padé approximants [99].

2.4.2 Properties

Since formal orthogonal polynomials play an important role in Padé approximation, continued fractions, and Lanczos-type methods for solving systems of linear equations, we will spend some time on their algebraic properties.

Let us come back to the expression (2.17) for the orthogonal polynomial P_k. Multiplying its first row by x^i and applying it the linear functional c, we see that for $i = 0, \ldots, k - 1$, two rows are identical. Thus $c(x^i P_k(x)) = 0$ for $i = 0, \ldots, k - 1$. This polynomial has degree k exactly if and only if $H_k(c_0) \neq 0$, a condition that we assume to hold for all k (which means that c is what we called *definite*). This condition also ensures that $c(x^k P_k(x)) \neq 0$, and thus, more generally, $c(Q(x)P_k(x)) \neq 0$ for any polynomial Q of degree equal to k. The normalization coefficient $(D_k)^{-1}$ can be arbitrarily chosen. However, for our purpose, two choices are privileged. For the Shanks transformation, we need to have $P_k(1) = 1$, which leads to

$$
D_k = \begin{vmatrix}
1 & 1 & \cdots & 1 \\
c_0 & c_1 & \cdots & c_k \\
c_1 & c_2 & \cdots & c_{k+1} \\
\vdots & \vdots & & \vdots \\
c_{k-1} & c_k & \cdots & c_{2k-1}
\end{vmatrix}.
$$

For Padé approximants and Lanczos-type methods, we must have $P_k(0) = 1$, which corresponds to $D_k = H_k(c_1)$. Obviously, when needed, we assume that these determinants are different from zero for all k.

The main result that will be proved now is that when they all exist and have degree equal to their index (that is, when c is definite), formal orthogonal polynomials, like the usual ones, satisfy a three–term recurrence relationship. In what follows, the variable x will be indicated only when necessary.

We will formulate and prove this result in two different ways.

The first is the *constructive way*: we will show that *if* the polynomials are orthogonal, *then* they satisfy a three–term recurrence relation, and we will show how to compute the coefficients of this relation. In this kind of proof, we don't assume a priori that orthogonal polynomials satisfy such a recurrence relation. This recurrence relation is unknown and we will find it. This is the way most researches are usually conducted.

The second is the *direct way*: we assume that a family of polynomials satisfies a three–term recurrence relation whose coefficients have a certain form, and *then* we will prove that they are orthogonal. In this kind of proof, we know a priori the form of the recurrence relation satisfied by the polynomials, and then we verify that these polynomials are orthogonal. Some results in mathematics are proved following such a scheme. This is, for example, the case for the proof of Fermat's last theorem, which says that the equation $x^n + y^n = z^n$ has no solution in integers for $n > 2$.

Theorem 2.16 (Constructive Proof) *Every family of formal orthogonal polynomials satisfies a three–term recurrence relationship of the form*

$$P_{k+1}(x) = (A_{k+1}x + B_{k+1})P_k(x) - C_{k+1}P_{k-1}(x), \quad k = 0, 1, \ldots, \quad (2.18)$$

with $P_{-1}(x) = 0$ and $P_0(x) = D_0 \neq 0$. The coefficients of this relation are given by

$$A_{k+1} = h_{k+1}/\beta_k,$$
$$B_{k+1} = -h_{k+1}\alpha_k/(h_k\beta_k),$$
$$C_{k+1} = \beta_{k-1}h_{k+1}/(\beta_k h_{k-1}),$$

with $\alpha_k = c(xP_k^2), \beta_k = c(xP_kP_{k+1})$, and $h_k = c(P_k^2)$.

Proof xP_k is a polynomial of degree $k + 1$. Thus, there exist $a_0, \ldots, a_{k+1}$ (depending on k), with $a_{k+1} \neq 0$, such that

$$xP_k = a_0P_0 + a_1P_1 + \cdots + a_{k+1}P_{k+1}.$$

Multiplying both sides by P_i and applying c gives

$$c(xP_iP_k) = a_0c(P_iP_0) + a_1c(P_iP_1) + \cdots + a_{k+1}c(P_iP_{k+1}).$$

For $i = 0, \ldots, k - 2 \geq 0$, xP_i is a polynomial of degree $k - 1$ at most, and thus $c(xP_iP_k) = 0$. Moreover, from the orthogonality conditions, we have $a_i = 0$ for $i = 0, \ldots, k - 2$. Now, for $i = k - 1, k, k + 1$, we get $a_i = c(xP_iP_k)/c(P_i^2)$, that is, $a_{k-1} = \beta_{k-1}/h_{k-1}$, $a_k = \alpha_k/h_k$, and $a_{k+1} = \beta_k/h_{k+1}$. Thus the recurrence relation is proved for $k \geq 1$.

We can write it as (2.18) with $A_{k+1} = 1/a_{k+1}$, $B_{k+1} = -a_k/a_{k+1}$, and $C_{k+1} = a_{k-1}/a_{k+1}$. The recurrence relation (2.18) can be started from $k = 0$, with $P_{-1}(x) = 0$. Doing that, we obtain $P_1(x) = (A_1x + B_1)P_0(x)$, which has the same form as $xP_0(x) = a_1P_1(x) + a_0P_0(x)$. Let us notice that since $P_{-1}(x) = 0$, the coefficient C_1 can be arbitrarily chosen (thus h_{-1} is not needed in the formula for C_1). ∎

Obviously, we have to obtain other expressions for the coefficients, since the expressions given in Theorem 2.16 involve P_{k+1}, the polynomial we want to compute.

Let t_k be the coefficient of x^k in P_k, and s_k the coefficient of x^{k-1}, that is, $P_k(x) = t_k x^k + s_k x^{k-1} + \cdots$. Thus $h_k = t_k c(x^k P_k)$, $\beta_k = c(x P_k P_{k+1}) = t_k c(x^{k+1} P_{k+1})$, and it follows, after some algebraic manipulations [94], that

$$A_{k+1} = \frac{t_{k+1}}{t_k}.$$

We have

$$C_{k+1} = \frac{\beta_{k-1}}{\beta_k} \frac{h_{k+1}}{h_k} = \frac{t_{k-1} t_{k+1}}{t_k^2} \frac{h_k}{h_{k-1}}.$$

Finally, $\alpha_k = t_k c(x^{k+1} P_k) + s_k c(x^k P_k)$, and we get

$$B_{k+1} = -\frac{\alpha_k t_{k+1}}{h_k t_k}.$$

Arbitrary nonzero values could also be assigned to the coefficients t_k, since the FOPs are defined up to a multiplying factor. In particular, we can take $t_k = 1$ for all k (monic FOP). Thus, the three coefficients of the recurrence relation can be computed.

Theorem 2.17 (Direct Proof) *Let us assume that for all k, P_k has degree k exactly and that $\{P_k\}$ forms a family of formal orthogonal polynomials. Then these polynomials satisfy the recurrence relation*

$$P_{k+1}(x) = (A_{k+1} x + B_{k+1}) P_k(x) - C_{k+1} P_{k-1}(x), \quad k = 0, 1, \ldots,$$

with $P_{-1}(x) = 0$ and $P_0(x) \neq 0$, and where the coefficients are given by

$$A_{k+1} c(x^k P_k) - C_{k+1} c(x^{k-1} P_{k-1}) = 0,$$
$$A_{k+1} c(x^{k+1} P_k) + B_{k+1} c(x^k P_k) - C_{k+1} c(x^k P_{k-1}) = 0.$$

Then the polynomials P_k form a family of formal orthogonal polynomials with respect to a linear functional c whose moments can be computed (see Theorem 2.19 below).

Proof We have to prove that the polynomial P_{k+1}, computed by the relation given above, satisfies the orthogonality conditions $c(x^i P_{k+1}) = 0$ for $i = 0, \ldots, k$. We multiply the recurrence relationship by x^i, for $i = 0, \ldots, k - 2$, and we apply the linear functional c. We obtain

$$c(x^i P_{k+1}) = A_{k+1} c(x^{i+1} P_k) + B_{k+1} c(x^i P_k) - C_{k+1} c(x^i P_{k-1}).$$

By the orthogonality condition, $c(x^{i+1} P_k) = 0$ for $i = 0, \ldots, k - 2$, $c(x^i P_k) = 0$ for $i = 0, \ldots, k - 1$, and $c(x^i P_{k-1}) = 0$ for $i = 0, \ldots, k - 2$. Thus $c(x^i P_{k+1}) = 0$ for $i = 0, \ldots, k - 2$.

Let us see what happens for $i = k - 1$ and $i = k$. In these cases, multiplying the recurrence relationship respectively by x^{k-1} and x^k and applying c, we obtain the two relations given in the theorem. So, if the three coefficients A_{k+1}, B_{k+1} and C_{k+1}

are chosen in order to satisfy them, then $c(x^i P_{k+1}) = 0$ for $i = 0, \ldots, k - 1$. Thus, thanks to the uniqueness property of FOPs, the polynomial P_{k+1} computed by the three–term recurrence relationship is the $(k + 1)$th member of the family of FOPs with respect to c. ∎

So, we see, from the previous theorem that we have two relations for determining the three coefficients involved in the recurrence relationship. The missing relation is given by the normalization condition. For example, if we want the polynomials to be *monic* (that is, the coefficient of x^k in P_k equal to 1), we must take $A_{k+1} = 1$, and we obtain

$$C_{k+1} = \frac{c(x^k P_k)}{c(x^{k-1} P_{k-1})},$$

$$B_{k+1} = \frac{C_{k+1} c(x^k P_{k-1}) - c(x^{k+1} P_k)}{c(x^k P_k)}.$$

We see that the quantity $c(x^k P_k)$, which appears in the denominator of B_{k+1}, has to be different from zero. From the determinantal formula (2.17), $c(x^k P_k) = (-1)^k (D_k)^{-1} H_{k+1}^{(0)}$. Since $(D_k)^{-1} \neq 0$, the condition $c(x^k P_k) \neq 0$ is equivalent to $H_{k+1}^{(0)} \neq 0$, that is, to the existence of P_{k+1}. We saw that in the expression of C_{k+1}, we have to divide by the quantity $c(x^{k-1} P_{k-1})$. Its value must also be nonzero, which is equivalent to the existence of P_k. So, we have proved the following theorem.

Theorem 2.18 *In the case of monic FOPs, a division by zero occurs in the computation of the coefficients of the recurrence relationship if and only if the polynomial to be computed does not exist.*

The occurrence of such a division by zero is called a *true breakdown*, and the functional c is said to be *indefinite*. Let us consider the case of the normalization condition $P_k(0) = 1$. Setting $x = 0$ in the three–term recurrence relationship, we find that we have the additional equation $B_{k+1} - C_{k+1} = 1$. So, we obtain a system of three equations in the three unknowns $A_{k+1}, B_{k+1}, C_{k+1}$. This system may be singular even if P_{k+1} exists, and so this recurrence does not allow us to compute P_{k+1}. This type of breakdown is called a *ghost breakdown*, since it does not correspond to the nonexistence of the polynomial to be computed, but to the impossibility of using a particular recurrence relationship for its computation.

The reciprocal of Theorem 2.16 is true. It is named the *Shohat–Favard theorem*

Theorem 2.19 *Let $\{P_k\}$ be a family of polynomials satisfying a recurrence relation of the form (2.18) with, for all k, $A_{k+1} C_{k+1} \neq 0$. Then $\{P_k\}$ is a family of formal orthogonal polynomials with respect to a linear functional c whose moments can be computed.*

Proof Since the polynomials are defined up to a multiplying factor, we do not restrict the generality by assuming that they are monic.

We have $c(P_0) = c(1) = c_0$, where $c_0 \neq 0$ is an arbitrary number. For $k \geq 0$, we must have $c(P_{k+1}) = 0$. Thus for $k = 0$, this condition is $c(P_1) = c(xP_0) + B_1 c(P_0) = c_1 + B_1 c_0 = 0$, which gives c_1. For $k = 1$, we have $c(P_2) = c(xP_1) + B_2 c(P_1) - C_2 c(P_0) = 0$. Thus we get $c(xP_1)$. But multiplying P_1 by x and using the recurrence relation gives $c(xP_1) = c(x^2 P_0) + B_1 c(xP_0) = c_2 + B_1 c_1$, which provides c_2.

By induction, we assume that $c_0, \ldots, c_k$ are known. Then $c(P_{k+1}) = c(xP_k) + B_{k+1} c(P_k) - C_{k+1} c(P_{k-1}) = 0$. Since $c_0, \ldots, c_k$ are known, $c(P_k)$ and $c(P_{k-1})$ can be computed. We have $c(xP_k) = c(x^{k+1}) + c(p) = 0$, where p is a polynomial of degree at most k. Therefore $c(p)$ is known, which leads to the value of $c(x^{k+1}) = c_{k+1}$. ∎

An important result is the *Christoffel–Darboux identity*:

Theorem 2.20 *For $k \geq 0$,*

$$\frac{t_k}{h_k t_{k+1}} [P_{k+1}(x)P_k(t) - P_{k+1}(t)P_k(x)] = (x - t) \sum_{i=0}^{k} h_i^{-1} P_i(x)P_i(t),$$

where t_k is the coefficient of the term of degree k of P_k, and $h_k = c(P_k^2)$.

Proof There are two different proofs of this identity. The classical one makes use of the three–term recurrence relation, while the second proof is a direct one.

Proof 1: We have

$$P_{i+1}(x) = (A_{i+1}x + B_{i+1})P_i(x) - C_{i+1}P_{i-1}(x),$$
$$P_{i+1}(t) = (A_{i+1}t + B_{i+1})P_i(t) - C_{i+1}P_{i-1}(t).$$

Multiplying the first relation by $P_i(t)$, the second one by $P_i(x)$, and subtracting it from the first one gives

$$P_{i+1}(x)P_i(t) - P_{i+1}(t)P_i(x) = A_{i+1}(x-t)P_i(x)P_i(t) - C_{i+1}[P_i(t)P_{i-1}(x) - P_i(x)P_{i-1}(t)].$$

But $C_{i+1} = (t_{i-1}t_{i+1}h_i)/(t_i^2 h_{i-1}) = (A_{i+1}h_i)/(A_i h_{i-1})$, and thus

$$(x - t)h_i^{-1} P_i(x)P_i(t) = h_i^{-1} A_{i+1}^{-1}[P_i(t)P_{i+1}(x) - P_i(x)P_{i+1}(t)]$$
$$+ h_{i-1}^{-1} A_i^{-1}[P_i(t)P_{i-1}(x) - P_i(x)P_{i-1}(t)].$$

Summing up these equalities for $i = 0, \ldots, k$ yields the result.

Proof 2: A direct proof of this identity (that is, not using the three–term recurrence relationship) exists. It is quite technical (five pages); it makes use of the notion of kernel described below and the determinantal identities of Sylvester and Schweins. We refer the interested reader to [101]. ∎

Letting t tend to x in the Christoffel–Darboux identity leads to

Theorem 2.21

$$\frac{t_k}{h_k t_{k+1}} [P_{k+1}'(x)P_k(x) - P_{k+1}(x)P_k'(x)] = \sum_{i=0}^{k} h_i^{-1} P_i^2(x).$$

Proof On the left-hand side of the Christoffel–Darboux identity (Theorem 2.20), add and subtract $P_{k+1}(t)P_k(t)$, divide both sides by $x - t$, and let t tend to x. ∎

Assuming that for all k, P_k has the exact degree k, we see that the rational function $(P_k(x) - P_k(t))/(x - t)$ is a polynomial of degree $k - 1$ in the variable x and also in the parameter t. Thus we can apply to it the functional c (which acts on x), and we consider the polynomial $Q_k \equiv Q_k^{(0)}$ of degree $k - 1$ in t defined by

$$Q_k(t) = c\left(\frac{P_k(x) - P_k(t)}{x - t}\right).$$

The family $\{Q_k\}$ is called the family of *associated polynomials* to $\{P_k\}$.

From the determinantal expression (2.17) of P_k, we immediately have the following result.

Theorem 2.22

$$Q_k(t) = \begin{vmatrix} 0 & c_0 & c_0 t + c_1 & \cdots & c_0 t^{k-1} + c_1 t^{k-2} + \cdots + c_{k-1} \\ c_0 & c_1 & c_2 & \cdots & c_k \\ \vdots & \vdots & \vdots & & \vdots \\ c_{k-1} & c_k & c_{k+1} & \cdots & c_{2k-1} \end{vmatrix}, \quad k = 1, 2, \ldots,$$

and $Q_0(t) = 0$.

Theorem 2.23 *The family of associated polynomials $\{Q_k\}$ satisfies the three–term recurrence relationship (2.18), with $Q_{-1}(x) = -1, Q_0(x) = 0$, and C_1 defined by $C_1 = A_1 c(P_0(x))$.*

Proof Let us write the recurrence relation of the polynomials P_k twice, once with the variable x, and once with the variable t. Subtract the first one from the second, and divide both sides by $x - t$. We get

$$\frac{P_{k+1}(x) - P_{k+1}(t)}{x - t} = A_{k+1}\frac{xP_k(x) - tP_k(t)}{x - t} + B_{k+1}\frac{P_k(x) - P_k(t)}{x - t}$$
$$-C_{k+1}\frac{P_{k-1}(x) - P_{k-1}(t)}{x - t}.$$

But

$$\frac{xP_k(x) - tP_k(t)}{x - t} = t\frac{P_k(x) - P_k(t)}{x - t} + P_k(x).$$

Applying the linear function c, and since $c(P_k(x)) = 0$ for $k > 0$, we obtain the recurrence relation for $k = 1, 2, \ldots$. We also have

$$Q_0(t) = c\left(\frac{P_0(x) - P_0(t)}{x - t}\right) = c(0) = 0,$$

$$Q_1(t) = c\left(\frac{(A_1 x + B_1)P_0(x) - (A_1 t + B_1)P_0(t)}{x - t}\right) = A_1 c(P_0(x)),$$

since $P_0(x) = P_0(t)$ (a constant). Using this expression for Q_1 and the recurrence relation for $k = 0$ yields

$$Q_1(t) = A_1 c(P_0(x)) = -C_1 Q_{-1}(t).$$

Since C_1 can be arbitrarily chosen, we can set $C_1 = A_1 c(P_0(x))$ and $Q_{-1}(t) = -1$. ∎

From the recurrence relation of both families, we have the following theorem.

Theorem 2.24

$$P_k(x)Q_{k+1}(x) - Q_k(x)P_{k+1}(x) = A_{k+1}h_k, \quad k \geq 0.$$

Proof Multiply by Q_k the recurrence relation of P_{k+1}, by P_k the relation giving Q_{k+1}, and subtract. We obtain

$$P_k Q_{k+1} - Q_k P_{k+1} = C_{k+1}(P_{k-1}Q_k - Q_{k-1}P_k).$$

Thus we get

$$P_k Q_{k+1} - Q_k P_{k+1} = C_{k+1}C_k \cdots C_1 (P_{-1}Q_0 - Q_{-1}P_0).$$

But $C_{i+1} = (A_{i+1}h_i)/(A_i h_{i-1})$ and $C_1 = A_1 c(P_0)$. Thus $C_{k+1}C_k \cdots C_1 = A_{k+1}h_k$. ∎

This result is important, since it shows that P_k, Q_k, P_{k+1}, and Q_{k+1} cannot have a common zero.

Using the same notation as above for the polynomials P_k, several interesting results follow [94].

Theorem 2.25 *The family $\{Q_k\}$ satisfies the Christoffel–Darboux identity (Theorem 2.20) and its consequence (Theorem 2.21).*

Moreover, a kind of Christoffel–Darboux identity, mixing both families, exists.

Theorem 2.26

$$1 + \frac{t_k}{h_k t_{k+1}}[P_{k+1}(x)Q_k(t) - Q_{k+1}(t)P_k(x)] = (x - t)\sum_{i=0}^{k} h_i^{-1} P_i(x)Q_i(t).$$

Proof The proof is quite similar to that of Theorem 2.20. Write the recurrence relation for $P_{i+1}(x)$, and multiply it by $Q_i(t)$. Write the recurrence relation for $Q_{i+1}(t)$, multiply it by $P_i(x)$, and subtract from the first one. Multiply the result by $(h_i A_{i+1})^{-1}$, and use the expression $C_{i+1} = (A_{i+1}h_i)/(A_i h_{i-1})$. Finally, sum these relations for $i = 0, \ldots, k$. We get

$$(x - t)\sum_{i=0}^{k} h_i^{-1} P_i(x)Q_i(t) = \frac{t_k}{h_k t_{k+1}}[P_{k+1}(x)Q_k(t) - Q_{k+1}(t)P_k(x)]$$

$$+ (h_{-1}A_0)^{-1}[Q_0(t)P_{-1}(x) - P_0(x)Q_{-1}(t)].$$

The second term in the right-hand side equals 1, since $C_1 = (h_0 A_1)/(h_{-1} A_0)$, and by definition of $C_1, Q_{-1}, Q_0, P_{-1},$ and P_0. ∎

If we let t tend to x in the preceding relation, we obtain the following result.

Theorem 2.27

$$P'_{k+1}(x)Q_k(x) - Q_{k+1}(x)P'_k(x) = A_{k+1}h_k \sum_{i=0}^{k} h_i^{-1} P_i(x)Q_i(x),$$

$$P_{k+1}(x)Q'_k(x) - Q'_{k+1}(x)P_k(x) = -A_{k+1}h_k \sum_{i=0}^{k} h_i^{-1} P_i(x)Q_i(x).$$

2.5 Continued Fractions

In this section, we consider three types of continued fractions: the *arithmetic* continued fractions in which the a_i and the b_i are numbers, the *analytic* continued fractions in which they are functions of the complex variable z, and the *interpolating* continued fractions. The first are used in arithmetic and number theory, the second serve to approximate functions, and the third type is devoted to rational interpolation. The generalities explained in Sect. 2.5.1 are valid for these three types.

A general reference on continued fractions is [418]. The interdependence of moments, Hankel determinants, the qd-algorithm, continued fractions, and formal orthogonal polynomials is discussed in [200]. The impact of continued fractions on numerical analysis is analyzed in [360].

2.5.1 Generalities

A *continued fraction* is an expression of the form

$$C = b_0 + \cfrac{a_1}{b_1 + \cfrac{a_2}{b_2 + \cfrac{a_3}{b_3 + \cfrac{a_4}{\ddots}}}}.$$

For evident typographical reasons, it will be written as

$$C = b_0 + \frac{a_1|}{|b_1|} + \frac{a_2|}{|b_2|} + \frac{a_3|}{|b_3|} + \cdots;$$

a_k and b_k are called the kth *partial numerator* and *partial denominator*, respectively, a_k/b_k is the kth *partial quotient*, and

$$C_n = b_0 + \frac{a_1|}{|b_1} + \cdots + \frac{a_{n-1}|}{|b_{n-1}} + \frac{a_n|}{|b_n}$$

is called the nth *convergent* of the continued fraction C (even if the continued fraction does not converge). We have $C = C_n + R_n$, where $R_n = \frac{a_{n+1}|}{|b_{n+1}} + \cdots$ is the *remainder*. The continued fraction is said to converge if the sequence (C_n) converges as n goes to infinity.

After reducing it to a common denominator, C_n can be written as $C_n = A_n/B_n$. It can be computed by the *forward recurrences*

$$A_k = b_k A_{k-1} + a_k A_{k-2},$$
$$B_k = b_k B_{k-1} + a_k B_{k-2}, \quad k = 1, 2, \ldots,$$

with

$$A_0 = b_0, \quad A_{-1} = 1,$$
$$B_0 = 1, \quad B_{-1} = 0.$$

The convergent C_n can also be obtained by a *backward recurrence*. Set $D_0 = b_n$, and compute

$$D_{i+1} = b_{n-i-1} + a_{n-i}/D_i, \quad i = 0, 1, \ldots, n - 1.$$

Then $D_n = C_n$.

Since the ratio $C_n = A_n/B_n$ is determined only up to a multiplying factor, the B_n can be arbitrarily chosen. In particular, they can be taken so that $\forall n$, $B_n/B_{n-1} = 1$. Thus, several continued fractions with the same convergents but different a_n and b_n are obtained. They are said to be *equivalent*. We consider the two continued fractions

$$C = b_0 + \frac{a_1|}{|b_1} + \frac{a_2|}{|b_2} + \frac{a_3|}{|b_3} + \cdots,$$

$$C' = b_0 + \frac{d_1 a_1|}{|d_1 b_1} + \frac{d_1 d_2 a_2|}{\lceil d_2 b_2} + \frac{d_2 d_3 a_3|}{\lceil d_3 b_3} + \cdots.$$

They are equivalent and $C'_n = A'_n/B'_n$ with $A'_n = d_1 \cdots d_n A_n$ and $B'_n = d_1 \cdots d_n B_n$. All the continued fractions equivalent to C can be obtained in that way.

We consider the series $S = u_0 + u_1 + \cdots$ with partial sums S_n. It is possible to define a continued fraction C whose convergents satisfy $C_n = S_n$. It is given by

$$a_n = -u_n/u_{n-1}, \quad b_n = 1 + u_n/u_{n-1}, \quad n = 1, 2, \ldots,$$

with $b_0 = u_0$ and $a_1/b_1 = u_1$.

Reciprocally, a continued fraction with convergents C_n can be transformed into a series whose terms are $u_n = C_n - C_{n-1}$ and $S_n = C_n$.

Let (C_n) be the sequence of convergents of the continued fraction C and let (C_{p_n}) be a subsequence. The continued fraction

$$C' = b'_0 + \frac{a'_1|}{|b'_1} + \frac{a'_2|}{|b'_2} + \frac{a'_3|}{|b'_3} + \cdots$$

whose convergents satisfy $C'_n = C_{p_n}$ is given by

$$a'_n = \frac{C_{p_{n-1}} - C_{p_n}}{C_{p_{n-1}} - C_{p_{n-2}}}, \quad b'_n = \frac{C_{p_n} - C_{p_{n-2}}}{C_{p_{n-1}} - C_{p_{n-2}}},$$

with $b'_0 = C_{p_0}$, $b'_1 = 1$, and $a'_1 = C_{p_1} - C_{p_0}$. This operation is called a *contraction* of the continued fraction. Usually, $p_n = 2n$.

We consider the *linear fractional transformation*

$$S_n(w) = \frac{A_n + w A_{n-1}}{B_n + w B_{n-1}}.$$

We have

$$S_n(0) = C_n, \quad S_n(a_{n+1}/b_{n+1}) = C_{n+1}, \quad \lim_{w \to \pm \infty} S_n(w) = C_{n-1}, \quad S_n(R_n) = C.$$

Since the sequence (R_n) is unknown, it can be replaced by another sequence (r_n), and we set $C^*_n = S_n(r_n)$. This procedure is called a *modification* of the continued fraction, and (r_n) is called a sequence of *converging factors*. This sequence can be chosen so that (C^*_n) converges to C faster than (C_n).

A useful property for proving the convergence of some continued fractions is that the convergents of the continued fraction $\frac{1|}{|1} + \frac{a_2|}{|1} + \frac{a_3|}{|1} + \cdots$ are identical to the partial sums of the series $C_1 + (C_2 - C_1) + (C_3 - C_2) + \cdots$ [563].

2.5.2 The Analytic Theory of Continued Fractions

The analytic theory of continued fractions is concerned with the convergence of continued fractions as a function of the complex variable z. Let us consider the continued fraction

$$C = b_0 + \frac{a_1 z|}{|1} + \frac{a_2 z|}{|1} + \frac{a_3 z|}{|1} + \cdots.$$

By the recurrence relations, we see that A_{2k-1}, A_{2k}, and B_{2k} are polynomials of degree k in z and that B_{2k-1} is a polynomial of degree $k-1$. The expansions of C_k and C_{k-1} in ascending powers of z agree up to the term of degree $k-1$ inclusively. It is possible to choose $b_0, a_1, a_2, \ldots$ so that the expansion of C_k agrees with that of a given series $f(z) = c_0 + c_1 z + c_2 z^2 + \cdots$ up to the term of degree k. This continued fraction is called the continued fraction *corresponding* to the series f. By a contraction of this continued fraction (with $p_k = 2k$ as explained above), we obtain a continued fraction whose convergent C'_k agrees with that of f up to the term of degree $2k$. This is the continued fraction *associated* to the series f. Thus, by the uniqueness property of Padé approximants, $C_{2k} = [k/k]_f(z)$ and $C_{2k+1} = [k+1/k]_f(z)$.

The other Padé approximants $[p/q]$ can also be related to continued fractions; see, for example, [94].

Let us now see how to obtain the partial numerators and denominators of the corresponding continued fraction of a formal power series $f(z) = c_0 + c_1 z + c_2 z^2 + \cdots$. The *qd-algorithm*, due to Heinz Rutishauser (1918–1970)[6] [536, 538], consists of the recursive rules

$$e_k^{(n)} = q_k^{(n+1)} - q_k^{(n)} + e_{k-1}^{(n+1)}, \quad q_{k+1}^{(n)} = \frac{q_k^{(n+1)} e_k^{(n+1)}}{e_k^{(n)}}, \qquad k = 1, 2, \ldots; n = 0, 1, \ldots,$$

with $e_0^{(n)} = 0$ and $q_1^{(n)} = c_{n+1}/c_n$. These rules relate numbers at the four vertices of a rhombus in Fig. 2.4.

$$\begin{array}{cccccc}
 & q_1^{(0)} & & & & \\
e_0^{(1)} & & e_1^{(0)} & & & \\
 & q_1^{(1)} & & q_2^{(0)} & & \\
e_0^{(2)} & & e_1^{(1)} & & e_2^{(0)} & \\
 & q_1^{(2)} & & q_2^{(1)} & & \ddots \\
e_0^{(3)} & & e_1^{(2)} & & e_2^{(1)} & \\
 & q_1^{(3)} & & q_2^{(2)} & & \ddots \\
e_0^{(4)} & \vdots & e_1^{(3)} & & e_2^{(2)} & \\
\vdots & & \vdots & q_2^{(3)} & \vdots & \ddots \\
 & & & \vdots & &
\end{array}$$

Fig. 2.4 The qd table

It can be proved that

$$q_k^{(n)} = \frac{H_k^{(n+1)} H_{k-1}^{(n)}}{H_k^{(n)} H_{k-1}^{(n+1)}}, \quad e_k^{(n)} = \frac{H_{k+1}^{(n)} H_{k-1}^{(n+1)}}{H_k^{(n)} H_k^{(n+1)}},$$

where $H_k^{(n)} \equiv H_k(c_n)$ is the Hankel determinant defined in Sect. 2.2.3.

The successive convergents $C_k^{(n)}$ of the continued fraction

[6] See Sect. 8.10 for his biography.

$$C^{(n)}(z) = c_0 + c_1 z + \cdots + c_n z^n + \left|\frac{c_{n+1} z^{n+1}}{1}\right. - \left|\frac{q_1^{(n+1)} z}{1}\right.$$

$$- \left|\frac{e_1^{(n+1)} z}{1}\right. - \left|\frac{q_2^{(n+1)} z}{1}\right. - \left|\frac{e_2^{(n+1)} z}{1}\right. - \cdots$$

are, in fact, the Padé approximants of the series f,

$$C_{2k}^{(n)}(z) = [n + k/k]_f(z), \quad C_{2k+1}^{(n)}(z) = [n + k + 1/k]_f(z),$$

which lie on the staircase of the Padé table

$$[n/0]_f$$
$$[n + 1/0]_f \quad [n + 1/1]_f$$
$$[n + 2/1]_f \quad [n + 2/2]_f$$
$$\ddots$$

Here $C^{(n)}$ is the continued fraction corresponding to the series f. By contracting it we obtain the continued fraction associated to f:

$$C'^{(n)}(z) = c_0 + c_1 z + \cdots + c_n z^n + \left|\frac{c_{n+1} z^{n+1}}{1 - q_1^{(n+1)} x}\right. - \left|\frac{q_1^{(n+1)} e_1^{(n+1)} z^2}{1 - (q_2^{(n+1)} + e_1^{(n+1)})z}\right.$$

$$- \left|\frac{q_2^{(n+1)} e_2^{(n+1)} z^2}{1 - (q_3^{(n+1)} + e_2^{(n+1)})z}\right. - \cdots .$$

Let us mention that the qd-algorithm was the basis for the development by Rutishauser of the LR-algorithm for the computation of the eigenvalues of a matrix [536] (see also [321]).

2.5.3 Interpolating Continued Fractions

Let us now turn to rational interpolation. Such rational functions can be constructed by means of continued fractions, as shown by Thorvald Nicolai Thiele (1838–1910)[7] [637], a Danish astronomer and director of the Copenhagen Observatory. Consider the continued fraction

$$C^{(n)}(x) = \alpha_0^{(n)} + \left|\frac{x - x_n}{\alpha_1^{(n)}}\right. + \left|\frac{x - x_{n+1}}{\alpha_2^{(n)}}\right. + \cdots,$$

with $\alpha_k^{(n)} = \varrho_k^{(n)} - \varrho_{k-2}^{(n)}$ for $k = 1, 2, \ldots$, and $\alpha_0^{(n)} = \varrho_0^{(n)} = f(x_n)$, and where the scalars $\varrho_k^{(n)}$ are recursively computed by

---

[7] See Sect. 8.15 for his biography.

$$\varrho_{k+1}^{(n)} = \varrho_{k-1}^{(n+1)} + \frac{x_{n+k+1} - x_n}{\varrho_k^{(n+1)} - \varrho_k^{(n)}},$$

with $\varrho_{-1}^{(n)} = 0$ and $\varrho_0^{(n)} = f(x_n)$. These quantities are the so-called *reciprocal differences* of f (see [550] for an alternative definition). They were used by Peter Wynn for defining his *ϱ-algorithm* [712] for rational extrapolation at infinity; see Sect. 5.1.5.

The kth convergent $C_k^{(n)}$ of this continued fraction satisfies the interpolation conditions $C_k^{(n)}(x_i) = f(x_i)$ for $i = n, \ldots, n + k$, and it can be proved that it has the form

$$C_{2k}^{(n)}(x) = \frac{\varrho_{2k}^{(n)} x^k + \cdots}{x^k + \cdots}.$$

Thus $\lim_{x \to \infty} C_{2k}^{(n)}(x) = \varrho_{2k}^{(n)}$, an essential property in rational extrapolation to infinity. Thus if f has the form

$$f(x) = \frac{S x^k + a_1 x^{k-1} + \cdots + a_k}{x^k + b_1 x^{k-1} + \cdots + b_k},$$

then for all n, $\varrho_{2k}^{(n)} = S$. Moreover, setting $S_n = f(x_n)$, we have

$$\varrho_{2k}^{(n)} = \frac{\begin{vmatrix} 1 & S_n & x_n & x_n S_n & \cdots & x_n^{k-1} & x_n^{k-1} S_n & x_n^k S_n \\ \vdots & \vdots & \vdots & \vdots & & \vdots & \vdots & \vdots \\ 1 & S_{n+2k} & x_{n+2k} & x_{n+2k} S_{n+2k} & \cdots & x_{n+2k}^{k-1} & x_{n+2k}^{k-1} S_{n+2k} & x_{n+2k}^k S_{n+2k} \\ 1 & S_n & x_n & x_n S_n & \cdots & x_n^{k-1} & x_n^{k-1} S_n & x_n^k \\ \vdots & \vdots & \vdots & \vdots & & \vdots & \vdots & \vdots \\ 1 & S_{n+2k} & x_{n+2k} & x_{n+2k} S_{n+2k} & \cdots & x_{n+2k}^{k-1} & x_{n+2k}^{k-1} S_{n+2k} & x_{n+2k}^k \end{vmatrix}}. \tag{2.19}$$

Let $C_k^{(n)}(x) = A_k^{(n)}(x)/B_k^{(n)}(x)$, and assume that $x_n, \ldots, x_{n+k} \in [a, b]$. Then

$$f(x) - C_k^{(n)}(x) = \frac{(x - x_n) \cdots (x - x_{n+k})}{(k + 1)! B_k^{(n)}(x) Q(x)} \frac{d^{k+1}}{d\xi^{k+1}} [f(\xi) B_k^{(n)}(\xi) Q(\xi)], \quad \xi \in [a, b],$$

where Q is a polynomial of the same degree as $B_k^{(n)}$. If f has no pole in $[a, b]$, one can take $Q(x) = B_k^{(n)}(x)$. This interpolation process was studied by Niels Erik Nörlund (1885–1981) [471], a Danish mathematician; see [188] for a more recent treatment.

Let us now consider the confluent reciprocal differences of a function f defined by

$$\varrho_{k+1}(t) = \varrho_{k-1}(t) + \frac{k + 1}{\varrho_k'(t)},$$

with $\varrho_{-1}(t) = 0$ and $\varrho_0(t) = f(t)$. This formula would be later named the *confluent form of the ϱ-algorithm* by Peter Wynn [721] and used as an extrapolation method for functions; see Sect. 5.1.2. *Thiele's expansion* of a function f is

$$f(t + h) = f(t) + \frac{h}{|\alpha_1(t)} + \frac{h}{|\alpha_2(t)} + \cdots,$$

with $\alpha_k(t) = \varrho_k(t) - \varrho_{k-2}(t)$ for $k = 1, 2, \ldots$. It is the analogue for continued fractions of the Taylor's expansion into a series. Since as Taylor expansion terminates when f is a polynomial, Thiele's expansion terminates when f is a rational function with a numerator and a denominator of the same degree, or when the degree of the denominator is that of the numerator minus one. The error of the convergent $C_k(t) = A_k(t)/B_k(t)$ is given by

$$f(t + h) - C_k(t + h) = \frac{h^{k+1}}{(k + 1)!} \frac{1}{B_k(t + h)Q(t + h)} \frac{d^{k+1}}{dx^{k+1}} [f(\xi)B_k(\xi)Q(\xi)],$$

where $\xi \in [x, x + h]$ and Q is a polynomial of the same degree as B_k.

Replacing t by 0 and h by x, we get

$$f(x) = f(0) + \frac{x}{|\alpha_1(0)} + \frac{x}{|\alpha_2(0)} + \cdots.$$

The successive convergents $C_k(x) = A_k(x)/B_k(x)$ of this continued fraction are such that $f(x) - C_k(x) = O(x^{k+1})$. Since Padé approximants are uniquely defined, we have

$$C_{2k}(x) = [k/k]_f(x), \quad C_{2k+1}(x) = [k + 1/k]_f(x),$$

and the error is given by

$$f(x) - C_k(x) = \frac{x^{k+1}}{(k + 1)!B_k(x)Q(x)} \frac{d^{k+1}}{d\xi^{k+1}} [f(\xi)B_k(\xi)Q(\xi)], \quad \xi \in [0, x].$$

If f has no pole in $[0, x]$, one can take $Q(x) = B_k(x)$. Thus, the preceding continued fraction can be obtained by the qd-algorithm with $n = 0$. In other words, there is a connection between the qd-algorithm and the confluent form of the ϱ-algorithm. See [90] for details.

Chapter 3
The Mathematical Landscape up to the Mid-Twentieth Century

The aim of this chapter is to give a picture of the knowledge already acquired around the middle of the twentieth century in the areas that interest us. Some of these domains had been explored for quite a long time, while some others were only recently opened, and other ones yet to be discovered. Before visiting each domain in particular (extrapolation methods, formal orthogonal polynomials, Padé approximants, continued fractions), we begin with a general presentation.

3.1 Overview

Linear sequence transformations have been known for a long time, at least two centuries since the paper of Leonhard Euler (1707–1783) dated from 1760 (but communicated in 1755) [224]. They emerge from the treatment of divergent series. In the late seventeenth century there was a controversy about what "sum," if any, could be assigned to the series $1 - 1 + 1 - 1 + 1 - 1 + \cdots$. This series was considered by the Italian priest Luigi Guido Grandi (1671–1742) in 1703. He noticed that inserting parentheses into it produced two different results: either $(1 - 1) + (1 - 1) + \cdots = 0$ or $1 + (-1 + 1) + (-1 + 1) = 1$. He gave a religious explanation of this phenomenon:

> By putting parentheses into the expression $1 - 1 + 1 - 1 + \cdots$ in different ways, I can, if I want, obtain 0 or 1. But then the idea of the creation *ex nihilo* is perfectly plausible.

However, he thought that the true value of the series was $1/2$ for various reasons [285]. As explained in [370], Gottfried Wilhelm von Leibniz (1646–1716), in a letter to the German philosopher and mathematician Christian Wolf (1679–1754), published in *Acta eruditorum* of 1713, agreed with Grandi's result, but he argued that since the partial sums take the values 0 and 1 with equal probability, the sum of the series should be $1/2$, as suggested by the theory of probability. He conceded that the argument was more metaphysical than mathematical, but said that there is more metaphysical truth in mathematics than is generally recognized. This argument was accepted by the Bernoullis. But Euler preferred to consider that this value arose from

© Springer Nature Switzerland AG 2020
C. Brezinski, M. Redivo-Zaglia, *Extrapolation and Rational Approximation*,
https://doi.org/10.1007/978-3-030-58418-4_3

the series expansion of $1/(1 + a) = 1 - a + a^2 - a^3 + \cdots$, with $a = 1$. According to Émile Borel (1871–1956) [72], Euler was thinking that if one happened to encounter this series, its value must surely be $1/2$. It is certainly a curious observation, since the application of the Aitken Δ^2 process to the sequence of its partial sums produces the constant sequence ($\varepsilon_2^{(n)} = 1/2$).

In [224], Euler also proposed a procedure for speeding up the convergence of slowly convergent series and to sum divergent ones. As explained in [670, pp. 151–152] (see also [671]), starting from a series whose terms are $(-1)^n a_n$, Euler considered its successive forward differences

$$\Delta^n a_0 = C_n^0 a_0 - C_n^1 a_1 + \cdots + (-1)^n C_n^n a_n.$$

Then, writing a_n instead of $(-1)^n a_n$, the series $\sum_{n \geq 0} a_n$ is transformed into the series $\sum_{n \geq 0} b_n$ with

$$b_n = (C_n^0 a_0 + C_n^1 a_1 + \cdots + C_n^n a_n)/2^{n+1}.$$

The partial sums t_n of this new series are

$$t_n = (C_n^0 a_0 + C_n^1 a_1 + \cdots + C_n^n a_n)/2^n.$$

This transformation fulfills the conditions of the Toeplitz's theorem, and thus it transforms any convergent series into a series that converges to the same limit. Other methods for the summation of divergent series bear the names of Abel, Cesàro, Hölder, among others. They were studied by Borel in [70] and in his book [72] which was first published in 1901, followed by a second edition in 1928. It contains a presentation of asymptotic series according to Henri Poincaré (1854–1912) [497] and the doctoral thesis of Thomas Jan Stieltjes (1856–1894) [622], an analysis of the analytic theory of continued fractions due to Stieltjes, and a description of the work of Henri Eugène Padé on the rational approximants named after him. Summation methods were used for the analytic continuation of functions defined by a power series expansion as explained, for example, in the book of Jacques Hadamard (1865–1963) and Szolem Mandelbrojt (1899–1983) in 1926 [302]. Let us mention that in 1953, Adriaan van Wijngaarden (1916–1987) proposed a summation method that, starting from the partial sums of an alternating series $S_{0,k} = \sum_{n=0}^{k} (-1)^n a_n$, forms other sequences by the rule $S_{j+1,k} = (S_{j,k} + S_{j,k+1})/2$ (Euler's transform corresponds to $S_{j,0}$ [667]). As we will see in Chap. 4, Wynn worked with him between 1960 and 1964 during his stay in Amsterdam. The work of van Wijngaarden was extended in 1966 by Peter Karl Henrici (1923–1987), who proposed a constructive version of a method due to Karl Weierstrass (1815–1897) [690] for continuing an analytic function, defined by its Taylor series at one point, along an arbitrary path [320].

However, due to their linear character, summation methods are not very powerful for accelerating convergence except in the case of the Richardson extrapolation. The idea of eliminating the first term (in h^2) in the expansion of the error of a discretization process by an extrapolation method is due to Lewis Fry Richardson in 1910 [520]. In [523], he extended this idea to the cancellation of the terms in h^2 and h^4 (see also the related paper by John Arthur Gaunt (1904–1944) [261]). As explained in Chap. 2, Romberg's method for improving the accuracy of the trapezoidal rule is strongly connected to the Richardson extrapolation method, although Werner Romberg (1909–2003)[1] derived it independently in 1955 [535].

The first nonlinear sequence transformation used for the purpose of convergence acceleration is the Aitken Δ^2 process [5] in 1926. Aitken used his method in 1937 for accelerating the convergence of the power method (Rayleigh quotients) for computing the dominant eigenvalue of a matrix [6]. However, the process does not seem to have been widely quoted until the mid-twentieth century. It appeared in [313] and [330, p. 571], which also contains an account of reciprocal differences, rational interpolation and approximation by continued fractions, and Padé approximation. Reciprocal differences for rational interpolation via continued fractions and the Thiele's formula were explained in several other textbooks of that time; see, for example, [451, 452, 472]. The Aitken process is also mentioned in [148, 336, 375, 479, 705], and in many books that were published around and after 1960. We note especially [280], where h^2 extrapolation is mentioned, as well as the Aitken Δ^2 process under the name of *exponential extrapolation*. Later, in 1970, the Aitken process, the Steffensen method, the Shanks transformation, and the ε-algorithm can be found in a book by Alston Scott Householder (1904–1993) [337].

Newton's method for finding a zero of the equation $f(x) = 0$ is well known. It consists in the iterations $x_{n+1} = x_n - f(x_n)/f'(x_n)$ from a given x_0. If the sequence (x_n) converges to x, and if $f'(x) \neq 0$, then $f(x) = 0$. If x is a simple zero of f, and if f is twice continuously differentiable in a neighborhood of x, then there exists a neighborhood of x such that for all x_0 in it, (x_n) converges quadratically to x. Obviously, the drawback of this method is the need for the mathematical expression for f'. In 1933, Johan Frederik Steffensen (1873–1961)[2] published a method for solving nonlinear equations [615]. For finding the fixed point x of F, he considered the iterations $x_{n+1} = F(x_n)$ and assumed that three consecutive iterates x_0, x_1 and x_2 are known. By linear interpolation with divided differences, one has

$$F(x) = F(x_0) + (x - x_0)[x_0, x_1]_F + (x - x_0)(x - x_1)[x, x_0, x_1]_F$$
$$= x_1 + (x - x_0)\frac{x_1 - x_2}{x_0 - x_1} + R_1,$$

where $R_1 = (x - x_0)(x - x_1)[x, x_0, x_1]_F$. Since $x = F(x)$, then, assuming that R_1 is constant, he obtained

$$x = x_0 - \frac{(\Delta x_0)^2}{\Delta^2 x_0} + R$$

with $R = -(x - x_0)(x - x_1)\dfrac{(\Delta x_0)^2}{\Delta^2 x_0}[x, x_0, x_1]_F$. If F has a continuous second derivative,

then $R = -(x - x_0)(x - x_1)\dfrac{(\Delta x_0)^2}{\Delta^2 x_0}F''(\xi)/2$, where ξ belongs to an interval containing

---

[1] See Sect. 8.8 for his biography.
[2] See Sect. 8.13 for his biography.

x_0, x_1, x_2, and x. Neglecting R, the preceding formula furnishes an approximation of x. Then Steffensen iterated his process from x, which thus can be compactly written as

$$x_{n+1} = x_n - \frac{(F(x_n) - x_n)^2}{F(F(x_n)) - 2F(x_n) + x_n} = x_n - \frac{[f(x_n)]^2}{f(x_n + f(x_n)) - f(x_n)},$$

where $f(x) = F(x) - x$. This formula shows that Steffensen's method consists in replacing, in Newton's method, $f'(x)$ by $[f(x + f(x)) - f(x)]/f(x)$. If $F'(x) \neq 1$, a condition equivalent to $f'(x) \neq 0$, the convergence of the method is quadratic under assumptions similar to those required for Newton's method, but it does not need the knowledge of the derivative of the function. It is, in fact, a cycling of the Aitken process with restarting, but Steffensen derived it directly. At the end of his paper, he made a curious remark. In the case that x_0, x_1, x_2 move away from the fixed point x, he proposed to set $y_0 = x_2, y_1 = x_1, y_2 = x_0$. Then y_0, y_1, y_2 will approach the fixed point, his process can be applied, and he obtained $x_0 - (\Delta x_0)^2/\Delta^2 x_0 = y_0 - (\Delta y_0)^2/\Delta^2 y_0$. Then he concluded that his method *may, in all cases, be interpreted as a method founded on a sequence approaching to the root, although there is no need to ascertain beforehand whether the sequence has this property or not.*

Steffensen's process was rediscovered independently in 1942 by the Belgian astronomer and physicist Georges Lemaître (1894–1966) [404], one of fathers of the Big Bang theory. For solving the nonlinear equation $x = F(x)$, starting from x_1, he linearized F and computed

$$x_2 = F(x_1) = \alpha + \beta x_1,$$
$$x_3 = F(x_2) = \alpha + \beta x_2,$$
$$x = F(x) = \alpha + \beta x.$$

Then he wrote that the system is compatible, that is,

$$\begin{vmatrix} x_2 & 1 & x_1 \\ x_3 & 1 & x_2 \\ x & 1 & x \end{vmatrix} = 0.$$

Expanding this determinant with respect to its last row, he obtained

$$x = \frac{x_1 x_3 - x_2^2}{x_1 - 2x_2 + x_3},$$

which is Aitken's formula, but he did not iterate with it. Even if he did not give a numerical example, Lemaître considered the method to be a convergence acceleration procedure, since he wrote: *We can therefore use the formula of rational iteration to accelerate convergence where it exists.* He called his method *l'itération rationnelle* (rational iteration). He curiously made the same remark as Steffensen that it is not necessary that the iterates x_1, x_2, x_3 converge to the solution since putting them in the reverse order does not change the formula. Applying the method to the sequence generated by $x_{n+1} = N/x_n$ leads to $x_{2n} = x_2 = N/x_1$, $x_{2n+1} = x_1$, and the result is $x = (x_1 + x_2)/2$. Then Lemaître applied his process to the solution of a first-order differential equation for accelerating the convergence of its iterates. In a second paper [405], he considered the differential equation $y' = 2y^2(y - x)$, whose solution has $y = x$ as an asymptote as x goes to infinity. Starting from an approximation y_n of the solution, he computed the next one by $1/y_{n+1} = 2\int_x^\infty (y_n - x)\, dx$. Having obtained three consecutive approximations y_n, y_{n+1}, and y_{n+2}, he used his rational iteration to get the following one, $y_{n+3} = (y_n y_{n+2} - y_{n+1}^2)/(y_n - 2y_{n+1} + y_{n+2})$. A discussion of the properties of this method follows. See [401] for a detailed analysis.

In 1949, Daniel Shanks obtained his sequence transformation, which generalized the Aitken process [567]. This is the starting point of all modern methods for accelerating the convergence of sequences. As we will see in Sects. 6.4 and 6.7 of Chap. 6, this transformation will also have an important impact on iterative methods for solving systems of linear and nonlinear equations.

In 1950, Peter Wynn was 19 years old. In this chapter we describe the state of knowledge of the main topics he, the researchers around him, his contemporaries, and his followers would soon begin to work. These topics are extrapolation methods, formal orthogonal polynomials, Padé approximation, and continued fractions. Around the same years, the transformation of series, rational approximants, and continued fractions by means different from those used by Wynn were developed by other researchers.

Among these contributors was Cornelius Lanczos (1893–1974), a Hungarian born physicist and mathematician, who proposed a process he named *economization of a power series*. He interpolated a function by a series in Chebyshev polynomials, and he suppressed the terms whose coefficients were below a required accuracy. Thus the series was reduced to its most economical form with *a smaller number of terms without sacrificing essentially in accuracy* [385]. In [388, pp. 457–463], the process was called the *telescoping method*. Lanczos's economization method was extended to continued fractions by Hans Jakob Maehly (1920–1961) in [426] where he described *three closely related methods for adjusting the coefficients of a truncated continued fraction (approximant) so that the maximum of the absolute value of the error, on a given interval, is nearly minimized*. Maehly obtained a PhD in physics on eigenvalue computations at ETH Zürich in 1951. Six weeks before his death, he joined the Applied Mathematics Division at the Argonne National Laboratory. However, before turning to new problems, he planned to complete his results on rational approximation. These results were obtained between 1958 and 1960 under a contract between Princeton University, where he had a position, and the Bureau of Ships and its Applied Mathematics Laboratory at the David Taylor Model Basin (where Shanks was working). Maehly's results were posthumously prepared and published by Christoph Johann Witzgall (b. 1929), a German-American mathematician who worked in the group of Friedrich Ludwig Bauer (1924–2015)[3] in Munich at the same period as Wynn [427]. An application of the work of Maehly

---

[3] See Sect. 8.3 for his biography.

and Ervand George Kogbetliantz (1888–1974) [372] was given by Kurt Spielberg (1928–2009) in 1960 [609, 610]. These two papers mention the paper of Wynn [720]. The economization of rational approximations, including Padé approximants, was later treated by Anthony Ralston (b. 1930) in 1963 [518].

3.2 Extrapolation Methods

As already said in the preceding section, in the nineteenth century and earlier, and until the 1920s, only linear summation methods were studied. They were mostly used to assign a sum to a divergent series. They bear the names of Cesàro, Hölder, Euler, et al., see [69, 494]. Under mild conditions given by a theorem due to Otto Toeplitz (1881–1940), they transform any convergent sequence into a sequence converging to the same limit [642]. This theorem is a byproduct of a result on the weak convergence of a sequence of linear bounded operators in a Banach space that is itself based on the Banach–Steinhaus theorem. However, their acceleration properties are quite limited, except in some special cases. In particular, Euler's method was evoked, or studied, or generalized by Wynn in several papers [715, 727, 735, 736, 773, 777]. It seems that no nonlinear sequence transformation was studied before the twentieth century, at least in Western countries (see [476] for an early use of the Aitken Δ^2 process in Japan).

3.2.1 Richardson's Extrapolation Method

The extrapolation method due to Richardson dates from 1910 [520] and 1927 [523]. It is a linear sequence transformation, but it was built for the purpose of accelerating convergence. He suggested to eliminate the first term of the error in the central differences formulas given by William Fleetwood Sheppard (1863–1936) [573] using several values of the step size. He wrote:

> [...] the errors of the integral and of any differential expressions derived from it, due to using the simple central differences of §1.1 instead of the differential coefficients, are of the form
> $$h^2 f_2(x, y, z) + h^4 f_4(x, y, z) + h^6 f_6(x, y, z) + \&tc.$$
> Consequently, if the equation be integrated for several different values of h, extrapolation on the supposition that the error is of this form will give numbers very close to the infinitesimal integral.

He called this procedure the *deferred approach to the limit* or h^2-*extrapolation*.

In his 1927 paper, Richardson extended this procedure for solving a differential eigenvalue problem of order six, and wrote:

> Confining attention to problems involving a single independent variable x, let h be the "step", that is to say, the difference of x which is used in the arithmetic, and let $\phi(x, h)$ be the solution of the problem in differences. Let $f(x)$ be the solution of the analogous problem

in the infinitesimal calculus. It is $f(x)$ which we want to know, and $\phi(x, h)$ which is known for several values of h. A theory, published in 1910, but too brief and vague, has suggested that, if the differences are "centered" then

$$\phi(x, h) = f(x) + h^2 f_2(x) + h^4 f_4(x) + h^6 f_6(x) \ldots \text{ to infinity } \ldots \qquad (1)$$

odd powers of h being absent. The functions $f_2(x)$, $f_4(x)$, $f_6(x)$ are usually unknown. Numerous arithmetical examples have confirmed the absence of odd powers, and have shown that it is often easy to perform the arithmetic with several values of h so small that $f(x) + h^2 f_2(x)$ is a good approximation to the sum to infinity of the series in (1).

If generally true, this would be very useful, for it would mean that if we have found two solutions for unequal steps h_1, h_2, then by eliminating $f_2(x)$ we would obtain the desired $f(x)$ in the form

$$f(x) = \frac{h_2^2 \phi(x, h_1) - h_1^2 \phi(x, h_2)}{h_2^2 - h_1^2}. \qquad (2)$$

This process represented by the formula (2) will be named the "h^2-extrapolation".

If the difference problem has been solved for three unequal values of h it is possible to write three equations of the type (1) for h_1, h_2, h_3, retaining the term $h^4 f_4(x)$. Then $f(x)$ is found by eliminating both $f_2(x)$ and $f_4(x)$. This process will be named the "h^4-extrapolation".

Let us mention that he referred to a paper by Nikolai Nikolaevich Bogolyubov (1909–1992) and Nikolai Mitrofanovich Krylov (1879–1955) of 1926, in which the deferred approach to the limit can already be found [67]. Richardson's paper [523] has a second part [261] written by John Arthur Gaunt [629], at that time bachelor of arts, and scholar of Trinity College, Cambridge.

In his analysis of Richardson's papers, John (Jack) Todd (1911–2007)[4] wrote [639]:

> Undoubtedly the process is a valuable practical tool, but there are certainly cases where it is unreliable. Richardson was fully aware of the possible failures and difficulties which might be encountered in its application and discussed various bad examples.

Applications of the Richardson extrapolation to the numerical solution of ordinary and partial differential equations are discussed in [250, 433, 709].

In 1925, Richardson published what can be called a descriptive paper on the approximate solution of differential equations by arithmetic methods [522]. Let us quote some parts of it, since they shed light on the personality of their author and his British sense of humor. The paper begins with a section entitled *I. The first obstacle may be one of sentiment*, and then he writes:

> It is said that in a certain grassy part of the world a man will walk a mile to catch a horse, whereon to ride a quarter of a mile to pay an afternoon call. Similarly, it is not quite respectable to arrive at a mathematical destination, under the gaze of a learned society, at the mere footpace of arithmetic. Even at the expenses of considerable time and efforts, one should be mounted on the swift steed of symbolic analysis.

> The following notes are written for those who desire to arrive by the easiest route, and who are not self-conscious about the respectability of their means of locomotion.

---

[4] See Sect. 8.16 for his biography.

Then Richardson describes the steps to be followed, one of them looking like, in a descriptive manner difficult to interpret, an extrapolation procedure. The paper ends with Section XIV:

XIV. Discontinuities. At discontinuities this series $Ah^2 + Bh^4 + \cdots$ may not be convergent. There are, so to speak, in the mathematical country, precipices and pit-shafts down which it would be possible to fall; but that need not deter us from walking about. Yet if we wish to explore these steep descents, pedestrianism must be supplemented by the acrobatics of the pure mathematician.

See [211] for the collected papers of Richardson.

3.2.2 Romberg's Method

In 1955, Werner Romberg published a method for improving the accuracy of the trapezoidal and the midpoint rules for computing an approximate value of a definite integral. Romberg made the basic observation that the trapezoidal rule $T(h)$ has an error in h^2. Using Richardson's deferred approach to the limit, he got

$$S(h) = (4T(h) - T(2h))/3,$$

where $S(h)$ is nothing else than Simpson's rule. Since this rule has an error in h^4, Richardson's process can be reapplied to give

$$C(h) = (16S(h) - S(2h))/15,$$

which is the closed Newton–Cotes formula of order 6. Continuing the process, the next step produces

$$R(h) = (64C(h) - C(2h))/63,$$

which is a new quadrature formula. More systematically, let T_m^i and U_m^i be respectively the trapezoidal and the midpoint rules computed from the step sizes $h_i, \ldots, h_{i+m}$. Romberg showed that when $h_{i+s} = h_i/2^s$ for $s = 1, 2, \ldots$, and $i = 0, 1, \ldots$, then for $i = 0, 1, \ldots$ and $m = 1, 2, \ldots$,

$$T_m^i = T_{m-1}^{i+1} + \frac{T_{m-1}^{i+1} - T_{m-1}^i}{4^m - 1} = \frac{4^m T_{m-1}^{i+1} - T_{m-1}^i}{4^m - 1}, \tag{3.1}$$

and a similar relation for the U_m^i. Moreover, $T_m^i = (T_m^{i-1} + U_m^{i-1})/2$. These values can be displayed in a double table in which the lower index indicates a column, and the upper one a descending diagonal. It is a triangular scheme similar to Fig. 2.1. Romberg observed that the leading terms in the expansions of the errors of T_m^i and U_m^i have the same order but opposite signs. Thus, if this term is dominant in each expansion, the exact value of the integral lies between T_m^i and U_m^i. If the step sizes h_i are in a geometric progression with a ratio larger than 2, the number of function evaluations becomes rapidly huge. If in Richardson's extrapolation method (2.5) one has $x_i = h_i^2 = (h/2^i)^2$, then (3.1) is recovered.

The basic idea behind Romberg's method was not new. It goes back to Archimedes method for computing approximations of π. If $T(1/n)$ is the perimeter of the regular n-gon inscribed in a circle of diameter 1, Archimedes computed $T(1/n)$ with $n = 6, 12, 24$, and 48, that is, the first column of Richardson's table shown in Fig. 2.1. In 1654, Christiaan Huygens (1629–1695) improved these values by computing the second column of the table [340]. In 1936, Karl Kommerell (1871–1962) obtained the third column, but did not pursue the subject further [373].

Romberg's method became widely known after its error and its convergence were studied [40, 41, 43, 539, 541, 620, 621, 626]. Many other references on it can be found on the internet. Thus Romberg's method found its way to textbooks, where it has long been presented without any reference to its author, a proof of true celebrity!

A complete history of Richardson's and Romberg's methods is given in [216, 221, 311, 361] (see also [686] for a mathematical analysis). Other historical accounts can be found in [109, 112].

3.2.3 The Aitken Process

The first nonlinear sequence transformation is the Δ^2 process of Aitken, which appeared in 1926 [5] (submitted on 19 December 1925, as he wrote in [9]). Let us give here only an account of its genesis, which is fully described in [135].[5] Aitken's goal was to accelerate the convergence of the method of Daniel Bernoulli (1700–1782) for computing the dominant zero of a polynomial $a_0 z^k + a_1 z^{k-1} + \cdots + a_k$ with $a_0 \neq 0$. Let (u_n) be the sequence recursively obtained from the homogeneous linear difference equation of order k

$$a_0 u_{n+k} + a_1 u_{n+k-1} + \cdots + a_k u_n = 0, \quad n = 0, 1, \ldots,$$

starting from almost arbitrary values for $u_0, \ldots, u_{k-1}$. Under some assumptions on $u_0, \ldots, u_{k-1}$, $\lim_{n \to \infty} u_{n+1}/u_n = z_1$, where z_1 is the zero of the polynomial of greatest modulus.

In Section 8 of his paper, Aitken wrote:

So far we have not obtained a very great degree of accuracy [. . .] We shall now proceed to derive from the primary sequences successive sequences of increasing approximative power.

For that purpose, he sets $Z_1(n) = u_{n+1}/u_n$ (we replaced his original variable t by n), and constructs the finite differences of first and second order $\Delta Z_1(n) = Z_1(n+1) - Z_1(n)$ and $\Delta^2 Z_1(n) = \Delta Z_1(n+1) - \Delta Z_1(n) = Z_1(n+2) - 2Z_1(n+1) + Z_1(n)$. He notices that if the zeros of the polynomial are such that $|z_1| > |z_2| \geq |z_3| \geq \cdots$, then $Z_1(n) = z_1 + O((|z_2|/|z_1|)^n)$, and that the first finite differences behave similarly, that is:

$\Delta Z_1(n)$ tends to become a geometric sequence [. . .] of common ratio z_2/z_1. Hence the *deviations* of $Z_1(n)$ from z_1 will also tend to become a geometric sequence with the same

---

[5] This analysis is adapted, with permission, from [135].

common ratio. Thus a further approximate solution is suggested, viz.

$$\frac{z_1 - Z_1(n+2)}{z_1 - Z_1(n+1)} = \frac{\Delta Z_1(n+1)}{\Delta Z_1(n)},$$

and solving for z_1 we are led to investigate the *derived* sequence

$$Z_1^{(1)}(n) = -\frac{\begin{vmatrix} Z_1(n+1) & Z_1(n+2) \\ Z_1(n) & Z_1(n+1) \end{vmatrix}}{\Delta^2 Z(n)}. \tag{8.2}$$

This formula is exactly the Aitken Δ^2 process. He then concluded that this new sequence converges to z_1 as $O((|z_3|/|z_1|)^n)$ or $O(((|z_2|/|z_1|)^2)^n)$ according to the largest of these ratios, and he added in a footnote

> Nägelsbach, in the course of a very detailed investigation of Fürstenau's method of solving equations, obtains the formulæ (8.2) and (8.4) [*his process and its reapplication*], but only incidentally.

In a paper of 1937 [6], Aitken used his process for accelerating the convergence of the power method (Rayleigh quotients) for computing the dominant eigenvalue of a matrix. A section is entitled *The δ^2-process for accelerating convergence*, and on pages 291–292, he wrote:

> For practical computation it may be remembered by the following *memoria technica*: product of outers minus square of middle, divided by sum of outers minus double of middle.

Aitken's papers contain almost all the ideas that led Rutishauser to the discovery of his *qd*-algorithm.

3.2.4 The Shanks Transformation

The transformation now called the Shanks transformation was introduced by Daniel Shanks (1917–1996). At that time, Shanks was working at the Naval Ordnance Laboratory in White Oak, Maryland. There, he published a memorandum of 42 pages, dated 26 July 1949, where his transformation is described and studied [567].[6]

Without having done any graduate work, Shanks wanted to present this work to the Department of Mathematics of the University of Maryland as a Ph.D. thesis. But he had first to complete the degree requirements before his work could be examined as a thesis. Hence, it was only in 1954 that he obtained his Ph.D., which was published in 1955 in the *Journal of Mathematical Physics* [568], a paper submitted on 9 June 1954. Dan Shanks considered this paper one of his best two (the second was his computation of π to 100,000 decimal places written with John William Wrench, Jr., (1911–2009) [570]).

---

[6] It can be downloaded from the address https://apps.dtic.mil/dtic/tr/fulltext/u2/a800123.pdf.

These determinants may be readily transformed into other interesting forms. Some of these are:

$$B = \frac{\begin{vmatrix} A_0 & A_1 & \cdots\cdots & A_K \\ \Delta A_0 & \Delta A_1 & ---\,\,\Delta A_K \\ \Delta^2 A_0 & \Delta^2 A_1 & \cdots & \Delta^2 A_K \\ \vdots & \vdots & & \vdots \\ \Delta^K A_1 & \Delta^K A_1 & \cdots & \Delta^K A_K \end{vmatrix}}{\begin{vmatrix} 1 & 1 & \cdots & 1 \\ & SAME\ AS\ (23) \end{vmatrix}} = A_K + \frac{\begin{vmatrix} (A_0 - A_K)(A_1 - A_K) & \cdots & 0 \\ \Delta A_0 & \cdots\cdots & \Delta A_K \\ \vdots & & \vdots \\ \Delta A_{K-1} & \cdots\cdots & \Delta A_{2K-1} \end{vmatrix}}{\begin{vmatrix} SAME\ AS\ (23) \end{vmatrix}}$$

(23a,b,c)

$$B = \frac{\begin{vmatrix} A_0 & A_1 & \cdots\cdots & A_K \\ A_1 & A_2 & \cdots\cdots & A_{K+1} \\ \vdots & \vdots & & \vdots \\ A_K & A_{K+1} & \cdots & A_{2K} \end{vmatrix}}{\begin{vmatrix} \Delta^2 A_0 & \Delta^2 A_1 & \cdots & \Delta^2 A_{K-1} \\ \vdots & \vdots & & \vdots \\ \Delta^2 A_{K-1} & \cdots & \Delta^2 A_{2K-2} \end{vmatrix}}$$

II. Two Nonlinear Sequence-to-Sequence Transforms

18. Let us return to the sequence of partial sums of the slowly convergent ln 2 = 1 - 1/2 + 1/3 - 1/4 + ... We have $A_0 = 1$, $A_1 = 1/2$, $A_2 = 5/6$, etc. Given the first seven partial sums, $A_0 \to A_6$,

11   NOLM 9994

Shanks's memorandum, 1949.
© Naval Ordnance Laboratory.

Let us analyze the memorandum [567] and the paper [568],[7] and highlight the main points that could have caught the attention of Peter Wynn and inspire his work. The abstract of Shanks's memorandum begins thus:

In mathematics, and in applied mathematics especially, one wishes to obtain accurate answers rapidly. One obstacle often met with is that the simplest and most obvious analysis gives

[7] Parts of this analysis are adapted extracts, with permission, from [135].

mathematical sequences which are slowly convergent or even divergent. The proper treatment of such sequences is therefore a general problem of real importance. This memorandum gives and discusses some methods of treating such sequences.

Then, in the introduction of his memorandum, Shanks explains the meaning of the words *transients* and *mathematical sequences* that appear in the title of [567], and what he will do with them:

1. In this paper we shall discuss an analogy between transients and mathematical sequences. By the term "physical transient" we mean a physical quantity, p, which, when expressed as a function of time, takes the form

$$p(t) = B + \sum_{i=1}^{K} a_i e^{\alpha_i t}.$$

It will appear below that it is useful to regard some mathematical sequences, A_n, as functions of n of the form

$$A_n = B + \sum_{i=1}^{K} a_i e^{\alpha_i n}$$

and because of this we may call such sequences "mathematical transients."

2. We shall be concerned here with the analysis of mathematical transients. By the term "analysis" we mean the determination of the "amplitudes," a_i; the "frequencies," α_i; and the "base line constant,"the B, of these transients. If all the α_i have negative real parts, the transient converges to its limit, B. If one or more α_i has a nonnegative real part the transient is divergent and has no limit. In such cases we may call B the "antilimit" of the transient.

Let us mention that all the mathematical formulas of this memo are handwritten by Shanks (see above). He then takes the example of a linear difference equation of order 3 with constant coefficients and a constant term $B = 0$, and shows how to determine the six unknowns a_i and $\alpha_i = \ln q_i / T$ for $i = 1, 2, 3$ from six successive values of $p(t_n) = a_1 q_1^n + a_2 q_2^n + a_3 q_3^n$ with $t_n = nT$ (T is the period of the transient). Then he removes the condition $B = 0$, and with the help of a seventh equation he is able to obtain B as a ratio of determinants. Then he says that the general formula is

$$B = \frac{\begin{vmatrix} A_0 & \cdots & A_K \\ \Delta A_0 & \cdots & \Delta A_K \\ \vdots & & \vdots \\ \Delta A_{K-1} & \cdots & \Delta A_{2K-1} \\ 1 & \cdots & 1 \\ \Delta A_0 & \cdots & \Delta A_K \\ \vdots & & \vdots \\ \Delta A_{K-1} & \cdots & \Delta A_{2K-1} \end{vmatrix}}{} = \frac{\begin{vmatrix} A_0 & \cdots & A_K \\ A_1 & \cdots & A_{K+1} \\ \vdots & & \vdots \\ A_K & \cdots & A_{2K} \\ \Delta^2 A_0 & \cdots & \Delta^2 A_{K-1} \\ \vdots & & \vdots \\ \Delta^2 A_{K-1} & \cdots & \Delta^2 A_{2K-2} \end{vmatrix}}{}.$$

He then applies this formula to the computation of an approximation of $\ln 2 = 0.69314718$ from the first nine partial sums of its series expansion, and he obtains 0.69314733. Instead of computing B starting from A_0, he says that it can be started from any A_N, and since the value of B will depend on N and K, he decides to denote it by B_{KN}, and to *designate the transform as* e_K.

After a discussion of the Aitken process and several numerical examples, some of them involving continued fractions, he studies in detail the relation of his transform to Padé approximants. Then comes an important result. Shanks proves that

$$e_k(A_n) = [n + k/k]_f(x),$$

where A_n denotes the nth partial sum of the series $f(x) = c_0 + c_1 x + c_2 x^2 + \cdots$. He also noticed that the other half of the Padé table (that is, when the degree of the denominator is greater than that of the numerator) can be obtained by applying his transform to the partial sums of the reciprocal series g of f, that is, the series such that $f(x)g(x) = 1$. When the series is the expansion of a rational function, he says that this function is exactly recovered. He also explains the connection with Thiele's reciprocal differences and continued fractions [637], which provide rational approximations, studied by Nörlund [471], which can be expressed as ratios of determinants similar to his transform.

Let us now examine the contents of the paper [568], where results of the memorandum and new ones are reported. We begin by quoting some sentences of the abstract, since they show that Shanks was perfectly aware of the effectiveness of his transform, of its numerous applications, and of its difficulties as well:

Examples are given of the application of these transformations [*Aitken's process, its iterated applications, and his own transformation*] to divergent and slowly convergent sequences. In particular the examples include numerical series, the power series of rational and meromorphic functions, and a wide variety of sequences drawn from continued fractions, integral equations, geometry, fluid mechanics, and number theory. Theorems are proven which show the effectiveness of the transformations both in accelerating the convergence of (some) slowly convergent sequences and in inducing convergence in (some) divergent sequences. The essential unity of these two motives is stressed. Theorems are proven which show that these transforms often duplicate the results of well-known, but specialized techniques. These special algorithms include Newton's iterative process, Gauss's numerical integration, an identity of Euler, The Padé Table, and Thiele's reciprocal differences. Difficulties which sometimes arise in the use of these transforms such as irregularity, non-uniform convergence to the wrong answer, and the ambiguity of multivalued functions are investigated. The concept of antilimit and of the spectra of sequences are introduced and discussed.

Shanks, after slightly changing his notation, begins by setting $B_{k,n}$ equal to the ratio

$$\frac{\begin{vmatrix} A_n & \cdots & A_{n+k} \\ \Delta A_n & \cdots & \Delta A_{n+k} \\ \vdots & & \vdots \\ \Delta A_{n+k-1} & \cdots & \Delta A_{n+2k-1} \end{vmatrix}}{\begin{vmatrix} 1 & \cdots & 1 \\ \Delta A_n & \cdots & \Delta A_{n+k} \\ \vdots & & \vdots \\ \Delta A_{n+k-1} & \cdots & \Delta A_{n+2k-1} \end{vmatrix}},$$

and he denotes it by $e_k(A_n)$, where (A_n) is the sequence to be transformed. To achieve greater homogeneity with the notation used in this book and, in particular, with those of Sects. 2.2.3 and 6.4, we prefer to number in ascending order the iterates used, starting from A_n, instead of the indices used originally by Shanks. Moreover, in what follows, the sequence (A_n) corresponds to our notation (S_n), and B indicates the limit S.

Shanks first noticed that his transformation can be written as

$$e_k(A_n) = \frac{c_n A_n + \cdots + c_{n+k} A_{n+k}}{c_n + \cdots + c_{n+k}},$$

where the coefficients c_i are themselves functions of the A_n, and thus the transformation is nonlinear. However, we have $e_k(aA_n + b) = ae_k(A_n) + b$, a property later named *quasilinearity*. An important property Shanks noticed is that for all $i = 0, \ldots, k$, one also has

$$e_k(A_n) = \frac{c_n A_{n+i} + \cdots + c_{n+k} A_{n+k+i}}{c_n + \cdots + c_{n+k}}.$$

Then he discussed the iterated application of the transformation, an idea taken up later by Wynn in [726, 753, 782]. After a historical account of the e_1 transform (which is identical to the Aitken Δ^2 process), he displayed the $e_k(S_n)$ in a double array that Wynn would call the ε-*array* [713], and pointed out its connection with the Padé table. He wrote:

> This paper includes an improved version of the author's memorandum; and a number of new topics such as non-uniform convergence, spectra of sequences, meromorphic functions, and Gauss Numerical Integration, which have not been previously published.

Drawing a smooth curve through the A_n, he noticed that the graph looks like the graph of a *physical transient*, that is, a physical quantity p that is a function of time of the form

$$p(t) = B + \sum_{i=1}^{k} a_i e^{\alpha_i t}, \quad \alpha_i \neq 0,$$

where the α_i are arbitrary complex numbers. Thus the sequence (A_n) suggested to him

> [...] the possibility of regarding (some) sequences (A_n) as if they were *mathematical transients*, that is, as if they were functions of n of the form

$$A_n = B + \sum_{i=1}^{k} a_i q_i^n, \quad q_i \neq 1, 0.$$

> [...] Again if (A_n) is a transient and one or more $|q_i| \geq 1$, A_n does not converge, but we will say that A_n "*diverges from S*." We will then call B the "*antilimit*" of (A_n) and we intend to compute B as a summation method for (A_n).

A part of the kernel of his transformation is recognized in this last expression.

Then, after explaining the heuristic motivation of the Aitken process and his transformation, he studies the iterated application of the Aitken process e_1^m (that is, applying again the process m times to the sequence given by the previous step), for which he gives several theoretical results. Then several numerical examples are presented including the relation between The Newton's method for computing the square root and the iterated Aitken process. Starting from the sequence of the convergents of the continued fraction $\sqrt{2}$, that is, $A_n = 1 + \dfrac{1|}{|2} + \dfrac{1|}{|2} + \cdots$, he proves that

$$e_1^m(A_n) = \frac{1}{2}\left(e_1^{m-1}(A_n) + \frac{2}{e_1^{m-1}(A_n)}\right).$$

By a counterexample, he showed that e_1 is not regular. However, if (A_n) and $e_1(A_n)$ converge, they both have the same limit, a result due to Lubkin [423]. The convergence of the repeated application of the e_1 transform is treated, and divergent series as well. After a discussion of the e_2 transform, Shanks comes back to the link, already given in [567], between his transformation and Padé approximants. He showed that if his transformation is applied to the partial sums $f_n(z)$ of a formal power series $f(z)$, then $e_k(A_n) = [n + k/k]_f(z)$. After mentioning some convergence results, he shows how to obtain the second half of the Padé table by applying the transformation to the partial sums of the reciprocal series g of f. Shanks gives several convergence results for these approximants, including the well-known theorem by Montessus de Ballore on the convergence of Padé approximants for meromorphic functions, and he discusses the block structure of the Padé table. The connection between the e_k transformation and some continued fractions is also discussed.

After that, Shanks computes $\ln(1 + z) = \int_0^z dq/(1 + q)$ by a Gaussian quadrature formula with k points, and proves that it gives the same result as $B_{k,k}$ when applied to the partial sums of the series for the logarithm, and also that the sequence $(B_{k,k})$ converges to the principal value of $\ln(1 + z)$ if z is not on the real cut $z \leq -1$. For this series, convergence outside *the shadow of branch points* is discussed, and semi-convergence (that is, convergence to the smallest term of an asymptotic series) also. The links between integration, sequence transformations, and continued fractions form the topics of several papers by Wynn [723, 768, 789, 802, 808, 811].

In the last section, Shanks assumes the existence of a "Padé surface," $B(x, y)$, that interpolates the discrete values $B_{k,n}$, and thus leads to an interpretation of the transformation e_x with x not necessarily an integer. He wrote:

> With this composite picture in mind [*a spiral*] one interprets e_1 geometrically as a trans-
> formation of one spiral into the next and may well ask whether there exists a continuous
> transformation T_x dependent upon a *continuous* parameter x which takes the m'th spiral
> into an $m + x$'th interpolating spiral for x not necessarily an integer. If so we would have an
> interpretation of e_1^x with x not necessarily an integer. Again, assume the existence of a "Padé
> Surface" - that is a continuous function $B(x, y)$ which interpolates the discrete points of an
> array (9) - and we would have an interpretation of the e_x transform with x not necessarily
> an integer. But to date these generalizations are merely speculative.

Let us mention that Wynn wrote several papers devoted to partial differential equations and the Padé surface [749, 785, 790, 799, 801]. In his unpublished handwritten notes recently discovered (see Sect. 5.10), Wynn dedicated a manuscript to spirals. It is analyzed in [122].

Finally, in the appendix, Shanks treated applications of his transformation to number theory (thus presaging Shanks's interest in this domain), to the detection of errors in numerical sequences and in power series, to Hardy's puzzle, and also to a typographical error. The last section of the paper is about remarks on the history of his transform, including Steffensen's method (without a reference) for the computation of a fixed point, thus providing a source for further references.

Let us report an interesting testimony, not appearing in the papers of Shanks. When Sylvester's determinantal identity is applied to a Hankel determinant, it gives the recurrence relation

$$H_{k+1}(u_n)H_{k-1}(u_{n+2}) = H_k(u_n)H_k(u_{n+2}) - (H_k(u_{n+1}))^2,$$

with $H_0(u_n) = 1$ and $H_1(u_n) = u_n$. Since $e_k(S_n) = H_{k+1}(S_n)/H_k(\Delta^2 S_n)$, this ratio can be computed by applying separately the preceding recurrence relation to its numerators and to its denominators. As Shanks himself explained to the first author of this book, he was proceeding that way for implementing his transform recursively.

Dan Shanks in December 1976.
© Claude Brezinski.

The work of Shanks is mentioned in a footnote of the paper where Lubkin presented his convergence acceleration method, which is quite close to Aitken's [423]. He writes:

> Since the first draft of this paper, the author has learned of a talk presented by D. Shanks at the Naval Ordnance Laboratory, White Oaks, Md. entitled "Mathematical sequences treated as transients," which basically considers each term of a series as the sum of corresponding terms of one or more geometric series. When a single geometric series is considered, his procedure reduces to that discussed in this paper. More complicated, but generally more effective, transformations result from the use of more than one component geometric series. Finally, he has found it possible to use an infinite number of component series, in which case the sum can be represented as a ratio of two determinants of infinite order, which generally converge with considerable rapidity. The author also understands that Otto Szász has had, for sometime past, a paper under preparation which embodies similar material.

We were unable to locate this paper of Otto Szász (1884–1952), maybe because he died before its completion.

Shanks's memo was immediately unclassified in 1949 when it was written, and it was known outside the Naval Ordnance Laboratory in White Oak, where Shanks was working at that time. In the paper where he presented his ε-algorithm [713], Peter Wynn cited Shanks's paper [568], but he also quotes the memo for an application about the fitting of certain types of statistical data. He mentioned that Shanks was partitioning data into groups, and he wrote: *This was one of the first problems upon which the transformation $e_m(S_n)$ [5] was used ([5] is the memorandum [567]).* Since this application was not repeated in Shanks's paper [568], it means that Wynn knew the original memo. In 1956, Wynn was working at the Scientific Computing Service, 23 Bedford Square, in London. Maybe these two governmental agencies were in contact. However, a question remains: how did Shanks's memo come into his hands?

In his paper, Wynn also mentioned that the same transformation was used by Robert J. Schmidt from Imperial College London [554] in a paper submitted 29 September 1939 but published only in 1941, for the purpose of solving a system of linear equations, and not for building a sequence transformation. Schmidt combined several iterates computed by the Gauss–Seidel method for obtaining a new approximation of the solution. This new iterate is obtained by solving a linear difference equation, and thus it is given as a ratio of determinants similar to Shanks's. Schmidt's paper is analyzed in [135]. We were unable to find information on him.

Wynn's interest in convergence acceleration methods may also be due to the historical roots of the Scientific Computing Service. This service was founded in 1937 by Leslie John Comrie (1893–1950), the superintendent of the Nautical Almanac since 1930, as the world's first private company for scientific computing. Then it was incorporated as Scientific Computing Service, Limited. Comrie is known mostly for his paper on the construction of mathematical tables by interpolation [177]. In this paper, he described the use of punched card equipment for interpolating tables of data, in contrast to the more inefficient and error-prone methods involving mechanical devices. This unconventional use of machines for calculation caused tensions with his superiors, and he was suspended in August 1936 as superintendent.

Jeffrey Charles Percy Miller (1906–1981) was the director of the Scientific Computing Service from 1946 to 1950, and continued as an advisory member after joining the University of Cambridge in 1950 as a lecturer. Originally, Miller was an astronomer, but due to some health problems, he was unable to satisfy the medical requirements for obtaining the position of assistant to Comrie, and he left astronomy. However, he continued to work closely with Comrie, who encouraged his interest in all aspects of numerical calculations, both theoretical and practical, and persuaded him to undertake the production of mathematical tables for special functions. In Cambridge, Miller extended his interests in numerical analysis, particularly with regard to his research students (see [542] for a biography). He is well known for a method that computes the most rapidly decreasing solution of a second-order linear difference equation. It was originally developed to compute tables of the modified Bessel's function, but it also applies to other situations. The general solution of a second-order linear difference equation contains two exponential terms f_n and g_n. If one of them decreases rapidly, Miller's algorithm provides a stable way for its computation [60, p. xvii]. Let f_n be the minimal solution, that is, $\lim_{n \to \infty} f_n / g_n = 0$. A relative error ε in either f_0 or f_1 induces a relative error proportional to $\varepsilon g_n / f_n$ in the computed values of f_n for $n \geq 2$. Thus the error is unbounded. Assume that $f_2, \ldots, f_M$ have to be computed from given f_0 and f_1. Miller's backward recurrence consists in selecting an integer N much larger than M, setting $f_N^{(N)} = 0$ and $f_{N-1}^{(N)} = 1$, and using the recurrence relation backward to compute $f_i^{(N)}$ for $i = N - 2, N - 3, \ldots, 0$. Then $\widetilde{f}_n^{(N)} = f_0 f_n^{(N)} / f_0^{(N)}$ provides an estimate of f_n. If these estimates are not good enough, a larger value of N has to be chosen. As proved by Walter Gautschi (b. 1927) [262], for a fixed value of n, $\lim_{N \to \infty} \widetilde{f}_n^{(N)} = f_n$ if and only if f_n is a minimal solution of the difference equation satisfying $f_0 \neq 0$. For a detailed analysis of Miller's algorithm, see [700].

The papers [718, 730, 736] of Wynn concern a method due to William Gee Bickley (1893–1969) and Miller in 1936 [61]. Before giving an account of it, let us quote the introduction of [61], which contains interesting observations, and an overview of the method:

> Many problems in pure and applied mathematics lead to infinite series, which must be summed numerically before the solution of the problem can be regarded as complete. Provided the series converge fairly rapidly, so that the desired accuracy can be attained by computing a reasonably small number of terms, no further difficulty arises. When, however, the convergence is slow, some indirect method of summation becomes necessary; several such are known, but in the case of a slowly convergent series of *positive* terms, none of the appropriate methods seem to be applicable unless the nth term is a simple function of n. For series with more complicated terms, we seem to be forced back upon methods which are arithmetical rather than analytical. Arithmetic alone will of course not suffice; we cannot, without some law, "continue" a series whose early terms are known. In the series which gave rise to the present investigation, for instance, such a "law" was known, but it did not enable the nth term of the series to be calculated independently of its predecessors. Indeed, each term of the series was of the form $C_n f_n(x)$, where the C_n and f_n had to be determined for $n \geq 2$ by means of recurrence formulæ. What could be obtained even here, and analytically, was the ratio of consecutive terms expanded as far as desired in powers of $1/n$, and such an expansion, or at least a few of its leading terms, will usually be obtainable. In what

follows we use such an expansion to determine, for any value of n, a multiplier such that the product of it and the nth term, added to the sum of the first n terms of the series, gives an approximation to the sum of the series considerably better than the partial sum.

Since this paper contains many interesting ideas, let us analyze it in detail. The authors consider the partial sums $S_n = \sum_{r=1}^{n} u_r$ of a series S. The series is said to be *slowly convergent* if the ratio $M(n) = (S - S_n)/u_n$ becomes large compared with unity for moderate values of n. Then they assume that u_n/u_{n-1} can be developed in the form

$$\frac{u_n}{u_{n-1}} = \varrho \left(1 - \frac{A_1}{n} + \frac{A_2}{n^2} + \frac{A_3}{n^3} + \cdots \right),$$

and they restrict themselves to the case $\varrho = 1$, because the analysis is entirely different if $\varrho \neq 1$. If the series converges, A_1 must be greater than 1, and in this case, they assume that $M(n)$ can be expressed in the form

$$M(n) = \alpha_{-1} n + \alpha_0 + \frac{\alpha_1}{n} + \frac{\alpha_2}{n^2} + \cdots .$$

Then, they set

$$S'_n = S_n + M(n) u_n,$$
$$S'_{n-1} = S_{n-1} + M(n-1) u_{n-1},$$

and wrote:

If we can determine $M(n)$ so that $S'_n = S'_{n-1}$ for all n, then the common value will be the sum sought. In general this is not accurately possible, but the new sequence S'_n must approach S as a limit [. . .].

Then they consider S'_n as the nth partial sum of a series with terms

$$u'_n = S'_n - S'_{n-1} = u_n(1 + M(n)) - u_{n-1} M(n-1) = u_{n-1} \sum_{i=0}^{\infty} \beta_i/n^i,$$

and they relate the new coefficients β_i to the α_i and the A_i, and they decide to choose the α_i so that $\beta_0, \beta_1, \ldots, \beta_r$ all vanish. Thus

$$u'_n = u_{n-1} \sum_{i=r+1}^{\infty} \beta_i/n^i,$$

and the series u'_n converges faster than the original series. The same process can be repeated several times. An improvement of the procedure is then described. Bickley and Miller continue:

In addition to the methods outlined above, there is another which has several advantages in practice, the main being that it is entirely computational and can be applied without any recourse to analysis. It consists essentially in the *elimination*, rather than the determination, of the α, and leads to a simple equation for the determination of the approximation to S.

This elimination consists in writing

$$\frac{n^r(S - S_n)}{u_n} = \alpha_{-1}n^{r+1} + \alpha_0 n^r + \cdots + \alpha_r + O(1/n).$$

Applying the operator Δ^{r+2} to this relation all the α_i disappear, and after having neglected the error term, they obtain

$$S \simeq \Delta^{r+2}(n^r S_n/u_n)/\Delta^{r+2}(n^r/u_n).$$

For $r = 0$, the Aitken Δ^2 process is recovered. For $r > 0$, this type of sequence transformation was later studied [692].

The authors claim that the method can be refined if numbers $\xi_1, \xi_2, \ldots$ can be found such that $M(n)$ takes the form

$$\alpha_{-1} + \alpha_0 + \frac{\gamma_1}{n + \xi_1} + \cdots + \frac{\gamma_r}{n + \xi_r} + \sum_{p=1}^{\infty} \frac{\alpha'_{2r+p}}{n^{2r+p}} = \alpha_{-1}n + \alpha_0 + \sum_{i=1}^{\infty} \alpha_i/n^i.$$

Using $\prod_{i=1}^r (n + \xi_i)$ instead of n^r as the multiplier allows one to *effectively eliminate twice as many of the original α without raising the order of the necessary differences*.

The optimal multiplier of degree s is

$$F_s(n) = n^s + p_1 n^{s-1} + \cdots + p_s = \prod_{i=1}^{s}(n + \xi_i),$$

and the authors write that we must have

$$\sum_{i=1}^{s} \gamma_i/(n + \xi_i) = \sum_{i=1}^{2s} \alpha_i/n^i + O(1/n^{i+1}).$$

Obviously, there is a typing error, and $O(1/n^{i+1})$ has to be replaced by $O(1/n^{2s+1})$. It follows that $\sum_{i=1}^{s} \gamma_i/(n+\xi_i)$ is the $[s-1/s]$ Padé approximant of the series $\sum_{i=1}^{2s} \alpha_i/n^i$ in the integer variable n.

Expanding the left-hand side in powers of $1/n$ and equating the coefficients, they deduce

$$\sum_{i=1}^{s} \gamma_i \xi_i^j = (-1)^j \alpha_{j+1}, \quad i = 1, \ldots, s.$$

Eliminating the γ_i, they finally obtain

$$\alpha_{s+i+1} + p_1 \alpha_{s+i} + \cdots + p_s \alpha_{i+1} = 0 \quad \text{for} \quad i = 0, \ldots, s - 1.$$

Eliminating the coefficients p_i from this set of equations shows that $F_s(n)$ is a constant multiple of

$$\begin{vmatrix} \alpha_{s+1} & \alpha_s & \cdots & \alpha_2 & \alpha_1 \\ \alpha_{s+2} & \alpha_{s+1} & \cdots & \alpha_3 & \alpha_2 \\ \vdots & \vdots & & \vdots & \vdots \\ \alpha_{2s} & \alpha_{2s-1} & \cdots & \alpha_{s+1} & \alpha_s \\ n^s & n^{s-1} & \cdots & n & 1 \end{vmatrix}.$$

This expression reveals that $F_s(n)$ is the formal orthogonal polynomial of degree s in the integer variable n belonging to the family of polynomials formally orthogonal with respect to the linear functional α defined by $\alpha(n^i) = \alpha_{i+1}$ for $i = 0, 1, \ldots$ (see Sect. 2.4).

It is obvious that these researches inspired Peter Wynn in his work. Indeed, many of the ingredients he will later use already appear herein.

3.3 Formal Orthogonal Polynomials

Formal orthogonal polynomials, that is, orthogonal with respect to a general linear functional on the vector space of polynomials (see Sect. 2.4), play a relevant role in several fields, and they were, in particular, used by Wynn in several of his papers [720, 724, 763], and implicitly play a role in some others.

In a paper published in 1938 [576], James Alexander Shohat (also known as Jacques Chokhate) (1866–1944) considered polynomials satisfying a recurrence relation of the form

$$\Phi_n(x) = (x - c_n)\Phi_{n-1}(x) - \lambda_n \Phi_{n-2}(x),$$

where the coefficients λ_n are no longer assumed to be positive, but only different from zero. They are called *polynômes orthogonaux généralisés* (the paper is in French), and they satisfy

$$\int_{-\infty}^{+\infty} \Phi_m(x)\Phi_n(x) \, d\psi(x) = 0, \quad m \neq n; \quad m, n = 0, 1, \ldots,$$

where ψ is a function of bounded variation in $(-\infty, +\infty)$. He gave their determinantal expression with respect to a given sequence (α_n) of moments such that their corresponding Hankel determinants do not vanish. He mentioned that the proof of this result, which uses the fundamental properties of algebraic continued fractions, is based on a theorem just presented by Ralph Philip Boas (1912–1992) at the Mathematical Seminar of the University of Pennsylvania. This theorem states that for any infinite sequence of real numbers (α_n) such that for all k, $H_k(\alpha_0) \neq 0$, there always exists an infinitude of functions ψ of bounded variation such that

$$\alpha_n = \int_{-\infty}^{+\infty} x^n \, d\psi(x), \quad n = 0, 1, \ldots.$$

In a remark, Shohat commented that such polynomials were considered for the first time by Harry Levern Krall (1907–1994), of Pennsylvania State University in a personal communication to him.

Section 1.6 of the book of Gabor Szegő (1895–1985) published in 1939 [632] is entitled *Linear functional operations*. However, since the author assumes that his linear functional is continuous, then by the representation theorem of Frigyes Riesz (1880–1956), it can be written in the form of an integral with respect to a measure of bounded variation, and the corresponding orthogonal polynomials are the usual ones.

In 1940, Yakov Lazarevich Geronimus (1898–1984) published two papers in Russian with almost the same title [267, 268]. We were unable to find the first one, but in the second one, he introduced polynomials with respect to a linear functional whose moments form an arbitrary sequence. A determinantal formula is given. If two sequences of moments are specified, he looked at necessary and sufficient conditions for expressing the polynomials of one family as a linear combination of the polynomials of the second family. The solution of this problem allows him to answer a question posed by Wolfgang Hahn (1911–1998) [303] on the conditions for a family of *classical orthogonal polynomials* and its derivatives to be simultaneously orthogonal (on the various definitions of these polynomials, see [18]). Although it is devoted almost entirely to the classical case, linear functionals related to orthogonality also appeared in the book by James Alexander Shohat and Jacob David Tamarkin (1888–1945), which was first published in 1943 [577].

In the Russian translation of Szegő's book published in 1961, there are appendices by Geronimus. They were translated, gathered, and published in 1977 [269]. He said:

Since many properties of orthogonal polynomials are obtained by purely formal methods, they are valid under considerably more general hypotheses than are used in the book.

We start with a sequence of nonsingular bilinear forms

$$H_m = \sum_{i,k=0}^{m} c_{ik} x^i \bar{x}_k \quad (m = 0, 1, \ldots)$$

or, what is the same thing, a sequence $\{c_{ik}\}_0^m \ (m = 0, 1, \ldots)$ of complex numbers called *moments* and subject to the conditions

$$D_n = |c_{ik}|_{i,k=0}^{n} \neq 0 \quad (n = 0, 1, \ldots, m; m = 0, 1, \ldots).$$

On the space of polynomials we define a linear functional $\mathcal{G}$ by the equations

$$\mathcal{G}\{z^i \bar{z}^k\} = c_{ik}, \quad (i, k = 0, 1, \ldots, m; m = 0, 1, \ldots).$$

The polynomials $P_n(z) = z^n + \cdots$ satisfying the conditions

$$\mathcal{G}\{P_i(z)\overline{P_k(z)}\} = \begin{cases} 0, & i \neq k, (i, k = 0, 1, \ldots, m; m = 0, 1, \ldots), \\ h_k = D_k/D_{k-1} \neq 0, & i = k, \end{cases}$$

are called *orthogonal with respect to the sequence* c_{ik} [. . .]

We have introduced a very general definition of orthogonality; let us show that it contains all the cases discussed in the book.

He then mentioned that when $z = \bar{z}$, we obtain orthogonality on the real axis, when $\bar{z} = 1/z$, we recover orthogonality on the unit circle; and that the general case corresponds either to orthogonality on an arbitrary curve or to orthogonality on a domain (see [107]).

Formal orthogonal polynomials were studied by Herman van Rossum (1917–2006)[8] in his thesis [662] defended on 6 July 1953 at the University of Utrecht, The Netherlands, under the supervision of Jan Popken (1905–1970). In the introduction, van Rossum wrote:

> Here the word "orthogonal" is taken in a much broader sense than is usually done. Let a sequence of real numbers $\mu_0, \mu_1, \mu_2, \ldots$ be given and consider a linear operator Ω (the so-called moment operator), which applies to any polynomial $p(x) = a_0 + a_1 x + \cdots + a_n x^n$, so that
>
> $$\Omega[p(x)] = a_0 \mu_0 + a_1 \mu_1 + \cdots + a_n \mu_n.$$
>
> Now, if a system of polynomials $p_0(x), p_1(x), p_2(x), \ldots$ has the property
>
> $$\Omega[p_m(x)p_n(x)] \begin{cases} = 0 \text{ if } m \neq n, \\ \neq 0 \text{ if } m = n, \end{cases}$$
>
> then it is said, that the polynomials form an orthogonal system with respect to the operator Ω (or: with respect to the sequence $\mu_0, \mu_1, \mu_2, \ldots$). If we used the word "orthogonal" above, then we meant it in this sense.

Then he discussed Padé approximants, J-fractions, and related them to these orthogonal polynomials. The denominators of Padé approximants on a diagonal or a staircase of the Padé table are proved to be orthogonal polynomials. But he said:

> Our treatment is far from complete. We only want to show, how this extensively studied classical theory, fits into our more general scheme. Especially it is interesting to see the fundamental importance of the theory of moments for the application of our general theory.

The theory is applied to some hypergeometric series. Van Rossum published several papers on this topic (see [94]), and he was followed in this way by Abraham van der Sluis (1928–2004) in his thesis defended at the University of Utrecht on 23 May 1956 [655]. In its introduction, one can read:

> In our days the theory of orthogonal polynomials, which started relatively late, has grown out to an extensive theory [. . .]
>
> However, in most of the work in this field the relations between the different sets of orthogonal polynomials seem to be more or less incidental, and their common origin is not clear. In our opinion this is mainly due to the starting-points of the authors [. . .] Hence it seems desirable to study the foundations and the structure of the theory from another point of view. This is important not only for the further development of this subject, but also in view of its connections with other fields of research in analysis, such as the problem of moments [. . .]
>
> It is our intention to develop a theory as general as possible and therefore we shall apply algebraic methods. As such it corresponds with the modern trend to apply algebraic methods to analysis. For instance the coefficients of the polynomials and the series that will be considered, can be taken from a ring R with a unit element. This makes the theory also applicable to e.g. matrix polynomials.

---

[8] See Sect. 8.17 for his biography.

> In our opinion the most unified theory is possible if we take the relation between orthogonal
> polynomials and Padé-fractions as a starting point.

Then the author discusses Padé approximants in the light of these orthogonal polynomials. He gives recurrence relations for their computation, and emphasizes the
orthogonality relations of the approximants. The case of hypergeometric series is
treated in detail.

As we will see in Chap. 6, formal orthogonal polynomials have important applications other than Padé approximation.

3.4 Padé Approximants

After their true discovery by Lambert and Lagrange (see Sect. 2.3.1), Padé approximants were used for accelerating the convergence of series; see, for example, the
paper of Ernst Eduard Kummer (1810–1893) dating from 1837 [380]. Let us now
describe the state of the art on the convergence of Padé approximants in the first part
of the twentieth century.

Henri Eugène Padé defended his doctoral thesis in 1892 under the guidance of
Charles Hermite (1822–1901) [481]. His work was analyzed by Eduard Burr van
Vleck (1863–1943) at a colloquium of the American Mathematical Society held
in Boston, 2–5 September 1903. At this occasion, he introduced the appellations
Padé table and *approximant* during the six lectures *Divergent Series and Continued
Fractions* he delivered as a "Colloquium lecturer" [665]. The proceedings appeared
only in 1905. In his paper, Van Vleck discussed the convergence of continued
fractions, and he analyzed in depth the works of many French mathematicians, all
published in French journals, and among them, the theorem of de Montessus of 1902.
This shows the speed of dissemination of knowledge on both sides of the Atlantic
even at a time when the internet did not exist, and there was no language barrier.
About Padé's thesis, he wrote:

> As this thesis is the foundation for a systematic study of continued fractions, it will be
> necessary to give a recapitulation of its chief results [...]
>
> The existence of [*Padé*] approximants was, of course well known before Padé, but no system-
> atic examination of them had been made except by Frobenius [251] [*our bibliography*], who
> determined the important relations which normally exist between them. Padé goes further,
> and arranges the approximants, expressed each in its lower terms, into a table of double
> entry [...]

At that time, several convergence results were already known. Obviously, they are
related to convergence results of continued fractions. In 1902, Robert de Montessus
de Ballore proved his famous theorem (Theorem 2.14 of this book) about meromorphic functions given above. It stated that if a series has k poles counted according to
their multiplicities and no other singularities in a disk D, then the sequence $([n/k])_n$
converges uniformly to the function represented by the series as n goes to infinity

except in neighborhoods of the poles included in D [456]. In 1927, Rowland Wilson (1895–1980) investigated the behavior of this sequence of approximants on the circle of convergence and at the included poles [698].

It seems that the first work on the convergence of sequences of Padé approximants for functions with branch points was undertaken by Samuel Dumas (1881–1938) in his thesis (in French) defended at the University of Zürich in 1908 under the supervision of Adolf Hurwitz (1859–1919) [214]. He proved that the poles of the $[n/n]$ approximants of the square root of a fourth-degree polynomial plus a second-degree polynomial chosen in order to have a function in $O(z^{-1})$ as z tends to infinity approach a certain locus S containing the four branch points as n goes to infinity. A special case was studied by Naoum Achyeser (1901–1980) in 1934 [2].

In his thesis, defended in 1927 under the supervision of Eduard Burr van Vleck, Hubert Stanley Wall (1902–1971) gave a complete analysis of the convergence behavior of the forward diagonal sequences of the Padé table for a Stieltjes series [679] (published as [680]). In 1931 and 1932, he extended this analysis to the cases in which the range of integration is $[a, b]$ with $-\infty \leq a < b < +\infty$ and $-\infty < a < b < +\infty$ [681, 682]. The second reference consists of only the following abstract:

Let $P(z)$ have a corresponding continued fraction $b_1/(1+(b_2z/1+(b_3z/1+\cdots)))$ in which $\lim_n b_n = b$. Then if $b = 0$ and $\lim_n (b_n/b_{n-1}) = \pi > 0$, every diagonal file of the Padé table converges to one and the same meromorphic limit. If $b \neq 0$ the files converge to a common limit over the entire plane except along the whole or a part of that segment of the real axis from $x = -b/4$ to $x = \infty$ which does not contain the origin, and except possibly at certain isolated points. Within the plane so cut the limit is holomorphic except at these isolated points, which are poles. (Received March 1, 1932.)

As explained by Doron Shaul Lubinsky (b. 1955) [422], the problem of spurious poles is important in the convergence theory:

Physicists such as George Allen Baker, Jr. (1932–2018)[9] in the 1960's endeavoured to surmount the problem of spurious poles. They noted that these typically affect convergence only in a small neighborhood, and there were usually very few of these "bad" approximants. Thus, one might compute $[n/n]$, $n = 1, 2, 3, \ldots, 50$, and find a definite convergence trend in 45 of the approximants, with 5 of the 50 approximants displaying pathological behavior. Moreover, the 5 bad approximants could be distributed anywhere in the 50, and need not be the first few. Nevertheless, after omitting the "bad" approximants, one obtained a clear convergence trend. This seemed to be a characteristic of the Padé method, and led to a famous conjecture.

In 1961, Baker, Gammel, and Wills conjectured that if the function f is mero-morphic in the unit disk D, then there exists an infinite subsequence of the sequence $([n/n]_f)$ that converges locally uniformly for $z \in D/\{\text{poles of } f\}$ [30]. In 1997, Herbert Stahl (1942–2013) reviewed partial results concerning this conjecture. He formulated weaker and more special versions of it, and he investigated their plau-sibility [611]. This conjecture was disproved in 2001 by Lubinsky [421], but it generated a number of related unresolved conjectures whose status is reviewed in

---

[9] See Sect. 8.2 for his biography.

[422]. Counter-examples to this conjecture and others were also given; for example, see [27]. Many other results relating to the convergence of Padé approximants can be found in the literature, in particular in [422].

3.5 Continued Fractions

As we already mentioned, continued fractions have a history almost as long as that of mathematics itself [102]. In the nineteenth century, all mathematics books had a chapter on the topic. They were studied for their own interest, in particular for their properties in the approximation of transcendental numbers, and they also served in the solution of various problems. For example, Georg Cantor (1845–1918) used them in his famous work on cardinality in 1878. For the state of knowledge in this field at the end of the nineteenth century, see [510].

Although most of the algebraic properties of arithmetic continued fractions (mostly used in arithmetic) had already been discovered by the beginning of the nineteenth century, some others were still to be obtained. Let us review some of them.

An infinite product can be transformed into an equivalent continued fraction (that is, whose convergents are identical to the partial products of the infinite one) and vice versa. Both transformations were studied in 1833 by Maritz Abraham Stern (1807–1894) [616]. In 1851, Joseph Liouville (1809–1882) proved the existence of transcendental numbers (such as, for instance, $\sum_{n=0}^{\infty} 10^{-n!}$), that is, numbers that are not the zero of a polynomial with integer coefficients, that there exist infinitely many such numbers, and he used continued fractions for approximating them [411] (the numbers e and π also are transcendental). A determinantal formula for the numerators and the denominators of the convergents of a general continued fraction was given by James Joseph Sylvester in 1853 [631]. In 1858, Louis Félix Painvin (1826–1875) gave the solution of the three-term recurrence relationship satisfied by the numerators and the denominators of the successive convergents (no reference known).

Any real positive number x can be expanded into a continued fraction of the form $x = a_0 + \cfrac{1}{\lvert a_1} + \cfrac{1}{\lvert a_2} + \cdots$ by an extension of Euclid's algorithm for the greatest common divisor. In 1877, Henry John Stephen Smith (1826–1883) proved that the convergents C_k of this continued fraction satisfy $|x - C_{k+1}| < |x - C_k|$, and that $C_k = P_k/Q_k$ with $Q_k > 1$ approximates x more accurately than any other fraction with a smaller denominator [602], a result already noticed by Christiaan Huygens but not proved by him. In 1895, Karl Theodor Vahlen (1869–1945) proved that of these two successive convergents, at least one of them satisfies $|x - P/Q| < 1/(2Q^2)$ [651]. This result, which is useful in proving the irrationality of a given number, was improved in 1903 by Émile Borel, who showed that the factor 2 can be replaced by $\sqrt{5}$ [73].

Let us now come to the important notion of *characteristic* of a rational number, which was later used by Wynn in some of his works. Let $x \in]0, 1[$. Let us denote by $\lfloor nx \rfloor$ the greatest integer less than or equal to nx, and let us set $g_1 = 0$ and $g_n = \lfloor nx \rfloor - \lfloor (n-1)x \rfloor$, for $n \geq 1$. This sequence, introduced in 1875 by Elwin Bruno Christoffel (1820–1900), is called the *characteristic* of x [171]. Its elements are 0 or 1. The characteristic of a rational number is finite and determines x uniquely. Christoffel gave the explicit relationship between the characteristic of x and its continued fraction expansion. When x is rational, the sequence (g_n) terminates, and its continued fraction expansion is finite. Christoffel extended this notion to irrational numbers in 1888 [173]. Characteristics of irrational numbers are infinite, but again determine the number uniquely. The characteristic of x can be obtained from its continued fraction expansion and vice versa. Thus irrational numbers can be considered symbols for distinguishing various characteristics. This concept did not achieve much attention before Wynn revived it.

The study of the convergence of continued fractions began in the nineteenth century. The first general results were obtained for arithmetic continued fractions. Philip Ludwig von Seidel (1821–1896) and Maritz Abraham Stern gave precise definitions of their convergence and divergence [565, 617]. They proved independently that the divergence of the series $\sum_{i=1}^{\infty} a_i$, where the a_i are strictly positive, is a necessary and sufficient condition for the convergence of the continued fraction $C = a_0 + \dfrac{1}{\lfloor a_1} + \dfrac{1}{\lfloor a_2} + \cdots$. Moreover, $|C - C_n| < |C_n - C_{n-1}|$ for $n \geq 2$ and $C_{2n-1} < C_{2n+1} < C_{2n+2} < C_{2n}$ for $n \geq 1$. Julius Worpitzky (1835–1895) in 1865 [701], and Ivan Vladislavovich Śleszyński (1854–1931) in 1889 [598] proved that the condition $|c_n| \leq 1/4$ for $n \geq 2$ is a sufficient condition for the uniform convergence of the continued fraction $\dfrac{c_1}{\lfloor 1} + \dfrac{c_2}{\lfloor 1} + \cdots$. Since Worpitzky's result was published in a quite unknown publication, it is not surprising that it did not attract attention. It was rediscovered in 1899 by Alfred Pringsheim (1850–1941) [511], the father-in-law of the Nobel Prize winner in literature Thomas Mann (1875–1955), and by van Vleck in 1901 [663]. Many other convergence results were obtained by Pringsheim,[10] and others; see [102].

Several convergence results on algebraic continued fractions were also obtained. In October 1863, Bernhard Riemann (1826–1866) proved the convergence of the continued fraction for the ratio of two hypergeometric series, a result found in his papers after his death [525]. The theory culminates at the end of the nineteenth century with the work of Thomas Joannes Stieltjes (1856–1894)[11] in 1894 [624]. He was the real founder of the analytic theory of continued fractions. He introduced the notion of Stieltjes integral, and proved the convergence of Gaussian quadrature methods for a finite interval of integration. His work was the starting point for the development of many ideas in mathematics such as the theory of equations, orthogonal polynomials, infinite matrices, quadratic forms in infinitely many variables, definite integrals, the

---

[10] See https://titurel.org/MathApprObit/PringsheimWerke.pdf for a list.
[11] See Sect. 8.14 for his biography.

moment problem, analytic functions, the summation of divergent series, and the spectral theory of operators (see [653] and [652] for analyses).

Let f be the Stieltjes integral with its series expansion

$$f(z) = \int_{-\infty}^{+\infty} \frac{d\mu(u)}{z - u} = \sum_{n=0}^{\infty} \alpha_n z^{-n-1}.$$

Stieltjes considered two types of continued fractions for f [624]. The *associated continued fraction*

$$\frac{\lambda_1}{|z - c_1} - \frac{\lambda_2}{|z - c_2} - \cdots$$

and the *corresponding continued fraction*

$$\frac{b_1}{|z} - \frac{b_2}{|1} - \frac{b_3}{|z} - \frac{b_4}{|1} - \cdots .$$

The nth convergent of the associated continued fraction has an error in $O(z^{-2n-1})$, while the error for the corresponding continued fraction is $O(z^{-n-1})$. These continued fractions were the topic of many papers such as [575] in which Shohat investigated the relation of the denominators of the odd convergents of the continued fraction

$$\int_{-\infty}^{+\infty} \frac{d\mu(u)}{z - u} = \frac{1}{|l_1 z} + \frac{1}{|l_2} + \frac{1}{|l_3 z} + \frac{1}{|l_4} + \cdots$$

with the corresponding family of orthogonal polynomials, or the paper of Jacob Sherman, a PhD student of Shohat, on the numerators of the continued fractions associated and corresponding to f [574]. Another related work was due to Hubert Stanley Wall [680]. An extension of Stieltjes's result to Gauss continued fractions was given by Edward Burr van Vleck [664]. In many of his works, Wynn referred to Stieltjes either only for quoting his results, or for his own researches [720, 724, 747, 766, 768, 780, 788, 810]. As general references, Wynn mentioned only the classical books of Oskar Perron (1880–1975) [493] and Hubert Stanley Wall [683] on continued fractions.

In 1900, Ivar Fredholm (1866–1927) gave a complete theory of integral equations of the form

$$-\lambda u(x) + \int_a^b K(x, y)u(y)\, dy = f(x),$$

where K is continuous in $[a, b] \times [a, b]$. The problem was taken up by David Hilbert (1862–1943), who founded the spectral theory of operators with a continuous spectrum [329], which led later to the theory of Hilbert spaces. Helge von Koch (1870–1925) [677] and Borel [71] extended the work of Stieltjes. A generalization of Riemann's result is due to van Vleck in 1900 [666], who in 1903, connected continued fractions with definite integrals of the type found by Stieltjes but with the range of integration over the entire real axis. The complete theory was obtained by

Ernst Hellinger (1883–1950) in 1922 using Hilbert's theory of infinite linear systems [317]. At about the same time, several other mathematicians reached the same goal by different methods: Rolf Nevanlinna (1895–1980) in 1922 by function-theoretic methods and asymptotic series [464], Torsten Carleman (1892–1949) by his theory of integral equations [160], and Marcel Riesz (1886–1969) by successive approx-imations [526]. The various papers of Padé on continued fractions must not be forgotten [483].

One of the most important problems studied at the beginning of the twentieth century was the *moment problem*, which consists, from a given sequence (c_n), in determining whether there exists a positive measure μ such that, for all n, $c_n = \int_I x^n \, d\mu(x)$. For $I = [0, +\infty)$, this problem is named after *Stieltjes*, after *Hamburger* for $I = (-\infty, +\infty)$, and after *Hausdorff* if I is a finite interval. The names of well-known mathematicians are attached to the solution of this problem in the 1920s [577]. The matrix moment problem is treated in [376].

Continued fractions whose coefficients obey a noncommutative law of multipli-cation appear for the first time in a paper by William Rowan Hamilton (1805–1865) [304], the discoverer of quaternions. In this paper, the author solved the difference equation $u_{n+1}(u_n + a) = b$, where a, b, and the u_n are quaternions.

Noncommutative continued fractions were then considered in 1913 by Joseph Henry Maclagen Wedderburn (1882–1948), a Scottish algebraist who proved that a finite division algebra is a field. He wrote [688]:

> The object of this note is to investigate the properties of simple continued fractions when the terms are not necessarily commutative with one another. The terms, for instance, may be the product of a function of x and differential operator D_x; or they may be matrices; or in fact any functional operator or hypercomplex quantity for which addition obeys the ordinary laws of algebra and multiplication is associative and distributive. In the demonstrations it is always assumed that an inverse exists, but, as the relations given are all integral identities, the truth of the theorems does not depend on the existence of an inverse: the use of the inverse could therefore no doubt have been avoided, as by studying Euclid's algorithm in place of continued fractions for instance, but it does not seem that there would be any material gain in doing so.

He does not seem to have been followed in this direction before Wynn took up the topic again in several of his papers.

The next appearance of noncommutative continued fractions is due to Herbert Westren Turnbull (1885–1961) [649, 650]. In the second of these papers, he wrote:

> An explicit, and apparently new, form is given for the rational reduction of a matrix to diagonal form, applicable to symmetric matrices and also to continuants [*another term for convergents*]. An account follows which co-ordinates several existent theories of generalized continued fractions, and concludes with a few determinantal theorems as corollaries.

Chapter 4
The Life of Peter Wynn

Peter Wynn in 1975.
© Claude Brezinski.

© Springer Nature Switzerland AG 2020
C. Brezinski, M. Redivo-Zaglia, *Extrapolation and Rational Approximation*,
https://doi.org/10.1007/978-3-030-58418-4_4

Peter Wynn was born on 1 September 1931 in Hoddesdon, Hertfordshire, England, 22 miles (35 km) north of London.[1] In [135], we wrote that he was born in Hertford, 7 km from Hoddesdon, on the basis of official documents found on the internet, but we recently had access to a curriculum vitæ typewritten by Wynn himself in which he mentioned his birthplace. The confusion is perhaps due to the fact that the birth was registered in Hoddesdon, where his father was still living in 1970, 54 Walton Street. His parents were John Wynn and Mabel G. North. They were married around April–June 1930 in Edmonton, an area of the London Borough of Enfield, 8.6 miles (13.8 km) north–east of Charing Cross.

Peter Wynn first received a bachelor of science degree in mathematics in 1953, and then a master of science degree in mathematics in 1954, both from the University of London. He began his career as a mathematician at the Admiralty Research Laboratory in Teddington in 1952–1953. Then he moved to the National Physical Laboratory, also in Teddington, for the period 1953–1955. At the time he published his paper on the ε-algorithm (in 1956, when he was only 25), he was attached to the Scientific Computing Service Ltd. in London (1955–1956) as a mathematician. This service was founded as a private venture in 1936, becoming a limited company in the following year, specialized in scientific calculations generally, and particularly those in which mechanical computation and mass production methods were employed. The managing director was Leslie John Comrie (1893–1950), an English astronomer and pioneer in mechanical computation. Comrie greatly influenced the development of scientific computation in the interwar period, and was elected Fellow of the Royal Society shortly before his death.

Since the scientific life of Wynn was most certainly greatly influenced by his belonging to the Admiralty Computing Service, let us open a parenthesis about it. Its historical roots and its operation are described in [641]:

> In 1942, in order to use more efficiently the scientific staff available in the Admiralty, the Director of Scientific Research set up, within the branch directed by Dr. J.A. Carroll, an Admiralty Computing Service to centralise, where possible, the computational and mathematical work arising in Admiralty Experimental Establishments.

> Mr. John Todd undertook the organisation and supervision of the Service. By agreement with the Astronomer Royal additional staff were attached to H.M. Nautical Almanac Office to carry out the computational work under the direction of the Superintendent, Mr. D.H. Sadler. In addition, arrangements were made to permit the employment of experts from the Universities and elsewhere as consultants.

> The work undertaken by Admiralty Computing Service was in general of one of two classes: heavy computation, or difficult mathematics [. . .]

> Shortly after the formation of Admiralty Computing Service it became apparent that research work in Admiralty (and other) Establishments would be greatly facilitated if their members were informed in certain mathematical and computational techniques not usually covered in undergraduate courses, and of which no adequate account was available in easily accessible literature. Accordingly the preparation of a series of monographs of an expository nature was begun [. . .]

---

[1] Parts of this biography are reprinted, with permission, from [135]. Additional material has been included.

Among the consultants employed during the war were Dr. N. Aronszajn, Professor W.G. Bickley, Dr. L.J. Comrie (Scientific Computing Service, Ltd.), Professor E.T. Copson, Dr. J. Cossar, Dr. A. Erdélyi, Professor P.P. Ewald, Dr. H. Kober, Dr. J. Marshall, Dr. J.C.P. Miller, Professor E. H. Neville [...]

Copies of the reports are only available for distribution to Government Departments and similar agencies but arrangements have been made for copies of some of the reports to be deposited with the Editors where they may be consulted. A very limited number of photostat copies of the unpublished tables is available for distribution or loan to institutions or individuals with a special computational requirement.

A part of this report is connected to the work Wynn was about to undertake

Summation of Certain Slowly Convergent Series. Stencilled typescript, on one side of 4 p.; undated but issued January 1943. 20 X 32.5 cm.

This note draws attention to a device which was apparently first applied in computational work by P.P. Ewald in his work on crystal-structure. Ann. Phys., v. 64, 1921, p. 253–287, Gesell, d. Wissen., Göttingen, math.-phys.K 1., n. s. v. 311, no. 4, 1938, p. 55–64. The analytical basis of the device is the Jacobi Imaginary Transformation of Theta-function Theory (see E. T. Whittaker & G. N. Watson, A Course in Modern Analysis, fourth ed., 1927, p. 475); there is a physical basis, too, which consists in replacing (taking the electrostatic analogy) point charges by Gaussian space-distributions. Applied to the series $S = \sum_{n=0}^{\infty}(n + 1/2)^{-1}e^{-1(n+1/2)/10}$ accuracy comparable with that obtained by summation of 50 terms of the original series may be obtained by taking a single term of one of the two infinite series into which S is transformed, and three terms of the other.

The year 1956 found Wynn as the technical director at the Computer Programming Service Ltd. In 1958, Wynn received a Research Appointment for one year at the *Technische Hochschule* in Munich, Germany, and another one at the *Johannes Gutenberg-Universität* in Mainz, where in 1959, he defended a Ph.D. thesis (Dr. Rer. Nat., D77) with the title *Über einen Interpolations-Algorithmus und gewisse andere Formeln, die in der Theorie der Interpolation durch rationale Funktionen bestehen* [719], under the supervision of Friedrich Ludwig Bauer, a German pioneer of computer science and numerical analysis, and one of the developers of Algol, and he became his assistant.

In 1960, he obtained a Ph.D. from the University of London with a thesis entitled on *Converging Factors for Continued Fractions*. Then from 1960 to 1964, he was a researcher at the *Mathematisch Centrum* in Amsterdam, Netherlands, participating in the early development of Algol with Adriaan van Wijngaarden.[2] In 1965, the University of London granted Peter Wynn the degree of doctor of science (mathematics), a degree conferred on a scholar who has a recognized international reputation. A candidate for this degree is usually required to submit a selection of his publications to the board of the appropriate faculty, which decides whether the candidate merits this accolade. The degree is only exceptionally and rarely awarded to a scholar under the age of forty (he was 34).

Next, Wynn held several researcher's positions in the United States and Canada. From 1964 to 1967, he had a research appointment at the Mathematics Research Center of the University of Wisconsin in Madison. He spent the summer of 1967 at

[2] See Sect. 8.18 for his biography.

Friedrich Ludwig Bauer around 1963.
Credit Repro Uli Benz / TUM. Archiv.

the *Eidgenössische Technische Hochschule* in Zürich, Switzerland, as an academic guest. In 1968–1969, he returned to the *Mathematisch Centrum* in Amsterdam. Then in 1969, he joined the Louisiana State University in New Orleans for two years as an associate professor. During his stay there, he had an M.Sc. student, Lonny Christopher Breaux [78]. Wynn also became friends with Manuel Phillip Berriozábal (b. 1931), who was a professor at the same university. His role in Wynn's legacy will be described in Sect. 5.10. Wynn spent the summer of 1971 as a visiting professor at the Mathematics Research Center of the University of Wisconsin in Madison. After that, he went to the *Centre de Recherches Mathématiques* of the *Université de Montréal* from 1971 to 1975 with a research appointment, before going to McGill University, also in Montreal, as a visiting professor. He moved to Mexico in 1981, first in Guanajato and, from around 1990, in Zacatecas, where he was a visiting professor at the *Universidad Autónoma de Zacatecas* for some time.

As he wrote to Prof. Naoki Osada[3] on 17 February 1997:

> Regarding your question as to where I have been and what I have been doing for the last twenty years, the answer is that I came to Mexico early in 1981 with the intention of taking a fine year holiday but a Mexican peso devaluation of 32,000% (from 25 pesos to the US dollars in 1981 to 8 new pesos (i.e. 8000 old ones) now; against the Japanese yen the devaluation has been greater) has allowed me to live as a tourist here since then. Zacatecas is in the tropics but 2500 meters above see level; it is not cold enough to snow in the winter time and the summer months are quite mild. Of course, one of the first Spanish words that one must learn to use properly is "Soborno"(="bribe") but that is a word which must be learned in many languages. I have worked at mathematics, discovering a number of wonderful things, and have published nothing. For some time now I have been honorary professor at the local university.

When writing the paper [135], the authors of this book tried to obtain information about Peter Wynn, who was living in Zacatecas, Mexico. The second author of this book, M.R.-Z., found that in his book *Topología de Conjuntos, un Primer Curso*, Juan Antonio Pérez, from the *Unidad Académica de Matemáticas, Universidad Autónoma de Zacatecas*, was thanking Wynn. She wrote him, and he answered that Wynn was found dead from a heart attack in his apartment in December 2017, and he added:

> There is not much to be said about Peter in Zacatecas. He was devoted to read, walking our narrow streets and visiting cafeterias where he used to speak with people, mostly about politics.
>
> I never heard about his personal life, family, relatives or something else. Peter used to go to my office when I was working at "Unidad Académica de Estudios Nucleares" (Nuclear Sciences). I am a topologist, so our chats about mathematics were quite limited.

On 15 April 2008, Juan Antonio Pérez gave us more details about Wynn's life in Zacatecas and his sad death:

> Professor Peter Wynn arrived in Zacatecas about 1995, when he asked for active mathematicians working for the local university, the Universidad Autónoma de Zacatecas (UAZ). After doing my PhD I came back to my home town, and I started working for the Nuclear Sciences Department of UAZ (Unidad Académica de Ciencias Nucleares). He arrived there looking [*for*] mathematics related people. From then on, we had frequent fruitful conversations on mathematics, culture and even politics. Wynn was no longer active at that time as a mathematician, so he was not interested in finding common themes for scientific collaboration, but he continued writing quite a lot of mathematical reviews [*141 for zbMATH*]. Peter lived in Zacatecas always on his own, having very few friends and walking daily through the beautiful roads of Zacatecas City, until his death on December last year (2017). He died alone at home, his death being discovered some days afterwards. His body had a very discrete burial and now he rests forever in Herrera Cemetery (Panteón de Herrera).

He also added that Wynn had a very valuable personal library and that since no relative of him was known, the police closed his house after his death. However, some friends and colleagues were able to collect his belongings, and his books will be donated to the UAZ, and his manuscripts will be classified in order to find out whether it is worth publishing them. After that, we never received more information.

---

[3] See his testimony in Sect. 7.9.

We were able to contact another colleague of Peter Wynn from the same department, Héctor René Vega-Carrillo, who wrote us on 5 April 2018:

Dr Peter Wynn was a kind and warm person, he came time to time to the Unidad Academica de Estudios Nucleares of the Universidad Autonoma de Zacatecas (Nuclear Studies school from the University Autonomous of Zacatecas) and talk with faculty about different topics. In our talks Dr Wynn was very concern with the political and social situation in Mexico, in science he was trying to look for better algorithms to reduce the computing time in calculations. We also have long talks about the ethics in research and how to improve the formation of new researchers. He was very interested in our method (named GLAPHI method) to improve the competencies to do research. Sometimes he arrives to the school late in the afternoon and he likes to come down into the Nuclear Measurements Laboratory to talk to late hours in the night, it was very pleasant to listen him to sing with a very strong voice as he moves from the 2nd to the 1st floor. With a large smile, and long and white hair he show up on my door asking what are we doing today?, something to shock the world?

It was my privilege to know Dr Peter Wynn and to have a large talk sessions.

Wynn knew English, German, Russian, Dutch, Spanish, and some French.

He is missed by many people, not only mathematicians, for different reasons related to his unique personality.

Chapter 5
The Works of Peter Wynn

We will now analyze the mathematical contents of the works (papers and reports) published by Peter Wynn.

The analysis of each paper or report is preceded by ▶, and ends with ◀. Its reference in the bibliography, which contains other information, is indicated after its title. A line with the symbol ∞ separates the commentaries about the paper above it from those related to the next paper. Quotations are in small roman characters.

For some papers it was difficult to produce a deep analysis because of the plethora of notations introduced by Wynn. Thus, only a rough idea of their contents was given.

We were unable to obtain a copy of only three of his reports: the *Centre de Recherches Mathématiques* of the *Université de Montréal* has no trace of [792], and the *School of Computer Science* of *McGill University* in Montreal does not possess [806] and [807].

The works have been divided into sections, each of them devoted to a particular topic, and in each section, we followed the chronological order.

5.1 Convergence Acceleration

We begin with the works of Wynn dedicated to sequence transformations for accelerating their convergence and the related algorithms for their implementation. Wynn was interested in several such transformations.

5.1.1 The Scalar ε-Algorithm

The papers on his ε-algorithm and its implementation are the most important ones. They also often contain a part dealing with continued fractions and Padé approximants, since these topics are strongly related.

© Springer Nature Switzerland AG 2020
C. Brezinski, M. Redivo-Zaglia, *Extrapolation and Rational Approximation*,
https://doi.org/10.1007/978-3-030-58418-4_5

5.1.1.1 The Algorithm and Its Properties

One of Wynn's most important contributions to convergence acceleration methods is certainly his discovery of the scalar ε-algorithm, which allows a recursive implementation of the Shanks transformation without computing the determinants involved in its mathematical expression (2.13). This is why we begin with its analysis.

▶ *P. Wynn, On a device for computing the $e_m(S_n)$ transformation, Math. Tables Aids Comput., 10 (1956) 91–96. [713]*

The ε-algorithm for implementing recursively the scalar Shanks sequence transformation [567, 568] was presented by Wynn in [713]. The paper appeared in April 1956, but it was written before 31 October 1955, since the paper [712], where [713] is quoted, was submitted at that date. At that time, he was 24 years old, and he was working at the *Scientific Computing Service*, Ltd., 23, Bedford Square, London, England. He wrote about the Shanks transformation:

> A consideration which militates against the use of the transformation is that the vast amount of labor spent in evaluating the determinants in (1) [*the ratio of determinants defining it*] serves only to produce one transformed result [. . .] It is proposed to show that the transformation (1) may be effected by a simple algorithm, in which transformed results for $n = \tilde{n}, \tilde{n} + 1, \ldots,$ $m = 1, 2, \ldots$ are progressively available for comparison.

Then he gave the rule of the ε-algorithm and related it to the ratio of determinants appearing in the Shanks transformation. The proof is quite technical, with manipulations of the rows and the columns of the determinants, and it makes use of the determinantal identities named after Sylvester and Schweins. Then he arranged the $\varepsilon_k^{(n)}$ into a double entry array (the ε-array) similar to the table for the powers of the operator Δ, where the index k denotes a column and n a descending diagonal. The rule of the ε-algorithm relates four quantities located at the vertices of a lozenge (see Fig. 2.2). He added: [. . .] *this simple calculation is easily programmed for a digital computer* [. . .]. Then a numerical example follows, and Wynn pointed out that the algorithm could be useful in the computation of the functions of mathematical physics, and also [. . .] *that the transformation may be applied to establish a criterion for the fitting of certain types of statistical data.* ◄

Wynn attributed the transformation not only to Shanks but also to Robert J. Schmidt [554], who, in a paper submitted on 29 September 1939, used it for the solution of a system of linear equations (see [135] for an analysis of this paper). We were unable to identify him. Wynn found the determinantal identities of Schweins and Sylvester in a book by Aitken [7]. He mentioned its 6th edition of 1949, Chapter V, page 108, formula (2) and page 49, formula (9).

The journal *Mathematical Tables and Other Aids to Computation* (MTAC), in which the paper appeared, was founded in 1943. It was published quarterly, and its first editor was Raymond Clare Archibald (1875–1955), a world-renowned historian of mathematics. The editorial board consisted of 24 members. In 1944, Derrick Henry Lehmer (1905–1991), a number theorist, was appointed as its second editor.

The journal continued to be published until 1959. Then it became *Mathematics of Computation* (see [341] for a detailed history).

∞

▶ *P. Wynn, On repeated application of the epsilon algorithm, Revue française de traitement de l'information, Chiffres, 4 (1961) 19–22. [726]*

In this paper, Wynn gives an example of the repeated application of the ε-algorithm. It consists in computing the sequence $(\varepsilon_{2k}^{(n)})$ for a fixed value of k and using it as a new sequence for reapplying the algorithm. Obviously, the procedure can be repeated several times. No theoretical results are presented. A numerical example involving the series expansion of the exponential integral is given. It was programmed on the Z22 computer, which was the seventh computer model developed by Konrad Zuse (1910–1995), a German computer pioneer. ◀

This paper was published in the journal *Chiffres*, edited by the *Association française de calcul* (AFCAL) founded in 1958 by Jean Kuntzmann (1912–1992), a professor at the University of Grenoble and a pioneer of numerical analysis and applied mathematics in France. It was followed, in 1962, by the *Association française de calcul et de traitement de l'information* (AFCALTI). The journal *Chiffres* is issued from these associations. Its first issue was published in 1958, and the journal continued under that name until 1963. Then, integrating people in operation research, the association took the name *Association française d'informatique et de recherche opérationnelle* (AFCET), with 4500 members in 1985, and the journal took the name of *Revue d'Informatique et de Recherche Opérationnelle* (R.I.R.O.), and later R.A.I.R.O., where the A stands for *automatique*. AFCET also published an internal journal, *AFCET-Interfaces*, with C.B.[1] as its editor-in-chief for some time. AFCET disappeared in 1998 because of bankruptcy. However, the journal still exists under this title but only publishes papers on operations research. This repeated application of the ε-algorithm would also be used in [782].

∞

In the next paper, Wynn discusses how to apply it to other difference operators. In the ε-algorithm, implicit use is made of the operator E defined by $Eu_n = u_{n+1}$.

▶ *P. Wynn, The epsilon algorithm and operational formulas of numerical analysis, Math. Comp., 15 (1961) 151–158. [725]*

It is well-known that difference operators are related to the derivative and integral operators. In order *to develop new methods for numerical integration and differentiation of continuous functions in terms of their tabular differences*, Zdeněk Kopal (1914–1993) transformed series of such operators into linearized rational functions of operators [374]. As stated by Wynn in [725] (submitted November 1959; revised June 1960), Kopal's aim was to replace the operational equation

---

[1] See his testimony in Sect. 7.13.

$$\left(\sum_{s=0}^{\infty} c_s d^s\right) F = f,$$

where F is a known function from which f is to be determined and d a finite difference operator, by the linearized rational approximant $U_{rr}(d)F = V_{rr}(d)f$, where U_{rr} and V_{rr} are respectively the numerator and the denominator of the $[r/r]$ Padé approximant of F. Wynn pointed out that

> Kopal was only able to find useful application of the technique when d was the backward difference operator, though his numerical results, which related to the forward integration of a differential equation, appeared to be very promising. However, the same effect over a very much larger range of problems may be achieved by recourse to another method.

This other method consisted in applying the scalar ε-algorithm to the partial sums $\varepsilon_0^{(m)} = \sum_{s=0}^{m} a_s p^s F$, where p is an associative and commutative operator such that $a_s p^s F = c_s x^s, s = 0, 1, \ldots$. Then he obtained $\varepsilon_{2s}^{(m)} = U_{s,m+s}(x)/V_{s,m+s}(x)$. The first numerical example concerns the computation of the derivative at $z = 0$ of the function $\exp(hz)$ for $h = 0.6$ by means of the formula

$$\left(\frac{d}{dz}F\right)_{z=0} = \sum_{s=0}^{\infty} \frac{(-1)^s}{s+1} \Delta^{s+1} F,$$

where

$$\frac{(-1)^s}{s+1} \Delta^{s+1} F = \frac{(-1)^s}{s+1} (e^h - 1)^{s+1}.$$

In this example, the $[r/r]$ Padé approximants are the successive convergents of the continued fraction

$$x^{-1} \log(1 + x) = \frac{1|}{|1} + \frac{1^2 x|}{|2} + \frac{1^2 x|}{|3} + \frac{2^2 x|}{|4} + \frac{2^2 x|}{|5} + \cdots.$$

The ε-algorithm *converges quite reasonably for $(e^{hz} - 1) > 1$, when the series diverges rapidly.* The second example is about the interpolation of the function $\log(0.6 + hz)$ for $h = 0.1$ and $z = 0.25$ with points of tabulation at unit intervals of z by use of Bessel's interpolation formula. The result is not spectacular, as he mentioned, but having experimented with the procedure in a large number of cases, he concluded that *in none of these was the accuracy of the transformed results worse than the original partial sums*. The last example concerns the Euler–Maclaurin integration formula, where the gain was about three decimal places. He concluded that it would be useful to relate the determinantal expression of the Shanks transformation to the solution of the operational equation of Kopal, *but this appears to be one of the cases in which a statement of the problem is not a great step forward to its solution.* This work was never continued. ◀

∞

As stated above, the scalar ε-algorithm allows one to compute the Padé approximants at one point of a function given by its power series expansion when it is applied to the sequence of its partial sums. Although the next paper deals with Padé approximants, Wynn obtained the result via his ε-algorithm. This is why we decided to discuss it here.

▶ *P. Wynn, Upon systems of recursions which obtain among the quotients of the Padé table, Numer. Math., 8 (1966) 264–269. [754]*

In this paper (submitted 5 May 1965), Wynn derived the *cross rule* for the scalar ε-algorithm. By two simple algebraic manipulations, Wynn eliminated the quantities with an odd lower index, which are intermediate computations, in the rule of his ε-algorithm, and he obtained a rule linking quantities with only even lower indices (we recall that $\varepsilon_{2k}^{(n)} = e_k(S_n)$). Similarly, the quantities with an even lower index can be eliminated and another cross rule relating quantities with only odd indices is obtained. It is

$$\frac{1}{\varepsilon_{k+2}^{(n)} - \varepsilon_k^{(n+1)}} + \frac{1}{\varepsilon_{k-2}^{(n+2)} - \varepsilon_k^{(n+1)}} = \frac{1}{\varepsilon_k^{(n+2)} - \varepsilon_k^{(n+1)}} + \frac{1}{\varepsilon_k^{(n)} - \varepsilon_k^{(n+1)}},$$

with the initial conditions $\varepsilon_{-2}^{(n)} = \infty$, $\varepsilon_{-1}^{(n)} = 0$, and $\varepsilon_0^{(n)} = S_n$ for $n = 0, 1, \ldots$. Thus this rule relates, in a table in which only the quantities with a lower index of the same parity have been kept, five quantities denoted by the cardinal points, and displayed as follows (C means "center"):

$$N = \varepsilon_k^{(n)}$$

$$W = \varepsilon_{k-2}^{(n+2)} \quad C = \varepsilon_k^{(n+1)} \quad E = \varepsilon_{k+2}^{(n)}$$

$$S = \varepsilon_k^{(n+2)}$$

and the cross rule is

$$\frac{1}{N - C} + \frac{1}{S - C} = \frac{1}{W - C} + \frac{1}{E - C},$$

with the initializations

$$\varepsilon_{-2}^{(n)} = \infty, \quad \varepsilon_{-1}^{(n)} = 0, \quad \varepsilon_0^{(n)} = S_n.$$

This relation leads to the singular rule $N + S = W + E$ when a division by zero occurs in the cross rule. ◀

This paper is an important one. Since the ε-algorithm is strongly related to Padé approximants, and because these approximants satisfy several recurrence relations obtained by Ferdinand Georg Frobenius, the cross rule was named the *missing identity of Frobenius* (see [102]). Later, it played a role in the proof of the convergence of the ε-algorithm for totally monotonic sequences [84, 91]. Moreover, thanks to this rule, Wynn was able to propose particular rules for avoiding (exact or near) singularities in the computation of the ε-array [743].

$$\infty$$

In the next paper, Wynn proposed an algebraic identity holding for the scalar ε-algorithm. He would come back to it several times, and he also proved a similar identity for the confluent form of his algorithm.

▶ *P. Wynn, Upon an invariant associated with the epsilon algorithm, MRC Technical Summary Report 675, University of Wisconsin, Madison, July 1966. [759]*

In this technical report, Wynn proved that if the scalar ε-algorithm is applied to a sequence (S_n) satisfying, for all n, $\sum_{i=0}^{k} a_i S_{n+i} = 0$ with $\sum_{i=0}^{k} a_i \neq 0$, then for all n,

$$\varepsilon_0^{(n)}\varepsilon_1^{(n)} - \varepsilon_1^{(n)}\varepsilon_2^{(n)} + \varepsilon_2^{(n)}\varepsilon_3^{(n)} - \cdots + \varepsilon_{2k-2}^{(n)}\varepsilon_{2k-1}^{(n)} = -\sum_{i=0}^{k} i a_i / \sum_{i=0}^{k} a_i.$$

The proof is based of the determinantal expressions for the scalars $\varepsilon_k^{(n)}$, and on Schweins's extensional identity. ◀

Notice that surprisingly, the result is the derivative at $x = 1$ of $\log(a_0 + a_1 x + \cdots + a_k x^k)$. Its meaning has still to be understood.

5.1.1.2 Error Propagation and Stability

Wynn was a genuine numerical analyst. He dedicated several papers to the propagation of rounding errors and the numerical stability of the scalar ε-algorithm and other nonlinear algorithms related to it. He gave particular rules to be used for avoiding such situations.

$$\infty$$

▶ *P. Wynn, On the propagation of error in certain non-linear algorithms, Numer. Math., 1 (1959) 142–149. [716]*

This paper (submitted on 26 February 1959) was written when Wynn was at the *Rechenzentrum of the Technische Hochschule*, Munich, Germany, but the address given at its end is *Institut für Angewandte Mathematik der Universität Mainz*. He acknowledged a grant from the *Deutsche Forschungsgemeinschaft*.

The rules of several nonlinear recursive algorithms relate quantities located at the four vertices of a lozenge. This is the case of the ε-algorithm but also of the qd-algorithm, the first and the second g-algorithms, the η- and the ϱ-algorithms. If a division by zero occurs in one of these algorithms, it has to be stopped, or the breakdown has to be bypassed. If there is a division by a number close to zero, the algorithm becomes numerically unstable. In this work, [...] *formulæ are given which describe the propagation of error, and methods are described which enable the numerical stability of the algorithms to be assessed.* By some graphics, Wynn showed in which region of the table containing the quantities computed by the algorithm an error in the initial quantities propagates. ◄

The many connections between the ε- and the qd-algorithms, and other algorithms such as the η and the g, are described in [38, 39, 42].

$$\infty$$

In the two following papers, Wynn was interested in sufficient conditions for the numerical instability of the qd- and the ε-algorithms.

► *P. Wynn, A sufficient condition for the instability of the q-d algorithm, Numer. Math., 1 (1959) 203–207. [717]*

This paper (submitted 25 March 1959, about 1 month after [716]) is entirely devoted to the stability of the qd-algorithm, as explained by Wynn:

> In the following note a simple expression is derived for the errors in certain quantities which are produced by means of the $q - d$ algorithm, resulting from a special distribution of error in the initial conditions. In any application of the algorithm this distribution and the resulting error growth exist only as possibilities. If however the algorithm is shown to be unstable to a certain extent in this case, the certainty that it can be more stable has been removed. This leads to the establishment of a criterion for the sufficiency of the instability of the algorithm.
>
> Three examples of the application of the algorithm are given for which an a priori formulation of the criterion is possible.

The first example is the transformation of an infinite series into a continued fraction in two different ways; the other one expresses the series as a Laplace integral. ◄

► *P. Wynn, A sufficient condition for the instability of the ε-algorithm, Nieuw Arch. Wiskd., 9 (1961) 117–119. [729]*

In this short paper, Wynn, extending to the ε-algorithm the results of [717], writes

> At the time at which the note referred to was written, corresponding differential forms of the ε-algorithm had not been discovered, but this has latterly been remedied and accordingly a similar treatment of the ε-algorithm is possible.

Wynn considers the application of the ε-algorithm to $\varepsilon_0^{(m)} = \sum_{s=0}^{m-1} \phi^{(s)}(a) z^{-s-1}$, $m = 1, 2, \ldots$, with $\varepsilon_0^{(0)} = 0$, and he uses the second confluent form of his algorithm [731] (at that time yet to appear) to derive differential–difference equations satisfied by the $\varepsilon_k^{(m)}$. For $z = 1$, if a is replaced by $a + \Delta a$, he derives the changes induced in the $\varepsilon_k^{(m)}$. He concludes that if the algorithm [...] *is applied to a series of terms which have approximately constant modulus and argument* [...] *it is a matter of numerical experience that the ε-process is unstable.* ◄

$$\infty$$

After having studied the sufficient conditions for the instability of the qd and the ε algorithms, Wynn comes to the problem of curing them by singular rules. He also shows how to cheaply program the algorithms by keeping only the last ascending diagonal in the table that contains the quantities. For that, he used the technique of a moving lozenge that needs only two temporary additional memories. This kind of technique would later attract much wider interest, not only for the ε-algorithm, but also for other recursive algorithms whose elements, represented in a two-dimensional array, are related by more complicated relations and are applied to vector and matrix sequences. It was systematically used in the Fortran and Matlab codes given in [127] and [133].

▶ *P. Wynn, Singular rules for certain non-linear algorithms, BIT, 3 (1963) 175–195. [743]*

After having described in [716] the propagation of rounding errors in nonlinear recursive algorithms, Wynn is interested in [...] *how this misfortune can be overcome*. He treats and solves the case in which two, and only two, consecutive quantities in a column of the arrays of the ε-, the ϱ-, and the qd-algorithms are equal or nearly equal (an *isolated singularity* as he denominates it), and he gives particular rules for jumping over the (near-)singularity, and continuing the application of the algorithm. Then he considers in detail how to program such algorithms by adding one by one the terms of the sequence to be accelerated, and climbing along an ascending diagonal of the array. It is his technique of the moving lozenge, proposed for the first time in [729], that avoids having to store the complete array. Then he gives the Algol procedure for the ε-algorithm, explains the strategy, and illustrates it by a numerical example. He also discusses the case of nonisolated singularities, and concludes that his rules are inadequate for that. ◀

The particular rules for isolated singularities were systematically used, in particular, in the simplified implementation of a generalization of the ε-algorithm for sequences of elements of a general vector space [132].

The case of nonisolated singularities was solved by Florent Cordellier (b. 1937) [180] (see also [181]). His rule allows one to turn around a square block of identical approximants in the Padé table and to continue to compute them. Denoting by C the common value of these approximants, by $N_i, i = 1, \ldots, N$, those on the north side of the block beginning with $i = 1$ at the northwest corner, by W_i the approximants on the west side from the northwest corner, by S_i those on the south side starting with $i = 1$ at the southeast corner, and by E_i those on the east side numbered from the southeast corner, the extended cross rule of Cordellier is

$$(N_i - C)^{-1} + (S_i - C)^{-1} = (W_i - C)^{-1} + (E_i - C)^{-1}, \quad i = 1, \ldots, N.$$

Jumping over such blocks using extended recurrence relationships between elements of adjacent families of formal orthogonal polynomials was treated by André Draux (b. 1943)[2] in his doctoral thesis [206].

5.1.1.3 Application to Special Sequences

Wynn published several papers on the application of the scalar ε-algorithm to special classes of sequences. Since some results appear in several papers, we decided to gather their analysis below without explaining separately the contents of each of them. They are

▶ *P. Wynn, On the convergence and stability of the epsilon algorithm, SIAM J. Numer. Anal., 3 (1966) 91–122. [755]*

▶ *P. Wynn, A note upon totally monotone sequences, Report CRM-139, Centre de Recherches Mathématiques, Université de Montréal, Montréal, November 1971. [776]*

▶ *P. Wynn, Sur les suites totalement monotones, C.R. Acad. Sci. Paris, 275A (1972) 1065–1068. [786]*

▶ *P. Wynn, Transformation de séries à l'aide de l'ε-algorithm, C.R. Acad. Sci. Paris, 275A (1972) 1351–1353. [787]*

▶ *P. Wynn, Accélération de la convergence de séries d'opérateurs en analyse numérique, C.R. Acad. Sci. Paris, 276A (1973) 803–806. [796]*

A sequence (S_n) is said to be totally monotonic (or monotone) if $\forall k, n \geq 0, (-1)^k \Delta^k S_n \geq 0$, where Δ is the usual forward difference operator. As proved by Felix Hausdorff (1868–1942) in 1921, a necessary and sufficient condition that there exist a Borel measure μ such that $S_n = \int_0^1 x^n \, d\mu(x)$ is that (S_n) be totally monotonic [307, 308].

In [755] (submitted on 16 September 1965 to the Mathematics Research Center, University of Wisconsin, Madison, Wisconsin), Wynn first proved that applying the ε-algorithm to such a sequence yields, for all k and n, $\varepsilon_{2k}^{(n)} \geq 0$ and $\varepsilon_{2k+1}^{(n)} \leq 0$. The result is illustrated by the partial sums of a Newton series and a Dirichlet series under some conditions. A sequence (S_n) is said to be totally oscillating if the sequence $((-1)^n S_n)$ is totally monotonic. For such a sequence, the ε-algorithm produces $(-1)^n \varepsilon_{2k}^{(n)} \geq 0$ and $(-1)^n \varepsilon_{2k+1}^{(n)} \leq 0$. If a totally oscillating sequence converges, then its limit is zero. Again, examples of Newton and Dirichlet series are given. The study is based on the sign of the Hankel determinants involved in the numerator and the denominator of the Shanks transformation. This sign goes back to the

---

[2] See his testimony in Sect. 7.3.

spectral theory of operators due to Stieltjes and Hilbert. Then four special classes of asymptotic sequences are studied in detail: monotonic and alternating Newton series, and monotonic and alternating Dirichlet series. The asymptotic behavior of the quantities $\varepsilon_k^{(n)}$ is given when k is fixed and n goes to infinity. The conclusion is that the convergence of series with monotone terms is not accelerated but only improved by a small factor by the algorithm, while series with alternating sign are accelerated. Then a detailed analysis of the numerical stability of these four classes follows as described by the author:

> In the present inquiry we have been able to derive estimates for the order of magnitude of the quantities being computed. Using these estimates we are able to obtain approximate values for the factors which magnify or diminish the small errors which are introduced at each stage of the calculation. We shall, that is, give a stability analysis for the cases being considered.

Numerical examples end the paper, and an appendix is devoted to the smoothing effect of the ε-algorithm. Let us mention that some new totally monotonic or oscillating sequences can be derived from other ones, as exhibited by Wynn in [776, 786, 787] and also, later, in [88]. In [796], Wynn showed that some sequences coming from some numerical methods are totally monotonic under certain assumptions: Newton's interpolation series, Newton's series for the derivative, Euler–Maclaurin formula, Newton–Gregory series formula. Numerical results are provided. ◄

These results were extended in [83, 84], where the convergence of the columns and the diagonals of the ε-array is proved for totally monotonic and totally oscillating sequences. Then in [91], the acceleration with respect to the initial sequence was demonstrated. An open problem is to know whether each column and each diagonal converges faster than the preceding one.

5.1.2 Confluent Forms

Confluent forms of the algorithms that transform a sequence into another one converging faster to the same limit are aimed at transforming a function into another one converging faster to its limit as its argument tends to infinity. These confluent forms were all obtained by a limiting process. The case of the confluent ε-algorithm was explained in Chap. 2. A similar treatment is possible for the other algorithms. Wynn devoted several papers to their derivation, their properties, and their applications.

▶ P. Wynn, *Confluent forms of certain non-linear algorithms*, Arch. Math., 11 (1960) 223–236. [721]

In the introduction of this paper (submitted 5 October 1959) Wynn wrote:

> Rutishauser, in his work upon the qd algorithm [. . .] derived by means of a formal limiting procedure, an infinitesimal analogue of the qd algorithm, and described some interesting properties of this confluent form of the algorithm. The following note extends the limiting technique to a number of other nonlinear algorithms, and properties of the confluent forms of the algorithms are derived.

The nonlinear algorithms considered obey a lozenge rule (or two such rules) relating quantities $\Phi_s^{(m)}$ displayed in an array. Their interpretation can differ according to whether the lower index is odd or even, such as in the ε-algorithm, where the $\varepsilon_{2k}^{(n)}$ and the $\varepsilon_{2k+1}^{(n)}$ are separated. The confluent form of such an algorithm is obtained by replacing the discrete variable m in $\Phi_s^{(m)}$ by a continuous one defined by $t = a + m\Delta t$. Thus, as m increases by 1, t increases by Δt. Then defining functions of t related to the $\Phi_s^{(m)}$, plugging them into the rule(s) of the algorithm, and letting Δt tend to zero leads to the confluent form of the algorithm. This was the procedure followed by Rutishauser in his paper [537]. Wynn explained that *Properties of the confluent forms of the algorithms may be deduced from properties of the algorithms themselves by means of formal limiting procedures.* He then applied the same idea to deduce the confluent form of the first and the second g-algorithms of Bauer [39], the η-algorithm [38], the ϱ-algorithm [712], and, of course, his ε-algorithm. For this algorithm, Wynn begins by replacing $\varepsilon_{2k+1}^{(n)}$ by $\varepsilon_{2k+1}(t)/\Delta t$, and $\varepsilon_{2k}^{(n)}$ by $\varepsilon_{2k}(t)$. As Δt tends to 0, the (first) confluent form of the ε-algorithm is obtained,

$$\varepsilon_{k+1}(t) = \varepsilon_{k-1}(t) + 1/\varepsilon_k'(t),$$

with $\varepsilon_{-1}(t) = 0$ and $\varepsilon_0(t) = f(t)$. Then he proved that if $f(t) = a + \sum_{i=0}^k a_i \exp(\lambda_i t)$, then $\varepsilon_{2k}(t) = a$ for all t. Thus, under certain conditions, and for a certain k, $\varepsilon_{2k}(t) = \lim_{t\to\infty} f(t)$. In the same way, as the scalar ε-algorithm transforms a sequence into a set of other sequences converging faster under some assumptions, its confluent form transforms a function into a set of other functions converging faster under some assumptions. Wynn also derived the confluent form of the ϱ-algorithm,

$$\varrho_{k+1}(t) = \varrho_{k-1}(t) + (k+1)/\varrho_k'(t),$$

with $\varrho_{-1}(t) = 0$ and $\varrho_0(t) = f(t)$. He gave the ϱ_k as ratios of determinants.

In the last section of the paper, Wynn explained that some of these results have only a minor interest:

The manner in which properties of the algorithms themselves may be transcribed into properties of the confluent forms, when differentiation is in the z-direction, and m is retained as the discrete variable is not so apparent. Indeed the following confluent forms of the algorithms are given without properties, and may have no other significance than that of exercises in the technique of deriving difference-differential relations from purely discrete recursions. ◄

We see that Wynn was lucid about his own work and did not try to cheat.

As already explained in Sect. 2.5.3, Thiele's interpolating continued fraction is connected to the ϱ-algorithm. When all interpolation points coincide, Thiele's continued fraction becomes its confluent form, and we obtain the so-called Thiele's expansion formula

$$f(t+h) = f(t) + \cfrac{h}{|\alpha_1(t)} + \cfrac{h}{|\alpha_2(t)} + \cdots,$$

with $\alpha_k(t) = \varrho_k(t) - \varrho_{k-2}(t)$, for $k = 1, 2, \ldots$. While Taylor's expansion terminated when f is a polynomial, Thiele's ends when f is a rational fraction. Replacing t by 0 and h by x, we obtain

$$f(x) = f(0) + \frac{x}{|\alpha_1} + \frac{x}{|\alpha_2} + \cdots,$$

with $\alpha_k = \alpha_k(0)$. The convergents C_k of this continued fraction are such that

$$C_{2k}(t) = [k/k]_f(t), \quad C_{2k+1}(t) = [k + 1/k]_f(t).$$

Thus, a connection also exists between the confluent form of the ϱ-algorithm and the qd-algorithm [94].

$$\infty$$

Theoretical properties of the first confluent form of the ε-algorithm and of the qd-algorithm are studied by Wynn in the following two papers.

▶ P. Wynn, *A note on a confluent form of the ε-algorithm*, Arch. Math., 11 (1960) 237–240. [722]

▶ P. Wynn, *Una nota su un analogo infinitesimale del q-d algoritmo*, Rend. Mat. Roma, 21 (1962) 77–85. [737]

These two papers are almost identical. If f satisfies a linear differential equation of a form similar to the difference equation defining the kernel of the Shanks transformation

$$a_0(f(t) - a) + a_1 f'(t) + \cdots + a_k f^{(k)}(t) = 0,$$

with $a_0 a_k \neq 0$, then $\varepsilon_{2k}(t) = a$ for all t, as shown by Wynn in [722]. In the same paper, he also proved that $\varepsilon_{2k}(t) = H_{k+1}^{(0)}/H_k^{(2)}$, $\varepsilon_{2k+1}(t) = H_k^{(3)}/H_{k+1}^{(1)}$, and $\varepsilon_{2k+2}(t) = \varepsilon_{2k}(t) - [H_{k+1}^{(1)}]^2/[H_k^{(2)} H_{k+1}^{(2)}]$ (the sign is wrong in the paper), where for all t,

$$H_k^{(n)} = \begin{vmatrix} f^{(n)} & \cdots & f^{(n+k-1)} \\ \vdots & & \vdots \\ f^{(n+k-1)} & \cdots & f^{(n+2k-2)} \end{vmatrix},$$

with $H_0^{(n)} = 1$. These relations correspond to those of the scalar algorithm. The paper ends with similar results for the confluent form of the qd-algorithm that he used in [737] for expressing the Laplace integral

$$F(z) = \int_0^\infty e^{-zt} \Phi(a + t) \, dt$$

as a continued fraction in two different ways. ◀

Observe that the kernel of the first confluent form of the ε-algorithm is a homogeneous linear differential equation of order k, and thus it is the continuous equivalent of the kernel of the scalar ε-algorithm.

$$\infty$$

Then Wynn proposed a second confluent form for his ε-algorithm and studied the connections between the two forms.

▶ *P. Wynn, Upon a second confluent form the ε-algorithm, Proc. Glasgow Math. Assoc., 5 (1962) 160–165. [731]*

A second confluent form of the ε-algorithm was presented in this paper (submitted 21 July 1961). By a different limiting process, Wynn arrived at the formulas $\varepsilon_{2k}(t) = H_{k+1}^{(-1)}/H_k^{(1)}$ and $\varepsilon_{2k+1}(t) = H_k^{(2)}/H_{k+1}^{(0)}$, where $f^{(-1)}(t) = 0$, and at the difference–differential relations

$$\varepsilon_{2k+2}(t) - \varepsilon_{2k}(t) = 1/\varepsilon_{2k+1}'(t),$$
$$\varepsilon_{2k+1}(t) - \varepsilon_{2k-1}(t) = 1/(\varepsilon_{2k}'(t) + f(t)),$$

with $\varepsilon_{-1}(t) = \varepsilon_0(t) = 0$. Using Schweins's identity, he then obtained the relation $\varepsilon_{2k+1}(t) - \varepsilon_{2k-1}(t) = [H_k^{(1)}]^2/[H_{k+1}^{(0)} H_k^{(0)}]$. Wynn claimed that under certain conditions, $\varepsilon_{2k}(t) = \int_t^\infty f(x)\, dx$, but we did not find any trace of this result in the paper. A generalization is given with an application to a particular continued fraction. ◀

$$\infty$$

Then Wynn studied the relations between the two confluent forms of the ε-algorithm.

▶ *P. Wynn, On a connection between the first and the second confluent forms of the ε-algorithm, Niew. Arch. Wisk., 11 (1963) 19–21. [745]*

The two confluent forms of the ε-algorithm are connected in this paper (submitted 29 October 1962). Denoting with an additional star the results due to the second form, and initializing the two forms from $f^{(n)}$, $n \geq 0$ (and thus adding n as an upper index), Wynn proved that $\varepsilon_{2k+1}^{(n)*}(t) = \varepsilon_{2k+1}^{(n-1)}(t)$ and $\varepsilon_{2k}^{(n)*}(t) = \varepsilon_{2k}^{(n-1)}(t) - f^{(n-1)}(t)$. No application is given. ◀

$$\infty$$

In the next paper, Wynn came again to the invariant he obtained for the scalar ε-algorithm [759], and showed that a quite similar invariant also holds for its first confluent form.

▶ *P. Wynn, Invariants associated with the epsilon algorithm and its first confluent form, Rend. Circ. Mat. Palermo, 21 (1972) 31–41. [783]*

If the scalar ε-algorithm is applied to a sequence (S_n) satisfying, for all n, $\sum_{\nu=0}^k c_\nu S_{n+\nu} = 0$, with $c_0 c_k \neq 0$, then, using manipulations on determinants, Wynn proved that, for all n,

$$\sum_{i=0}^{2k-2}(-1)^i\varepsilon_i^{(n)}\varepsilon_{i+1}^{(n)} = -\sum_{v=1}^{k}vc_v\Big/\sum_{v=1}^{k}c_v.$$

Similarly, applying the first confluent form of the ε-algorithm to a function f satisfying, for all t, an irreducible differential equation of the form $\sum_{v=0}^{k}c_vf^{(v)}(t) = 0$, the same kind of proof led Wynn to the relation, for all t,

$$\sum_{i=0}^{2k-2}(-1)^i\varepsilon_i(t)\varepsilon_{i+1}(t) = -c_1/c_0.$$

Again, no application is evoked. ◄

$$\infty$$

Wynn devoted several papers to the use of the confluent forms of the ε-algorithm in the computation of infinite integrals. These methods use the derivative of the integrand at the finite endpoint of the integral. In the first of these papers, he introduced a new recursive algorithm, the ω-algorithm, which, depending on its initializations, allows one to implement various discrete or confluent transformations.

▶ P. Wynn, *Upon some continuous prediction algorithms. I, II, Calcolo, 9 (1973) 197–234; 235–278. [793]*

Together, the two papers (parts I and II, published consecutively in the same volume) total 82 pages. Their detailed analysis is quite complicated. This is why we give only the abstract and then present an interesting algorithm they contain.

Three difference–differential processes operating upon an initial value function $\Phi(\mu)$ and yielding a sequence of approximations to $\lim_{\mu=\infty}\Phi(\mu)$ are described. The processes are confluent analogues of an algorithm related to Romberg's extrapolation procedure, of the ϱ-algorithm, and of the ε-algorithm. It is shown that these processes can be modified to yield approximations to the integral $\int_{\mu}^{\infty}\psi(\mu')\,d\mu'$, and that in certain circumstances this integral can be evaluated by a process of repeated differentiation involving the function $\psi(\mu)$. Discrete algorithms for approximating $\lim_{\mu=\infty}\Phi(\mu)$, for which it is assumed that values of the successive derivatives of the function $\Phi(\mu)$ are available, are described. It is shown that these algorithms can be applied to the evaluation of the integral $\int_{\mu}^{\infty}\psi(\mu')\,d\mu'$, it being assumed that the values of the successive derivatives of the function $\psi(\mu)$ are available. A number of examples in which closed expressions for the transformed estimates can be derived are discussed. Algol procedures for implementing the discrete algorithms are given.

Confluent forms of Richardson's extrapolation process when $x_n = 1/(n+1)$ are given and discussed (without mentioning the name or a reference), as well as confluent forms of the ϱ-algorithm that are expressed by ratios of Hankel determinants, and related to Padé approximants. Then Wynn turns to the first confluent form of the ε-algorithm and proposes an application to integration using the derivatives of the integrant. Then he defines a quite interesting new algorithm, the ω-algorithm, whose rules are

$$\omega_{2k+1}^{(n)} = \omega_{2k-1}^{(n+1)} + \omega_{2k}^{(n)}/\omega_{2k}^{(n+1)},$$

$$\omega_{2k+2}^{(n)} = \omega_{2k}^{(n+1)}(\omega_{2k+1}^{(n)} - \omega_{2k+1}^{(n+1)}),$$

with $\omega_{-1}^{(n)} = 0$ and $\omega_0^{(n)} = a_n$, where (a_n) is a given sequence. The ω with an odd lower index are intermediate results that can be eliminated. We have $\omega_{2k}^{(n)} = H_{k+1}(a_n)/H_k(a_{n+2})$. According to its initialization, this algorithm allows one to implement several transformations. Thus, if $a_n = \Delta^n S_i$ for a fixed value of i, then $\omega_{2k}^{(0)} = e_k(S_i) = \varepsilon_{2k}^{(i)}$ and $\omega_{2k}^{(1)} = 1/\varepsilon_{2k+1}^{(i)}$. This algorithm can also be used for implementing the confluent form of the ε-algorithm. Indeed, if $a_n = f^{(n)}(t)$, then $\omega_{2k}^{(0)} = \varepsilon_{2k}(t)$. When $a_n = f^{(n)}(t)/n!$, we obtain the confluent form of the ϱ-algorithm, that is, $\omega_{2k}^{(0)} = \varrho_{2k}(t)$ and $\omega_{2k}^{(1)} = 1/\varrho_{2k+1}(t)$. A singular rule and Algol procedures for implementing this algorithm are given. Let us mention that this algorithm can also be used with $a_n = \Delta^n S_i/n!$, with the divided differences $a_n = [x_i, \ldots, x_{i+n}]$ of a function f such that $f(x_i) = S_i$, or with $a_n = n![x_i, \ldots, x_{i+n}]$ for a fixed value of i. ◄

Although it seems to be quite interesting, the ω-algorithm has not been extensively studied.

∞

A convergence theory for the integration methods described in the preceding paper [793] is given in a report and a paper. Since they are quite similar, although written with a gap of 2 years and published with one of 4 years, they are analyzed together.

▶ *P. Wynn, A convergence theory of some methods of integration, Report CRM-193, Centre de Recherches Mathématiques, Université de Montréal, Montréal, May 1972. [789]*

▶ *P. Wynn, A convergence theory of some methods of integration, J. Reine Angew. Math., 285 (1976) 181–208. [808]*

Apart from its mathematical content, the difficulty of [808] (submitted 22 March 1974 but previously published as a report in 1972 [789]) lies in the numerous special notations introduced and their typography. This is also the opinion of Harry I. Miller in his MR 0415119. The paper is too dense to be summarized. The introduction describes its purpose:

Recently [*preceding paper and Ref. [793] of this book*] the author has introduced three methods for evaluating an integral over a semi-infinite interval in terms of the derivatives of the integrand at the finite end-point. The methods operate in very much the same way as the Euler–Maclaurin process, except that an infinite sum of values of the integrand over the range of integration is not required, and that two of the methods are nonlinear. They can be extended to the determination of integrals over a finite interval and can also be adapted to the estimation of the value of a given function as its argument tends to infinity.

Wynn makes use of the three difference–differential processes introduced in [793]. One of them is related to the confluent form of the ε-algorithm, and another one to that of the ϱ-algorithm. They are respectively

$$\varepsilon_{2r+1}(\mu) = \varepsilon_{2r-1}(\mu) + 1/(\psi(\mu) + \varepsilon'_{2r}(\mu)),$$
$$\varepsilon_{2r+2}(\mu) = \varepsilon_{2r}(\mu) + 1/\varepsilon'_{2r+1}(\mu),$$

with $\varepsilon_{-1}(\mu) = \varepsilon_0(\mu) = 0$, and

$$\varrho_{2r+1}(\mu) = \varrho_{2r-1}(\mu) + (2r + 1)/(\psi(\mu) + \varrho'_{2r}(\mu)),$$
$$\varrho_{2r+2}(\mu) = \varrho_{2r}(\mu) + (2r + 2)/\varrho'_{2r+1}(\mu),$$

with $\varrho_{-1}(\mu) = \varrho_0(\mu) = 0$. The methods operate in a way similar to that of the Euler–Maclaurin formula, except that an infinite sum of values of the integrand in the interval of integration is not needed. Another process related to the ω-algorithm is also presented. ◄

5.1.3 Nonscalar ε-Algorithms

Wynn extended the rule of his scalar ε-algorithm to treat vector or matrix sequences. The paper would later have many applications in the numerical solution of various problems; see Chap. 6, in particular Sects. 6.4, 6.6, and 6.7.

► *P. Wynn, Acceleration techniques for iterated vector and matrix problems, Math. Comp., 16 (1962) 301–322. [732]*

In this important paper (submitted on 1 August 1961), Wynn extended his scalar ε-algorithm to sequences of vectors or matrices. In the second case, the usual inverse is used. For vectors, it is necessary to define an inverse. Obviously, the inverse of the inverse has to give back the original vector, and for a vector y of dimension 1, one has to recover $y^{-1} = 1/y$. Wynn discussed several possible definitions, and he decided to use that of Klaus Samelson (1918–1980),[3] $y^{-1} = \bar{y}/(\bar{y}, y)$, which is nothing else than the pseudo-inverse. Then he gave, without proof, the following partial result on the kernel of the transformation $(S_n) \longmapsto (\varepsilon_{2k}^{(n)})_n$ when k is fixed. If for all n, $\sum_{i=0}^{k} c_i S_{n+i} = b$, with $\sum_{i=0}^{k} c_i \neq 0$, then for all n, $\varepsilon_{2k}^{(n)} = b/\sum_{i=0}^{k} c_i$. Several numerical applications are given, and he explained that his technique of the moving lozenge has to be used for avoiding waste of storage for problems of large dimensions. A note on programming, where he discusses the numerical instability and its treatment by his particular rules, ends the paper. ◄

This is a relevant paper, since it introduces the vector ε-algorithm. This work was done when Wynn was at the *Mathematisch Centrum*, 2e Boerhaavestraat 49,

---

[3] See Sect. 8.11 for his biography.

STICHTING
MATHEMATISCH CENTRUM
2e BOERHAAVESTRAAT 49
AMSTERDAM (O.)

Telefoon 747272 (3 lijnen)

Bankier: Amsterdamsche Bank N.V.
Bijkantoor Sarphatistraat

Postgiro 462890 - Gem. Giro M. 2138

Uw ref.:

Onze ref.: PW/JL

Onderwerp:

JUL 27 1961

Amsterdam, 5th July 1961.

Dr. H. Polachek,
Chairman, Editorial Committee,
Mathematics of Computation,
Applied Mathematics Laboratory,
David Taylor Model Basin,
W a s h i n g t o n, 7 D.C.
U.S.A.

Dear Dr. Polachek,

I am sufficiently encouraged by the tone of your *)
kind letter of September 22nd, 1960 to submit yet
another paper:
"Acceleration Techniques for Iterated Vector and
Matrix Problems" for your consideration with a
view to publication in "Mathematics of Computation".

For your possible interest I enclose a copy of
"Continued Fractions whose Elements Obey a Non-Commutative Law of Multiplication".

My view is that this is too long for publication in
your journal, and you may well share my opinion.

I must ask you to believe that at one time I really
was a modest and unassuming lad, but now of course
Numerical Analysis has quite ruined me, and in the
event I incline to the view that the paper which I
have submitted is worthy of the high standards and
distinguished appeal of your journal: but if you are
going to throw the thing out could you possibly do
so reasonably quickly so that I can bang it through
somewhere else without much loss of time.

Yours sincerely,

P. Wynn

*) under separate cover

Letter of Peter Wynn to Harry Polachek.
© Claude Brezinski.

Amsterdam-O., The Netherlands. On 5 July 1961, he submitted this paper to Harry
Polachek (1913–2002), at that time editor-in-chief of the journal *Mathematics of
Computation*. His letter is reproduced above for its last paragraph, which is worth
reading.

Let us mention that particular rules for avoiding breakdown and near-breakdown
in it were obtained by Florent Cordellier [179]. He was working on them for several
years already. C.B. was discussing Cordellier's results in his letters to Wynn, who

replied (17 November 1973), after six pages of comments: *Though how he would have found this out mystifies me*. This letter certainly helped Cordellier to rework his paper, which was finally published in 1977 after further direct comments to him by Wynn.

The vector and the matrix ε-algorithms lack an algebraic theory. In contrast to the scalar one, they were not created for implementing a sequence transformation whose kernel was defined by a closed formula, but they were directly derived from the rule of the scalar ε-algorithm.

The kernel of the vector ε-algorithm was obtained by John Bryce McLeod (1929–2014)[4] [443] in 1971 by a difficult proof involving Clifford algebra and an isomorphism between complex vectors of dimension n and $2^n \times 2^n$ real matrices. The kernel still has the form (2.12). The proof holds for complex vectors, but the coefficients in this difference equation have to be real. Later, the vectors $\varepsilon_{2k}^{(n)}$ were expressed, as in the scalar case, by ratios of determinants, but of dimension $2k + 1$ instead of $k + 1$, as given by Peter Russell Graves-Morris (b. 1941) and Chris Jenkins [287, 288]. The proof of this result was achieved via the relation of this algorithm to Pfaffians and vector Padé approximants. Their result is valid for complex coefficients in the difference equation (2.12) defining the kernel. See [529] for a review of the algebraic foundations of the vector ε-algorithm. They can also be expressed as ratios of designants (see Sect. 2.1.3), a generalization of determinants in a noncommutative algebra, a work due to Ahmed Salam [546, 548]. Due to the noncommutativity of the product, we have two expressions, one involving left designants and the second one the right designants. They are

$$
{}^l\varepsilon_{2k}^{(n)} =
\begin{vmatrix}
1 & \cdots & 1 \\
\Delta S_n & \cdots & \Delta S_{n+k} \\
\vdots & & \vdots \\
\Delta S_{n+k-1} & \cdots & \Delta S_{n+2k-1}
\end{vmatrix}_l^{-1}
\begin{vmatrix}
S_n & \cdots & S_{n+k} \\
\Delta S_n & \cdots & \Delta S_{n+k} \\
\vdots & & \vdots \\
\Delta S_{n+k-1} & \cdots & \Delta S_{n+2k-1}
\end{vmatrix}_l ,
$$

and

$$
{}^r\varepsilon_{2k}^{(n)} =
\begin{vmatrix}
S_n & \cdots & S_{n+k} \\
\Delta S_n & \cdots & \Delta S_{n+k} \\
\vdots & & \vdots \\
\Delta S_{n+k-1} & \cdots & \Delta S_{n+2k-1}
\end{vmatrix}_r
\begin{vmatrix}
1 & \cdots & 1 \\
\Delta S_n & \cdots & \Delta S_{n+k} \\
\vdots & & \vdots \\
\Delta S_{n+k-1} & \cdots & \Delta S_{n+2k-1}
\end{vmatrix}_r^{-1} .
$$

Both expressions are equal to S for all n if the sequence (S_n) satisfies (2.12). Moreover, the left and right $\varepsilon_{2k+1}^{(n)}$ are the inverses of the $\varepsilon_{2k}^{(n)}$ applied to the sequence (ΔS_n), a property similar to that holding for the scalar ε-algorithm.

However, the algebraic theory was not entirely satisfactory, and that is why a new approach was taken up later. It will be described in Sect. 6.4.

∞

---

[4] See Sect. 8.5 for his biography.

Ten years later, Wynn produced a large report for extending the algorithms to various sets of elements.

▶ *P. Wynn, The abstract theory of the epsilon algorithm, Report CRM-74, Centre de Recherches Mathématiques, Université de Montréal, Montréal, February 1971.* [774]

This report from the *Centre de Recherches Mathématiques* of the *Université de Montréal* has 132 pages. It is a purely theoretical work. After recalling what the scalar ε-algorithm is, Wynn writes that its *convergence theory derives in great part from the theory of associated and corresponding continued fractions, and hence from the theory of orthogonal polynomials*. Then he recalls the connection with the qd-algorithm and with Padé approximation. He mentions that since the operations involved in the rule of the algorithm are only additions, subtractions, and the formation of inverses, it can be extended to numbers of more general types. This was already done to the case of continued fractions whose coefficients are elements of a noncommutative associated ring with inverse [746]. The ε-algorithm has also been applied to sequences of matrices and vectors in [732]. However, in this latter case, Wynn points out the lack of a theory of determinants over a field. As already mentioned above, such a theory was later built by Ahmed Salam [546, 547, 548]. Then Wynn explains the purpose of this work:

Although the theories referred to in the preceding paragraphs are quite extensive, it is eminently desirable that the ε-algorithm should be applied to numbers of still further types, and that the powerful convergence acceleration properties of the algorithm should be made available for the treatment of a far wider range of problems. In this paper we first offer a brief catalogue of the types of number to which the ε-algorithm (and, indeed, any algorithm employing solely rational operations) might feasibly be applied. We then give a careful analysis of certain elementary results in the theory of the ε-algorithm, with particular reference to the minimal necessary arithmetic properties (such as associativity and commutativity of addition, and so on) of the numbers concerned, discussing in each case which of the more general types of number are adumbrated by each of the results. We also consider, in its most abstract setting, the problem of exponential extrapolation. We indicate how algebraic isomorphisms may be used to extend to types of number to which they appear to have no direct application certain special results in the theory of the ε-algorithm, and conclude with an account of variants of the ε-algorithm in which addition and subtraction do not occur.

The next section is about these systems of numbers: distributive rings, linear algebras, extended Cayley numbers, algebraic extensions of linear algebras, formal power series, rings with star inverse, complementary systems, and nonassociative number systems. Then a section is devoted to number species: elements of a field, elements of a neo-field, elements of an associative division algebra, elements of an associative division ring, elements of an associative algebra with inverse, elements of an associative ring with inverse, and more. After that, Wynn discusses the arithmetic operations used in a nonscalar ε-algorithm. He observes that obviously, the results that are given concern only the numbers that exist, since some of them may not exist due to a singularity. The following section presents theorems of three types about the ε-algorithm:

Firstly it is shown that if the initial values are modified in a certain way, then a corresponding array can be constructed whose members are simply related to those of the original array; secondly it is shown that similar results hold with respect to modification of the operations; thirdly it is shown that numbers belonging to subsequences taken from an ε-array also satisfy some relationships.

Then a section is devoted to isomorphisms that allow one to deduce results proved for a certain type of numbers to another type. This is what J. Bryce McLeod did for obtaining the kernel of the vector ε-algorithm [443]. Special versions of the algorithm are constructed that require only multiplication and the formation of an inverse. Finally, Wynn interprets in purely geometric terms the application of the vector ε-algorithm to vectors with three real components. Although he also mentioned earlier that the ε-algorithm could be applied to sequences of quaternions over a formally real field, he does not link these two problems. ◄

5.1.4 Euler Transformation

Wynn devoted several papers to Euler transformation (see Sect. 3.1 for an account of it). Let us analyze them.

► P. Wynn, *Central difference and other forms of the Euler transformation*, *Quart. J. Mech. Appl. Math.*, 9 (1956) 249–256. [715]

Euler transformation consists in transforming the series

$$s = \sum_{n=0}^{\infty} (-a)^n v_n \quad \text{into} \quad s = \sum_{n=0}^{\infty} (-a)^n \Delta^n v_0 / 2^{n+1},$$

which converges faster. In [715], Wynn delayed it until the mth term,

$$s = \sum_{n=0}^{m-1} (-a)^n v_n + \frac{(-a)^m}{1+a} \sum_{s=0}^{\infty} \left(\frac{-a}{1+a}\right)^s \Delta^s v_m,$$

and developed alternative formulas based on the backward and the central difference operators. ◄

In the following paper, Wynn compared the results obtained by Euler transformation, a linear transformation, with those furnished by the ε-algorithm, which is a nonlinear one.

► P. Wynn, *A comparison between the numerical performances of the Euler transformation and the ε-algorithm*, *Revue française de traitement de l'information*, *Chiffres*, 4 (1961) 23–29. [727]

Wynn considers the generalized Euler transformation

$$\sum_{s=0}^{\infty} v_s x^s = \frac{1}{1-x} \sum_{s=0}^{\infty} \left(\frac{x}{1-x}\right)^s \Delta^s v_0.$$

He shows that *The class of expansions upon which the Euler transformation successfully operates is therefore a subclass of those upon which the ε-algorithm successfully operates*. Thus, the latter method is *stronger than the Euler transformation*. Numerical examples support this claim. ◄

∞

Then Wynn studied a variant of Euler transformation in a paper published in two parts.

▶ *P. Wynn, The numerical transformation of slowly convergent series by methods of comparison, Part I, Revue française de traitement de l'information, Chiffres, 4 (1961) 177–210. [728]*

▶ *P. Wynn, The numerical transformation of slowly convergent series by methods of comparison. Part II, Revue française de traitement de l'information, Chiffres, 5 (1962) 65–88. [741]*

In these papers, as stated in the abstract, Wynn gives a formal theory of the transformation

$$\sum_{s=0}^{\infty} c_s v_s x^s \sim \sum_{s=0}^{m-1} c_s v_s x^s + \sum_{s=0}^{\infty} x^{m+s} \Phi_m^{(s)}(x) \Delta^s v_m,$$

where

$$\Phi_m(x) \sim \sum_{s=0}^{\infty} c_{m+s} x^s.$$

The case in which Φ_0 satisfies a linear differential equation and the coefficients v_s obey a linear recurrence relation in s are treated in detail.

In the first paper [728], Wynn is more specific, and he says that if $\Phi(x) \sim \sum_{s=0}^{h} c_s x^s$, then $\Theta(x) \sim \sum_{s=0}^{h} c_s v_s x^s \sim \sum_{s=0}^{h} x^s \Phi^{(s)}(x) \Delta^s v_0 / s!$ He calls this transformation the Euler–Gudermann transformation [293].

Then Wynn delays the application of this transformation and generalizes it to the form

$$\sum_{s=0}^{h} c_s v_s x^s \sim \sum_{s=0}^{m-1} c_s v_s x^s + \sum_{s=0}^{h-m} u_m^{(s)} \Delta^s v_m, \quad h > m,$$

where $u_m^{(s)} = x^{m+s} \Phi_m^{(s)}(x)/s!$ and $\Phi_m \sim \sum_{s=0}^{h-m} c_{m+s} x^s$. He shows that if the functions $u_0^{(s)}$ and $u_m^{(0)}$ have been computed for $s = 0, 1, \ldots$ and $m = 0, 1, \ldots$, respectively, then

$u_m^{(s)} = u_{m+1}^{(s)} + u_{m+1}^{(s-1)}$ for $m, s = 1, 2, \ldots$. The partial sums of the transformation are displayed in a double entry table, and computed recursively.

If Φ_0 satisfies a linear differential equation of the form $\sum_{n=0}^{l} p_n^0(x) \Phi_0^{(n)}(x) = f_0(x)$, where p_n^0 and f_0 are polynomials, recurrences for the $u_0^{(s)}$ and the $u_m^{(0)}$ are established. Several particular cases are treated and numerical examples are given. The example of the Wilson integral was suggested to Wynn by Frank William Jones Olver (1924–2013).

In the second part of the paper [741], a second version of a higher order, and a variant of the Euler–Gudermann transformation are presented. An integral transform of the transformation is given. It consists in multiplying the function to be transformed by a weight function and integrating the result with respect to x along a suitable contour. A convergence theory of the transformation is established. In the last section, Wynn points out that since in certain cases, the sequence $(u_s^{(0)} \Delta^s v_0)$ may well be dominated by a term of the form μ^s, *the Euler–Gudermann transformation appears to offer a promising point of application for the ε-algorithm. This is indeed substantiated by numerical experience.*

In these papers, the most complicated formulas are handwritten by Wynn! ◄

Christoph Gudermann (1798–1852) was a German mathematician who worked on geometry and special functions. He was a student of Gauss and the teacher of Weierstrass.

$$\infty$$

In the next two papers, after several years, Wynn came back to Euler transformation.

▶ *P. Wynn, A note on the generalised Euler transformation, The Computer Journal, 14 (1971) 437–441; Errata 15 (1972) 175. [773]*

In this paper, the author considered the series $\sum_{s=0}^{\infty} u_s$, where it is assumed that its terms behave like those of a geometric progression with ratio z, that is, $u_s = z^s v_s$, and that v_s is approximately constant. Defining the difference operators E and Δ by $E^m v_s = v_{s+m}$ and $\Delta = E - I$, we have

$$\sum_{s=0}^{\infty} u_s = \sum_{s=0}^{\infty} (zE)^s v_0 = \left(\frac{I}{I - zE} \right) v_0 = \frac{1}{1-z} \sum_{s=0}^{\infty} \left(\frac{z}{1-z} \right)^s \Delta^s v_0.$$

This transformation can be delayed, thus leading to

$$\sum_{s=0}^{\infty} u_s = \sum_{s=0}^{m-1} u_s + \frac{z^m}{1-z} \sum_{s=0}^{\infty} \left(\frac{z}{1-z} \right)^s \Delta^s v_m.$$

Denoting by $S_m^{(r)}$ the partial sums of this infinite series after the first r terms, Wynn proved that they obey the recurrence

$$S_m^{(r+1)} = \frac{1}{1-z} S_{m+1}^{(r)} - \frac{z}{1-z} S_m^{(r)}.$$

Then he discussed the choice of the parameter z. It can be determined either by analysis or by numerical estimation. In this second case, the ε- or the ϱ-algorithm can be applied to the sequence (u_{s+1}/u_s) for estimating z. A numerical example and the corresponding Algol procedures are given. ◄

$$\infty$$

Another paper on Euler transformation is the following one.

▶ *P. Wynn, A transformation of series, Calcolo, 8 (1971) 255–272. [777]*

The problem Wynn treated here *is that of transforming certain series with real terms of the same sign into series whose terms oscillate in sign.* Starting from the function

$$f(z) = \int_0^1 \frac{d\sigma(\rho)}{1 - z\rho},$$

where σ is bounded and nondecreasing in $[0, 1]$, Wynn considered the series $\mathcal{F}(z) = \sum_{\nu=0}^{\infty} t_\nu z^\nu$, where $t_\nu = \int_0^1 \rho^\nu \, d\sigma(\rho)$. The generalized Euler transformation applied to f leads to

$$f(z) = \sum_{\nu=0}^{\infty} \left\{ \sum_{\tau=0}^{k-1} z^\tau \Delta^\nu t_{k\nu+\tau} + \frac{z^k}{1-z} \Delta^\nu t_{k(\nu+1)} \right\} \left(\frac{z^{k+1}}{1-z} \right)^\nu, \tag{6}$$

where k is an integer. For $k = 0$, this expression simply becomes the usual Euler transformation

$$f(z) = \frac{1}{1-z} \sum_{\nu=0}^{\infty} \Delta^\nu t_0 \left(\frac{z}{1-z} \right)^\nu.$$

Then Wynn wanted to *determine the orders of magnitude of the terms of expansion (6), and derive conditions that are sufficient to ensure convergence of this expansion.* A quite long analysis follows in which restrictions on σ are imposed. Then *We also consider the conditions under which it is profitable to apply the transformation (6). The absolute values of the successive terms of the series (6) are, as we have shown, dominated by the terms of a geometric progression* [...]

Finally, Wynn discussed the optimal choice of k in (6), and the signs of its terms. Since the transformation can be delayed, he gave a recursive algorithm for computing the $S_m^{(r)}$, which now also depend on k. A numerical example ends the paper. ◄

5.1.5 Other Transformations

At the same time he produced his ε-algorithm, Wynn was also interested in other transformations for accelerating the convergence of scalar sequences. Let us review these contributions.

▶ *P. Wynn, A note on Salzer's method for summing certain convergent series, J. Math. and Phys., 35 (1956) 318–320. [711]*

This paper was submitted on 19 July 1955, when Wynn was at the National Physical Laboratory in Teddington, Middlesex, England. In it, he gave an alternative formulation of an acceleration method proposed by Herbert Ellis Salzer (1915–2006) [551]. For summing a series, this method consists in assuming that its partial sums S_n behave like a polynomial in $1/n$, interpolating a certain number of them by a polynomial, and extrapolating the polynomial at $n = +\infty$. For that, Wynn used Lagrange's interpolation formula and gave an estimation for the remainder. Numerical examples concluded the paper. This is the idea underlying Richardson's extrapolation method [520, 523] after replacing a polynomial in $1/n$ by a polynomial in an auxiliary variable x_n. However, Richardson did not present it that way in his papers.
◀

Salzer discussed this extrapolation procedure again in [552], and in a joint paper with Genevieve M. Kimbro [553], where a polynomial in $1/n^2$ is also considered and the method related to Richardson's. Let us anticipate by saying that the convergence of Richardson's method would be extensively studied by Pierre-Jean Laurent in several papers and his doctoral thesis in 1964 [395] and in [394] (see Sect. 6.2 for details).

∞

The next paper is important, since Wynn introduced a peculiar form of reciprocal differences and used them for extrapolation at infinity by a rational function in n. His ϱ-algorithm is a particular form of the general reciprocal differences because, it is not known why, he chose to restrict it to the case $x_n = n$ as its auxiliary sequence.

▶ *P. Wynn, On a procrustean technique for the numerical transformation of slowly convergent sequences and series, Math. Proc. Cambridge Philos. Soc., 52 (1956) 663–671. [712]*

In this paper (submitted on 31 October 1955), instead of extrapolating by a polynomial in $1/n$ at 0, Wynn extrapolated the partial sums $S_n = \sum_{i=0}^{n} u_i$ by a rational function in n at infinity. He wrote:

> By means of Thiele's interpolation formula, the partial sums S_n may be regarded as rational functions of n. The reciprocal differences of S_n for the successive arguments $n = m, m + 1, \ldots$ are formed by tabulating the sequences $\rho_s(S_{m+r})$ $(s = 0, 1, \ldots; r = 0, 1, \ldots)$, where

$$\rho_0(S_{m+r}) = S_{m+r}, \quad \rho_1(S_{m+r}) = \frac{1}{u_{m+r+1}},$$

and $\rho_s(S_{m+r}) = \rho_{s-2}(S_{m+r+1}) + \dfrac{s}{\rho_{s-1}(S_{m+r+1}) - \rho_{s-1}(S_{m+r})}$ $\quad (s = 2, 3, \ldots)$.

Then, he considered the continued fraction

$$S_n = S_m + \dfrac{n-m}{\lfloor \rho_1(S_m) \rfloor} + \dfrac{n-m-1}{\lfloor \rho_2(S_m) - S_m \rfloor} + \dfrac{n-m-2}{\lfloor \rho_3(S_m) - \rho_1(S_m) \rfloor} + \cdots$$

The successive convergents of the even part of this continued fraction are ratios of polynomials of the same degree which, as n tends to infinity, tend to $\rho_{2n}(S_m)$, which is taken as an approximation of the limit of (S_n). Thus, setting $\varrho_{2n}^{(m)} = \rho_{2n}(S_m)$, Wynn obtained the so-called ϱ-algorithm defining the set of sequence transformations $(S_n) \longmapsto \{(\varrho_{2k}^{(n)})\}$. These quantities are expressed as ratios of determinants (see (2.19)). Numerical examples are given. ◄

Instead of extrapolating at infinity by a rational extrapolation in n, an obvious generalization is to extrapolate at infinity by a rational function in x_n where the auxiliary sequence (x_n) tends to infinity [81]. The convergence of this algorithm was later studied by Naoki Osada [475].

∞

In [76] and other papers [75, 77], John William Bradshaw (1878–1967) proposed a method for transforming two types of slowly convergent series and continued fractions. It consisted in modifying the series by the term-by-term addition of another series whose terms b_n have to behave like $n^{-\beta-\alpha}$ if the successive terms of the initial series behave like $n^{-\beta}$. The b_n may be derived from a system of linear equations, they are rational functions, and they may be established as the convergents of a continued fraction.

Let us present a short biography of this researcher. John William Bradshaw was born on 14 June 1878 at Batavia, Illinois, USA. In the fall of 1900, he entered the graduate school at Harvard, where he received his master's degree in 1902. The following 2 years were spent in Strasbourg (at that time in Germany, now in France), where he held for the 2 years the Parker and Kirkland Travelling Fellowships. He received his doctoral degree from the University of Strasbourg in 1904. Upon his return to America, he became instructor in mathematics at the University of Michigan, a position that he held until his appointment to an assistant professorship. He also became registrar of the university in April 1906, a position that he held until October 1908. He died in 1967.

Another method, due to Bickley and Miller [61], is applicable to a larger class of series. It is based on the notion of converging factor C_n defined as $R_n = u_n C_n$, where R_n is the remainder term of the series $\sum_{i=0}^{\infty} u_n = \sum_{i=0}^{n-1} u_n + R_n$; see Sect. 3.2.4 for details.

In the next paper, Wynn examined these two methods and confirmed that nonlinear sequence transformations are more powerful than linear summation processes.

► *P. Wynn, A note on a method of Bradshaw for transforming slowly convergent series and continued fractions, Amer. Math. Monthly, 69 (1962) 883–889. [730]*

In this paper, Wynn concluded that

From the preceding remarks it can be seen that Bradshaw's continued fractions may be obtained from the Bickley-Miller expansions by expanding $u_n C_n$ as a series in inverse powers of n and applying the $q - d$ algorithm relationship to the coefficients of this series. In general this procedure is more efficient than that proposed by Bradshaw, for the derivation of each of his rational functions necessitates the solution of a completely independent set of linear equations; the Bickley-Miller method and the $q - d$ algorithm are however recursive procedures in which the coefficients in the approximation of one degree assist in the computation of those in the next.

Examples are given. ◄

$$\infty$$

Then comes a paper on two techniques for obtaining converging factors for the acceleration of slowly convergent series.

► *P. Wynn, On a connection between two techniques for the numerical transformation of slowly convergent series, Koninkl. Nederl. Akad. Weten., 65A (1962) 149–154. [736]*

In [61], the authors proposed methods for obtaining the converging factors of series such that $S \sim \sum_{n=0}^{\infty} u_n$ when either $u_{n+1}/u_n = x(1 + A_1/n + A_2/n^2 + \cdots)$ or $u_{n+1}/u_n = x(1 - A_1/n + A_2/n^2 + \cdots)$. As Wynn explains in [736], the converging factor defined by $u_n C_n \sim \sum_{s=0}^{\infty} u_{n+s}$ is, for series of the first type, expanded as $C_n = \alpha_0 + \alpha_1/n + \alpha_2/n^2 + \cdots$, and the α_i are obtained by substitution in the difference equation $u_n C_n = u_n + u_{n+1} C_{n+1}$. This method can be applied to the asymptotic series $z e^z Ei(-z) \sim \sum_{n=0}^{\infty} n!(-z)^{-n}$. With the change of variable $z = (n + h)\beta$, where $\beta = e^{-i\theta}$ and h is small, there follows $u_{n+1}/u_n = -e^{-i\theta}(1 + \sum_{s=0}^{\infty}(-h)^s n^{-s-1})$, from which the coefficients of the converging factor may be derived. For large $|z|$, an improvement in the accuracy can be obtained by applying Euler transformation to the series involving the exponential integral. John Barkley Rosser (1907–1989)[5] [35] pointed out that if the terms thus obtained are expanded in inverse powers of n and rearranged, one obtains a series for C_n identical to that obtained by the method of Bickley and Miller.

Then Wynn writes that the first purpose of this paper is to *point out that this equivalence is general and not confined to a specific example.* For series of the second type given above, the appropriate converging factor is $C_n = \alpha_{-1} n + \alpha_0 + \alpha_1 n^{-1} + \cdots$, and the expressions for the α_i are given. Then Wynn gives *a further suitable method for accelerating the convergence of slowly convergent series.* It is too technical to be explained here. He concludes thus:

The results of this note show how numerically equivalent techniques may be applied, which demand only the previous determination of $x = \lim_{n \to \infty} u_{n+1}/u_n$ or $A_1 = \lim_{n \to \infty} n(1 - u_{n+1}/u_n)$. ◄

---

[5] See Sect. 8.9 for his biography.

$$\infty$$

In the next paper, Wynn considered a procedure that consists in modifying the sequence to be accelerated before applying it a sequence transformation. In a paper published in 1965 (thus certainly developed when Wynn was working with him), van Wijngaarden suggested a simple linear process for transforming a series of positive terms into an alternating one [668]. Let $s = \sum_{i=1}^{\infty} a_i$. Then $s = \sum_{j=1}^{\infty} (-1)^{j+1} v_j$, where $v_j = \sum_{i=1}^{\infty} b_{j,i}$ with $b_{j,i} = 2^{i-1} a_{j2^{i-1}}$. Obviously, to be useful, the series giving the v_j has to be either known or rapidly converging. For example, if $a_i = i^{-c}$ with $c > 1$, then $v_j = j^{-c}(1 - 2^{1-c})^{-1}$. The theoretical aspects of this technique were studied by James Wilson Daniel [190], who showed that it can be applied repeatedly, and compared it with the results given by Wynn in [712] and [753]. This process was recently extended [10, 357] and combined with an acceleration method proposed by Avram Sidi (b. 1947)[6] [578].

▶ *P. Wynn, Accelerating the convergence of a monotonic sequence by a method of intercalation, MRC Technical Summary Report 674, University of Wisconsin, Madison, January 1967. [762]*

Numerical experiments show that an acceleration method, such as the ε-algorithm, is often more effective on alternating sequences than on series with monotone terms. In [762], Wynn considers a series of positive terms $\sum_{\nu} u_{\nu}$ whose sum U is required. Let $V = \sum_{\nu} v_{\nu}$ be another series of positive terms whose sum in known. He explains that the process of intercalation consists in considering the sequence $W = u_0 - v_0 + u_1 - v_1 + \cdots$. Obviously $U = V + W$. Then, using the four types of series studied in [755], he shows, by numerical experiments, that the improvement brought by applying the ε-algorithm is virtually negligible for series with monotone terms, while those with alternating terms are accelerated. ◀

In a letter dated 17 January 1997 to Prof. Naoki Osada, Wynn wrote:

> You should not worry too much about never having heard of van Wijingaarden's condensation method. It is used as a standard technique in research centers in Holland and Germany, but is equally unknown to Smith and Ford in their SIAM J Numer Anal 16 (1979) 223–240 survey paper and, in their books, to Wimp, to Brezinski and Redivo-Zaglia, and to Walz (You may enjoy reading the Math of Comp 39 (1982) 736–738 review of Wimp's book and the Zbl. reviews of the other two books.) Incidentally, the method works well on series of the form $\sum 1/n \log(n)$ and $\sum 1/n \log(\log(n)), \ldots$ which occur in lifting factors in aerofoil theory.

Wynn added that the documentation for the implementation of van Wijngaarden's method and a C program can be found in pages 291–297 of [398] and a numerical application on page 667.

---

[6] See his testimony in Sect. 7.10.

5.1.6 Varia

In three papers, Wynn applied the ε-algorithm to sequences that are interconnected. He thus obtained ε-arrays that are also connected. He called them *hierarchies of arrays*. These hierarchies are still to be further studied, and they still lack theoretical results, in particular about their convergence and acceleration properties.

▶ *P. Wynn, Upon a hierarchy of epsilon arrays, Technical Report 46, Louisiana State University, New Orleans, October 1970. [772]*

The aim of the report (dated October 1970) was to

[...] establish the existence of a hierarchy of arrays of numbers produced by means of the ε-algorithm from a system of initial value sequences; the members of each initial value sequence are the successive first differences of the members of the preceding sequence and, at the same time, the successive sums of the members of the following sequence. It is shown that the members of each array may be simply expressed in terms of those of the two neighboring arrays in the hierarchy.

To be more specific, let $(S_m^{(0)} = S_m)$ be a sequence of numbers, and consider the sequences $(S_m^{(i)})$ derived from the relationships

$$S_m^{(i)} = \sum_{v=0}^{m} S_v^{(i-1)}, \quad i = 1, 2, \ldots,$$

$$S_m^{(i-1)} = S_{m+1}^{(i)} - S_m^{(i)}, \quad i = 0, -1, -2, \ldots.$$

Let $\{^{(i)}\varepsilon_r^{(m)}\}$ be the numbers obtained by applying the ε-algorithm from the initial values

$$^{(i)}\varepsilon_{-1}^{(m)} = 0, \quad ^{(i)}\varepsilon_0^{(m)} = S_m^{(i)}.$$

Wynn gave the relations between the $^{(i)}\varepsilon_r^{(m)}$ and the $^{(i-1)}\varepsilon_r^{(m)}$ when r is respectively odd and even. He discussed the case of possible breakdowns in these arrays. He proved, by a succession of lemmas, that if the sequence (S_m) satisfies the irreducible linear recurrence $\sum_{v=0}^{h} c_v S_{m+v} = H$ for $m = 0, 1, \ldots$, and if $C = \sum_{v=0}^{h} c_v$, then if C and H are different from zero,

$$\begin{aligned}
&^{(0)}\varepsilon_{2h}^{(m)} = H/C, \quad ^{(0)}\varepsilon_{2h+1}^{(m)} = \infty, \qquad && m = 0, 1, \ldots, \\
&^{(i)}\varepsilon_{2(h+i)}^{(m)} = \infty, \qquad && i = 1, 2, \ldots; m = 0, 1, \ldots, \\
&^{(i)}\varepsilon_{2h}^{(m)} = 0, \quad ^{(i)}\varepsilon_{2h+1}^{(m)} = \infty, \qquad && i = -1, -2, \ldots; m = 0, 1, \ldots.
\end{aligned}$$

The cases $C \neq 0, H = 0$, and $C = 0, H \neq 0$ are also treated. The proofs take several pages. Then Wynn considered the case of the confluent form of the ε-algorithm. Starting from a prescribed function $S^{(0)}(\mu)$, he defined functions derived from it by

$$S^{(i)}(\mu) = \int_a^\mu S^{(i-1)}(\mu') \, d\mu', \quad i = 1, 2, \ldots,$$

$$S^{(i-1)}(\mu) = dS^{(i)}(\mu)/d\mu, \quad i = 0, -1, -2, \ldots,$$

and applied to them the confluent form. Relations between the various arrays similar to those obtained for the discrete case are derived. ◄

Wynn did not explain the motivations or the interest of these results. They seem to be purely speculative.

∞

The preceding work was continued in the following paper 2 years later.

► *P. Wynn, Hierarchies of arrays and function sequences associated with the epsilon algorithm and its first confluent form, Rend. Mat. Roma, 5, Serie VI (1972) 819–852.* [784]

In this paper, submitted 15 May 1972 and accepted 9 June 1972 Wynn came back to what he proposed in [772]. The results are essentially the same. There is an additional section on the connection with the theory of the Padé table, and a motivation, but they concern only the standard scalar ε-algorithm and not the hierarchies of arrays built as explained in the paper. However, Wynn made an interesting remark. It is well known that if the ε-algorithm is applied to the partial sums of the reciprocal series g of f, that is, the series formally defined by $f(t)g(t) = 1$, then $\varepsilon_{2k}^{(n)} = [n + k/k]_g(t) = 1/[k/n + k]_f(t)$. Wynn noticed that *it is unnecessary to determine the coefficients and the partial sum of the reciprocal series*. In fact the inverses of the partial sums of the reciprocal series are obtained by extending the ε-array to negative values of the upper index. Thus, if $\varepsilon_{2m}^{(-m)} = 0$ and if $\varepsilon_0^{(m)}$ is the partial sum of f up to the degree $m - 1$ inclusively, then $\varepsilon_{2m}^{(1-m)}$ is the inverse of the corresponding partial sum of g.

He concluded that determinantal proofs of some of his results may also be given, and that the method of proof he adopted can be extended to numbers and functions that are elements of a ring with an associative inverse, as done in [774], but that this *is not at present possible, since the theory of determinants over such systems of numbers has not yet been established.* ◄

As already mentioned, such a theory was later built by Ahmed Salam using designants (see Sect. 2.1.3), which generalize determinants in a noncommutative algebra [546, 547, 548].

∞

Let us mention that the risk of having an important propagation of rounding errors due to cancellation in subtractions is less with alternating series than with monotone ones. The method of intercalation can be viewed as a "preconditioning" or a preprocessing of the series. In the next paper, Wynn gives an application of this technique to the ε-algorithm.

▶ *P. Wynn, Convergence acceleration by a method of intercalation, Computing, 9 (1972) 267–273. [782]*

Wynn came back to the intercalation process in this paper, and gave two examples that showed that this artifice can be quite effective. He applied the ε-algorithm to the sequence $(\varepsilon_0^{(n)} = S_n)$ to be accelerated, and obtained results, denoted by $_0\varepsilon_0^{(0)}, _0\varepsilon_0^{(1)}, _0\varepsilon_2^{(0)}, _0\varepsilon_2^{(1)}, \ldots, _0\varepsilon_{2k}^{(0)}, _0\varepsilon_{2k}^{(1)}, \ldots$ that is, the staircase of elements along the first two descending diagonals of the ε-array (see Fig. 2.2). Then, he reapplied the algorithm to this sequence with the initializations $_1\varepsilon_0^{(2n)} = _0\varepsilon_{2n}^{(0)}, _1\varepsilon_0^{(2n+1)} = _0\varepsilon_{2n}^{(1)}, n = 0, 1, \ldots$, and so on. He thus constructed nested ε-arrays. This process can be understood as a method of intercalation (on this procedure, see the analysis of [762] in Sect. 5.1.5), since two sequences are mixed together before reapplying the algorithm, one term of the sequence with the upper index 0 alternating with one of the sequence with the upper index 1. The numbers $_\tau\varepsilon_0^{(0)}$ and $_\tau\varepsilon_0^{(1)}$, that is, the first two terms of all sequences, are the same in all the ε-arrays $\{_\tau\varepsilon_k^{(n)}\}$ produced in this way. Starting with m terms of the original sequence, Wynn said that *it suffices to give the numerical values of $\varepsilon_0^{(1)}, _0\varepsilon_0^{(m')}, _1\varepsilon_0^{(m')}, _2\varepsilon_0^{(m')}, \ldots$; the numbers $\varepsilon_0^{(1)}, _0\varepsilon_0^{(m')}$ taken together indicate the rate of convergence of the original sequence, the number $_1\varepsilon_0^{(m')}$ (equal to $_0\varepsilon_{m'/2}^{(0}$ if m' is event and to $_0\varepsilon_{m'/2+1}^{(0}$ if m' is odd) indicates the improvement in convergence obtained after one application of the ε-algorithm, the number $_2\varepsilon_0^{(m)}$ indicates the further improvement resulting from one repetition, and so on.* Numerical examples show the effectiveness of the procedure.
◀

Recall that Wynn applied this technique of nested ε-arrays to totally monotonic sequences in [787], and also in [796], where other such sequences are produced. These two papers were analyzed in Sect. 5.1.1.3.

5.1.7 Applications

Wynn studied several applications of the scalar and the vector ε-algorithm to the solution of various problems. As we will see below, his ideas were taken up again later and extended. The first paper is a kind of review in which he explains how to program lozenge algorithms.

▶ *P. Wynn, Acceleration technique in numerical analysis with particular reference to problems in one independent variable, in Information Processing 1962, Proc. IFIP Congress 62, Munich, 27 August–1 September 1962, C.M. Popplewell ed., North-Holland, Amsterdam, 1963, pp. 149–156. [742]*

The beginning of this paper is interesting:

I propose to discuss certain methods for accelerating the convergence of iterative processes
for solving computational problems. In certain respects this paper is an account of the work
which has recently been carried out, and it is perhaps fair to state at the outset that in order
to invest it with some coherence and unity I have borne in mind a remark of Herodotus: In
human affairs nothing very much happens at all and certainly the right thing never happens
at the right time; the conscientious historian will remedy this.

Wynn begins by describing the idea behind the construction of an extrapolation
method from its kernel as explained in Sect. 2.2. He called such techniques *procrustean*. This word, already used in the title of [712], describes situations in which
different properties are fitted to an arbitrary standard. In the Greek myth, Procrustes
was a son of Poseidon, and he was known to amputate the excess of length. Accordingly, this paper is a kind of survey on the ϱ-, the ε-, and the qd-algorithms, the
Padé table, and continued fractions. It seems that it is in Section 12 of this paper that
Wynn described for the first time his *moving lozenge* technique for implementing
rhombus rules. It consists in storing only the last ascending diagonal of the ε-array.
When a new term of the sequence to be transformed is added, the new ascending
diagonal can be computed from the old one with the storage of only two additional
quantities. Let us mention that this saving in storage requirement is particularly
important for sequences of vectors or matrices; see [132]. The following sections
deal with nonscalar cases. Applications are given in particular to the linear Lichtenstein–Gershgorin integral equation. The integral was discretized by, as he said, *the
operational formula of Stone Age Numerical Analysis* consisting of the trapezoidal
rule. ◀

The trapezoidal rule was used again in [136], but for a nonlinear Fredholm
integral equation, and its convergence was accelerated by a generalization of the
Shanks transformation that will be explained in Sect. 6.4.

This paper was published in the proceedings of an IFIP congress. Let us briefly
give some information about this series of congresses. IFIP stands for *International
Federation for Information Processing*. It is a nongovernmental organization, created
in 1960 and recognized by the United Nations. It gathers more than 50 national and
international scientific societies and academies, bringing together more than one
million scientists on computer science and networks. The first international IFIP
congress was held in Paris, 15–20 June 1959, in the Unesco building. France was
represented by the *Association Française de Calcul*, a society editing the journal
Chiffres, in which Wynn published several papers. The congress was attended by
1772 participants, among them many well-known computer scientists and numerical
analysts [3]. Wynn was present. Only eleven plenary talks on numerical analysis
were given. The second IFIP congress took place in Munich from 27 August to
1 September 1962. Wynn was an invited speaker there in the session *Differential
and Integral Equations*. Among the 2300 attendees, one finds Friedrich Bauer,
Germund Dahlquist, Noël Gastinel, Alston Householder, Jean Kuntzmann, Hans
Maehly, Heinz Rutishauser, Klaus Samelson, Eduard Stiefel, and many other well-
known persons. IFIP will celebrate its 60th anniversary in 2020.

$$\infty$$

The next paper treats the solution of systems of linear equations. Wynn proposed several conjectures, some of which were later proved to be wrong.

▶ *P. Wynn, Upon a conjecture concerning a method for solving linear equations, and certain other matters, MRC Technical Summary Report 626, University of Wisconsin, Madison, April 1966. [756]*

In this report of 122 pages Wynn discussed the conjecture that if the matrix ε-algorithm (with the pseudo-inverse replacing the inverse) is applied to a sequence (S_n) of rectangular matrices satisfying, for all n, $\sum_{i=0}^{k} a_i(S_{n+i} - S) = 0$, then for all n, $\varepsilon_{2k}^{(n)} = S$. Here the a_i are scalars and S is a rectangular matrix of the same dimension as the S_n. The conjecture was already known to be true for vectors, as proved by McLeod [443] (called by Wynn *the minor conjecture*), and for square matrices, as demonstrated by himself [746].

For trying to establish this conjecture, Wynn settled on an important artillery but he justified it by the sentence: *As with most mathematical conjectures the importance of the result lies not so much in itself as in that of the theory which must be developed in order to prove it.*

The minor conjecture is first introduced by its connection with the iterative solution of systems of linear equations. Then Wynn offered a survey of special cases in which it can be proved by recourse to existing results. Particular cases are studied. For proving the conjecture, he then tried to adapt the proofs of these existing results, for example determinantal identities, continued fractions, and Padé approximants. He also tried to relate the vector ε-algorithm to the conjugate gradient method of Hestenes and Stiefel [327]. He did not succeed, since the conjugate gradient method is, in fact, related to the topological ε-algorithm, as later proved in [94, pp. 184–190] (see Sect. 6.6). Then Wynn discussed the proof of McLeod [443], which uses an isomorphism between vectors and elements of a Clifford algebra. He also introduced nonlinear difference–differential recurrences for treating the case of differentiable rectangular matrices depending on a parameter. ◀

Finally, the main conjecture was not proved, and a counterexample of it was given by Thomas Nall Eden Greville[7] (1910–1998) in 1968 [290].

∞

The next paper is quite important, since Wynn's work in this direction was later pursued and gave rise to a procedure for reducing the Gibbs phenomenon for Fourier series, as will be seen after the analysis.

▶ *P. Wynn, Transformations to accelerate the convergence of Fourier series, in Gertrude Blanch Anniversary Volume, B. Mond, G. Blanch eds., Wright Patterson Air Force Base, 1967, pp. 339–379; also as MRC Technical Summary Report 673, University of Wisconsin, Madison, July 1966. [764]*

---

[7] See Sect. 8.4 for his biography.

The aim of Wynn is to describe a number of methods for accelerating the convergence of Fourier series. He first explains the various reasons why Fourier series either diverge or converge too slowly, thus rendering them useless for computational purposes. Then, he discusses linear methods: a method of Krylov, Abel summability, analytic continuation, and Toeplitz methods. As explained by J.G. Herriot in his review MR 0215553:

> Since discontinuities in the function and its derivatives cause the corresponding Fourier series to converge slowly, Krylov suggested the subtraction of Fourier series corresponding to functions with known singularities in order to speed up convergence. Although the Cesàro methods of summability play a significant role in the theory of Fourier series, they cannot be recommended for numerical purposes.

Then Wynn comes to nonlinear methods: corresponding and associated continued fractions, the continued fraction transformation of a Fourier series (his Section III.2), transformations based on the theory of the Padé table. The ε-algorithm is also applied to the series to show its marked superiority over the other procedures. The paper ends with an extension to boundary value problems such as Dirichlet problem for the unit circle.

The most interesting result is at the beginning of Section III.2. Let us quote him

> Given, for example, the series
>
> $$C \sim \sum_{s=0}^{\infty} c_s \cos(s\theta)$$
>
> we adjoin to it the conjugate series
>
> $$i \sum_{s=0}^{\infty} c_s \sin(s\theta)$$
>
> and have now expressed the required sum as the real part of the formal sum of the power series
>
> $$f(e^{i\theta}) = \sum_{s=0}^{\infty} c_s e^{is\theta}$$
>
> for
>
> $$C = \text{Re} \{ \lim_{z=e^{i\theta}} f(z) \}.$$
>
> We transform the power series (106) [*that is, the series* $f(e^{i\theta})$] into a corresponding or associated continued fraction, expecting the later to converge more rapidly than the series (106), and extract C from this continued fraction as the real part of a suitable convergent.

But no numerical experiments on this process are provided. ◄

Numerical comparisons show that a generalization of summation by parts gives better results for some particular Fourier series [368]. However, Wynn's idea was pursued later with some success. In [294], the ε-algorithm was applied to the partial sums of Fourier and Legendre series of functions with jumps, or whose first derivative presented discontinuities. The numerical computations showed that the ε-algorithm was able to detect the location of such discontinuities and, moreover, that it attenuated the Gibbs phenomenon. In [111], Wynn's idea was taken up again. The ε-algorithm was applied to the partial sums of the series $f(e^{i\theta})$, and then only the real parts of the $\varepsilon_{2k}^{(n)}$ were kept. This procedure greatly attenuated the Gibbs phenomenon, or even suppressed it almost completely in some cases. It was extended to Chebyshev series. Analytic estimates of the error of this method for a particular case of hypergeometric functions were given in [48], and also the rate of convergence of the columns of the ε-array. Fourier–Padé approximants have also been used for smoothing the Gibbs phenomenon [212], as well as Hermite–Padé approximants [46].

∞

The next report is devoted to the solution of systems of nonlinear equations that arise in estimating unknown parameters in theoretical models based on experimental data.

▶ *P. Wynn, A numerical method for estimating parameters in mathematical models, Report CRM-443, Centre de Recherches Mathématiques, Université de Montréal, Montréal, August 1974. [800]*

As Wynn explained in the abstract, a mathematical model is often described by a system of equations, of any type and arbitrary complexity, with a vector valued dependent variable $y(x)$ defined for values of x over a prescribed real interval. The equations, and possibly the initial conditions on y, depend on a vector p of unknown parameters that have to be determined. They are given directly as the solution of a system of nonlinear equations or, by the intermediary of a transfer function, as the best fit to a set of experimental measurements. Wynn's idea was to treat this problem by the vector ε-algorithm.

Using his notation, the problem is formulated as the solution of $m(y(p, x), p, x) = 0 \in \mathbb{R}^M$, with $y \in \mathbb{R}^M$ and $p \in \mathbb{R}^N$. The solution y is defined for x belonging to some finite or infinite interval, or a subset of points of it, and initial and/or boundary values, which may or may not depend on p, are given at some points. Obviously, conditions have to be imposed so that a solution exists and is unique. If the solution is directly observable and the transfer function is simply $t(p, y, x) = y$, with y independent of p and x, the problem reduces to that of achieving the best fit between this solution and given data at points of a prescribed set. A list of applications follows.

As Wynn wrote:

> [...] the suggestion naturally prompts itself that we should attempt to solve the equation $z(p) = 0$ [*a simplification he introduced in the notation*] by iteration, using a scheme of the form $p^{(\tau+1)} = p^{(\tau)} + qz(p^{(\tau)})$ where q is a real scalar. One implementation of such a scheme is as follows: starting with an initial approximation p' of p and a prescribed scalar q', we determine $p'' = p' + q'z(p')$. We then set $q = -(p_I'' - p_I')/(z_j(p'') - z_J(p'))$, where $|p_I'' - p_I'| = \max_v |p_v'' - p_v'|$, $|z_J(p'') - z_J(p')| = \max_v |z_v(p'') - z_v(p')|$, take $p^{(0)} = p''$ and iterate as described.

Then Wynn used the method proposed by one of us (C.B.) [79, 80] and, independently, by Eckart Gekeler [263, 264] that consists in computing $2k + 1$ vectors $p^{(\tau)}$, $\tau = 0, \ldots, 2k$, applying to them the vector ε-algorithm, and restarting the iterates from $\varepsilon_{2k}^{(0)}$. Under certain assumptions on z and taking $k = N$, the dimension of the system, the sequence thus obtained converges quadratically to the solution of the nonlinear system $z(p) = 0$. Let us mention that there was a gap in the proofs given by these two authors, as well as in that of Stig Skelboe [595, 596, 597], and that a correct one was given only in 1992 by Hervé Le Ferrand [399] in the case of the topological ε-algorithm.

Wynn modified the parameter q after each cycle of iterates according to a procedure similar to that described above. He gave a numerical example, and discussed the accuracy of the derived solutions. He then explained how to use the procedure for the numerical integration of implicit differential equations. He gave a numerical example showing that the procedure works even when the model and the data are incompatible.

An important problem in iterative methods is the choice of the starting iterate. Wynn discussed the choice of q and the initial values proposed in [138] in the case of a boundary value problem for a system of ordinary differential equations with conditions given at the endpoint.

Then he turned to a comparison with linear methods, to the treatment of models of undetermined forms, to the prediction of the equilibrium position of a model, to growth functions, and to the determination of (possibly improper) integrals. ◄

In solving a system of nonlinear equations $f(x) = 0$ by an iterative method of the form $x_{n+1} = x_n + \alpha f(x_n), n = 0, 1, \ldots$, an open problem is the choice of the parameter α. If these iterations diverge, this has, in theory, no effect on the quadratic convergence of the restarted ε-algorithm. However, if the divergence is fast, rounding errors can affect the convergence. Obviously, the scalar α can be replaced by a sequence of scalars (α_n), or even by a nonsingular matrix, or a sequence of matrices. In the case of a system of linear equations $f(x) = b - Ax$, such iterates correspond to a method due to Richardson in 1910 [520], a paper in which he also discussed the choice of the parameters α_n.

5.2 Rational Approximation

Wynn worked on two types of rational approximation: interpolation by rational functions using Thiele's method, and Padé approximation. Both topics are strongly related to continued fractions.

5.2.1 Thiele Interpolation

The reference [719] is Wynn's doctoral thesis defended at the Natural Sciences Faculty of the Johannes Gutenberg University in Mainz, Germany, in 1959. His dissertation advisor was Prof. Friedrich Ludwig Bauer, a mathematician and a pioneer of computer science in Germany, in particular a developer of Algol.

▶ P. Wynn, *Über einen Interpolations-algorithmus und gewisse andere Formeln, die in der Theorie der Interpolation durch rationale Funktionen bestehen*, Numer. Math., 2 (1960) 151–182. [719]

The paper was submitted on 31 July 1959. It begins with the expression of the Legendre's interpolation polynomial using divided differences. He then presents Aitken's and Neville's recursive schemes. He then proceeds to interpolation by a rational function through reciprocal differences and the Thiele's interpolation formula. Rational expressions are given. A result about the Schweins's determinantal formula is given. Then, in order to avoid the calculation of reciprocal differences and the use of the continued fraction expansions, an algorithm is described that exhibits a relation between two interpolating fractions with different degrees in their numerators. A recursive method for the evaluation of these fractions is given. The recursive formulas are derived by first expressing the fractions as quotients of determinants and then introducing certain other quotients of determinants.

The scheme is also formulated in the form of an Algol program (with a few misprints that the reader can easily correct). It is pointed out that this scheme requires a smaller number of arithmetic operations than the Thiele interpolation formulas. A numerical example concerning the Bessel function J_0 is given. Particular rules are discussed. There is a comparison of the results obtained with those furnished by polynomial interpolation. Then it is pointed out that a considerable saving in the calculation can be effected when there is an even number of symmetrically distributed arguments.

Let us quote the end of the review MR 0128597 of this paper by J.G. Herriot

Formulas are given for solving the corresponding interpolation problem of constructing a rational approximation to f when $2n$ or $2n + 1$ derivatives of f are given at a single point. A discussion of the error propagation in the algorithmic scheme is included. A second algorithm, of "central" type analogous to the Neville method for polynomial interpolation, is also described. The consequences of letting the arguments $x_0, x_1, \ldots$ tend to a common value is investigated. ◄

In 1961, Josef Stoer (b. 1934) defended his thesis on rational interpolation under Friedrich Bauer and Klaus Samelson [625]. In his paper, the previous paper of Wynn is quoted and discussed. Rational functions are superior to polynomials for interpolation and extrapolation, as in Richardson's method, because rational functions are able to approximate functions with poles. Polynomials give good results only if the nearest pole is rather far outside a circle around the data points in the complex plane, while rational functions can have quite a good accuracy even in the presence of nearby poles. On this topic, see the papers by Roland Zdeněk Bulirsch (b. 1932) and Stoer [149, 150]. The application of rational extrapolation to the solution of ordinary differential equations by the modified midpoint rule was pointed out by William Bryan Gragg, Jr. (1936–2016) [282, 283].

∞

The second paper Wynn wrote on Thiele's method was published 11 years after his thesis. In between, several papers on rational interpolation appeared by various authors, including Stoer and Luc Wuytack (b. 1943) [702]. They are mentioned by Wynn himself. For an introduction to the topic, see [188]. In this paper, difference–differential equations between Padé approximants are given.

▶ *P. Wynn, Difference–differential recursions for Padé quotients, Proc. London Math. Soc., S. 3, 23 (1971) 283–300. [778]*

Wynn derived two general difference–differential recursions relating adjacent Padé approximants. In his *Theorem 1*, he considered the approximants of the series $\sum_{\nu=0}^{\infty} f^{(\nu)}(\mu) z^{\nu}/\nu!$, where $z = z' - \mu$. *Theorem 2* concerned the approximants of the series $\sum_{\nu=0}^{\infty} \varphi^{(\nu)}(\mu) z^{\nu}$. These results are too technical to be given here, but Wynn's commentaries are interesting. He wrote (references are those of our bibliography)

> The results of Theorems 1 and 2 are rather deep and, starting from first principles, it would be a matter of extraordinary difficulty to discover them; in this instance it may interest the reader to learn how they were in fact reached.

> The background to Theorem 1 is as follows: there exists a classical method, due to Thiele [636] [. . .] for the recursive construction of interpolatory rational functions whose values agree with those of a given function for prescribed values of the argument; in my Mainz dissertation [719] I showed that the same rational functions can be constructed by means of a process which, in terms of the algebraic operations involved, is somewhat more economical than Thiele's method; in his dissertation [625], Stoer showed that my process can considerably be improved, and in a recent paper [389], Larkin has shown that Stoer's formulæ can be refined still further.

Theorem 2 derives from the theory of the ε-algorithm. ◀

5.2.2 Padé Approximation

Although it is difficult to separate papers on continued fractions from those on Padé approximation (see Sect. 2.3), Wynn devoted eight papers to the second topic.

In the first of these papers, assuming the convergence a diagonal sequence of Padé approximants, he derived results on the convergence of other diagonal sequences.

▶ *P. Wynn, Upon the diagonal sequences of the Padé table, MRC Technical Summary Report 660, University of Wisconsin, Madison, May 1966. [758]*

Here, Wynn was interested in the convergence of diagonal sequences of Padé approximants. He wrote that it can be easily derived that if $\lim_{r\to\infty}[m + r + 2/r] = p_{m+2}$ exists, and if the sequence $([m + r + 1/r])$ is such that there exists $\delta > 0$ and r' such that $|[m + r + 1/r] - p_{m+2}| \geq \delta$ for $r \geq r'$, then $\lim_{r\to\infty}[m + r/r] = p_m$ exists. In this result, the nature of the sequence $([m + r + 1/r])$ is unrestricted. In the second theorem, he asserted that if $\lim_{r\to\infty}[m + 2r + 2/2r] = p'_{m+2}$, $\lim_{r\to\infty}[m + 2r + 3/2r + 1] = p''_{m+2}$, and $\lim_{r\to\infty}[m + r + 1/r] = p_{m+1}$ exist with $p_{m+1} \neq p'_{m+2}$ and $p_{m+1} \neq p''_{m+2}$, then $\lim_{r\to\infty}[m + 2r/2r] = p'_m$ and $\lim_{r\to\infty}[m + 2r + 1/2r] = p''_m$

exist, and that these limits satisfy his cross rule. No proofs and no hints for them are given. ◄

∞

Monotonicity and optimality properties of the diagonal Padé approximants for the Stieltjes series are studied in the next paper.

▶ P. Wynn, *Upon the Padé table derived from a Stieltjes series*, SIAM J. Numer. Anal., 5 (1968) 805–834. [766]

This paper is concerned with the Padé table of the series $\sum_{s=0}^{\infty}(-1)^s c_s z^s$ where $c_s = \int_0^\infty u^s \, d\psi(u)$ with ψ bounded and nondecreasing.

It is shown that under certain conditions, when z is real and positive, the Padé quotients along both forward and backward diagonals form monotonic sequences; an optimal property of the quotients lying upon the principal diagonal is proved. Some new convergence results are derived. The Padé quotients are compared with the transformed sums produced by certain linear methods.

The proofs are obtained by means of the continued fractions corresponding to the series and its reciprocal. The optimality property states that among the Padé approximants that can be computed from the partial sums of the series up to the term of degree $2r - 1$ inclusively, $[r/r]$ provides the most accurate approximation of $F(z) = \int_0^\infty d\psi(u)/(1 + zu)$. If one more term of the series is used, the best approximants are either $[r/r + 1]$ or $[r + 1/r]$. Then Wynn considers the uniform convergence of the Padé approximants lying on the diagonal of the Padé table in two cases: when ψ is constant for $u > b$ and has one point of increase at $u = b$, and when ψ has points of increase for arbitrary large values of u. Numerical results are given for a particular case. In the next section, Wynn examines weighted means of the partial sums of the series. It is assumed that the sum of the weights is equal to 1 in order to preserve the limit. The conclusion is that Padé approximants give better results. Then Wynn discusses the case of other series, and concludes *that our proof of the optimal property of the principal diagonal breaks down if the Padé table is constructed from a series which is not of Stieltjes type*. ◄

Other results on the best Padé approximant are given in [270, 273].

∞

The following paper is based on a criterion of Lev Semyonovich Pontryagin (1908–1988) for polynomials in two independent variables [502]. It concerns the Padé approximants of the exponential function.

▶ P. Wynn, *Zur Theorie der mit gewissen speziellen Funktionen verknüpften Padéschen Tafeln*, Math. Z., 109 (1969) 66–70. [769]

Consider the polynomial $Q(z, w) = \sum_{s=0}^{m} \sum_{t=0}^{n} q_{st} z^s w^t$. The term $q_{st} z^s w^t$ is called the *principal term* of Q if $q_{st} \neq 0$ and if for each other term $q_{ij} z^i w^j$ with $q_{ij} \neq 0$, we have either $s > i$ and $t > j$, or $s = i$ and $t > j$, or $s > i$ and $t = j$. Clearly, not every polynomial has a principal term (see, for example, [49, pp. 440–441]).

Following this result, if the polynomial Q has no principal term, then the function $Q(z, e^z)$ has an unbounded number of zeros with an arbitrary large positive real part. In this paper, Wynn deduced from this result that if $j > i$ (respectively $j < i$), then there is an unrestricted number of values of z with an arbitrarily large positive (respectively negative) real part for which $[j/i]_{e^z} = e^z$. A complementary result immediately follows for the Padé approximants of e^{-z}, since $e^{-z} e^z = 1$. Then Wynn gave conditions such that a function f satisfies $[j/i]_f(z) = f(z)$ for an unrestricted number of values of z with arbitrarily large positive real part. He then considered the case of a polynomial Q in three variables. ◄

∞

A convergence result for a special class of functions is given in the following paper.

► P. Wynn, *Upon a convergence result in the theory of the Padé table*, Trans. Amer. Math. Soc., 165 (1972) 239–249. [788]

Wynn considers the series expansion in ascending powers of z of the function

$$f(z) = \sum_{\nu=1}^{n} \frac{M_\nu}{1 + b_\nu z} + \int_a^b \frac{d\sigma(s)}{1 + zs},$$

with $M_\nu > 0$ and $b_1 > \cdots > b_n > b > 0$. Let $([p_j/q_i])$, $i, j \geq n$ be an infinite progressive sequence of Padé approximants of f such that either $p_{i+1} \geq p_i$ and $q_{j+1} > q_j$, or vice versa, and let $\mathcal{D}$ be the open disk $|z| < 1/b$ cut along $(-1/b, -1/b_1]$. Wynn proves that this sequence of Padé approximants converges uniformly to f in $\mathcal{D}$. The proof makes use of results by Montessus de Ballore, Nevanlinna, Markoff, Padé, Stieltjes, and van Vleck. ◄

∞

The next paper is a review of recent results on continued fractions, Padé approximation, the ε-algorithm, and some applications. It corresponds to the talk given by Wynn at the congress *Padé Approximation and Related Matters* held in Boulder, USA, 19–22 June 1972.

► P. Wynn, *Some recent developments in the theories of continued fractions and the Padé table*, Rocky Mountain J. Math., 4 (1974) 297–324. [801]

After an introduction in which Wynn defines his notation, the following section deals with interpolation by means of Thiele's continued fraction and the ϱ-algorithm. Starting from Newton's interpolation series, a determinantal formula for such interpolatory rational functions was given by Jacobi [348] in 1845. When all interpolation points tend to the same value, this formula becomes that of Padé approximants. Difference–differential recurrences relating Padé approximants are given. Several connections are recalled, in particular with the scalar ε-algorithm. Applications to the solution of systems of nonlinear equations are mentioned. The convergence theory of Padé approximants is discussed. There is a section on confluent prediction algorithms and integration methods. The partial differential equation of the Padé surface is given. Continued fractions whose coefficients belong to a ring are studied, and, in particular, vector continued fractions. Finally, extensions to nonassociative number systems are presented. ◄

$$\infty$$

The next paper gives the error of the Padé approximants for a Stieltjes series.

► P. Wynn, *Extremal properties of Padé quotients*, Acta Math. Hungar., 25 (1974) 291–298. [798]

Wynn considered the series $f(\lambda) = \sum_{\nu=0}^{\infty} t_\nu \lambda^{-\nu-1} = \int_\alpha^\beta d\sigma(\varrho)/(\lambda + \varrho)$ where $t_\nu = (-1)^\nu \int_\alpha^\beta \varrho^\nu d\sigma(\varrho)$ and σ is not a simple step function with a finite number of jumps in $-\infty \le \alpha < \beta \le +\infty$. Let C_r be the successive convergents of the continued fraction associated with the series f, let X denote the set of real numbers $X_1, \ldots, X_r$, and let

$$M(X, r, m) = \int_\alpha^\beta \left(1 + \sum_{\nu=1}^r X_\nu(\lambda + \varrho)^\nu\right) \frac{\varrho^m d\sigma(\varrho)}{\lambda + \varrho}.$$

Wynn proved that $f(\lambda) - C_r(\lambda) = \min_X(M(X, r, m)) > 0$ when $\lambda \in (-\alpha, \infty)$, and $C_r(\lambda) - f(\lambda) = \min_X(-M(X, r, m)) > 0$ when $\lambda \in (-\infty, -\beta)$. Results of the same type are given for the continued fraction corresponding to the series. Since the associated and corresponding continued fractions are Padé approximants, these results are then translated in terms of these approximants. Convergence results are also established. ◄

$$\infty$$

The next paper comes from a congress on Padé and rational approximation organized by Edward B. Saff and Richard S. Varga in Tampa, Florida, USA, 15–17 December 1976.

► P. Wynn, *The transformation of series by the use of Padé quotients and more general approximants*, in Padé and Rational Approximation. Theory and Applications, E.B. Saff and R.S. Varga eds., Academic Press, New York, 1977, pp. 121–144. [810]

The paper begins with a caustic remark:

> Much of what has recently been written on the theory of continued fractions was known and better understood fifty years ago.

Wynn's purpose was to provide a synopsis of the development of selected topics of the theory as initially presented, together with the original references. The author quoted the works of Borel, Carleman, Watson, F. Nevanlinna, Hamburger, Stieltjes, Hardy, Le Roy, F. Bernstein, and Wall. Then he linked the successive convergents of the continued fractions associated to the series $\sum f_\nu \lambda^{-\nu-1}$ and $\sum f_\nu z^\nu$ with quadrature formulas of the form $R_i = \sum M_\nu^{(i)} g(t_\nu^{(i)})$ for the approximation of $R = \int_{-\infty}^{+\infty} g(t) \, d\sigma(t)$, where $g(t) = (\lambda - t)^{-1}$ and $g(t) = (1 - zt)^{-1}$ respectively. He explained that the methods of proof deriving from the theory of quadrature suffice to demonstrate the convergence of these continued fractions under appropriate conditions, and he referred to the work of Marcel Riesz for that. Then, according to Wynn, these results can be trivially extended to functions of the form $G(z, t) = \sum G_\nu(z)t^\nu$, and to approximants derived from more general Padé approximants denoted by $W_{i,j}(z)$. For these approximants, he gave the reference [806], a report from McGill University that we were unable to locate. He said that the use of these approximants *allows a given series to be transformed by the use of approximating functions whose behavior is far more in accord with that of its sum or formal sum*. Then, he treated the case of the Euler–Maclaurin series that is transformed by these approximants and the usual Padé approximants. The last part of the paper is devoted to the methods he introduced in [768, 793, 808, 811] for evaluating an integral over a semi-infinite interval in terms of the derivatives of the integrand at the finite endpoint. As a *bonne bouche*, as Wynn wrote, the method may be combined with the Euler–Maclaurin formula for approximating the sum of a series with respect to the derivatives of its first term. Use of the ϱ- and the ε-algorithms is made. ◄

∞

The following paper, which appeared in 1981, is the last research paper published by Wynn. At that time, he was associated to the *Centro de Investigacion en Matematicas* in Guanajuato, Mexico, the town where he was living. The two other papers he published the same year are of a historical nature [812, 813].

► P. Wynn, *The convergence of approximating fractions*, Bol. Soc. Mat. Mexicana, 26 (1981) 57–71. [814]

In this paper, Wynn was interested in the uniform convergence of the forward diagonal sequences of the Padé approximants of a formal power series with coefficients $c_i = \int_\alpha^\beta t^i \, d\sigma(t)$, where σ is bounded and nondecreasing in $[\alpha, \beta]$. Various cases for α and β are discussed, but the notation used is too heavy to be reproduced here. Then, as Wynn explained in his introduction:

> Two methods for transforming possibly divergent power series have been extensively studied. The first of these was introduced by Le Roy, and is an extension of the integral transformation subsequently investigated in detail by Borel [. . .] An extension of Le Roy's method of transforming series is introduced. After conditions upon the rate of growth of the above moments have been imposed, the summability of the series by use of Le Roy's method and its extension is demonstrated, the convergence of certain sequences of approximating fractions obtained from the series is established, and the consistency of these two diverse methods of defining the sum of the series considered is proved. The results derived are refinements and extensions of theorems due to F. Bernstein, Hamburger and Wall.

Wynn also gave other references by himself. ◄

Édouard Louis Emmanuel Julien Le Roy (1870–1954) was a French philosopher and mathematician. He was a friend of the philosopher Henri Bergson (1859–1941), to whom he succeeded as a professor at the *Collège de France*, an institution founded by the French king François I in 1530, and as a member of the *Académie française*;

see [442]. In [402], Le Roy studied the properties that the coefficients of a possibly divergent series must possess so that the series could be analytically continued. An interesting paper on the theory of summable series in which the methods of Borel and Le Roy are presented among many others, was written by Robert Daniel Carmichael (1879–1967) [162], whose Ph.D. thesis is considered the first significant American contribution to the knowledge of differential equations.

5.2.3 The Padé Surface

Wynn published four papers dealing with the partial differential equations related to lozenge algorithms such as the ε-algorithm. The quantities computed by these algorithms depend on two integer indices, k (lower index) and n (upper index), and they are assigned to initial conditions. They are displayed in a two-dimensional table, and the rule(s) of the algorithm connect four quantities lying at the vertices of a lozenge in this array (similar to the ε-array of Fig. 2.2). Moreover, they can be expressed by determinantal expressions that differ in form according to whether their lower index k is even or odd. Thus, for some of them, their rule can be separated into two rules according to the parity of k.

∞

Let us analyze in detail the first of these papers.

▶ *P. Wynn, Partial differential equations associated with certain non-linear algorithms, Z. Angew. Math. Phys., 15 (1964) 273–289. [749]*

Wynn began by describing how partial differential equations are obtained from a lozenge rule:

> [. . .] it means finally that by contracting the lozenge into a point, partial differential equations corresponding to these algorithms may easily be derived [. . .]

> It transpires that in all cases which are considered in detail the lozenge algorithms are first order finite difference approximations to the derived partial differential equations. It is emphasised at the outset that in this treatment the partial differential equations are derived from the finite difference equations, and not conversely as is more often the case.

> The purposes of this paper are twofold. In the following section we shall discuss some properties of the algorithms being considered and introduce certain definitions of which subsequent use will be made. In the last section we place on record the partial differential equations which have been referred to, and derive certain properties of the solutions to these equations.

In the array, the quantities computed by the lozenge algorithm

> [. . .] lie at certain points, the positions of which are determined by n and k, in the plane of the paper. We now introduce new coordinates x and y (which are assumed to be continuous) and an interval h, so that the quantities lying in the above large lozenge do so at the following points in the x–y plane

$$x, y - 2h$$

$$x - h, y - h \qquad\qquad x + h, y - h$$

$$x, y$$

$$x - h, y + h \qquad\qquad x + h, y + h$$

$$x, y + 2h$$

This transformation of coordinates corresponds to the substitutions $x = a + 2kh$, $y = b + 2(n + k)h$, where a and b are two constants.

The quantities computed by the algorithm (which depend on n and k) are replaced by continuous functions depending on x and y (and also on n, k, and h), and then Wynn let h tend to zero, and thus he obtained the partial differential equation corresponding to the algorithm. He pointed out the nonuniqueness of the substitutions, and discussed some properties of lozenge algorithms such as their centro-symmetry, and those of the partial differential equations derived from them.

Then Wynn studied particular cases. For the ε-algorithm, he set $\varepsilon_{2k-1}^{(n)} = h^{-1}{}_1\varepsilon(x, y)$ and $\varepsilon_{2k}^{(n)} = h^{-1}{}_2\varepsilon(x, y)$, which led him, as h tends to zero, to ${}_1\varepsilon_x'{}_2\varepsilon_y' = 1$ and ${}_2\varepsilon_x'{}_1\varepsilon_y' = 1$. The functions ${}_1\varepsilon(x, y)$ and ${}_2\varepsilon(x, y)$ separately satisfy the partial differential equation $(1/\varepsilon_y')_y = (1/\varepsilon_x')_x$. He called this equation *the partial differential equation of the Padé surface*.

Partial differential equations for the qd-algorithm, the first and the second g-algorithms, the η-algorithm of Bauer [39], and the ϱ-algorithm are derived. Their properties are studied. The conclusion is:

> We remark that a theory of the types of initial and boundary conditions which are necessary for a solution to these equations to exist, has been constructed; but at the present time this is somewhat speculative and incomplete, and we do not examine this aspect of the theory here.
>
> When considering questions relating to the existence and uniqueness of the solutions of a partial differential equation it is often of great assistance if explicit solutions to a finite difference equation approximation to the partial differential equation can be given. In the case of all the algorithms of this paper, if the initial conditions are chosen in a certain manner, determinantal formulæ for the solutions of the algorithmic relationships can be derived (in certain cases such formulæ have been given). This property may well facilitate further research, and makes the algorithms of this paper particularly interesting. ◄

∞

Wynn came back to the partial differential equation of the Padé surface in [785], and gave some of its particular solutions. Two other papers on the same topic follow [790, 799].

► P. Wynn, *A note on a partial differential equation*, Report CRM-22, Centre de Recherches Mathématiques, Université de Montréal, Montréal, 1972. [785]

Results about the solution of the partial differential equation of the Padé surface are given in this report. If g is twice differentiable, then $\varepsilon(x, y) = g(\alpha xy + \beta x + \gamma y + \delta)$, with either $\alpha \neq 0$ or $\beta\gamma \neq 0$, is a solution. If f is an analytic function, then $\varepsilon(x, y) = f(x + iy)$ and $f(x - iy)$ are solutions that are in common with those of the Laplace equation $f_{xx}'' f_{yy}'' = 0$. If g is twice differentiable, $\varepsilon(x, y) = g(x + y), g(x - y)$ and $g(xy)$ are solutions. Writing the partial differential equation in the form $(\varepsilon_y')^2 \varepsilon_{xx}'' =$

$(\varepsilon_x')^2\varepsilon_{yy}''$, then in a certain region of the x–y plane, the ratio $\varepsilon_x'/\varepsilon_y'$ is real, finite, and nonzero, which means that it is hyperbolic in this region. Wynn also gave results on the solutions when there are transformations of the dependent or independent variables. He also provided a method for constructing one of the functions $_1\varepsilon$ and $_2\varepsilon$ when the other one is known. The paper ends by exhibiting the partial differential equation $(1/e_x')_x' = (1/e_y')_y'$ as a necessary condition for the solution of a problem in the calculus of variations. ◄

∞

► *P. Wynn, The partial differential equation of the Padé surface, Report CRM-197, Centre de Recherches Mathématiques, Université de Montréal, Montréal, June 1972.* [790]

Other solutions of the partial differential equation of the Padé surface are given in this report. For example, $1/(x-y)$ and $e^{\beta x+\gamma y+\delta}$. A self-conjugate inversion formula is obtained. However, this paper does not contain many new results. ◄

∞

► *P. Wynn, Sur l'équation aux dérivées partielles de la surface de Padé, C.R. Acad. Sci. Paris, 278A (1974) 847–850.* [799]

Since the ε-algorithm is related to Padé approximants, Wynn simply reformulated his results on the solutions of the partial differential equation of the Padé surface obtained via the ε-algorithm in terms of Padé approximants. ◄

5.3 Continued Fractions

As seen in Sect. 5.1, continued fractions (see Sect. 2.5) were a favorite topic of Wynn. He devoted many reports and papers to them, their development, and their extensions.

5.3.1 Algebraic Continued Fractions

► *P. Wynn, A comparison technique for the numerical transformation of slowly convergent series based on the use of rational functions, Numer. Math., 4 (1962) 8–14.* [733]

In the paper [728] (already analyzed in Sect. 5.1.4), Wynn was interested in the transformation of the series $\varphi_0(x) \sim \sum_{s=0}^{\infty} c_s x^s$ into the series $\sum_{s=0}^{\infty} c_s v_s x^s$. In this paper, he proposed to develop $(\sum_{s=0}^{\infty} c_s v_s x^s)/(\sum_{s=0}^{\infty} c_s x^s)$ into a continued fraction. He started from

The partial differential equation of the Padé surface

by P. Wynn

1. *Derivation of the equation.* One may obtain the values of the quotients $\{R_{i}(z)\}$ of the normal Padé table [1] (see [2,3]) derived from the series $\sum_{\nu=0}^{\infty} t_{\nu} z^{\nu}$ ($t_{\nu} \in \mathbf{Z}$ ($\nu = 0,1,...$), $z(\neq 0) \in \mathbf{Z}$ ($\overline{\mathbf{Z}}$ [z] is the complete [the finite part of the] complex plane)) by use of the ε-algorithm: set

$$\varepsilon_{-1}^{(m)} = \varepsilon_{2m}^{(-m)} = 0 \quad (m = 1,2,...), \quad \varepsilon_{0}^{(m)} = \sum_{\nu=0}^{m-1} t_{\nu} z^{\nu} \quad (m = 0,1,...), \text{ and determine}$$

$$\varepsilon_{r+1}^{(m)} = \varepsilon_{r-1}^{(m+1)} + 1/(\varepsilon_{r}^{(m+1)} - \varepsilon_{r}^{(m)}) \quad (r = 0,1,...; \ r' = [r/2], \ m = -r', -r'+1,...)$$

then $\varepsilon_{2r}^{(m)} = R_{m+r}(z)$ $(r = 0,1,...; \ m = -r, -r+1,...)$ [4,5]. We introduce new coordinates: set $x = a + 2rh$, $y = b + 2(m+r)h$ $(h(\neq 0), a, b \in \mathbf{Z})$; set $P(x,y) = \varepsilon_{2r}^{(m)}$, $Q(x+h, y+h) = 4h^2 \varepsilon_{2r+1}^{(m)}$. Two of the complete system of ε-algorithm relationships may then be written

$$[\{Q(x+h,y+h) - Q(x-h,y+h)\}/2h][\{P(x,y+2h) - P(x,y)\}/2h] = 1$$

$$[\{P(x,y) - P(x-2h,y)\}/2h][\{Q(x-h,y+h) - Q(x-h,y-h)\}/2h] = 1.$$

First page of a manuscript by Peter Wynn published as [799].
© Claude Brezinski.

$$\varphi_0(x) \frac{\alpha_{1,0} + \alpha_{1,1}x + \alpha_{1,2}x^2 + \cdots}{\alpha_{0,0} + \alpha_{0,1}x + \alpha_{0,2}x^2 + \cdots},$$

with $\alpha_{0,s} = c_s$ and $\alpha_{1,s} = c_s v_s$ for $s = 0, 1, \ldots$. Using a method described in [816, p. 31] (a book that he translated from Russian), he obtained the continued fraction

$$\sum_{s=0}^{m-1} \alpha_{1,s} x^s + \left(\varphi_0(x) - \sum_{s=0}^{m-1} \alpha_{0,s} x^s\right) \left(\frac{\alpha_{1,0}^{(m)}}{\alpha_{0,0}^{(m)}} + \frac{\alpha_{2,0}^{(m)} x}{\alpha_{1,0}^{(m)}} + \frac{\alpha_{3,0}^{(m)} x}{\alpha_{2,0}^{(m)}} + \cdots\right),$$

where the coefficients are obtained by the recurrence

$$\alpha_{r,s}^{(m)} = \alpha_{r-1,0}^{(m)} \alpha_{r-2,s+1}^{(m)} - \alpha_{r-2,0}^{(m)} \alpha_{r-1,s+1}^{(m)}$$

from the initial values $\alpha_{0,s}^{(m)} = \alpha_{0,m+s}$ and $\alpha_{1,s}^{(m)} = \alpha_{1,m+s}$, for $s = 0, 1, \ldots$. The successive convergents $C_r^{(m)} = A_r^{(m)} / B_r^{(m)}$ are computed by

$$A_r^{(m)} = \alpha_{r-1,0}^{(m)} A_{r-1}^{(m)} + \alpha_{r,0}^{(m)} x A_{r-2}^{(m)},$$
$$B_r^{(m)} = \alpha_{r-1,0}^{(m)} B_{r-1}^{(m)} + \alpha_{r,0}^{(m)} x B_{r-2}^{(m)},$$

with

$$A_0^{(m)} = \sum_{s=0}^{m-1} c_s v_s x^s, \qquad\qquad\qquad\qquad B_0^{(m)} = 1,$$
$$A_1^{(m)} = \alpha_{0,0}^{(m)} A_0^{(m)} + \alpha_{1,0}^{(m)} \left(\varphi_0(x) - \sum_{s=0}^{m-1} c_s x^s \right), \qquad B_1^{(m)} = \alpha_{0,0}^{(m)}.$$

An Algol program and three numerical examples are given. The first one occurs in the conductivity of dense stars, the second one is about the transformation of the series for the modified Bessel function I_0, and the third example concerns the asymptotic development of an integral. These results are compared with those obtained in [728]. Determinantal formulas in terms of the coefficients c_s and $c_s v_s$ are given for the quantities $\alpha_{r,0}^{(m)}$. ◄

It is difficult to guess why Wynn was interested in such a problem.

∞

The next paper is a review on continued fractions. It was presented by Wynn at the *SIAM Symposium on Approximation* organized at the Oak Ridge National Laboratory in Gatlinburg, Tennessee, 21–26 October 1963. In 1961, the proposal for a series of Gatlinburg Symposia was initiated by Alton Scott Householder (1904–1993). He was then director of the Mathematics Division at the Oak Ridge National Laboratory and professor at the University of Tennessee. This series of meetings still continues and they are known as the Householder Symposia.

► *P. Wynn, On some recent developments in the theory and application of continued fractions, SIAM J. Numer. Anal., Ser. B, 1 (1964) 177–197. [751]*

With his British sense of humor, Wynn wrote in the introduction:

In this talk I propose firstly to deal with those aspects of the study of continued fractions which are of direct application in the theory of approximation, and secondly to sketch some developments which are of recent origin and as yet incompletely worked out. It has taken more than two hundred years for the theory of continued fractions to attain its present condition, and I do not propose to speak for quite that length of time.

Wynn first explained how a continued fraction results from a sequence of bilinear transformations of the form

$$f_0 = b_0 + f_1, \qquad f_r = \frac{a_r}{b_r + f_{r+1}}, \qquad r = 1, 2, \ldots.$$

Thus

$$f_0 = b_0 + \frac{a_1|}{|b_1} + \frac{a_2|}{|b_2} + \cdots + \frac{a_n|}{|b_n + f_{n+1}},$$

$$= \frac{A_n + f_{n+1}A_{n-1}}{B_n + f_{n+1}B_{n-1}}, \quad n = 1, 2, \ldots,$$

with

$$A_n = b_n A_{n-1} + a_n A_{n-2}, \quad n = 1, 2, \ldots; \quad A_{-1} = 1, \quad A_0 = b_0,$$
$$B_n = b_n B_{n-1} + a_n B_{n-2}, \quad n = 1, 2, \ldots; \quad B_{-1} = 0, \quad B_0 = 1.$$

He gave the example of the Riccati differential equation, which, after bilinear transformations as explained above, leads to a sequence of functions (f_r) satisfying $f'_r + \alpha_r f_r^2 + \beta_r f_r + \gamma_r = 0$. Thus the solution of a Riccati equation can be directly expanded in the form of a continued fraction. For example, $\tan x$ satisfies $y' = 1 + y^2$ and

$$\tan x = \frac{x|}{|1} - \frac{x^2|}{|3} - \frac{x^2|}{|5} - \cdots .$$

It was this continued fraction that Johann Heinrich Lambert (1728–1777) used to prove in 1761 that $\tan x$ is irrational if x is a nonzero rational number [384]. Since $\tan \pi/4 = 1$, Lambert concluded that $\pi/4$, and therefore also π are irrational [102, pp. 110–111].

Then Wynn gave two other applications of bilinear transformations to Nevanlinna and Richards–Goldberg continued fractions. He explained how continued fractions can be derived from power series. He considered the series $f_0(z) \sim \sum_{s=0}^{\infty} c_s z^{-z-1}$. Let f_1 be the formal power series defined by $f_0(z) = c_0/(z - c_1 c_0^{-1} - f_1(z))$, and so on. Thus

$$f_0(z) = \frac{c_0|}{|z - \alpha_0} - \frac{\beta_0|}{|z - \alpha_1} - \cdots - \frac{\beta_{r-2}|}{|z - \alpha_{r-1} - f_r(z)}.$$

The successive denominators p_r of this continued fraction satisfy the three-term recurrence relationship

$$p_r(z) = (z - \alpha_{r-1})p_{r-1}(z) - \beta_{r-2}p_{r-2}(z),$$

which shows that they form a family of formal orthogonal polynomials with respect to a sequence of moments. The successive denominators of the continued fraction are the associated polynomials that also form a family of orthogonal polynomials. The infinite continued fraction f_0 is associated to the series f_0. Then Wynn discussed the convergence theory of continued fractions. He took the example of Grommer fractions. Applying the ε-algorithm to the partial sums of the series f_0, he explained that the successive convergents of the continued fraction lie on a diagonal of the ε-array, and that three types of convergence can be distinguished: *regular convergence* when the convergents in all the diagonals converge to the same limit; *irregular convergence* when the convergents in each diagonal converge to different limits; and the *highly irregular convergence behaviour* of R. Nevanlinna when the even and odd members of each diagonal converge to different limits. Numerical examples are given. The last parts of the paper deal with noncommutative continued fractions, noncommutative determinants, series with vector coefficients, and an application to the acceleration of Picard iterates for solving a differential equation and a partial differential equation.

In his conclusions, the first point is a hope that noncommutative continued fractions will play a role similar to the usual ones in a further development of the theory of linear operators. And he stated:

> The second point (it is rather prosaic) is this: numerical analysis is very much an experimental science. ◄

He was completely right in asserting that.

∞

Wynn delivered several lectures on continued fractions and wrote the corresponding notes.

► *P. Wynn, Four lectures on the numerical application of continued fractions, in Alcune Questioni di Analisi Numerica, A. Ghizzetti ed., Series: C.I.M.E. Summer Schools, Vol. 35, Springer, Heidelberg, 1965, pp. 111–251. [752]*

► *P. Wynn, Five lectures on the numerical application of continued fractions, Orientation Lecture Series 5, MRC, University of Wisconsin, Madison. [804]*

These reports correspond to courses on continued fractions given by Wynn. The first one consisted of lectures delivered at the *Centro Internazionale Matematico Estivo* (C.I.M.E.) held in Perugia, Italy, 7–16 September 1964 [752]. The second one [804] (175 pages) corresponds to a course given in various U.S. Army research establishments (no date mentioned in it). They are quite similar. ◄

∞

In the next long paper, Wynn builds two constructive definitions for a divergent integral. They are based on the transformation into continued fractions of the series representing the integral.

► *P. Wynn, Upon the definition of an integral as the limit of a continued fraction, Arch. Rat. Mech. Anal., 28 (1968) 83–148. [768]*

In the paper, Wynn considers integrals of the form $\int_t^\infty \varphi(\xi)\,d\xi$, where φ is a complex valued function of bounded modulus and t a complex variable. He assumes that the path of integration runs from the point t along a path parallel to and in the same direction as the positive real axis. He writes:

> Perhaps the crudest possible method of interpreting the value of the above integral is to regard it as being the limit of the value of the infinite sum

$$\sum_{r=0}^{\infty} h\varphi(t + rh) \qquad (2)$$

as h tends to zero [...]

There is an extensive literature [...] (and the accompanying references) concerning linear methods for the transformation of divergent integrals. These methods have been developed in analogy with the application of Toeplitz matrices in the theory of divergent series [...]

In this paper we shall be concerned with two constructive definitions of the integral, both of which are based upon the continued fraction transformation of the series (2). Since the continued fraction is a non-linear mechanism, the fundamental ideas underlying the new methods are quite distinct from those upon which the more familiar linear transformations are based; the new theory of integration must therefore be established in its own right, independent of the existing linear theory. We attempt, in the following, to give as complete a treatment of the convergence theory of our process of integration as is possible in the light of extant knowledge of continued fractions.

Section 2 is devoted to continued fractions associated to a series. If the continued fraction associated to the series $\sum_{r=0}^{\infty} h\varphi(t + rh)z^{-r-1}$ either terminates, converges strictly, or diverges essentially, then Wynn says that the function φ is *continued fraction integrable*. Assuming that φ possesses derivatives of all orders at t, Wynn derived the following determinantal expression for its convergents: $\varepsilon_{2r}(t) = H_{r+1}^{(-1)}(\varphi^{(v)}(t))/H_r^{(1)}(\varphi^{(v)}(t))$, for $r = 1, 2, \ldots$. According to similar conditions on the behavior of the sequence $(\varepsilon_{2r}(t))$, Wynn says that φ is continued fraction integrable. If the limit of this sequence is equal to the integral, the function φ is said to be *regularly continued fraction integrable*. More definitions of the same kind are given. Then Wynn shows that the functions ε_{2r} can be obtained via the second confluent form of the ε-algorithm. He also gives the determinantal expression for $\varepsilon_{2r+1}(t)$. He proved that if φ satisfies the linear differential equation $\sum_{s=0}^{n} a_s \varphi^{(s)}(t) = 0$, where n cannot be replaced by a smaller integer, then $\varepsilon_{2n+1}(t)$ is identically infinite for all t. If $\varphi(t) = \theta(t)e^{-zt}$, Wynn proves that $\varepsilon_{2r}(t)$ is equal to e^{-zt} multiplied by the rth convergent of the associated continued fraction derived from the formal power series $\sum_{s=0}^{\infty} \theta^{(s)}(t)z^{-z-1}$. Wynn wrote:

We have, in a simple and most direct manner, reduced the convergence behavior of the sequence $\{\varepsilon_{2r}(t)\}$ to that of continued fractions derived from power series. We are thus able to annex to the theory of continued fraction integration a formidable corpus of work built up by Tschebyscheff, Markoff, Stieltjes, Hamburger, Carleman, Nevanlinna, and many others. The termination of the second confluent ε-algorithm process, described in the previous section, can now be understood in a new light. After the introduction of the variable z (this can take any finite value, its position in the complex plane not being, as in later developments, critical) the function $\theta(t)$ satisfies a linear homogeneous differential equation with constant coefficients: (88) [*the power series above*] is the power series expansion of a rational function, and the associated continued fraction derived from it terminates; the continued fraction does no more than reconstruct the original rational function.

Then, in order to investigate functions φ of far greater complexity than those dealt with before, Wynn states certain results in the theory of associated continued fractions that will later be used to establish the consistency of his integration process. Formal orthogonal polynomials and their associates are introduced, and the moment problem is discussed. Then:

The time has now come to show how we propose to exploit this corpus of results in the convergence theory of continued fraction integration.

Convergence results for the associated continued fraction are established. Then, Wynn investigates the Markoff theorem about this convergence, and he studies the Hausdorff, the Hamburger, and the determinate and indeterminate Stieltjes moment problems.

An important property of the associated continued fraction provides a sequence of nested circular regions within which the value of the continued fraction must lie. Wynn exploits this theory to give bounds on the error of $\varepsilon_{2r}(t)$. The continued fraction integrability of certain functions simply related to φ is deduced from that of φ.

The third part of the paper is concerned with rational function integration. An associated continued fraction can be deduced for each series $F_m(z) \sim \sum_{s=0}^{\infty} c_{m+s} z^{-s-1}$, $m = 0, 1, \ldots$, and its convergents are denoted by $C_n^{(m)}(z)$. Wynn considers the approximations obtained by keeping n fixed and letting m tend to infinity. Under certain assumptions, the sequence $(C_n^{(m)})$ converges to $F_0(z)$ as m goes to infinity. An analogous process in the construction of an infinite integral consists in evaluating the finite integral $\int_t^{t'} \varphi(\xi) \, d\xi$ and in approximating the integral $\int_{t'}^{\infty} \varphi(\xi) \, d\xi$ by means of $\varepsilon_{2n}(t')$ obtained by the second confluent form of the ε-algorithm. He studies the conditions under which $\int_t^{t'} \varphi(\xi) \, d\xi + \varepsilon_{2n}(t')$ converges, as t' tends to infinity, to a limit which he calls the *rational functional integral of* φ. Then Wynn recalls, in terms of continued fractions, the famous theorem of Robert de Montessus de Ballore [456] that states that if $\sum_{s=0}^{\infty} c_s z^{-z-1}$ is the formal power series expansion of a function F that is regular for $|z| > R$ except for a finite number ν of poles, counted according to their multiplicities, lying outside this circle, then the sequence $(C_\nu^{(m)}(z))$, with ν fixed, as m tends to infinity, converges uniformly to F in any domain $|z| > R$ that does not contain a pole of F. He deduces from this theorem that if φ is infinitely differentiable for $t \geq t'$, and the function $G(z) = e^{-zt} \int_0^{\infty} \varphi(t + \xi) e^{-z\xi} \, d\xi$ is regular for $\mathrm{Re}(z) \geq R' < \infty$ and has a finite number of poles in the infinite strip $0 \leq \mathrm{Re}(z) \leq R'$, then φ is regularly continued fraction integrable. Moreover, if the integral exists in the normal sense, both integrals are equal. Then, on the basis of the work of Hadamard about the asymptotic behavior of Hankel determinants of a Taylor series [301], Wynn gives an estimation of the rate of convergence of $\int_t^{t'} \varphi(\xi) \, d\xi + \varepsilon_{2n}(t')$ as t' tends to infinity, and an example.

In the next part, Wynn shows how the theories of continued and rational fraction integrations can be exploited to yield the convergence theories of two nonlinear difference–differential processes for predicting the limit of a function $S(t)$ as t tends to infinity. These processes are the scalar ε-algorithm and its first confluent form. He shows that $\varepsilon_{2n}^{(m)} = C_n^{(m)}(z)$ when applied to the partial sums of the series F, and thus *the ε-algorithm may be regarded as being a device for predicting the limit of the sequence (S_m) from the values of its first few members.*

A similar conclusion holds for its first confluent form, which leads him to define the continued fraction and the rational function limits of a function S. If for all t, $\sum_{r=0}^{n} a_r S^{(r)}(t) = a_0 S$ with $a_n \neq 0$, then for all t, $\varepsilon_{2n}(t) = S$. This property defines the continued fraction limit of the function S. If $S(t)$ asymptotically behaves like a sum of exponential functions, the confluent form defines its rational function limit. He concludes that:

> Finally, we venture to suggest that the continued fraction and rational function limits are mathematically somewhat simpler than the limit in the normal sense.

He shows that the convergence theory of the first confluent form of the ε-algorithm as applied to $S(t)$ can be derived from that of the second confluent form as applied to $\varphi(t) = dS(t)/dt$.

Then Wynn treats integration by part, and show that:

> [...] the formula for integrating by parts in the theory of continued fraction integration exactly paraphrases that which occurs in the theory of normal integration, with the provision that the limit on the left hand side is a continued fraction limit; a corresponding result in the theory of rational function integration may also be derived.

The conclusion of this long paper is:

> In this paper we have described four new theories: those of the continued fraction integral, of the rational function integral, of the continued fraction limit, and of the rational function limit; the most extensive treatment was devoted to the first of these, which is, perhaps, the most important [...]

> As is natural with the development of a new theory, a number of possible further developments can immediately be envisaged; we wish, on this occasion, to mention two.

> The first concerns a possible enrichment of the theory itself.

The second development is about orthogonal polynomials, and he writes:

> Although we have not worked out the details in their entirety, it appears that there is a theory of orthogonal polynomials derived from a system of moments each of which is expressed by an integral which is divergent in the usual sense; one derives root location theorems, an analogue of Bessel's inequality, and other counterparts of the conventional theory without great difficulty; these matters will be dealt with in forthcoming papers.

Wynn thanks Arthur Erdélyi (1908–1977), a leading expert on special functions, orthogonal polynomials, and hypergeometric functions, for many constructive criticisms of this paper. ◀

∞

Wynn devoted a paper to the analysis of the work of Elwin Bruno Christoffel on continued fractions [812], and a second one to its subsequent developments [813].

▶ P. Wynn, *The work of E.B. Christoffel on the theory of continued fractions*, in *E.B. Christoffel: The Influence of His Work on Mathematics and the Physical Sciences*, P.L. Butzer, F. Fehér, eds., Birkhäuser Verlag, Basel, 1981, pp. 190–202. [812]

The first paper is an analysis of the work of Christoffel on continued fractions. The abstract of the paper is as follows:

> Christoffel's principal contributions to the theory of continued fractions concerned firstly the expression of a real number by means of a characteristic having a close connection with the regular continued fraction for this number, secondly the expression in simple

closed form of the convergents of the continued fraction expansion of quotients of Bessel's functions of contiguous orders, and thirdly the continued fraction expansion of functions defined as integral transforms of weight functions with respect to which the denominators and the numerators of the convergents are respectively orthogonal and associated orthogonal polynomials.

Let us also quote what Wynn writes at the beginning of the introduction:

The theory of continued fractions stands in a curious relationship to the rest of mathematics. Most mathematical theories are developed by processes of consolidation based upon linear transformations. Euclid's algorithm (from which continued fractions of all known types ultimately derived) is however a recursive division process. The transformations involved are nonlinear and, since it is inconvenient to subsume them within a framework of linear relationships, the theory of continued fractions is often overlooked [...]

Nevertheless, as Hadamard remarks, *pour inventer il faut penser à côté*; and an attempt to apply the nonlinear operations of Euclid's algorithm in the study of any subject often leads to thinking to one side. The theory of continued fractions has been pre-eminently an avenue to new and unexpected results, and one that has been ventured upon by all the great innovative mathematicians. It is the natural province of the mathematician who for some time remains aloof from organized academic activities; [...] Christoffel spent a considerable time in isolation.

A sentence particularly well adapted to his own situation!

Wynn begins by explaining that Christoffel used three types of division process for transforming a series into a continued fraction: operations upon real numbers, transformation of algebraic forms, and manipulations of formal power series. Then Wynn discusses what Christoffel called the *characteristic* of a number, a quite technical notion that was evoked in Sect. 3.5, and comments on his work on continued fractions derived from the quotient of two power series.

Next, Wynn recalls the main results obtained by Christoffel. In his inaugural dissertation [169] at the University of Berlin in 1856, and later in [170], Christoffel studied the relation between the continued fraction transformation of the series with coefficients $1/(i + 1)$ and Gaussian quadrature formulas. Then, in [172], he dealt with the function $f(t) = \int_{-1}^{1} (t - x)^{-1} \omega(x) \, dx$, where ω has a constant sign in $[-1, 1]$, and the series with moments $c_i = \int_{-1}^{1} x^i \omega(x) \, dx$. If P_n is a polynomial of degree n, he considered the polynomial Q_n of degree $n - 1$ defined by

$$Q_n(t) = \int_{-1}^{1} \frac{P_n(t) - P_n(x)}{t - x} \omega(x) \, dx.$$

Let $r_n(t) = P_n(t) f(t) - Q_n(t) = \sum_{i=0}^{\infty} d_i t^{-i-1}$. Christoffel showed that if $d_i = 0$ for $i = 0, \ldots, n$, then the polynomials P_n form a family of orthogonal polynomials with respect to the weight function ω on the interval $[-1, 1]$, that its zeros t_i are real, distinct and lie in $[-1, 1]$, and that the formula

$$\int_{-1}^{1} f(x) \omega(x) \, dx = \sum_{i=1}^{n} \frac{Q_n(t_i)}{P_n'(t_i)} f(t_i)$$

is exact if f is a polynomial of degree strictly less than $2n$. This is a Gaussian quadrature formula, and its coefficients were later called the *Christoffel numbers*. Moreover, he proved the formula known as the Christoffel–Darboux formula, also given by Gaston Darboux (1842–1917) 1 year later [194].

In the conclusions of his paper, Wynn writes that:

> [...] the theory of Christoffel's characteristic has simply not been further developed; it is surprising that at a time when (or so the writer believes) general standards of competence and discernment were far higher than those of today, such a potentially fruitful idea was not immediately taken up [...] Christoffel's results must be rederived at a higher level. A requirement for the successful execution of such a programme is, of course, that the mathematician who carry it out will be as gifted in their day as Christoffel was in his. ◄

A quite severe judgment!

∞

► *P. Wynn, Remark upon developments in the theories of the moment problem and of quadrature, subsequent to the work of Christoffel, in E.B. Christoffel: The Influence of His Work on Mathematics and the Physical Sciences, P.L. Butzer, F. Fehér, eds., Birkhäuser Verlag, Basel, 1981, pp. 731–734. [813]*

In this second paper, Wynn qualifies Christoffel as *one of the many omnific mathematicians who have made significant contributions to the theory of continued fractions*. Then he says that his work initiated developments in the theories of the Stieltjes and Hamburger moment problems and of quadratures, and he reviews some of these works. ◄

5.3.2 Converging Factors

Wynn published several papers on converging factors for series and for continued fractions. The case of continued fractions was already explained in Sect. 2.5. Let us describe them for series. We consider the (possibly asymptotic) series $S = \sum_{i=0}^{\infty} u_i$ and its partial sums $S_n = \sum_{i=0}^{n} u_i$. The remainder term R_{n+1} is defined by $S = S_n + R_{n+1} = S_n + (R_{n+1}/u_{n+1})u_{n+1}$. The converging factor C_{n+1} is defined by $C_{n+1} = R_{n+1}/u_{n+1}$. Since the converging factor may itself be developed into an asymptotic series, it is convenient to discuss this quantity rather than the remainder term. However, since it is unknown, an approximation $\widetilde{C}_{n+1}$ of it has to be found so that $S_n + \widetilde{C}_{n+1}u_{n+1}$ will be a better approximation of S than S_{n+1}. As stated in [686], *the applicability of extrapolation methods is strongly connected with the existence of an asymptotic expansion of the sequence under consideration*. This is exactly the point of view developed in his book by Guido Walz [686] (see also [19], where examples are given).

∞

The first paper below is about converging factors for continued fractions.

▶ *P. Wynn, Converging factors for continued fractions, I, II, Numer. Math., 1 (1959)*
272–307; 308–320. [718]

Wynn begins by explaining the origin of converging factors for continued fractions. In November 1873, when considering the continued fraction

$$\frac{\pi}{2} = 1 + \frac{1|}{|1} + \frac{1 \cdot 2|}{|1} + \frac{2 \cdot 3|}{|1} + \cdots,$$

James Whitbread Lee Glaisher (1848–1928) noted that

$$\frac{\pi}{2} = 1 + \frac{1|}{|1} + \frac{1 \cdot 2|}{|1} + \frac{2 \cdot 3|}{|1} + \cdots + \frac{n^2 - n|}{|1} + \frac{n|}{|1}$$

was a much better approximation of $\pi/2$ than the nth convergent of the original
continued fraction [275]. His explanation of this phenomenon was that the tail
$u_n = \frac{n(n + 1)|}{|1} + \frac{(n + 1)(n + 2)|}{|1} + \cdots$ of the continued fraction satisfies the difference
equation $u_n(u_{n+1} + 1) = n(n+1)$, which has, as a first approximation, the solution $u_n = n$. Thus he looked for a better approximation of u_n in the form $u_n = \sum_{s=-1}^{\infty} \alpha_s n^{-s}$.
This series is described as a converging factor in analogy with similar expressions
for accelerating slowly convergent series.

In his paper, Wynn extends this idea to continued fractions whose tails have the
form

$$u_n = \frac{a_n|}{|b_n} + \frac{c_n|}{|d_n} + \cdots + \frac{y_n|}{|z_n} + \frac{a_{n+1}|}{|b_{n+1}} + \frac{c_{n+1}|}{|d_{n+1}} + \cdots .$$

Such a tail satisfies the Riccati difference equation $p(n)u_n + q(n)u_{n+1} + r(n)u_n u_{n+1} = s(n)$, where p, q, r, and s are polynomials in n. He approximates u_n by the series
$\sum_{s=-k}^{\infty} \alpha_s n^{-s}$, where the coefficients α_s are determined by a system of recurrences.
To this end, it is necessary to expand both u_{n+1} and $u_n u_{n+1}$ in inverse powers of n.
Then the Riccati equation allows one to determine all the coefficients by equating
coefficients of identical powers of n. All the details of the procedure are given. A
modification of the method is introduced to simplify the algebraic manipulations
involved. An equivalent transformation of the continued fraction is introduced. Then
Wynn shows how to incorporate the converging factor into the continued fraction
and how to obtain an indication of the accuracy achieved. He treats in detail the
cases of $\pi/2$, $\log_2 2$, Bessel's functions, the hypergeometric function, the confluent
hypergeometric function, certain asymptotic series, and, in *Part II* of the paper,
further continued fraction expansions.

The coefficients of the series expansion of converging factors may be regarded
as coefficients of generating functions. Following the method of Bickley and Miller
[61] (see Sect. 3.2.4) for finding the generating function for the converging factor of
an infinite series, *which, if successful, determines the coefficients at one blow*, as he
writes, Wynn shows how to find an integral equation for the generating function of

the converging factor of a continued fraction. The paper ends with several numerical results. ◄

∞

In the next paper, Wynn approximates the tail of a continued fraction by a series. He presents some conjectures without proofs, but gives numerical examples and the corresponding Algol programs.

► *P. Wynn, Note on a converging factor for a certain continued fraction, Numer. Math., 5 (1963) 332–352. [744]*

The paper is devoted to *a device for accelerating the numerical convergence of a certain class of continued fractions. The method used is of considerable theoretical interest in itself.* However, it is very technical and difficult to grasp. As he already explained in the preceding paper [718]:

This computational device consisted in essence of the replacement of the tail

$$u_n = \frac{a_n}{\lfloor b_n} + \frac{c_n}{\lfloor d_n} + \cdots + \frac{y_n}{\lfloor z_n} + \frac{a_{n+1}}{\lfloor b_{n+1}} + \frac{c_{n+1}}{\lfloor d_{n+1}} + \cdots$$

of the continued fraction (the form of whose coefficients, apart from the first three, is periodic; the functions $a_n, b_n, \ldots, z_n$ being $2p$ in number)

$$C = \xi_0 + \frac{\xi_1}{\lfloor \xi_2} + \frac{a_1}{\lfloor b_1} + \cdots + \frac{y_1}{\lfloor z_1} + \frac{a_2}{\lfloor b_2} + \cdots,$$

by a series of the form

$$u_n = \sum_{s=-k}^{\infty} \alpha_s n^{-s}.$$

In the case that $a_n, b_n, \ldots, z_n$ are rational functions of their index, the coefficients α_s are recursively determined from the difference equation

$$u_n = \frac{a_n}{\lfloor b_n} + \frac{c_n}{\lfloor d_n} + \cdots + \frac{y_n}{\lfloor z_n + u_{n+1}}.$$

The process is illustrated by the continued fraction for $\pi/2$, for which $k = 1$, and there are two possible values for α_{-1}: $+1$ and -1. The convergents of C satisfy

$$C_{p(n-1)+2} = \frac{b_n A_{p(n-1)+1} + a_n A_{p(n-1)}}{b_n B_{p(n-1)+1} + a_n B_{p(n-1)}}.$$

Making use of the converging factor obtained as explained above in [718] (first paper analyzed in this section) from $\alpha_{-1} = 1$, the quantity

$$C_{p(n-1)+2} = \frac{A_{p(n-1)+1} + u_n A_{p(n-1)}}{B_{p(n-1)+1} + u_n B_{p(n-1)}}$$

is a considerably better approximation of C. Using the converging factor obtained from $\alpha_{-1} = -1$:

[...] brought to life a ghost function with which the continued fraction C may be associated. A number of conjectures regarding this function were made in the original treatment. Here we do not pursue this matter further, other than allowing for its investigation in the Algol programme.

Then the procedure is developed for a ratio of hypergeometric functions. A parameter h is introduced to overcome a difficulty. Thus the coefficients α_i now depend on h, and the converging factor can be expressed as a series in h, namely $u_n = \sum_{s=-1}^{\infty} \alpha_s(h) n^{-s}$. Then Wynn applies the ε-algorithm to the partial sums of u_n with $\varepsilon_0^{(0)} = 0$, $\varepsilon_0^{(m)} = \sum_{s=-1}^{m-2} \alpha_s(h) n^{-s}$, for $m = 1, 2, \ldots$, and $\varepsilon_1^{(m)} = n^{m-1}/\alpha_{m-1}(h)$. He obtains a continued fraction whose convergents provide better estimates than partial sums of u_n. Algol programs and numerical results end the paper. ◄

∞

A special case of converging factor is treated in the next paper, where the ε-algorithm is also used.

► P. Wynn, *Converging factors for the Weber parabolic cylinder function of complex argument*, Proc. Kon. Nederl. Akad. Weten., IA, IB, 66 (1963) 721–736; 737–754. [748]

The paper is divided into two parts. In the introduction, Wynn affirms that:

[...] considerable interest therefore attaches to general and efficient methods for computing numerical values to great accuracy of the higher functions of Mathematical Physics. One such method is the application of the converging factor. [...] The converging factor is an important numerical device for hastening the convergence of slowly convergent series and increasing the accuracy obtainable by use of an asymptotic series.

Let $S \sim u_0 + u_1 + u_2 + \cdots$ and $R_n \sim u_n + u_{n+1} + u_{n+2} + \cdots$. The converging factor C_n is defined by $R_n = u_n C_n$. In [450], Miller gave a method for developing it into a series either of the form $C_n \sim \sum_{r=0}^{\infty} \beta_r z^{-r}$ or of the form $C_n \sim \sum_{r=0}^{\infty} \delta_r n^{-r}$ for the cases in which either the function S satisfies a linear differential equation in z or the terms u_r satisfy a linear difference equation in r. He took the example of the Weber's parabolic cylinder functions but restricted to real values of the argument. In his paper, Wynn extends the computations to the complex domain, and proposes a recursive technique for obtaining the coefficients of the series for the converging factor. Weber parabolic cylinder functions are the solutions of the differential equation $f''(z) - (z^2/4 + a)f(z) = 0$ or $f''(z) + (z^2/4 - a)f(z) = 0$. Wynn first establishes the difference equation satisfied by the remainders R_n, and then he develops R_n into a series whose coefficients are computed. A comparison with the results of Miller [450] and those of John Robinson Airey (1868–1937) [4] is given. He then transforms the series into a continued fraction by applying the ε-algorithm. In the second part of the paper the formalism is summarized, the corresponding Algol programs are given, and numerical results are presented. ◄

∞

The converging factor for a continued fraction can be itself transformed into a continued fraction, as showed by Wynn in the next paper.

▶ *P. Wynn, On the computation of certain functions of large argument and parameter, BIT, 6 (1966) 228–259. [760]*

Let us give some extracts of the paper, since it contains Wynn's ideas on the development of numerical analysis:

> It has often been suggested that in the last 20 years Numerical Analysis has established itself as one of the major mathematical disciplines. During the same period of time digital computers have attained widespread use [. . .]

> In the more recent past formal languages (such as Fortran and Algol) have been developed for the instruction of digital computers, but the existence of these languages has not had quite the effect on Numerical Analysis that one might have expected.

> The existence of these languages does not seem basically to have influenced the manner in which problems in Numerical Analysis are posed, and the way in which they are solved.

> The primary purpose of this paper is to provide an example, capable of widespread extension, of the way in which the existence of a formal language (in this case Algol) quite fundamentally influences the posing and solution of a computational problem.

Wynn was completely right, as proved by the impact of computers on the development of numerical analysis and the emergence of the discipline *scientific computing*.

In this paper, the author is concerned with the derivation and transformation of an asymptotic series. He took the example of the incomplete gamma function, and transformed a related asymptotic series

> [. . .] in such a way as to make it more suitable for numerical computation. The device used to accomplish this which will be studied here, was introduced by J.R. Airey and called by him the Converging Factor [. . .]

> The converging factor may also be transformed into a corresponding continued fraction by means of the ε-algorithm.

The reference to the paper of John Robinson Airey is [4]. Airey's method consists in expanding the converging factor into an asymptotic series and transforming it, by a change of variables, into a power series whose coefficients

> [. . .] may also be derived by further processes of formal expansion and recognition [. . .]

> Thus the converging factor is a most powerful computational device. It was extensively used by Airey in the tabulation of a number of important functions of Mathematical Physics.

> But the processes of formal expansion and recognition, while easily performed by a mathematician of Airey's calibre, are perhaps not entirely suited to use with a digital computer.

Then Algol programs are given and illustrated by numerical results. In his conclusion, Wynn discussed the numerical stability of the computation and the propagation of errors, and wrote:

> We have not examined the propagation of error, and have simply proceeded in a quite reprehensible manner to make use of prolific systems of formal recursions. ◀

5.3.3 Noncommutative Continued Fractions

Wynn dedicated several papers to noncommutative continued fractions. In the first paper [746], he established their theory. He did not cite Wedderburn [688] or any other previous contributor to this topic. On the contrary, the book by Perron [493] was mentioned several times.

▶ *P. Wynn, Continued fractions whose coefficients obey a noncommutative law of multiplication, Arch. Rat. Mech. Anal., 12 (1963) 273–312. [746]*

The aim of this paper (submitted on 13 August 1962) was to establish a theory of continued fractions whose coefficients obey a noncommutative law of multiplication. Wynn wrote:

> [...] the theory has already found application in the acceleration of slowly convergent iterative processes in numerical analysis and has therefore some relevance to this subject, but in any case it is of considerable interest as a self-contained intellectual discipline [...]

> The indicated domain of inquiry is, it would appear, completely unstudied, and therefore all the results to he given are original, but many of them are quite transparent adaptations of existing results in the conventional theory of continued fractions [...]

In the first section, Wynn defined noncommutative continued fractions and the nature of their coefficients. Various formulas are given in the second section. Then, due to the noncommutativity of the multiplication, the notions of pre and post orthogonal polynomials are introduced. This led Wynn to define continued fractions associated with and corresponding to a formal series with noncommutative coefficients. These continued fractions are computed by the backward recurrence given in Sect. 2.5. Since, in this formula, a_{n-i}/D_i has to be replaced either by $a_{n-i}D_i^{-1}$ or by $D_i^{-1}a_{n-i}$, two types of continued fractions, called pre and post, are obtained. The same types also arise from the forward recurrence, since C_n can be defined either by $C_n = A_n B_n^{-1}$ or by $C_n = B_n^{-1}A_n$. Then a noncommutative version of the qd-algorithm is proposed. It allows one to construct the continued fractions mentioned above.

In the following section, Wynn presented noncommutative versions of certain nonlinear sequence transformations such as the ε-algorithm. He proved that if this algorithm is applied to a sequence (S_n) of elements of a subset with division of an associative algebra, and satisfying for all n, $\sum_{i=0}^{k} a_i(S_{n+i} - S) = 0$, then for all n, $\varepsilon_{2k}^{(n)} = S$. This is nothing else than the same relation as that defining the kernel of the scalar Shanks transformation. Thus in particular, the result holds for square matrices. Then Wynn presented some continued fractions related to functions satisfying systems of three-term recurrence relationships or homogeneous linear differential equations of the second order.

Interpolatory continued fractions are described in the next section. Inverse and reciprocal differences are used. Rational extrapolation by the ϱ-algorithm is treated.

The last section deals with confluent forms. In his conclusions, Wynn raised two problems: the variable z has to be a scalar, and the difficulty in the construction of a convergence theory for noncommutative continued fractions. In a first appendix, the author discussed what an extension of determinants to the landscape considered could be. In a second appendix, Wynn showed that the vector ε-algorithm provides the exact solution of a system of linear equations when applied to the iterates of a relaxation method. An application to the discretization of a partial differential equation by finite differences ends this long paper. ◄

∞

An isomorphism between complex vectors and real matrices was used by McLeod [443] to obtain the kernel of the vector ε-algorithm. In the next paper [757], Wynn wanted to extend this isomorphism to complex vectors and complex matrices, and for that, he referred to the preceding paper [746], since he would use its material. Let us mention that Wynn's paper precedes McLeod's by 5 years, and that it is only quoted as "to appear."

► *P. Wynn, Complex numbers and other extensions to the Clifford algebra with an application to the theory of continued fractions, MRC Technical Summary Report 646, University of Wisconsin, Madison, May 1966. [757]*

After an introduction to Clifford algebra and the inverse of a vector, Wynn wrote:

During the time that the author was developing the use of this successful but entirely inde-fensible numerical technique, he also built up a comprehensive formal theory of continued fractions whose coefficients obey a noncommutative law of multiplication.

Let us quote his abstract, since the paper is too theoretical to be summarily analyzed:

This isomorphism is such that the sum and difference of two vectors correspond to the sum and difference of the equivalent matrices: furthermore the inverse of a vector (this is described in the text) corresponds to the inverse of its equivalent matrix. This isomorphism is first extended so as to concern vectors whose components are quaternions, and then still further generalized to deal with what are called hyper-quaternion vectors. The isomorphism is then applied to the investigation of certain matrix continued fractions: the resulting theory establishes the existence of the vector-valued Padé table.

If the matrix ε-algorithm is applied to the partial sums of a formal power series with matrix coefficients, the matrices $\varepsilon_{2k}^{(m)}$ are the matrix Padé approximants $[m+k/k]$ of the matrix series. For a series with vector coefficients, Wynn first transformed its partial sums (which are vectors of dimension n) into $2^{n+1} \times 2^{n+1}$ matrices, and applied to them the matrix ε-algorithm. Then the matrices obtained by the algorithm are transcribed into their vector equivalents. He wrote:

It is the existence of this second method which finally removes the vector ε-algorithm from any suspicion of being a mere adventitious computational trick, and offers a perfectly secure interpretation of the vector ε-array. ◄

Let us mention that vector Padé approximants were later introduced in their doctoral theses by Jeannette van Iseghem (b. 1943) [658] and Marcelis Gerrit de Bruin (b. 1944) [145], and developed in their subsequent papers.

∞

We saw that noncommutative continued fractions were already treated by Wynn in [746] (the first paper considered in this section). The notion of their convergence will now be addressed with its specificities.

▶ *P. Wynn, A note on the convergence of certain noncommutative continued fractions, MRC Technical Summary Report 750, University of Wisconsin, Madison, May 1967.* [765]

Wynn begins by observing that up to that time, the theory of such continued fractions had been entirely formal. Namely, if one starts from some initial conditions and performs a finite number of simple rational operations, one obtains a certain mathematical object, and that starting from certain other initial conditions, the same object is obtained. However, no conditions that all goes well in both cases are given, and questions of convergence when an infinite number of such rational operations is performed are not considered at all. In this report, Wynn proved the existence of such objects and also gave some simple convergence criteria. The coefficients of the continued fractions are members of a normed ring. For scalar continued fractions, the forward and the backward recurrences are equivalent; they work or break down together. Wynn showed that this is no longer true for noncommutative continued fractions, and he explained why. For discussing the notion of convergence, two real norms have to be associated with every coefficient: an infimum norm λ and a supremum norm Λ. The normed space of elements has to be complete, i.e., if for any sequence (X_n) of elements, and any prescribed quantity $\varepsilon > 0$, there exists N such that $\Lambda(X_n - X_{n+1}) \leq \varepsilon$ for $i = 0, 1, \ldots$, then there exists X such that $\lim_{n \to \infty} \Lambda(X_n - X) = 0$. In these circumstances, (X_n) is said to converge uniformly to X. Wynn proved that if

$$\Lambda(c_2) + \Lambda(c_3) + \sum_{i=4}^{r} \Lambda(c_i)(1 + \Lambda(c_2))(1 + \Lambda(c_3)) \cdots (1 + \Lambda(c_{i-2})) < 1$$

for $r = 4, 5, \ldots$, the pre and post continued fractions $c_0 + \dfrac{c_1|}{|I} + \dfrac{c_2|}{|I} + \cdots$ both exist and converge uniformly. The infimum norm λ is nowhere used. ◀

∞

Wynn was working on noncommutative continued fractions for some years when he published the next paper [767]. It is, in fact, the completion under the form of a research journal of his preceding publications [746, 751, 757, 765]. However, it is an important paper, since it gathers results scattered in the literature. Its ultimate goal was to establish results on the vector ε-algorithm. It also served as the starting point for several researchers who quoted it. It is dedicated to Alexander Markowich Ostrowski (1893–1986) on his 75th birthday.

▶ *P. Wynn, Vector continued fractions, Linear Algebra Appl., 1 (1968) 357–395.* [767]

The aim of Wynn in this paper was to [. . .] *construct the formal theory underlying a numerical process* [the vector ε-algorithm] *for accelerating the convergence of vector sequences.* He mentioned that such sequences occur in the numerical solution of linear algebraic equations when one is discretizing a differential or an integral equation. He first recalled the rule of the vector ε-algorithm and gave a numerical example showing that it is

> [. . .] a formidable resource of numerical analysis. However, the vector ε-algorithm not only lacks a rigorous convergence theory (this is true of many methods of numerical analysis, and should not worry anyone) but also, at least until now, has lacked even the support of a formal theory.

Then Wynn recalled some known facts concerning continued fractions, in particular those derived from a formal power series. In particular, the rth convergent of the continued fraction associated to a series agrees with it up to the term of degree $2r$ inclusively. Since it is a rational fraction with a numerator of degree $r - 1$ and a denominator of degree r, it is the Padé approximant $[r - 1/r]$ of the power series. The numerators of the successive convergents form a family of formal orthogonal polynomials with respect to the sequence of coefficients of the series. A similar result holds for the denominators. Delaying the construction of the associated continued fraction leads to the Padé approximants $[n + r - 1/r]$. By introducing the reciprocal series, Wynn explained how the other half of the Padé table can be obtained. Then he went to the ε-algorithm and explained that its application to the partial sums of the power series produces the Padé approximants. He mentioned that the ε-algorithm is related to other algorithms allowing one to compute recursively the coefficients of the recurrence relations of the families of orthogonal polynomials, a bridge highlighted by his thesis advisor F.L. Bauer [38] (see also [39, 42]).

After this recall of known results in the scalar case, Wynn came to noncommutative continued fractions whose elements belong to a division ring. He settled the landscape and discussed the fact that there are two types of such continued fractions, pre and post, since the multiplication law used in the backward recurrence is not commutative. Pre and post families of formal orthogonal polynomials are thus produced, and the pre and the post multiplication of the numerator by the inverse of the denominator thus leads to two types of Padé approximants. He mentioned that Bauer's bridge can be extended to the noncommutative case. Since the ε-algorithm does not involve multiplication, it allows one to construct pre and post associated continued fractions starting with differing initial conditions. Thus he stated: *There is no question of having a pre- or post-Padé table: the operator valued Padé table is unique.*

The next section is devoted to the vector ε-algorithm. Wynn recalled the difficulties encountered in the establishment of its theory, which has to use Clifford's algebra and McLeod's isomorphism between vectors and matrices [443] for obtaining the difference equations defining the vector sequences belonging to its kernel. In order to overcome the disadvantages of this theory, Wynn proposed a different isomorphism using anticommuting matrices, and proved the corresponding theoretical results. The remaining part of the paper concerns an extension of the theory to l_2 vectors having an infinite number of components. It is very technical, but it is quite clearly explained in the abstract of the paper:

> It is shown that if a given power series has coefficient vectors lying in Hilbert space, then the vector-valued Padé quotients derived from this series also lie in Hilbert space. Properties of the matrices occurring in the vector-matrix isomorphism are examined; in particular, it is shown that the numerical values of certain norms of a vector are equal to those of the corresponding norms of its companion matrix. The concept of a functional Padé table is introduced, and one of its properties is derived.

There is a first appendix on hyper-quaternion vectors. A second appendix is devoted to a functional form of the ε-algorithm and a functional Padé table. Wynn considered infinite vectors of the form $z = (z(0), z(h), z(2h), \ldots)^T$. In the rule of the vector ε-algorithm, the $\varepsilon_{2k}^{(n)}$ are unchanged, but the $\varepsilon_{2k+1}^{(n)}$ are replaced by $h\varepsilon_{2k+1}^{(n)}$. Letting h tend to zero, every member of the ε-array becomes a function (in L_2), and the algorithm evolves into a functional one,

$$\varepsilon_{k+1}^{(n)}(\xi) = \varepsilon_{k-1}^{(n+1)}(\xi) + (\varepsilon_k^{(n+1)}(\xi) - \varepsilon_k^{(n)}(\xi))^{-1},$$

with inversion defined by $z(\xi)^{-1} = z(\xi)/\int_0^\infty z(t)\bar{z}(t)\,dt$. Then Wynn said that breakdown can occur only when the two functions in the inversion process are identical, and that all functions $\varepsilon_k^{(n)}$ that can be constructed by this functional ε-algorithm belong to L_2. ◄

Padé approximants for operators were later treated by Annie Cuyt (b. 1956)[8] [185].

5.4 Formal Orthogonal Polynomials

Polynomials orthogonal on a finite or infinite interval with respect to a positive measure are interesting in themselves. They have been treated in a number of books such as, for example, [166, 632]. Their extension to a more general linear functional is related to the numerators and the denominators of Padé approximants, and thus, also to the convergents of certain continued fractions. Wynn was one of the pioneers in this domain, which has applications even in numerical linear algebra, as will be seen in Sect. 6.6.

∞

The next two papers are analyzed together, since a procedure to get the results of [720] more easily is given in [724]. For the other results, the paper [724] is quite similar to the first one.

▶ *P. Wynn, The rational approximation of functions which are formally defined by a power series expansion, Math. Comp., 14 (1960) 147–186. [720]*

---

[8] See her testimony in Sect. 7.2.

▶ *P. Wynn, L'ε-algoritmo e la tavola di Padé, Rend. Mat. Roma, (V) 20 (1961) 403–408. [724]*

Wynn introduced formal orthogonal polynomials in [720], a paper submitted on 5 November 1959. He began by relating Padé approximants and continued fractions by means of the *qd*-algorithm. He wrote:

> The various questions associated with the central problem of obtaining rational functions from power series expansions are unified and greatly clarified by an appeal to the classical theory of orthogonal polynomials [. . .]

> Since, however, the moments c_s are the central feature of the problem in hand, and the introduction of the integral and weight function $\phi(t)$ (in view of the difficulty of envisaging an appropriate form for $\phi(t)$) may even serve as an impediment to understanding, the present treatment will slightly be varied. Accordingly, define a process I by

$$(3.1.4) \qquad I(t^s) = c_s \quad s = 0, 1, \ldots,$$

and construct a sequence of orthogonal polynomials $p_n^{(m)}(z)$, $n = 0, 1, \ldots$, $m = 0, 1, \ldots$, from the relations

$$(3.1.5) \qquad I(t^{m+s} p_n^{(m)}) = \begin{cases} 0 & 0 \le s \le n-1 \\ \omega_n & s = n, \ldots \end{cases}$$

He chose ω_n so that $p_n^{(m)}$ was monic. Then he constructed a second family of polynomials, the associated polynomials, as

$$o_n^{(m)}(z) = I\left(\frac{t^m (p_n^{(m)}(z) - p_n^{(m)}(t))}{z - t} \right), \quad n = 1, 2, \ldots; m = 0, 1, \ldots,$$

where the linear functional I acts on the variable t, z being a parameter. Let us continue to quote Wynn:

> Note in passing that formally

$$(3.1.10) \qquad I\left(\frac{t^m}{z - t} \right) = \sum_{s=0}^{\infty} c_{m+s} z^{-s-1}$$

> Consider the quotient $o_n^{(m)}(z)/p_n^{(m)}(z)$, there follows [. . .]

$$\frac{o_n^{(m)}(z)}{p_n^{(m)}(z)} = I\left(\frac{t^m}{z - t} \right) - \frac{1}{p_n^{(m)}(z)} \left(\frac{t^m p_n^{(m)}(t)}{z - t} \right)$$

$$= [\ldots]$$

$$= \sum_{s=0}^{\infty} c_{m+s} z^{-s-1} + P_{2n+1}(z)$$

where $P_{2n+1}(z)$ is a power series beginning with a term in z^{-2n-1}.

Thus, these rational functions are nothing else than Padé approximants of the formal power series. A third family of polynomials is also defined. He gave a determinantal formula for $p_n^{(m)}$, and asserted that these three families of orthogonal polynomials satisfy the same three-term recurrence relation but with different initializations. The coefficients of these recurrences are then obtained via a result due to Stieltjes [623], thus leading to a matrix interpretation of the orthogonality conditions, or by the qd-algorithm of Rutishauser [536, 538] (see also [318]). Two recurrence relations between families of polynomials with the upper indices m and $m + 1$ are given, and after some manipulations, the usual three-term recurrence relation is recovered.

Then Wynn showed that his ε-algorithm allows one to compute recursively the upper half of the Padé table, and he explained how to construct the other half using the theory of continued fractions and the Frobenius identities. The procedure is quite complicated, but he said that in a paper to appear [724] (the second paper considered in this section), he would show how to obtain it via the ε-algorithm by applying the algorithm to the partial sums of the reciprocal series. The occurrence of blocks of identical approximants in the Padé table is discussed.

Wynn wrote:

> A systematic study of feasible methods for deriving rational approximations may thus be conducted by describing each of these stages in terms of the numbers and types of arithmetic operations involved and, subsequently, designing the methods by combining suitable steps. The arithmetic operations involved in each method may be assessed by adding together those involved in the composite stages, and a comparison of the methods thus be made.

The paper ends with a discussion of checking and correcting the approximations obtained. F.L. Bauer, H. Rutishauser, and E.L Albasiny are acknowledged. ◄

An interesting paper on the connections between these topics is due to William Bryan Gragg, Jr. [284].

∞

The next paper highlights the connection between associated and corresponding continued fractions and formal orthogonal polynomials.

► *P. Wynn, A general system of orthogonal polynomials, Quart. J. Math. Oxford (2), 18 (1967) 81–96. [763]*

Let us give an extract of the introduction of this paper (submitted on 8 September 1966):

> The original purpose of the work upon which this paper is based was to examine the coefficients of the continued fractions which correspond to certain series [. . .]

> There is a strong connexion between the transformation of a power series into its associated and corresponding continued fractions and the derivation of a system of orthogonal polynomials from a given set of moments: indeed the two processes are in a sense formally equivalent.

> After a little additional effort we discovered, as a bonus for our work on continued fractions, the coefficients of what may be regarded as a global set of orthogonal polynomials derived from a global set of moments. Again all the classical orthogonal polynomials may be derived as special cases of the general result, and further special cases lead to new systems.

> We do not, in this paper, treat such matters as a global weight function or a global generating function: we are concerned with an algebraic formulation of the condition of orthogonality.

Wynn begins by explaining what are the corresponding and the associated continued fractions derived from a formal power series in inverse powers of the variable, and he shows how their coefficients can be computed by the qd-algorithm of Rutishauser. The coefficients of the power series are related to the quantities computed by the qd-algorithm in some particular cases. ◄

This paper is less important than [720], and it does not bring decisive results. As seen in Sect. 2.4, formal orthogonal polynomials can be introduced directly without the machinery of continued fractions.

∞

Asymptotic results on the behavior of orthogonal polynomials at infinity are given in the next paper.

► P. Wynn, *Über orthonormale Polynome und ein assoziiertes Momentproblem*, *Math. Scand., 29 (1971) 104–112.* [780]

Let $\{P_k\}$ be the family of orthonormal polynomials with respect to a bounded and nondecreasing function on $(-\infty, +\infty)$. Wynn proves that if the Hamburger moment problem is not determined, then $\lim_{k\to\infty} |P_k(z)| = 0$ uniformly on every bounded open domain of the z-plane that does not contain a point of the real axis. A similar result holds for a bounded and nondecreasing function on $(0, +\infty)$ outside the positive real axis, provided that the Stieltjes moment problem is undetermined. In his second theorem he considers a bounded and nondecreasing function on the compact interval $[\alpha, \beta]$, where, of course, both moment problems are determined. Here he shows that $\lim_{k\to\infty} |P_k(z)| = \infty$ uniformly on every bounded open domain of the z-plane that does not contain a point of $[\alpha, \beta]$. Finally he compares his findings with other well-known asymptotic results given under more restrictive assumptions. ◄

This work was completed during a stay at the *Eidgenössische Technische Hochschule* in Zürich, where Wynn was invited by Eduard Stiefel (1909–1978), director of the *Institut für Angewandte Mathematik*.

5.5 Computation of Special Functions

The celebrated handbook [1] edited by Milton Abramowitz (1915–1958) and Irene Ann Stegun (1919–2008), of the United States National Bureau of Standards, first appeared in 1964. It contains definitions, identities, approximations, plots, and tables of values of numerous functions used in all fields of applied mathematics and mathematical physics.

Thus, it was not inappropriate that in 1961, Wynn was interested in the tabulation of indefinite integrals.

► P. Wynn, *On the tabulation of indefinite integrals, BIT, 1 (1961) 286–290.* [723]

He wrote:

> The tabulation of an auxiliary function which is related to an indefinite integral of the form $\int_\alpha^x \phi(t)\,dt$ may often be most economically carried out by the numerical integration of a differential equation which is satisfied by the auxiliary function. The choice of auxiliary function and auxiliary variable z where $x = g(z)$ is in the main determined by facilities for interpolation in the resulting table, and is logically that which provides the greatest information about the required indefinite integral in the smallest possible space.

The auxiliary function must be simply related to the integral, and the numerical solution of the differential equation that the function satisfies must be easy. Moreover:

> A number of efficient methods have been developed for the numerical solution of linear second order differential equations from which the term involving the first derivative is absent.

Wynn proved that the function

$$y(z) = \theta(z) \int_\alpha^{g(z)} \phi(t)\,dt + h(z)$$

satisfies a linear differential equation of order 2 with no term in y'. Several choices of the function θ are discussed. The exponential integral is given as an example. ◄

∞

When computing tables of values for mathematical functions, it is a priority to control the accuracy. Wynn tackled this problem in the following two papers.

▶ *P. Wynn, Numerical efficiency profile functions, Koninkl. Nederl. Akad. Wet., 65A (1962) 118–126. [734]*

▶ *P. Wynn, The numerical efficiency of certain continued fraction expansions, IA, IB, Koninkl. Nederl. Akad. Wet., 65A (1962) 127–137; 138–148. [735]*

The first paper [734] is an explanation of how to control the efficiency of a recursive process for computing a function of one argument z. For that purpose, Wynn proposes to construct a two-argument table that indicates the number n_{r_i,s_j} of iterations of the process for which a relative error less than or equal to $p^{-s_j}/2$ is attained at the point z_{r_i}, where p is a given radix. Both sequences (r_i) and (s_j) must be monotone, and n is a monotonic function that increases with both of its arguments.

Then Wynn explains that [...] *if the relative error in the computation of the function value must not exceed $p^{-s}/2$, the argument value is $z' \in (z_r, z_{r+1}]$ and $s' \in (s_{r'}, s_{r'+1}]$ then the index of the required partial sum or convergent is $n_{r+1,r'+1}$.* Having such a table is not required if a numerical efficiency profile function $n(z, s)$ of continuous variables z and s can be given such that $n(z_r, s_{r'}) \geq n_{r,r'}$. The profile function must be chosen so that $n(z_r, s_{r'}) - n_{r,r'}$ is as small as possible. It is a linear programming problem solved by the simplex method. A complete algorithm is given together with a numerical example involving a continued fraction for $\log(1 + z)$.

The paper [735] begins with a review of methods for deriving continued fractions for functions used in mathematical physics: Gauss's method, application of the qd-algorithm, Euler's method, expansion of a Laplace transform, and the use of reciprocal differences. Then Wynn gives many examples of the numerical efficiency table he introduced in [734]: Bessel's and hypergeometric functions, confluent hypergeometric functions, asymptotic series, slowly convergent series, and theta functions. He concludes:

> It is therefore possible that the continued fractions which are described in this note are archetypal for certain of those which will find numerical application in the future. ◄

∞

The next paper is concerned with the location of the zeros of certain hypergeometric functions by means of continued fractions.

► *P. Wynn, On the zeros of certain confluent hypergeometric functions, Proc. Amer. Math. Soc., 40 (1973) 173–183. [795]*

In the introduction, Wynn noticed that:

> [...] the location of the zeros of certain special confluent hypergeometric functions have been obtained by investigating the behaviour of integral expressions involving the functions in question and, in the case of real argument and parameters, by the use of Sturm sequences.

His aim was to introduce a new method based on continued fractions, and to prove that:

> [...] the zeros of certain confluent hypergeometric functions lie in a fixed half-plane and that those of certain combinations of these functions lie on the imaginary axis.

Thus he established that for $-1/2 < \alpha < \infty$, the zeros of $_1F_1(\alpha; 2\alpha + 1; z) = 0$ have a positive real part, and that those of $_1F_1(\alpha + 1; 2\alpha + 1; z) = 0$ have a negative real part. He deduced that the zeros of the Bessel functions $I_{\alpha-1/2} + I_{\alpha+1/2} = 0$ have a positive real part, and that if α is an integer, this equation has an unbounded number of zeros. Then he proved that the zeros of the equations

$$_1F_1(\alpha; 2\alpha; z) = {_1F_1}(\alpha; 2\alpha + 1; z) \quad \text{and} \quad {_1F_1}(\alpha; 2\alpha; z) = {_1F_1}(\alpha + 1; 2\alpha + 1; z)$$

are identical, purely imaginary, symmetrically distributed about the origin and unbounded in number. Finally, let $C_n(z)$ be the convergents of the continued fraction associated with the function $g(z) = {_1F_1}(\alpha; 2\alpha; z)/{_1F_1}(\alpha; 2\alpha + 1; z)$. The zeros $iy_\nu^{(n)}$, $\nu = 1, 2, \ldots$, of the equation $g(z) = C_n(z)$ have the same properties as those described above, and furthermore, they interlace $y_\nu^{(n)} < y_\nu^{(n+1)} < y_{\nu+1}^{(n)}$, $\nu = 1, 2, \ldots$. ◄

5.6 Algol Procedures

The language Algol (acronym for *Algorithmic Language*) was originally jointly developed in the mid and late 1950s by a committee of European and American computer scientists at a meeting held in 1958 at the *Eidgenössische Technische Hochschule* (ETH) in Zürich, among them Friedrich Ludwig Bauer (Wynn's thesis advisor), Heinz Rutishauser, Klaus Samelson, Hermann Bottenbruch, and others. The first version of the language, developed in Germany, was Algol 58. It uses the *normal form method* of describing programming languages. This method was due to the American computer scientist John Backus (1924–2007), who directed the team that invented and implemented Fortran. Then the next developments, resulting in Algol 60 and Algol 68, came from the *Centrum voor Wiskunde en Informatica* in Amsterdam, with Edsger Wybe Dijkstra (1930–2002), Jaap Anton Zonneveld (1924–2016), and Adriaan van Wijngaarden. Recall that Wynn was a member of this center from 1960 to 1964. Thus he really participated in the early developments of Algol. For a history of this language, consult [540].

In various papers, Wynn gave Algol programs. In this section, we present his papers that are more specifically dedicated to this language.

∞

The first one is devoted to programming with complex numbers.

▶ *P. Wynn, An arsenal of Algol procedures for complex arithmetic, BIT, 2 (1962) 232–255. [740]*

In this paper, the purpose of Wynn was

[...] to present a complete system of Algol procedures for carrying out arithmetic operations upon complex numbers, to describe the convention which governs their use, and to give some examples of their application.

These applications were the computation of the confluent hypergeometric function and Weber's parabolic cylinder function, the use of the ε-algorithm to series of complex terms, the qd-algorithm, and continued fractions. In these examples, a great deal of time was devoted to estimating the relative error of the successive convergents of the continued fractions. He concluded:

The above Algol programmes may not therefore be regarded as anything more than a first step in the construction of a library of Algol procedures for the evaluation of functions in the complex plane. ◀

∞

The paper [742] resulted from a talk Wynn gave at the IFIP congress *Information Processing 62* held in Munich, 27 August–1 September 1962 (for the analysis of this paper, see Sect. 5.1.7). The corresponding Algol programs were published only in the following paper.

▶ *P. Wynn, General purpose vector epsilon-algorithm Algol procedures, Numer. Math., 6 (1964) 22–36. [750]*

It seems that Wynn did not appreciate that these procedures were not included in the proceedings of the IFIP congress, since he wrote in the first paragraph of [750]:

> At the 1962 IFIP Congress in Munich the author gave an invited expository talk on accelera-
> tion techniques in Numerical Analysis. It had been his intention to include in the proceedings
> of this Congress two general purpose *algol* procedures together with a number of short pro-
> grammes illustrating their use. In this way work in what is at this time a critical domain
> of inquiry in Numerical Analysis would have been thrown open to as large as possible a
> forum of experimentation. Due to restrictions which were imposed upon space it was not
> possible to publish these procedures in the Congress proceedings; it is the purpose of this
> note to cause them to be published here. Before giving these procedures, a short explanation
> is embarked upon.

Then Wynn recalls the scalar and the vector ε-algorithms, and how to program them by his ascending diagonal and moving lozenge techniques. Numerical results concerning the iterative solution of the Lichtenstein–Gershgorin integral equation follow the Algol procedures. He concludes that:

> It will be realised that the given vector ε-algorithm procedures have quite general application
> in Numerical Analysis. ◄

∞

Wynn presented the repeated application of the ε-algorithm in [726] (see Sect. 5.1.1.1). In the next paper he explained how to program it, and gave the corresponding Algol procedures.

► *P. Wynn, A note on programming repeated application of the epsilon-algorithm, Revue française de traitement de l'information, Chiffres, 8 (1965) 23–62; Errata 156. [753]*

Wynn comes again to the repeated application of the ε-algorithm. In the first application of the algorithm, he starts from $_0\varepsilon_0^{(m)} = S_m$ as the set of initial values. Then the rth application of the algorithm can start from $_r\varepsilon_0^{(m)} = {}_{r-1}\varepsilon_{2m}^{(0)}$, which he calls the *associated repeated application* of the ε-algorithm, or from $_r\varepsilon_0^{(2m)} = {}_{r-1}\varepsilon_{2m}^{(0)}$ and $_r\varepsilon_0^{(2m+1)} = {}_{r-1}\varepsilon_{2m}^{(1)}$, called the *corresponding repeated application* (notice a typing error in the paper: the lower index is m instead of $2m$). Then he shows how to program these applications and gives the Algol procedures. He comments that:

> In a number of cases it would seem that the first mode is slightly the more powerful: however
> the second mode is more flexible and the programme for effecting it is shorter.

Some numerical examples are given. ◄

The corresponding repeated application of the ε-algorithm is what Wynn called the intercalation method in Sect. 5.1.5, and it leads to the hierarchies of ε-arrays studied in Sect. 5.1.6.

∞

The last of Wynn's papers containing Algol programs is about the computation of continued fractions and effecting the ε-algorithm.

▶ *P. Wynn, An arsenal of Algol procedures for the evaluation of continued fractions and for effecting the epsilon algorithm, Revue française de traitement de l'information, Chiffres, 9 (1966) 327–362. [761]*

The paper gives Algol programs for the computation of continued fractions and the implementation of the ε-algorithm. After an introduction, a procedure for the evaluation of the continued fraction

$$\frac{c_0}{|z} - \frac{q_1}{|z} - \frac{e_1}{|z} - \frac{q_2}{|z} - \frac{e_2}{|z} - \cdots$$

is given, assuming that the coefficients q_i and e_i are known in a closed form. It may be used for the evaluation of the incomplete gamma function of large argument.

Then it is shown how to compute

$$\frac{c_m}{|z} - \frac{q_1^{(m)}}{|z} - \frac{e_1^{(m)}}{|z} - \frac{q_2^{(m)}}{|z} - \frac{e_2^{(m)}}{|z} - \cdots,$$

assuming that $q_1^{(m)}$ (given by the qd-algorithm) can be simply expressed, or that c_m is known in a closed form. Application of the ε-algorithm is also discussed. Wynn concludes:

> During the computations of the qd and ε-algorithms considerable instability, resulting in a loss of precision, can occur. No measures to combat this have been taken, although at the cost of a considerable complication in the programmes it is possible to do so [*he cites his paper [743]*].

This means that he has not coded his particular rule for avoiding instability. ◀

5.7 Abstract Algebra

For his work on noncommutative continued fractions, Wynn had to study the algebra of series with noncommutative coefficients, in particular with coefficients in a ring and vector coefficients. Thus he began to be interested in abstract algebra. The corresponding papers and reports show that he acquired quite a deep knowledge of the field.

∞

The first of these papers deals with the computation of the inverse (also called the reciprocal) of a series f, that is, the series g formally satisfying $fg = 1$ or $gf = 1$.

▶ *P. Wynn, Upon the inverse of formal power series over certain algebra, Report CRM-53, Centre de Recherches Mathématiques, Université de Montréal, Montréal, November 1970. [771]*

Let $f(\mu, z) = \sum_{\nu=\mu}^{\infty} t_\nu z^\nu$ be a formal power series. Its reciprocal is the series $g(-\mu, z)$, which is formally defined by $f(\mu, z)g(-\mu, z) = 1$. It exists if and only if $t_\mu \neq 0$. Wynn showed that if the coefficients t_ν are elements of an associative ring with a multiplicative inverse, and if t_μ is invertible, then the pre- and post-reciprocal series (that is, $gf = 1$ and $fg = 1$) exist and are identical. The result has applications to quaternions, square matrices, lower and upper triangular matrices, and certain classes of infinite matrices. The same result holds for series with coefficients in a flexible ring. Applications are to Cayley's numbers defined over a field, and to the flexible rings he constructed in [770] (the next paper). Lastly, Wynn proved that if the coefficients of a formal power series are self-involutive members of a distributive ring, and if the pre- and post-reciprocal series of a given series are identical, then the coefficients of the reciprocal series are also self-involutive. This result may be applied to square matrices. ◄

$$\infty$$

An involution is a function f that is its own inverse, that is, $f(f(x))) = x$ for all x in the domain of f. Wynn studies them in a ring, in particular a flexible ring, that is, where $a \cdot (b \cdot a) = (a \cdot b) \cdot a$.

► *P. Wynn, Upon a recursive system of flexible rings permitting involution, Report CRM-50, Centre de Recherches Mathématiques, Université de Montréal, Montréal, November 1970. [770]*

The paper contains four parts. In the first one, Wynn gives a general theory of flexible rings permitting involution, and he investigates the conditions under which the norm of the product of two elements is equal to the product of their norms, and those under which the inverse of the product of two elements is equal to the reversed product of the inverses of the elements. In the second part, he shows that the linear algebras of a recursive system constructed by Albert [11, 12], which puts the algebras of real or complex numbers, quaternions, Cayley's and extended Cayley's numbers in a unified setting, are noncommutative Jordan algebras. Then, in the third part, Wynn derives a recursive system of rings in analogy with Albert's system, he shows that each of these rings is flexible, and he applies to them the theory of the first part. In the fourth part, Wynn describes a method for the exponential extrapolation of sequences of elements of a flexible ring. It consists in determining S from the members of the sequence $S_n = S + \alpha\beta^n, n = 0, 1, \ldots$, where S, α, and β are elements of a field. Wynn shows that for all n, $\varepsilon_2^{(n)} = S$, and that this result is still valid in a flexible ring. ◄

Abraham Adrian Albert (1905–1972) was an American mathematician known for his work on finite-dimensional division algebras over number fields, and as the founder of Albert algebras (see [364] for a photo, a biography, an analysis of his work, and a complete list of his publications). Cayley numbers, also known as octonions, were discovered by John Thomas Graves (1806–1870) in December 1843,

2 months after the discovery of quaternions by Hamilton. Graves communicated his discovery to Hamilton in a letter dated 4 January 1844, but it was published only in 1848, after having been rediscovered by Arthur Cayley (1821–1895) in 1845. Since then they have been called Cayley numbers; see [371].

∞

In the following paper, Wynn came back to reciprocal series of a series with complex rectangular matrix coefficients. He studied their existence and uniqueness.

▶ *P. Wynn, Upon the generalized inverse of a formal power series with vector valued coefficients, Compo. Math. 23 (1971) 453–460. [779]*

Wynn studied particular cases of obtaining the reciprocal series of a series with complex matrix coefficients. Let $f(z) = \sum_{\nu=0}^{\infty} T_\nu z^\nu$ be a formal power series, and $g(z) = \sum_{\nu=0}^{\infty} \widehat{T}_\nu z^\nu$ its reciprocal series. If the coefficients T_ν of f are square matrices, g is formally defined by $f(z)g(z) = I$, and it exists and is uniquely determined if and only if T_0 is invertible. If the coefficients T_ν are rectangular matrices, Wynn defined g by four relations similar to those defining the pseudo-inverse of a matrix, namely

$$f(z)g(z)f(z) = f(z), \quad g(z)f(z)g(z) = g(z), \quad f(z)g(z) = h(z), \quad g(z)f(z) = m(z),$$

where h and m are series with Hermitian matrix coefficients. He concluded that g is not, in general, uniquely determined by these relations. However, this is possible in the two particular cases in which the T_ν are either row or column matrices of finite dimension with $T_0 \neq 0$. The reciprocal series has coefficients that are, respectively, column or row matrices. In both cases, the recurrence relations for computing them are given. ◀

∞

The report [791] (dated September 1972) and the paper [805] (submitted 28 August 1974) are essentially the same. The ring of series whose coefficients belong to a distributive ring is studied.

▶ *P. Wynn, The algebra of certain formal power series, Report CRM-216, Centre de Recherches Mathématiques, Université de Montréal, Montréal, September 1972. [791]*

▶ *P. Wynn, The algebra of certain formal power series, Riv. Mat. Uni. Parma, (4) 2 (1976) 155–176. [805]*

The set of all formal power series in X with coefficients in a commutative ring **R** form another ring symbolized by **R**[[*X*]], and called the ring of formal power series in the variable X over **R**. In the case that **R** is a distributive ring, Wynn denoted it by **P**{**R**} in [791] and [805], and he showed that:

[...] it shares to a large extent the multiplicative properties of the coefficients; in particular that if the elements of **R** multiply commutatively, so do those of **P**{**R**}, that if **R** is a ring, $\mathbf{P}\{\mathbf{R}\}$ is also a ring, and that the same holds with regard to Lie, flexible, alternative commutative and noncommutative Jordan rings. It is also shown that if $\mathbf{R}$ possesses a unit element, $\mathbf{P}\{\mathbf{R}\}$ also possesses a unit element, that $\mathbf{P}\{\mathbf{R}\}$ has a center if $\mathbf{R}$ has one, that is $\mathbf{R}$ is without divisors of zero, the same is true of $\mathbf{P}\{\mathbf{R}\}$, [...]

and so on. Then he treated the case of the derivation and the involution. A transformation $\mathcal{D}$ operating on elements of a distributive ring $\mathbf{R}$ is called a derivation if $\mathcal{D}(A + B) = \mathcal{D}(A) + \mathcal{D}(B)$ and $\mathcal{D}(AB) = A(\mathcal{D}(B)) + (\mathcal{D}(A))B$ for $A, B \in \mathbf{R}$. Wynn proved that

$$\mathcal{D}(A(z)B(z)) = A(z)\mathcal{D}(B(z)) + \mathcal{D}(A(z))B(z).$$

Then he considered the case of involution, and proved similar results.
Wynn concluded [805] thus:

It is clearly possible to apply the theory of this paper recursively, considering power series in one variable whose coefficients are power series in another, and to do so a denumerably infinite number of times. ◄

One can doubt the usefulness of such a recursive use of this theory. It can be considered a pure mathematical speculation.

∞

In [794], Wynn goes one step further in his study of involution rings. He studies possibly nonlinear involutions permitting involution. An application to $\varepsilon_2^{(n)}$ is given.

► P. Wynn, *Distributing rings permitting involution*, Report CRM-281, Centre de Recherches Mathématiques, Université de Montréal, Montréal, 1973. [794]

As already mentioned above, an involution is an application that is its own inverse, that is, $f(f(x)) = x$ for all x in the domain of f. In this report, Wynn gave

[...] a systematic theory of (possibly nonlinear) involutions permitted by distributive (i.e. nonassociative) rings. In particular, it is shown that two numbers which belong to a distributive ring permitting an involution with central trace and norm and satisfy a set of identities of flexible and alternative type generate an associative ring. Properties of the inverse defined in terms of an involution are reviewed. The theory is applied to that of a hierarchy of recursively defined rings [...] The problem of locating the center of a spiral over certain systems of numbers is considered.

It is well known that if the scalar ε-algorithm is applied to a sequence of complex numbers $S_m = S + a\lambda^m, m = 0, 1, \ldots$, then $\varepsilon_2^{(m)} = S, m = 0, 1, \ldots$ (equivalent to the Aitken Δ^2 process). The points S_m lie on a spiral in the complex plane that is contracting if $|\lambda| < 1$, and expanding otherwise. The computation of $\varepsilon_2^{(m)}$ from S_m, S_{m+1}, and S_{m+2} is a process for locating the center S of the spiral. This interpretation is extensively discussed in a manuscript left by Wynn to his friend Manuel Berriozábal and found after his death. It is analyzed in [122]. The same result holds for quaternions and Cayley numbers over a field. Wynn proved that if the S_m are elements of a power associative ring (so that the powers λ^m are unambiguously defined) and if the inverse of elements of this ring behaves in a suitable manner, then the same result still holds. ◄

5.8 Miscellaneous

Wynn produced a certain number of papers and reports almost completely independent of his main interests. It is not known why and how he was led to these works.

$$\infty$$

The first of these papers concerns a method of order 3 for solving a special nonlinear equation.

▶ *P. Wynn, On a cubically convergent process for determining the zeros of certain functions, Math. Tables Aids Comput., 10 (1956) 97–100. [714]*

Wynn presented a cubically convergent method for finding the zero of a function ϕ satisfying $p\phi'' + q\phi' + r\phi = s$, where p, q, r, and s are functions of x. It is based on a method he attributed to Herbert William Richmond (1863–1948) [524] that does not make use of ϕ''. It consists in the iterations

$$x_{n+1} = x_n - \frac{2\phi\phi'}{2(\phi')^2 - \phi s + r\phi^2 + q\phi\phi'},$$

all functions being computed at $x = x_n$.

An application to the zeros of the Bessel functions J_0 and Y_0 is given. ◀

$$\infty$$

In the next paper, Wynn used Newton's method for fitting experimental data by a sum of exponential functions.

▶ *P. Wynn, A note on fitting certain types of experimental data, Stat. Neerl., 16 (1962) 145–150. [738]*

The paper is concerned with finding the $2h + 1$ parameters Z, A_i, and α_i, $i = 1, \ldots, h$, such that $\Theta = \sum_{i=1}^{n}(f_i - f(t_i))^2$ is minimal, where the f_i are observed values at the points t_i, and $f(t) = Z + \sum_{s=1}^{h} A_i e^{-\alpha_i t}$. The method used by Wynn is simply Newton's method. He gave expressions for the derivatives of Θ with respect to the parameters. A variant consists in keeping, for all iterations, the Jacobian computed at the starting point. This variant can be accelerated by the vector ε-algorithm. A numerical example is given. Wynn pointed out that when the points y_i are equidistant, the problem can be solved by a method due to the French engineer Gaspard Clair François Marie Riche, baron de Prony (1755–1839) in 1795 [513]. ◀

Prony's method has many applications in digital filtering and in signal processing, and it is also related to the z-transform, and to Padé approximation as explained in [691] and [110]. The z-transform is a functional transformation of sequences that can be considered equivalent to the Laplace transform for functions. While the Laplace transform is useful in solving differential equations, the z-transform plays a central role in the solution of difference equations. By changing z into z^{-1}, it is identical to the method of generating functions introduced by the French mathematician François Nicole (1683–1758) and developed by Joseph Louis Lagrange.

In signal processing, a function of time $f(t)$, called the input signal, is transformed into an output signal $h(t)$. This transformation is realized via a system G called a digital filter. The input signal f can be known for all values of the time t, and in that case we speak of a continuous signal and a continuous filter, or it can be known only at equally spaced values of t, $t_n = nT$ for $n = 0, 1, \ldots$, where T is the period, and in that case, we speak of a discrete signal and a discrete filter. The z-transform of a discrete signal is given by

$$F(z) = \sum_{n=0}^{\infty} f_n z^{-n},$$

where $f_n = f(nT)$. Corresponding to the input sequence (f_n), we have the output sequence $(h_n = h(nT))$. If we set

$$H(z) = \sum_{n=0}^{\infty} h_n z^{-n},$$

then the system G can be represented by its transfer function $G(z)$ such that

$$H(z) = G(z)F(z).$$

In other words, setting $G(z) = \sum_{n=0}^{\infty} g_n z^{-n}$, we have

$$h_n = \sum_{k=0}^{n} f_k g_{n-k}, \quad n = 0, 1, \ldots.$$

Thus if (f_n) and (h_n) are known, then (g_n) can be computed. An important problem in the analysis of digital filters is the identification of the transfer function when (f_n) and (h_n) are known (that is, when (g_n) is known). If the filter is linear, then G is a rational function of z, and if not, it can be approximated by a rational function $R(z) = P(z)/Q(z)$.

Setting

$$P(z) = a_0 z^s + a_1 z^{s-1} + \cdots + a_s,$$
$$Q(z) = z^s + b_1 z^{s-1} + \cdots + b_s,$$

the b_i are the solution of the system

$$\begin{pmatrix} g_s & g_{s-1} & \cdots & g_1 \\ \vdots & \vdots & & \vdots \\ g_{2s-1} & g_{2s-2} & \cdots & g_s \end{pmatrix} \begin{pmatrix} b_1 \\ \vdots \\ b_s \end{pmatrix} = - \begin{pmatrix} g_{s+1} \\ \vdots \\ g_{2s} \end{pmatrix},$$

and the a_i are then given by

$$a_0 = g_0,$$
$$a_1 = g_1 + b_1 g_0,$$
$$\vdots$$
$$a_s = g_s + b_1 g_{s-1} + \cdots + b_s g_0.$$

Thus, comparing to the relations defining the Padé approximants (see Sect. 2.3.1), we see that

$$R(z) = [s/s]_G(z),$$

which shows the equivalence between the z-transform method and Padé approximation.

As proved in [691], the z-transform is also equivalent to the method for interpolation by a sum of exponential functions due to Gaspard de Prony [513]. The problem is to find the A_j and the a_j, for $j = 1, \ldots, s$, such that

$$\sum_{j=1}^{s} A_j e^{a_j t_i} = g_i, \quad i = 1, \ldots, 2s,$$

where the g_i are given numbers and $t_i = iT$. Thus, setting $z_j = e^{a_j T}$, the preceding system can be written as

$$\sum_{j=1}^{s} A_j z_j^i = g_i, \quad i = 1, \ldots, 2s.$$

We set

$$Q(z) = b_s + b_{s-1} z + \cdots + b_1 z^{s-1} + z^s = \prod_{j=1}^{s}(z - z_j),$$

and we first calculate the b_i. Setting $b_0 = 1$, we multiply the first equation by b_s, the second one by b_{s-1}, and so on up to the $(s + 1)$th equation, which is multiplied by b_0. Then these equations are summed. We begin the same process again by multiplying the second equation by b_s, the third one by $b_{s-1}, \ldots$, the $(s + 2)$th by b_0, and then summing them. And so on. We finally obtain

$$\sum_{i=0}^{s} b_{s-i} \sum_{j=1}^{s} A_j z_j^{i+k} = \sum_{i=0}^{s} b_{s-i} g_{i+k}, \quad k = 1, \ldots, s,$$

which can also be written as

$$\sum_{i=0}^{s} b_{s-i} g_{i+k} = \sum_{j=1}^{s} A_j z_j^k \sum_{i=0}^{s} b_{s-i} z_j^i = \sum_{j=1}^{s} A_j z_j^k Q(z_j) = 0.$$

Thus the b_i are solution of the system

$$\sum_{i=0}^{s-1} b_{s-i} g_{i+k} = -g_{s+k}, \quad k = 1, \ldots, s,$$

which is exactly the system solved above to find the denominator of the Padé approximant $R(z) = [s/s]_G(z)$. Once the b_i have been obtained, the z_i are the zeros of the polynomial Q and $a_i = T^{-1} \ln z^i$ for $i = 1, \ldots, s$. Then the A_i can be uniquely determined from the z_i by solving a nonsingular subsystem extracted from the initial one or with the help of Padé approximants. We know that

$$[s/s]_G(z) = g_0 + z[s - 1/s]_{G_1}(z),$$

where $G_1(z) = g_1 + g_2 z + g_3 z^2 + \cdots$. Thus the partial fraction decomposition of $[s - 1/s]_{G_1}$ also gives the A_i since

$$[s - 1/s]_{G_1}(z) = \sum_{i=1}^{s} A_i (z - z_i)^{-1}$$

if the z_i are distinct. If it is not the case, it means that at least two A_i are identical, and thus s has to be replaced by a lower value. This shows the complete equivalence between Prony's method and Padé approximation in the z-transform domain.

A survey of the application of Padé approximation and continued fractions to model reduction problems and an extensive bibliography are given in [153]. The method is also particularly suited for the analysis of damped oscillations of an electrical transmission line [259]. It has applications to power system ringdown analysis as initiated in 1990 by J.F. Hauer et al. [306], and it is used for testing harmonics and interharmonics in electrical power systems [710]. In music, the *prony* is a unit for measuring the interval between two musical tones.

∞

The following paper of Wynn concerns the solution of a partial differential equation.

▶ *P. Wynn, Note on the solution of a certain boundary-value problem, BIT, 2 (1962) 61–64. [739]*

Wynn is interested in the iterative solution of the linear partial differential equation

$$D_x D_y f(x, y) = g(x, y, f, D_x f, D_x^2 f, D_x D_y f, D_y^2 f)$$

in a rectangle, where D_x and D_y denote the partial derivatives with respect to x and y. The partial derivatives are estimated by finite differences. This leads to a system of linear equations with a compound matrix, and he gives the inverse of this matrix in a closed form. ◀

∞

The next paper is devoted to the asymptotic expansion of an integral. In a particular case, the ε-algorithm is used, and the Algol procedure is given. This paper is certainly related to those in which Wynn was interested in the computation of special functions.

▶ P. Wynn, *A numerical study of a result of Stieltjes, Revue française de traitement de l'information, Chiffres*, 6 (1963) 175–196. [747]

Wynn uses the method of steepest descent to obtain an asymptotic expansion of the integral

$$(1) \qquad\qquad F(\rho) = \int_{\alpha^*}^{\alpha^{**}} G(t) e^{\rho H(t)} \, dt,$$

in which H is assumed to have a single maximum in the whole complex plane

[...] at $t = \alpha_H \in [\alpha^*, \alpha^{**}]$ and that the integral (1) may be so transformed that the path of integration lies along the contour (the steepest path) $Im\ H(t) = \text{constant} = Im\ H(\alpha_H)$.

The case of the exponential integral is treated in detail, and the ε-algorithm is applied to its expansion. Algol procedures and numerical results are given. ◀

$$\infty$$

The signs of the Hankel determinants of a very special function considered by Pólya are studied in [781].

▶ P. Wynn, *On an extension of a result due to Pólya, J. Reine Angew. Math.*, 248 (1971) 127–132. [781]

This paper is devoted to an extension of a result due to George Pólya (1887–1985) [501]. Wynn considered the function

$$g(\xi) = \sum_{m=0}^{n} \frac{\Psi_0^{(m)} e^{-\Psi_2^{(m)}(\xi)}}{\Psi_1^{(m)}(\xi) \left(\prod_{\tau=1}^{s_m} \varphi_\nu^{(m)}(\xi) \right) \left(\prod_{\tau=1}^{l_m} \varrho_\tau^{(m)}(\xi) \right)} = \sum_{\nu=0}^{\infty} \frac{t_\nu}{\nu!} \xi^\nu,$$

where the functions appearing in this formula have to satisfy conditions that are too complicated to be reproduced here. He proved that $H_{r+1}(t_{2m}) > 0$ and $H_{r+1}(t_{2m+1}) < 0$ for $r, m = 0, 1, \ldots$, and added:

The motivation of the above theorem was as follows: Pólya proved his result by a rather long but elementary argument using a theorem of Laguerre upon the zeros of integral functions; I attempted to shorten the proof by recourse to the Bernstein–Widder theorem and in so doing was led, by way of Lemma 1, to a more general result concerning the functions [*technical notation*]. (It should perhaps be stated that the Bernstein–Widder theorem was discovered after the publication of Pólya's result).

In passing, Wynn also gave a negative answer to a question raised by Pólya in his paper. An application to continued fraction ends the paper. ◀

The Bernstein–Widder theorem states that a necessary and sufficient condition that f satisfies, for all $n \geq 0$, $(-1)^n f^{(n)}(t) \geq 0$, for all $t \in [0, +\infty[$ is that $f(t) = \int_0^\infty e^{-tx} \, d\sigma(x)$ with σ bounded and nondecreasing in this interval [57, 695].

∞

The following report [775] was published later as a paper [803]. They are analyzed together.

▶ *P. Wynn, Upon a class of functions connected with the approximate solution of operator equations, Report CRM-103, Centre de Recherches Mathématiques, Université de Montréal, Montréal, June 1971.* [775]
▶ *P. Wynn, Upon a class of functions connected with the approximate solution of operator equations, Ann. Mat. Pura Appl., 104 (1975) 1–29.* [803]

Wynn explains, in the introductions of both papers, that the function e^z is real for real values of z, that it maps the open left half-plane $\mathrm{Re}(z) < 0$ onto the interior of the unit circle, and that it generates a well determined corresponding continued fraction with convergents $q_r(z)/p_r(z)$ all of whose roots of p_{2r} are in the open right half-plane $\mathrm{Re}(z) > 0$. Wynn then proves that functions of the form

$$f(z) = 1 + \cfrac{az}{1 - az/2 + z^2 \displaystyle\int_0^\infty \frac{d\phi(t)}{1 + z^2 t}}, \tag{5.1}$$

where ϕ is bounded and nondecreasing, and such that all moments $c_v = \int_0^\infty t^v \, d\phi(t)$ exist, possess the same properties as the exponential function: they map the open half plane $\mathrm{Re}(z) < 0$ into the interior of the unit disk, they satisfy the equation $f(z)f(-z) = 1$ in their domain of definition, and generate a corresponding continued fraction whose even-order convergents satisfy the two preceding properties. Furthermore, a function f is real for real values of z and possesses the three preceding properties if and only if it has the representation given above. Some functions derived from this first one also have these properties. An example involving a ratio of functions $_1F_1$ with different arguments is given. Such functions play a part in the solution of operator equations and in their approximate solution. A subclass is studied. Wynn writes:

> The properties of the function e^z [...] play an important rôle in the theory of the approximate solution of partial differential equations and in the stability theory of electrical networks (indeed, it was a discussion with Professor I.J. Schoenberg concerning such matters which caused the author to carry out the investigations upon which the theory of this paper is based).

These functions, introduced by Isaac Jacob Schoenberg (1903–1990) in connection with smoothing operations, are characterized. Using the theory of functions of the above type, it is shown that if when $-1/2 < \alpha < \infty$, $_1F_1(\alpha; 2\alpha + 1; z) = 0$, then $\mathrm{Re}(z) > 0$. This quite theoretical and difficult paper is the product of a short stay in Zürich, where Wynn was invited by Eduard Stiefel. ◀

∞

In continuation of the study of the functions introduced by Schoenberg in [556, 557], we have the following paper.

▶ *P. Wynn, On the intersection of two classes of functions, Rev. Roumaine Math. Pures Appl., 19 (1974) 949–959. [797]*

In this paper, Wynn comes back to the functions studied in [775] and [803] (analyzed in this section). Functions f of the form

$$f(z) = ae^{\gamma z} \frac{\prod_{\nu=1}^{j}(1 + \alpha_\nu z)}{\prod_{\nu=1}^{i}(1 - \beta_\nu z)}, \qquad \alpha_\nu, \beta_\nu > 0,$$

were introduced by Schoenberg in connection with smoothing [556, 557]. The series expansions of these functions are often called *Pólya frequency series*. Wynn proves that such functions can be expressed in the form (5.1) (see the analysis of the preceding paper) if and only if $a = 1$, $i = j$, and $\alpha_\nu = \beta_\nu$ for $\nu = 1, \ldots, i$. Moreover, if C_{2r}, $r = 0, 1, \ldots$, are the successive convergents of the associated continued fraction generated by a nonrational such function, then for a fixed $n \geq 0$, there exists an infinite sequence of positive real numbers $(y_\nu^{(n)})$ for which $f(z) = C_{2r}(z) = \sigma$, $|\sigma| = 1$ when $z = \pm i y_\nu^{(n)}$, and $y_\nu^{(n)} < y_\nu^{(n+1)} < y_{\nu+1}^{(n)}$, for $\nu = 1, 2, \ldots$. ◀

Let us mention that the convergence of the Padé approximants of Pólya frequency series was extensively studied by Robert J. Arms and Albert Edrei (a student of George Pólya) in [21].

$$\infty$$

The next paper (submitted on 17 July 1977) was published as an extra issue of the journal *Calcolo*. It has 103 pages, and cannot be found on the web site of this journal.[9]

▶ *P. Wynn, The evaluation of singular and highly oscillatory integrals by use of the anti–derivative, Calcolo, 15, Fasc. IV bis, (1978) 1–103. [811]*

Since the paper is quite long, with many particular notations, let us only give its abstract.

Four methods of producing, from a function $\psi(\mu)$ all of those derivatives exist at the point μ, a sequence of functions converging under certain conditions to an anti-derivative $I(\psi, \mu)$ satisfying the equation $dI(\psi, \mu)/d\mu = -\psi(\mu)$, are described. Convergence results together with a priori error bounds are given. The anti–derivative finds its main application in the evaluation of singular integrals over a semi-infinite interval in terms of the derivatives of the integrand at the finite end point, and its use neither presupposes knowledge of the nature of the singularities of the integrand nor requires their exact location to be known. It is shown how the anti-derivative may be used to evaluate integrals over a finite range, and that its use for this purpose is in certain cases more effective than that of the Euler–Maclaurin formula. The latter formula and the anti-derivative are used in conjunction in a method for approximating the sum of an infinite series in terms of the derivatives of its first term. It is shown how the anti-derivative may be used to predict the limiting value of a function as its argument tends to infinity in terms of the values of its derivatives for a finite argument value. Numerical illustrations of the theory are provided throughout. ◀

---

[9] A copy of it was sent by Wynn to Naoki Osada, who offered it to us.

Let us mention that the anti-derivative of a function is nothing else than its primitive function equipped, by Wynn, with a minus sign. The integration methods used in this paper arose from the study of algorithms for predicting $\lim_{\mu \to \infty} S(\mu)$ in terms of the derivatives of the function S at a finite argument value. By a simple transformation of the variable, the point at infinity can be brought to any finite distance. This work is based on the results of [768, 793, 808], where Wynn introduced a number of methods for evaluating an integral over a semi-infinite interval. This means that Wynn is using the confluent forms of algorithms to estimate the limit of a sequence. The theoretical results of the paper are illustrated by examples for which the anti-derivative may be expressed in a closed form. Applications to the computation of improper integrals were given by Lonny Christopher Breaux [78], an M.Sc. student of Wynn at Louisiana State University in New Orleans in 1971.

It was shown that the first confluent form of the ε-algorithm allows one to recover the principal part of the asymptotic expansion of a semi-infinite integral [202, p. 96]. Indeed, let $f(t) = \int_a^t g(x)\,dx$. Applying the first confluent form of the ε-algorithm gives $\varepsilon_2(t) = f(t) - g^2(t)/g'(t)$, and we thus obtain $\int_t^\infty g(x)\,dx \sim g^2(t)/g'(t)$ [81, pp. 115–116].

∞

▶ *P. Wynn, The calculus of finite differences over certain systems of numbers, Calcolo, 14 (1977) 303–341. [809]*

The purpose of this paper is given by the abstract:

A recursive process of interpolation over functions whose arguments belong to certain systems of numbers is described. The process can, in particular, be applied to functions of many variables and, for the examples considered, is both more flexible and more powerful than either the use of many dimensional divided differences or multivariate Lagrange interpolation. Recursive processes of differentiation, integration, and confluent interpolation over functions whose arguments belong to certain further systems of numbers are developed from the interpolation procedure.

After recalling the known multidimensional extensions of divided differences, Newton's and Lagrange's interpolation formulas, and their use in approximate differentiation and integration, Wynn writes:

In this paper, what appears to be a new variant of the calculus of finite differences is embarked upon. A method of interpolation over a certain system of numbers is derived and then, after introducing suitable restrictions upon the system, processes of approximate differentiation, integration, and confluent interpolation are established. The methods can, in particular, be applied to functions of many variables and, concerning their implementation in this case, we remark that it is not required that the function concerned should be evaluated at all intersections of a rectangular grid; furthermore, the constructed sets of approximating functions are extended after each function evaluation; it is not required that an additional group of further function values should be made available at each stage.

The paper is too technical to be described here in detail. For interpolation in two variables, Wynn begins by extending the Neville–Aitken scheme to dimension two by means of the scalar product of vectors in $\mathbb{R}^2$. The case of an arbitrary dimension is then considered. An extension of Lagrange's interpolation formula is also proposed.

An application to Picard iterates for solving $x = \phi(x)$ in $\mathbb{R}^n$ is given. The next sections deal with inverse interpolation, differentiation, and multiple integration. Wynn also studies the confluent case in which all points in the Neville–Aitken scheme tend to a finite value. Further processes of interpolation, differentiation, and integration are developed. They are based on the former functions, which

> [...] can be made to serve as the basis of an independent method of interpolation which can in turn be developed for the purposes of approximate differentiation and integration.

Remarks about programming are given. In his conclusion, Wynn writes:

> In providing a certain amount of theory and giving illustrations which exhibit the algorithms described in this paper in a favorable light, the author has conformed to that behaviour expected of inventors of numerical methods. For his part, the reader may by now be persuaded to the view that these algorithms open the way to further flexible and powerful techniques for the treatment of problems in numerical analysis; and this may even be true. However, the methods of this paper are not without attendant disadvantages, and a useful purpose is served by describing some of them.

These disadvantages are discussed at length, and Wynn shows that he is fully aware of the shortcomings of his work:

> Lastly, it is eminently to be desired that a newly discovered theory should exhibit a number of strong connections with existing ones. In the present context, it would greatly facilitate the theoretical investigations needed to set the numerical methods of this paper upon a firm base, if it could be shown that the expressions occurring in the formulæ involved were natural extensions of, for example, Bernoulli, Euler, and Bernstein polynomials. If this were true, convergence results from classical finite difference theory could be appropriated wholesale to the new theory. In fact, the methods of this paper are peculiarly lacking in desirable connections; in particular, there are no polynomials. ◄

5.9 Translations

In 1963, Wynn translated into English two books on continued fractions written in Russian.

► A. Ya. Khinchin, Continued Fractions, Translated from Russian by Peter Wynn, P. Noordhoff N.V., Groningen, 1963. [815]

► A.N. Khovanskii, The Application of Continued Fractions and Their Generalizations to Problems in Approximation Theory, Translated from Russian by Peter Wynn, P. Noordhoff N.V., Groningen, 1963. [816]

The first one is a book (first Russian edition published in 1935) by Aleksandr Yakovlevich Khinchin (1894–1959). It has 95 pages. The author was one of the most significant contributors to the Soviet school of probability theory. The book contains three chapters: properties of the apparatus, the representation of numbers by continued fractions, the measure theory of continued fractions.

The second translation concerns the book of Alexey Nikolaevitch Khovanskii (1916–1996) . It has 212 pages. The original Russian edition dates to 1956. In its preface, Wynn writes:

> The decision to translate the following book was taken, firstly because it contains a considerable amount of new material relating to the numerical application of continued fractions which will be of interest to the Western reader, and secondly because it offers an introduction to the analytic theory of continued fractions which, in the reasoned and systematic form given, is not available in the English language.

The four chapters of the book are: certain problems in the theory of continued fractions, continued fraction expansions of certain functions, further methods for obtaining rational function approximations, generalized continued fractions. The book originally contained 17 references, but Wynn added a bibliography of 109 works and 10 supplementary references.

5.10 Wynn's Legacy

When we were in the last stage of preparation of this book, new documents written by Peter Wynn came into our hands. Let us explain how they were discovered. They will be described in a forthcoming publication [122].

On 14 January 2020, C.B. received a message from Francis Alexander (Sandy) Norman III, an associate professor of mathematics and associate chair at the University of Texas at San Antonio, Department of Mathematics, a specialist in mathematical education with a Ph.D. from the University of Georgia. After reading the remembrance of Peter Wynn that appeared in [115], he informed us that he [. . .] *"inherited" from a retired colleague several rather heavy boxes of Wynn's papers which had been left with my colleague for safekeeping (or convenience) some years earlier* [. . .], and he asked us if we were interested in their contents.

The colleague Sandy was speaking about is Manuel Philip Berriozábal, a professor at the University of Texas at San Antonio (UTSA) since 1976. When Wynn was living in Mexico, he was obliged to return each year to the U.S. to renew some official documents. It was during his professorship at the Louisiana State University in New Orleans that Manuel Berriozábal met Peter Wynn and became a close friend. Wynn continued to visit him and his wife, Maria Antonietta, after they moved to San Antonio, and sometimes, he brought documents, asking them to keep them for him.

Of course, we answered Sandy that we were very much interested in the contents of these four boxes, and he began to look at them, to sort them, to scan a lot of material, and to send it to us. Almost all the documents are handwritten. There are lists of things to do, of references, of projects, and so on. But most of them are drafts of mathematical papers and/or parts of books. Some are easy to analyze, since they are in quite a good shape, almost ready to be published, but many of them contain only preparatory calculations on unnumbered pages, full of notation difficult to grasp (Wynn was fond of introducing his own notation). The topics covered run from Padé approximation, rational interpolation and approximation, the ε-algorithm

Francis Alexander Norman.
© Francis Alexander Norman.

and other algorithms, to pure algebraic developments whose aims are not explained. These documents have to be carefully analyzed to determine whether the material they contain is new, valuable, and can open the way to new research. We plan to try to give separately an analysis of these documents [122].

Maria Antonietta and Manuel Philip Berriozábal.
© M.A. and M.P. Berriozábal.

Chapter 6
Commentaries and Further Developments

In this chapter, we comment on and discuss some of the most important further developments obtained in the domains considered. More extensions are presented by the authors of the testimonies reproduced in Chap. 7.

6.1 Summation Methods

Although it is not our purpose to develop here extensions of summation methods, let us give only one example to show that such methods are related to optimization and best approximation.

We consider an absolutely convergent series of positive terms $u = \sum_{i=0}^{\infty} u_i$ in l_1, the set of sequences such that $\|u\|_1 = \sum_{i=0}^{\infty} |u_i| < +\infty$, and we look for an approximation v_N of u of the form $v_N = \sum_{i=0}^{N} A_i u_{n_i}$. The following two problems were considered and solved by Jacques Baranger (b. 1940) [32, 33]: the n_i being given, find the A_i minimizing $\|u - v_N\|_1$, and secondly, find the n_i and the A_i minimizing the same norm. In the subspace H_a of l_1 equipped with the scalar product $(u, v) = \sum_{i=0}^{\infty} \Delta u_i \Delta v_i / a^i$, $0 < a < 1$, the solution of the first problem is given by

$$A_0 = C_0 + 1/(1-a),$$
$$A_i = C_{i+1} - C_i, \qquad i = 1, \ldots, N-1,$$
$$A_N = n_N - C_{N-1},$$

where

$$C_i = \frac{n_i a^{n_i} - n_{i+1} a^{n_{i+1}}}{a^{n_i} - a^{n_{i+1}}}, \qquad i = 0, \ldots, N.$$

In particular, for $n_i = i$, the solution of the first problem is $A_0 = A_1 = \cdots = A_{N-1} = 1$, $A_N = 1/(1-a)$, and the l_1-norm of the error is equal to $a^{N+1}/(1-a)^3$. A sufficient condition for the sequence (v_N) to converge faster than the sequence (s_n)

© Springer Nature Switzerland AG 2020
C. Brezinski, M. Redivo-Zaglia, *Extrapolation and Rational Approximation*,
https://doi.org/10.1007/978-3-030-58418-4_6

of the partial sums of the series u is that $\lim_{n\to\infty} u_{n+1}/u_n = a$. Replacing a by its approximation $\Delta s_n/\Delta s_{n-1}$ allows one to recover the Aitken Δ^2 process.

For the second problem, the n_i are those that maximize

$$\Phi_N = \frac{1}{1-a}\left[a^{n_0}\left(n_0 + \frac{1}{1-a}\right)^2 + \sum_{i=0}^{N-1} \frac{(n_{i+1} - n_i)^2}{a^{-(n_{i+1}-n_i)} - 1}\right],$$

and the A_i are given by the same relations as for the first problem.

Let us mention that the Euler's transformation was studied by several researchers [254, 255, 256, 407, 466, 467], some of them mentioning the papers of Wynn [753, 773, 787].

6.2 Richardson's Extrapolation Method

In his papers [520, 523], Richardson eliminated the terms in h^2 and h^4 in a power series expansion of the error in even powers of h. It seems that Richardson's method, or at least a similar idea, rapidly entered into the toolbox of numerical analysts. Obviously, it was not possible to go through all the old journals to find references, and thus we limited ourselves to some of them.

A process called the ρ^r *correction process* for eliminating an error term of the form $\rho^r h^r C_r$ was proposed by Gertrude Blanch (1897–1996), an American mathematician, born in Russia, in a paper dealing with the numerical solution of parabolic partial differential equations by finite differences [64]. In 1964, she received the Federal Woman's Award for her pioneering work in numerical analysis and computation. Wynn dedicated his paper [764] to her in an anniversary volume in 1967.

Linear and quadratic extrapolation was used in [68] in the solution of Schrödinger's equation in one dimension by a method using central differences.

Richardson's deferred approach to the limit (h^2 extrapolation) was taken up by Michael Robert Osborne (b. 1934) in 1960 (submitted 23 April 1959) for a difference approximation to a differential equation eigenvalue problem [477]. The process was extended by Mario Giorgio Salvadori (1907–1997), a well-known Italian-born structural and civil engineer [549]. He was concerned with problems in which the approximation was a purely even function of h having more than a single term. He tabulated 3-point coefficients for (h^2, h^4)- and (h^4, h^6)-extrapolation, and 4-point coefficients for (h^2, h^4, h^6)- and (h^4, h^6, h^8)-extrapolation. He gave applications to differentiation, integration, boundary value problems, and eigenvalues of a differential operator, and concluded that: *It is well to remember that the extrapolations may "over-shoot" the mark and that it is not known in general, whether they are approximations from above or from below.*

Richardson's idea was generalized by Pierre-Jean Laurent in his doctoral thesis [395]. Let v be a function such that

$$v(x) = v(0) + xv'(0) + \frac{x^2}{2!}v''(0) + \cdots + \frac{x^{n-1}}{(n-1)!}v^{(n-1)}(0) + x^n S(x),$$

where S is bounded in a given real interval. Laurent started from a strictly decreasing sequence (h_k) of positive abcissas tending to zero as k goes to infinity, and he wanted to build an extrapolation process of the form

$$L_n(v) = \sum_{k=1}^{n} A_k^n v(h_k)$$

for obtaining an approximate value of $v(0)$. He chose the coefficients A_k^n such that L_n is exact for any polynomial P_{n-1} of degree $n - 1$, that is,

$$L_n(P_{n-1}) = P_{n-1}(0).$$

Thus the process consisted in interpolating $v(h_1), \ldots, v(h_n)$ by a polynomial P_{n-1} of degree $n - 1$ and then extrapolating its value at $x = 0$. The coefficients A_k^n are given by Lagrange's formula for an interpolation polynomial:

$$A_k^n = \prod_{\substack{j=1 \\ j \neq k}}^{n} \frac{1}{1 - h_k/h_j}, \qquad k = 1, \ldots, n.$$

They can be recursively obtained by the relation

$$A_k^{n+1} = \frac{h_{n+1}}{h_{n+1} - h_k} A_k^n,$$

which comes directly from the expression of the coefficients A_k^n.

Then Laurent proposed another way of obtaining $P_{n-1}(0)$. It consists in computing recursively the value at $x = 0$ of the interpolation polynomials by the Neville–Aitken formula. This led to Richardson's extrapolation method as given by formula (2.5). Newton's formula using divided differences can also be used. The convergence and the stability of the process are studied (see also [394]). Applications to numerical integration in one and two dimensions, to approximate derivation, to the integration of differential, partial differential, and integral equations are studied. If (2.5) is applied to the trapezoidal rule for computing a definite integral with the step sizes $h_k = h_0/2^k$, then Romberg's method is recovered [535]. Later, he extended this result to a linear combination of functions forming a Chebyshev system [396, pp. 312–315].

The case of an asymptotic expansion of the error of the form $a_1 h^{p_1} + a_2 h^{p_2} + \cdots$ in a discretization process was considered by Hans Jörg Stetter (b. 1930) in 1965 [618]. The idea was pursued by Victor Pereyra, and called the *deferred corrections* method [489, 491, 492]. In [619], Stetter exposed *the common structural principle of all these techniques and* [exhibit] *the principal modes of its implementation in a discretization context*, and he called it the *defect correction* principle. This principle was explicitly given in [490] under the form of a recursive algorithm, called the *method of successive extrapolations for functional equations*, which is very much in the style of the general extrapolation algorithm described below.

The idea of deferred corrections was applied to a number of problems even recently; see, for example, [191, 192, 193, 237, 238, 249, 296, 366, 703, 704] and the references given in these papers.

The case of a kernel whose sequences have the form (our notation) $S_n = S + a_1 x_n + \cdots + a_k x_{n+k-1}$, where (x_n) is a known sequence, was considered in [289] together with its confluent form. It is the G-transformation, which can be viewed as a generalization of Shanks's, which is recovered for the choice $x_n = \Delta S_n$, and which thus can be written as a quite similar ratio of determinants. A recursive algorithm for its implementation can be found in [514]. It can be related to the ε-algorithm if $x_n = \Delta S_n$.

As we saw in Chap. 2, the quantities involved in Richardson's extrapolation and the Shanks transformation are expressed as a ratio of determinants and, in both cases, an algorithm allows one to compute them recursively. A similar situation arises for many other sequence transformations, where in each case, the quantities involved are expressed as a ratio of determinants. Let us mention that a theoretical study of such rational sequence transformations was already conducted by Renato Pennacchi (b. 1926) in 1968 [487, 488].

Let us say a few words about Renato Pennacchi. After his doctoral degree in mathematics and some years spent at Rome University, he joined IBM, where he became manager of the IBM Pisa Scientific Center in 1967. His major professional and research interest have been in system theory and numerical analysis. The first four IBM scientific centers, as autonomous organizations, were established in 1964, and their success was extended all over the world. These centers contributed greatly, in collaboration also with university professors and students, to the development of scientific solutions and applications for important projects in an era of great changes in technology. In 1969, under his impulse, IBM decided to extend the activity of the scientific centers in Italy, and he was able to create two other centers (Bari and Venice), becoming responsible for the whole Italian network. In 1979, the last two centers were absorbed by the IBM Rome center, and due to a change in management, Pennacchi took a position at the *Centro Software Applicativo di Roma*, where he created a group of advanced technologies ("Ad Tech") whose aim was to study the evolution of the software products managed by the laboratory through implementation of prototypes, leveraging leading edge technologies and approaches. In these prototypes, behind which there was his vision and his intuitions, we can see ideas subsequently implemented in products successfully deployed worldwide. For instance, one of these products was used by NASA, and its graphical interface, representing the flux of data of an application, appeared on the cover of one issue of the NASA magazine. Let us give a quotation he loved to repeat (translation from Italian):[1]

> When you can't solve a problem, stand upside down, even physically, if you can: you will see it from another angle and you will find the solution more easily.

---

[1] *Quando non riesci a risolvere un problema, mettiti a testa in giù, se puoi anche fisicamente: lo vedrai da un'altra angolazione e ne troverai più facilmente la soluzione.*

As we saw, for each transformation expressed by a mathematical formula (ε-algorithm, ϱ-algorithm, Richardson's method, etc.), a special recursive algorithm for its implementation had to be found, if possible. Thus, there was a real need for a general theory of such sequence transformations and for a single recursive *general extrapolation algorithm* for their implementation. This work was performed independently between 1973 and 1980 by five different researchers; see [109].

It seems that the first appearance of such a general algorithm is due to Claus Schneider in a paper received on 21 December 1973 [555]. Let S be a function depending on a parameter h and such that $\lim_{h \to \infty} S(h) = S$. The approximations $S(h_i)$ being given for $i = 0, 1, \ldots$, with $h_0 > h_1 > \cdots > 0$, Schneider looked for a function $S'(h) = S' + a_1 g_1(h) + \cdots + a_k g_k(h)$ satisfying the interpolation conditions $S'(h_i) = S(h_i)$ for $i = n, \ldots, n+k$, where the g_j are given functions of h. Obviously, the value of the unknown S' obtained depends on k and n. Assuming that for all j, $g_j(0) = 0$, we have $S' = S'(0)$. Denoting by ϕ_k^n the extrapolation functional on the space of functions f defined at the points h_i and at the point 0, and satisfying $\phi_k^n f = f(0)$, we have

$$\phi_k^n S' = c_0 S(h_n) + \cdots + c_k S(h_{n+k})$$

with $c_0 + \cdots + c_k = 1$. The interpolation conditions become

$$\phi_k^n E = 1,$$
$$\text{and } \phi_k^n g_j = 0, \qquad j = 1, \ldots, k,$$

where E is defined by $E(h) \equiv 1$. Schneider wanted to express the functional ϕ_k^n in the form $\phi_k^n = a\phi_{k-1}^n + b\phi_{k-1}^{n+1}$, which leads to the two conditions

$$\phi_k^n E = a + b = 1$$

and

$$\phi_k^n g_k = a\phi_{k-1}^n g_k + b\phi_{k-1}^{n+1} g_k = 0.$$

The values of a and b follow immediately, and he obtained

$$\phi_k^n = \frac{[\phi_{k-1}^{n+1} g_k]\phi_{k-1}^n - [\phi_{k-1}^n g_k]\phi_{k-1}^{n+1}}{[\phi_{k-1}^{n+1} g_k] - [\phi_{k-1}^n g_k]}.$$

Thus the quantities $\phi_k^n S'$ can be recursively computed by this scheme with the initializations $\phi_0^n S' = S(h_n)$. The auxiliary quantities $\phi_k^n g_j$ appearing in this formula can be computed recursively by the same scheme with $\phi_0^n g_j = g_j(h_n)$. As we will see below, this scheme is exactly the algorithm called the E-algorithm in [95] and the B-H protocol in [699]. In a footnote, Schneider mentioned that this representation for ϕ_k^n was suggested by Professor Wolfgang Börsch-Supan, his advisor at the Johannes Gutenberg Universität in Mainz.

In 1976, Günter Meinardus (1926–2007) and Gerald D. Taylor considered a quite different problem in a paper on best uniform approximation from a linear subspace of functions [445]. Let $a \le h_1 < h_2 < \cdots \le b$, $M_{n,k} = \{h_n, \ldots, h_{n+k}\}$, and assume that for all k, $V_k = \mathrm{span}(g_1, \ldots, g_k)$ satisfies the Haar condition.

Using the standard theory of Haar subspaces, a linear functional L_n^k on $C[a, b]$ based on $M_{n,k}$ can be defined by

$$L_n^k(f) = \sum_{i=n}^{n+k} c_i f(h_i),$$

with, for consistency, $L_n^0(f) = f(h_n)$ for all n, and where the coefficients c_i, which depend on n and k, are such that $c_n > 0$, $c_i \ne 0$ for $i = n, \ldots, n+k$, $\mathrm{sgn}(c_i) = (-1)^{i-n}$, and

$$\begin{cases} \displaystyle\sum_{i=n}^{n+k} |c_i| = 1, \\[2em] \displaystyle\sum_{i=n}^{n+k} c_i g_j(h_i) = 0, \qquad j = 1, \ldots, k. \end{cases}$$

Moreover, $|L_n^k(f)| = \inf_{g \in V_k} \max_{h \in M_{n,k}} |f(h) - g(h)|$.

Then, using Chebyshev's alternation theorem and Gaussian elimination, they proved that there exists a unique decomposition of the linear functional $L_n^k(f)$ of the form

$$L_n^k(f) = \sum_{j=n}^{n+k-r} \lambda_j L_j^r(f), \quad r = 0, \ldots, k,$$

where the real numbers λ_j, depending on n, k, and r, are all different from zero, with $\mathrm{sgn}(\lambda_j) = (-1)^{j+r}$, and the sum of their absolute values equals 1. For $r = k - 1$, they obtained the recursive scheme

$$L_i^k(f) = \frac{L_{i+1}^{k-1}(g_k) L_i^{k-1}(f) - L_i^{k-1}(g_k) L_{i+1}^{k-1}(f)}{L_{i+1}^{k-1}(g_k) - L_i^{k-1}(g_k)}$$

with $L_i^0(f) = f(h_i)$, $i = n, \ldots, n + k$ (notice that there are two typing errors in their formula: they wrote $L_{i+1}^k(f)$ in the numerator instead of $L_{i+1}^{k-1}(f)$, and the sign is minus instead of plus in the denominator). This is the same scheme as that of Schneider.

Newton's formula for computing the interpolation polynomial is well known. It is based on divided differences. These formulas can be generalized to interpolation by a linear combination of functions forming a complete Chebyshev system. We seek

$$P_k^{(n)}(x) = a_0 g_0(x) + \cdots + a_k g_k(x)$$

satisfying the interpolation conditions

$$P_k^{(n)}(x_i) = f(x_i), \qquad i = n, \ldots, n+k,$$

where the x_i are distinct points and the g_i are given arbitrary functions. The $P_k^{(n)}$ can be computed recursively by an algorithm that generalizes the Neville–Aitken scheme for polynomial interpolation. Such an algorithm was obtained by Günter Mühlbach (b. 1941) in 1976 [458], and it was called the Mühlbach–Neville–Aitken algorithm (MNA for short). Its rule is

$$P_k^{(n)}(x) = \frac{g_{k-1,k}^{(n+1)}(x)P_{k-1}^{(n)}(x) - g_{k-1,k}^{(n)}(x)P_{k-1}^{(n+1)}(x)}{g_{k-1,k}^{(n+1)}(x) - g_{k-1,k}^{(n)}(x)}$$

with $P_0^{(n)}(x) = f(x_n)g_0(x)/g_0(x_n)$. The $g_{k,i}^{(n)}$ are computed recursively by a similar recurrence relation,

$$g_{k,i}^{(n)}(x) = \frac{g_{k-1,k}^{(n+1)}(x)g_{k-1,i}^{(n)}(x) - g_{k-1,k}^{(n)}(x)g_{k-1,i}^{(n+1)}(x)}{g_{k-1,k}^{(n+1)}(x) - g_{k-1,k}^{(n)}(x)},$$

with $g_{0,i}^{(n)}(x) = g_i(x_n)g_0(x)/g_0(x_n) - g_i(x)$. If $g_0(x) \equiv 1$, and if for all $i > 0$, $g_i(0) = 0$, the quantities $P_k^{(n)}(0)$ are the same as those obtained by the E-algorithm given below, and the MNA reduces to it. Let us mention that, in fact, the MNA is closely related to the work of Henri Marie Andoyer (1862–1929) that goes back to 1906 [17]; see [106] for detailed explanations.

Let us now come to the work of Tore Håvie (1930–2005). We already mentioned Romberg's method for accelerating the convergence of the trapezoidal rule. It is based on the existence of the Euler–Maclaurin expansion for the error. This expansion holds only if the integrand has no singularity in the interval of integration. In the presence of singularities, the expansion of the error is no longer a series in h^2 (the step size) but a more complicated one depending on the singularity. Thus, Romberg's scheme has to be modified to incorporate the various terms appearing in the expansion of the error. Several authors worked on this question, treating several types of singularities; see, for example, the paper of Leslie Fox (1918–1992) published in 1967 [248]. Tore Håvie began to study this question under Romberg (Romberg emigrated to Norway and came to Trondheim in 1949, where Håvie was working), and he published several papers on modifications of Romberg's method in which the integrand presents various types of singularities. In 1978, Håvie wrote a report, published 1 year later [309], in which he treated the most general case of an error expansion of the form

$$S(h) - S = a_1 g_1(h) + a_2 g_2(h) + \cdots,$$

where $S(h)$ denotes the approximation of the definite integral S obtained by the trapezoidal rule with a step size h, and the g_i are the known functions (forming an asymptotic sequence as h tends to zero) appearing in the expansion of the error. Let

$h_0 > h_1 > \cdots > 0$, $S_n = S(h_n)$, and $g_i(n) = g_i(h_n)$. Håvie set

$$E_1^{(n)} = \frac{g_1(n + 1)S_n - g_1(n)S_{n+1}}{g_1(n + 1) - g_1(n)}.$$

Replacing S_n and S_{n+1} by their expansions, he obtained

$$E_1^{(n)} = S + a_2 g_{1,2}^{(n)} + a_3 g_{1,3}^{(n)} + \cdots$$

with

$$g_{1,i}^{(n)} = \frac{g_1(n + 1)g_i(n) - g_1(n)g_i(n + 1)}{g_1(n + 1) - g_1(n)}.$$

The same process can be repeated for eliminating $g_{1,2}^{(n)}$ in the the expansion of $E_1^{(n)}$, and so on. Håvie gave an interpretation of this algorithm in terms of Gaussian elimination for solving the system

$$E_k^{(n)} + b_1 g_1(n + i) + \cdots + b_k g_k(n + i) = S_{n+i}, \qquad i = 0, \ldots, k,$$

for the unknown $E_k^{(n)}$.

In 1980, one of us (C.B.) took up the same problem, but from the point of view of extrapolation [95]. Let (S_n) be the sequence to be accelerated. Interpolating it by a sequence of the form $S_n' = S + a_1 g_1(n) + \cdots + a_k g_k(n)$, where the g_i are known sequences that can depend on the sequence (S_n) itself, leads to

$$S_{n+i} = S_{n+i}', \qquad i = 0, \ldots, k.$$

Solving this system directly for the unknown S (which, since it depends on n and k, is denoted by $E_k^{(n)}$) gives

$$E_k^{(n)} = \begin{vmatrix} S_n & \cdots & S_{n+k} \\ g_1(n) & \cdots & g_1(n + k) \\ \vdots & & \vdots \\ g_k(n) & \cdots & g_k(n + k) \end{vmatrix} \Big/ \begin{vmatrix} 1 & \cdots & 1 \\ g_1(n) & \cdots & g_1(n + k) \\ \vdots & & \vdots \\ g_k(n) & \cdots & g_k(n + k) \end{vmatrix}.$$

We see that $E_k^{(n)}$ is given as a ratio of determinants whose form is quite similar to the ratio involved in Richardson's extrapolation and in the Shanks transformation. Indeed, for the choice $g_i(n) = \Delta S_{n+i}$, the ratio appearing in the Shanks transformation results, while when $g_i(n) = x_n^i$, we obtain the ratio expressing the quantities involved in Richardson's process. Other algorithms may be similarly included in this framework.

The problem was to find a recursive algorithm for computing the $E_k^{(n)}$. Applying Sylvester's determinantal identity separately to the numerator and the denominator of the preceding determinantal expression, the author obtained the two rules of the so-called E-algorithm,

$$E_k^{(n)} = \frac{g_{k-1,k}^{(n+1)} E_{k-1}^{(n)} - g_{k-1,k}^{(n)} E_{k-1}^{(n+1)}}{g_{k-1,k}^{(n+1)} - g_{k-1,k}^{(n)}},$$

with $E_0^{(n)} = S_n$.

The auxiliary quantities $g_{k,i}^{(n)}$ are computed recursively by the quite similar rule

$$g_{k,i}^{(n)} = \frac{g_{k-1,k}^{(n+1)} g_{k-1,i}^{(n)} - g_{k-1,k}^{(n)} g_{k-1,i}^{(n+1)}}{g_{k-1,k}^{(n+1)} - g_{k-1,k}^{(n)}}$$

with $g_{0,i}^{(n)} = g_i(n)$.

This derivation is closely related to Håvie's, since Sylvester's identity can be proved using Gaussian elimination. The E-algorithm is, in fact, a particular case of the MNA; see [310] for a unified treatment. A new derivation of this algorithm, via annihilation operators and error estimates, was proposed in [124]. The algorithm was extended to the vector case [95].

The convergence and acceleration results proved for this algorithm show that for accelerating the convergence of a sequence, it is necessary to know the expansion of the error $S_n - S$ with respect to some asymptotic sequence $(g_1(n)), (g_2(n)), \ldots$, and it can be proved, under certain assumptions, that $\forall k$,

$$\lim_{n \to \infty} \frac{E_{k+1}^{(n)} - S}{E_k^{(n)} - S} = 0.$$

These results were refined by Avram Sidi [582, 584, 586]. Other acceleration results for this algorithm were obtained by Ana Cristina Matos and Marc Prévost (b. 1952) [441], Prévost [506], and Pascal Mortreux and Prévost [457]. These results show that the study of the asymptotic expansion of the error of the sequences to be accelerated is fundamental, see the book of Guido Walz [686]. A more economical algorithm than the E-algorithm was given by William Frank Ford (b. 1934) and Avram Sidi in [246]. The connection between the E-algorithm and the ε-algorithm was studied by Berhnard Beckermann (b. 1961) [45]. A general ε-algorithm connected to the E-algorithm was obtained by Carsten Carstensen [163]. See [100] for a more detailed review on the E-algorithm. Another quite general sequence transformation was introduced in [334, 335]. A complete theory of triangular computational schemes, with many applications, was given in [144].

Convergence acceleration algorithms can also be used for predicting the unknown terms of a series or sequence. This idea, introduced by Jacek Gilewicz (1937–2016) [270], was studied by Sidi and Levin [592], C.B. [97], and Denis Vekemans [674].

6.3 The Aitken Process

The Aitken Δ^2 process was widely used, commented, and extended. Many modifications and extensions of it can be found in the literature, and it is impossible to mention all of them here.

In a paper dated 1955 [527], James D. Riley, a colleague of Shanks at the U.S. Naval Ordnance Laboratory, proposed an iterative method for the regularization of a system of linear equations with a positive definite, symmetric, but possibly ill-conditioned matrix. Instead of solving $Ax = b$, he considered the system $(A + kI)y = b$, where k is a small positive constant. Setting $C = A + kI$, he has $A = C - kI$, thus $A^{-1} = C^{-1} + kC^{-2} + k^2C^{-3} + \cdots$, and it follows that $x = A^{-1}b = y + kC^{-1}y + (kC^{-1})^2 y + \cdots$. Starting from an approximation x_0 of x, he solved $Cz_0 = b - Ax_0$, and set $x_1 = x_0 + z_0$. Therefore $x - x_1 = kC^{-1}(x - x_0)$, and since the eigenvalues of kC^{-1} are less than one, x_1 is a better approximation of x than x_0. The process can be repeated, and he concluded that

> For accelerating the convergence of the sequence (x_i), the δ^2-process of Aitken and its extension by Shanks are especially appropriate. (See Forsythe [247], p. 309–310 [*our bibliography*]).

He mentioned that according to a memorandum by Harry Polachek [498], *an approximate solution can be improved without limit* since a significant inverse of C is available (however, Polachek took $C = A^{-1}$). No numerical results are given.

When solving a fixed-point problem by iterations of the form $x_{n+1} = f(x_n)$, applying the Aitken process to the sequence (x_n) leads to a second-order approximation of the fixed point x, since the coefficient of the first term of the error has been approximated. This is why the sequence produced by the Aitken process converges faster to x than (x_n). Developing the error of this new sequence and approximating the coefficient in the first term of its error gives a third-order approximation of the fixed point. The process can be iterated again, thus leading to an extension of the process proposed by Kjell Jørgen Overholt (1925–2016) [480].

Transformations that can be considered as an extension of the Aitken process were presented in 1973 by David Levin[2] [408]. Recall that the kernel of the Aitken process is the set of sequences of the form $S_n - S = a\Delta S_n$, Δ being the usual forward difference operator, while that of the transformation E_1, the first column of the E-algorithm, described in the preceding section, has a kernel with sequences of the form $S_n - S = ag(n)$, where $(g(n))$ is a known sequence. In his paper, Levin considered a kernel of the form

$$S_n - S = (a_1 + a_2(n + b)^{-1} + \cdots + a_k(n + b)^{-(k-1)})g(n), \quad n = 0, 1, \ldots.$$

Obviously, the corresponding transformation can be implemented via the E-algorithm with the choice $g_i(n) = (n + b)^{-(i-1)}g(n)$, but the algorithm is not optimal. On multiplying both sides of the preceding relation by $(n+b)^{k-1}/g(n)$, the right-hand

---

[2] See his testimony in Sect. 7.8.

side becomes a polynomial of degree $k - 1$ in n. Thus applying to it the operator Δ^k gives 0, and Levin's transform is then defined by

$$L_k^{(n)} = \frac{\Delta^k\big(S_n(n + b)^{k-1}/g(n)\big)}{\Delta^k\big((n + b)^{k-1}/g(n)\big)}.$$

According to the choice of the auxiliary sequence $(g(n))$, the three transformations of Levin are recovered. His u-transform corresponds to $g(n) = (n + b)\Delta S_{n-1}$, the choice $g(n) = \Delta S_{n-1}$ leads to his t-transform, and his v-transform is obtained with $g(n) = -\Delta S_{n-1}\Delta S_n/\Delta^2 S_n$.

These transforms are quite powerful, as experimentally showed by Smith and Ford in [600]. Many other choices are also possible, for instance $g(n) = \Delta S_n$, which provides the best simple remainder estimate for a sequence with strictly alternating differences ΔS_n [599]. Using the expression for the operator Δ^k and setting $L_k^{(n)} = N_k^{(n)}/D_k^{(n)}$, a recursive algorithm for its numerators and denominators has been found [233]:

$$N_{k+1}^{(n)} = N_k^{(n+1)} - \frac{(n + b)(n + k + b)^{k-1}}{(n + k + b + 1)^k} N_k^{(n)},$$

$$D_{k+1}^{(n)} = D_k^{(n+1)} - \frac{(n + b)(n + k + b)^{k-1}}{(n + k + b + 1)^k} D_k^{(n)},$$

with $N_0^{(n)} = S_n/g(n)$ and $D_0^{(n)} = 1/g(n)$. The corresponding subroutine is given in [234]. When $b = 1$, the asymptotic behavior of $L_k^{(n)}$ was studied by Sidi [582]. Levin's transforms can be generalized by replacing $(n + b)^{-i}$ in their kernel by x_n^{-i}, where (x_n) is a given sequence, and the operator Δ^k by the divided differences operator δ^k. Thus a slight generalization of the so-called *generalized Richardson extrapolation process* due to Sidi [579] is obtained. If $x_n = n$, the method reduces to Drummond's formula [213]. This transformation can be written in the form $D_k^{(n)} = \Delta^k(S_n/\Delta S_n)/\Delta^k(1/\Delta S_n)$, where now the D stands for Drummond (see [126] for a study). For $k = 1$, we recognize the Aitken process. If $k = 2$, we obtain the u_2-transform of Levin or the W transformation of Lubkin [423].

Ernst Joachim Weniger (b. 1949)[3] gave other results on these transformations, and derived new ones in [693]. A Levin-type transformation tailored for Fourier series was obtained by Herbert Hans Heinrich Homeier (b. 1957) [333], who gave a recursive algorithm for its implementation in [334]. This author published several other papers on this transform; they can be found on the internet. Other transformations were given by Levin and Sidi in [409].

Several other researchers proposed extensions of the Aitken process. A generalization for a kernel of the form $S_n = S + n^{-k}(c_0 + c_1 n^{-1} + c_2 n^{-2} + \cdots)$ was given in 1981 [62]. The case of a kernel given by $S_n = S + (a + bx_n)\lambda^n$, where a, b, and λ are unknown and (x_n) is a known sequence, was considered in [125] for the scalar and vector cases. Finally, let us mention still another extension for sequences of the form

---

[3] See his testimony in Sect. 7.11.

$S_n = S + a_n \lambda^n$, where the sequence (a_n) and the parameter λ are unknown [155]. Miscellaneous algorithms related to the Aitken process are reviewed in [123, pp. 131–143].

As shown in [692], transformations can be obtained by applying an *annihilation operator* to an estimate of the error. This is an important idea, which proved to be quite fruitful, and was, in fact, implicitly used in the derivation of Levin's transforms. The testimonies of David Levin, Avram Sidi, and Ernst Joachim Weniger are given in Chap. 7.

6.4 The ε-Algorithm and the Shanks Transformation

Let us now review results on, and extensions of, the scalar and the vector ε-algorithm, and those related to the Shanks transformation.

Several applications of the ε-algorithm can be found in different fields. Let us mention some of them. In [425], James Ross Macdonald (b. 1923), a physicist who was instrumental in building up the Central Research Laboratories of Texas Instruments, discussed applications of the ε-algorithm to the acceleration of slowly converging series and iterations, to divergent series, to the limit of iterated vector and matrix sequences, to the solution of differential and integral equations, to a new way of carrying out numerical integration, to extrapolation, and to the fitting of a curve to a polynomial or to a constant plus a sum of exponentials. Another application concerns fluid mechanics, more precisely the acceleration of Goldstein's expansion of the Oseen drag of a sphere in powers of the Reynolds number, was proposed by Milton van Dyke (1922–2010) [657].

Extensions of Wynn's results on the convergence of the scalar ε-algorithm were given in [83, 84, 88, 91] and in [585, 588].

The Shanks transformation, the ε-algorithm, and also Padé approximants have been used in various applications. An important domain in which extrapolation methods, in particular the ε- and the E-algorithms and Levin's transforms, were used is numerical quadrature, for example for Sommerfeld integrals, which appear frequently in dipole radiation problems [147], and for the acceleration of methods for evaluating oscillatory integrals [63]. See [516] for a review, and [496] for software. Let us mention only one more, since it was presented at the congress held in Boulder, USA, 19–22 June 1972, which Wynn was attending (and C.B. too). In [167], it is shown that the ε-algorithm is quite effective in accelerating quadrature methods, and that the approximation of integrand factors by Padé approximants is a useful methods for performing certain classes of integrals. See also [188]. Other references, which are too numerous to be quoted here, can easily be found from these or from the homepages of the authors of the papers [204, 222, 223, 266, 362, 378].

By embedding the successive terms $S_n, \ldots, S_{n+k}$ in Euclidean $(k + 1)$-space, Tucker derived the Shanks transformation geometrically [648].

A generalization of the scalar ε-algorithm was proposed in [669] in 1979. It consists in the introduction of a parameter α in the numerator of one of the ratios of Wynn's cross rule. Properties and numerical examples of this modification are given. The process was further studied in [34].

The ε-algorithm was extended to the computation of multivariate Padé approximants by Annie Cuyt [183, 184], and to a noncommutative algebra by André Draux [208] in 1988, who wrote:

> In the case of a noncommutative algebra, the epsilon algorithm is deduced from the Padé approximants at $t = 1$, and from the use of the cross rule; their algebraic properties are a consequence of those verified by the Padé approximants. The computation of the coefficients is particularly studied. It is shown, that it does not exist any non-invertible needed elements if and only if the Hankel matrices $M_k(\Delta^l S_n) = (\Delta^l S_{n+i+j})_{i,j=0}^{k-1}$, for $l = 1, 2$ and 3, have an inverse. Some results of convergence and convergence acceleration are also given.

Draux mentioned the papers [746, 754, 774] of Wynn. In [209], he studied the convergence of these approximants.

A function f is said to be totally monotonic if for all $t \geq 0$, $(-1)^k f^{(k)}(t) \geq 0$. A necessary and sufficient condition is that there exists α bounded and nondecreasing in $[0, \infty)$ such that for all t in this interval,

$$f(t) = \int_0^\infty e^{-xt}\, d\alpha(x).$$

The convergence of the confluent form of the ε-algorithm for such functions was proved in [82].

A confluent form of the topological ε-algorithm was proposed in [87], and extended in [131].

We will now discuss the nonscalar cases.

Given several manifolds and the corresponding perpendicular projection matrices, Leonard Duane Pyle (1930–1993) derived, in 1967, a closed formula for the perpendicular projection matrix of their intersection using results taken from the theory of generalized inverse together with an application of the rectangular matrix ε-algorithm [515].

A geometric interpretation of the rule of the scalar and vector Aitken Δ^2 processes and ε-algorithm were given by Alain Berlinet in [53]. Strangely, Bézier curves appear in this study. Here is the abstract:

> The paper aims at answering the following question, in the scalar as well in the vector case: What do the famous Aitken's Δ^2 and Wynn's ε-algorithm exactly do with the terms of the input sequence? Inspecting the rules of these algorithms from a geometric point of view leads to change the question into another one: By what kind of geometric object can the parallel (or harmonic) sum be represented? Thus, the paper begins with geometric considerations on the parallel addition and the parallel subtraction of vectors, including equivalent definition, and properties derived from the new point of view. It is shown how the parallel sum of vectors is related to the Bézier parabola controlled by these vectors and to their interpolating equiangular spiral. In the second part, observing that Aitken's Δ^2 and Wynn's ε-algorithm may be defined through hybrid sums, mixing standard and parallel sums, the consequences are drawn regarding the way one can define and analyze these algorithms. New explanatory

rules are derived for the ε-algorithm, providing a better understanding of its basic step and of its cross-rule.

The parallel sum of two numbers a and b is denoted by $a : b$ and defined as one-half their harmonic mean, that is, $a : b = ab/(a + b) = [a^{-1} + b^{-1}]^{-1}$. For vectors, the Samelson inverse is used, and one has $a : b = (\|b\|^2 a + \|a\|^2 b)/\|a + b\|^2$. The conclusion of the paper is:

> The vector Aitken's Δ^2 or VEA-1 transforms a sequence $x = (x_n)$ into a sequence Tx such that $(Tx)_n$ is the center of the similitude transforming simultaneously x_n into x_{n+1}, and x_{n+1} into x_{n+2}. In the case where x_n, x_{n+1} and x_{n+2} are collinear, the similitude is an homothety. If x_n, x_{n+1} and x_{n+2} are non collinear, $(Tx)_n$ is the focus of the parabola controlled by x_n, x_{n+1} and x_{n+2} and there is a unique equiangular spiral interpolating x_n, x_{n+1} and x_{n+2} and having $(Tx)_n$ as asymptotic point.

Moreover, the cross-rule of the ε-algorithm is the algebraic form of a geometric transformation, which is explained.

In our commentaries on Wynn's paper [732] (see Sect. 5.1.3), we mentioned that the vector and matrix ε-algorithms lack a simple algebraic theory. This was due to the fact that they were directly derived by defining the inverse of a vector or a matrix in the rule of the scalar ε-algorithm. They were not obtained as recursive algorithms for the implementation of a generalization of the Shanks transformation to the vector or matrix cases.

Faced with these difficulties, in 1975 C.B. decided to take up the problem from the beginning [86]. Let (S_n) be a sequence of elements of a topological vector space E on $\mathbb{C}$ (the topology is necessary for considering convergence problems), and assume that for all n, it satisfies the following linear homogeneous difference equation of order k:

$$a_0(S_n - S) + a_1(S_{n+1} - S) + \cdots + a_k(S_{n+k} - S) = 0 \in E, \tag{6.1}$$

where $S \in E$, $a_0, \ldots, a_k \in \mathbb{C}$, and with the conditions $a_0 a_k \neq 0$ (otherwise, k has to be replaced by a smaller integer) and $a_0 + a_1 + \cdots + a_k = 1$ (a normalization condition that does not restrict the generality). The problem was to build a generalization of the Shanks transformation $(S_n) \longmapsto \{(e_k(S_n))\}$ such that for all n, $e_k(S_n) = S$ if (S_n) satisfies (6.1).

Obviously, if the scalars $a_0, \ldots, a_k$ are known, this transformation is given by

$$e_k(S_n) = a_0 S_n + a_1 S_{n+1} + \cdots + a_k S_{n+k} \in E. \tag{6.2}$$

But in practice, the coefficients a_i are unknown, and the problem was to compute them. Writing (6.1) for the indices n and $n + 1$ and subtracting, we obtained

$$a_0 \Delta S_n + a_1 \Delta S_{n+1} + \cdots + a_k \Delta S_{n+k} = 0 \in E.$$

This relation in E did not allow us to compute the a_i, and it had to be transformed into a scalar relation.

Let us see how this problem was solved. Let y be an element of E^*, the algebraic dual space of E, that is, the vector space of linear functionals on E. Denoting by $\langle \cdot, \cdot \rangle$

the duality product between E^* and E and applying this product to the preceding relation yields

$$a_0\langle y, \Delta S_n\rangle + a_1\langle y, \Delta S_{n+1}\rangle + \cdots + a_k\langle y, \Delta S_{n+k}\rangle = 0 \in \mathbb{C}. \tag{6.3}$$

Writing this relation for the indices $n, n+1, \ldots, n+k-1$ and adding the normalization condition leads to a system of $k+1$ linear equations in the $k+1$ unknowns $a_0, \ldots, a_k$. Then from (6.2), we obtain the (first) *topological Shanks transformation*, defined by

$$e_k(S_n) = \frac{\begin{vmatrix} S_n & S_{n+1} & \cdots & S_{n+k} \\ \langle y, \Delta S_n\rangle & \langle y, \Delta S_{n+1}\rangle & \cdots & \langle y, \Delta S_{n+k}\rangle \\ \vdots & \vdots & & \vdots \\ \langle y, \Delta S_{n+k-1}\rangle & \langle y, \Delta S_{n+k}\rangle & \cdots & \langle y, \Delta S_{n+2k-1}\rangle \end{vmatrix}}{\begin{vmatrix} 1 & 1 & \cdots & 1 \\ \langle y, \Delta S_n\rangle & \langle y, \Delta S_{n+1}\rangle & \cdots & \langle y, \Delta S_{n+k}\rangle \\ \vdots & \vdots & & \vdots \\ \langle y, \Delta S_{n+k-1}\rangle & \langle y, \Delta S_{n+k}\rangle & \cdots & \langle y, \Delta S_{n+2k-1}\rangle \end{vmatrix}}. \tag{6.4}$$

The numerator is the linear combination of $S_n, \ldots, S_{n+k}$ obtained by expanding it with respect to its first row by the usual rule for computing a determinant. A second topological Shanks transformation can be defined by replacing the first row in the numerator by $S_{n+k}, \ldots, S_{n+2k}$.

The formula (6.4) can be written under the form of a Schur complement after replacing, in the numerator and in the denominator, each column (beginning with the second one) by its difference from the preceding column,

$$e_k(S_n) = S_n - [\Delta S_n, \ldots, \Delta S_{n+k-1}] \begin{pmatrix} \langle y, \Delta^2 S_n\rangle & \cdots & \langle y, \Delta^2 S_{n+k-1}\rangle \\ \vdots & & \vdots \\ \langle y, \Delta^2 S_{n+k-1}\rangle & \cdots & \langle y, \Delta^2 S_{n+2k-2}\rangle \end{pmatrix}^{-1} \begin{pmatrix} v_n \\ \vdots \\ v_{n+k-1} \end{pmatrix},$$

with $v_{n+i} = \langle y, \Delta S_{n+i-1}\rangle$ for $i = 0, \ldots, k-1$, and where $[\Delta S_n, \ldots, \Delta S_{n+k-1}] \in E^k$.

A recursive algorithm for implementing these transformations had then to be designed. The first step was to define an inverse in E satisfying all the usual properties an inverse has to have. After several unsuccessful attempts, it was found that the answer was to define the inverse of a couple $(y, u) \in E^* \times E$ as $(u^{-1}, y^{-1}) \in E^* \times E$ with $y^{-1} = u/\langle y, u\rangle \in E$ and $u^{-1} = y/\langle y, u\rangle \in E^*$. The corresponding recursive algorithm, named the (first) *topological ε-algorithm* (TEA1), is

$$\begin{cases} \varepsilon_{2k+1}^{(n)} = \varepsilon_{2k-1}^{(n+1)} + \dfrac{y}{\langle y, \varepsilon_{2k}^{(n+1)} - \varepsilon_{2k}^{(n)}\rangle} \in E^*, & k, n = 0, 1, \ldots, \\[3ex] \varepsilon_{2k+2}^{(n)} = \varepsilon_{2k}^{(n+1)} + \dfrac{\varepsilon_{2k}^{(n+1)} - \varepsilon_{2k}^{(n)}}{\langle \varepsilon_{2k+1}^{(n+1)} - \varepsilon_{2k+1}^{(n)}, \varepsilon_{2k}^{(n+1)} - \varepsilon_{2k}^{(n)}\rangle} \in E, & k, n = 0, 1, \ldots, \end{cases}$$

with $\varepsilon_{-1}^{(n)} = 0 \in E^*$ and $\varepsilon_0^{(n)} = S_n \in E, n = 0, 1, \ldots$, and one has

$$\varepsilon_{2k}^{(n)} = e_k(S_n) \quad \text{and} \quad \varepsilon_{2k+1}^{(n)} = y/\langle y, e_k(\Delta S_n)\rangle.$$

There exists a second topological ε-algorithm (TEA2) for implementing the second topological Shanks transformation [86]. Its rules are quite similar to those above. The corresponding Fortran subroutines are given in [123], and the Matlab functions in [133]. The principal difficulty of these algorithms lies in the storage and the computation of elements of E^* when E is not $\mathbb{C}^p$. When the S_n are square matrices, the element $y \in E^*$ can be defined, for example, by $\langle y, u\rangle = \text{tr}(u)$, where $u \in \mathbb{C}^{p\times p}$, and tr denotes the trace of a matrix. With this choice, the linear functionals $\varepsilon_{2k+1}^{(n)}$ are a combination of the linear functional tr, which makes them difficult to store. But of course, other functionals are possible, but all difficult to implement with these rules.

In 2014, the authors of this book, C.B. and M.R.-Z.,[4] noticed that a relation that was given in [86] could be used to simplify the two topological ε-algorithms, a property that was not noticed at that time. It was

$$e_{k+1}(S_n) = e_k(S_{n+1}) - \frac{\langle y, e_k(\Delta S_n)\rangle \langle y, e_k(\Delta S_{n+1})\rangle}{\langle y, \Delta e_k(S_n)\rangle \langle y, \Delta e_k(S_{n+1})\rangle} \Delta e_k(S_n)$$

$$= e_k(S_{n+1}) - \frac{e_k(\langle y, \Delta S_n\rangle) e_k(\langle y, \Delta S_{n+1}\rangle)}{\Delta e_k(\langle y, S_n\rangle) \Delta e_k(\langle y, S_{n+1}\rangle)}) \Delta e_k(S_n).$$

Thus they obtained two *simplified topological ε-algorithms* (STEA1 and STEA2) [129], the STEA2 being the most attractive one for the implementation reasons described in [132]. Each of these algorithms can be given in four equivalent forms. The most frequently used of them, for STEA2, is the following one, where the ε belonging to E are denoted by $\widehat{\varepsilon}$:

$$\widehat{\varepsilon}_{2k+2}^{(n)} = \widehat{\varepsilon}_{2k}^{(n+1)} + \frac{\varepsilon_{2k+2}^{(n)} - \varepsilon_{2k}^{(n+1)}}{\varepsilon_{2k}^{(n+1)} - \varepsilon_{2k}^{(n)}} (\widehat{\varepsilon}_{2k}^{(n+1)} - \widehat{\varepsilon}_{2k}^{(n)}),$$

with $\widehat{\varepsilon}_0^{(n)} = S_n \in E, n = 0, 1, \ldots$, and where the ε are scalars computed by applying the scalar ε-algorithm to the sequence $\varepsilon_0^{(n)} = \langle y, S_n\rangle$. It has to be noted that this algorithm has only one rule instead of two, that it is a triangular one and not a rhombus scheme, and that only elements of E with a lower even index appear in it. The corresponding software in Matlab is in the public domain [133]. It is an implementation optimized from the point of view of the computational cost and memory storage. These features are very important, since they allow one to use this algorithm for vectors and matrices of large dimensions, and also for tensors. In fact, for proceeding in the computation, the technique used computes the terms by ascending diagonal, instead of by columns. This technique follows the idea first given by Wynn in [743] (see the analysis in Sect. 5.1.1.2). The terms of the sequence to be accelerated are introduced one by one into the algorithm, and the last ascending

---

[4] See their testimony in Sect. 7.13.

diagonal of the ε-array is computed up to the desired element without fixing, a priori, its indices k and n. Thus in STEA2 (which is cheaper that the STEA1), adding a new term of the sequence makes use of only $k + 2$ previously stored elements. All the details are explained in the papers [129] and [132] and in the documentation provided with the software [133]. A different implementation of the topological ε-algorithms is given in [590, pp. 115–117]. The cost is comparable, but here, the indices k and n have to be fixed from the beginning, thus forcing all the computations to be redone if one of the indices is changed.

Other generalizations of the Shanks transformation (6.2) with the same kernel given by (6.1) can be obtained using again (6.3). Instead of using one element $y \in E^*$ and writing (6.3) for the indices $n, \ldots, n + k - 1$, we can fix the index n and write (6.3) for k linearly independent elements $y_i \in E^*$, $i = 1, \ldots, k$. This choice leads to the *Modified Minimal Polynomial Extrapolation* method (MMPE) [86], which can also be expressed as a ratio of two determinants:

$$e_k(S_n) = \frac{\begin{vmatrix} S_n & S_{n+1} & \cdots & S_{n+k} \\ \langle y_1, \Delta S_n \rangle & \langle y_1, \Delta S_{n+1} \rangle & \cdots & \langle y_1, \Delta S_{n+k} \rangle \\ \vdots & \vdots & & \vdots \\ \langle y_k, \Delta S_n \rangle & \langle y_k, \Delta S_{n+1} \rangle & \cdots & \langle y_k, \Delta S_{n+k} \rangle \end{vmatrix}}{\begin{vmatrix} 1 & 1 & \cdots & 1 \\ \langle y_1, \Delta S_n \rangle & \langle y_1, \Delta S_{n+1} \rangle & \cdots & \langle y_1, \Delta S_{n+k} \rangle \\ \vdots & \vdots & & \vdots \\ \langle y_k, \Delta S_n \rangle & \langle y_k, \Delta S_{n+1} \rangle & \cdots & \langle y_k, \Delta S_{n+k} \rangle \end{vmatrix}}.$$

This transformation can be also written as a Schur complement:

$$e_k(S_n) = S_n - [\Delta S_n, \ldots, \Delta S_{n+k-1}] \begin{pmatrix} \langle y_1, \Delta^2 S_n \rangle & \cdots & \langle y_1, \Delta^2 S_{n+k-1} \rangle \\ \vdots & & \vdots \\ \langle y_k, \Delta^2 S_n \rangle & \cdots & \langle y_k, \Delta^2 S_{n+k-1} \rangle \end{pmatrix}^{-1} \begin{pmatrix} \langle y_1, \Delta S_n \rangle \\ \vdots \\ \langle y_k, \Delta S_n \rangle \end{pmatrix}.$$

The MMPE can be recursively implemented by the $S\beta$-algorithm given by Khalide Jbilou in his thesis [351]. It is studied in detail in [353]. This algorithm was obtained as a byproduct of the *Recursive Projection Algorithm* (RPA) and its *Compact* form (CRPA) given in [96].

Let us now derive it directly by Sylvester's determinantal identity. The possibility of such a proof was mentioned in [123, p. 227], but it is given here for the first time. This proof and the algorithm are valid in a general vector space E. As explained in Sect. 2.1.1, denoting respectively by D, C, NW, SE, NE, and SW the determinants in this identity in their order of appearance from left to right, recall that Sylvester's identity (2.1) can be written

$$D \cdot C = NW \cdot SE - NE \cdot SW.$$

Theorem 6.1 *For $k = 1, 2, \ldots$ and $n = 0, 1, \ldots,$ we have*

$$e_k(S_n) = \frac{e_{k-1}^{(n)} - a_k^{(n)} e_{k-1}^{(n+1)}}{1 - a_k^{(n)}},$$

$$\beta_k^{(n)} = \frac{\beta_{k-1}^{(n)} - a_k^{(n)} \beta_{k-1}^{(n+1)}}{1 - a_k^{(n)}},$$

with $e_0(S_n) = S_n$ and $\beta_0^{(n)} = \Delta S_n$ for $n = 0, 1, \ldots,$ and where

$$a_k^{(n)} = \langle y_k, \beta_{k-1}^{(n)} \rangle / \langle y_k, \beta_{k-1}^{(n+1)} \rangle.$$

Proof We set

$$\beta_k^{(n)} = \begin{vmatrix} \Delta S_n & \cdots & \Delta S_{n+k} \\ \langle y_1, \Delta S_n \rangle & \cdots & \langle y_1, \Delta S_{n+k} \rangle \\ \vdots & & \vdots \\ \langle y_k, \Delta S_n \rangle & \cdots & \langle y_k, \Delta S_{n+k} \rangle \end{vmatrix} \Bigg/ \begin{vmatrix} 1 & \cdots & 1 \\ \langle y_1, \Delta S_n \rangle & \cdots & \langle y_1, \Delta S_{n+k} \rangle \\ \vdots & & \vdots \\ \langle y_k, \Delta S_n \rangle & \cdots & \langle y_k, \Delta S_{n+k} \rangle \end{vmatrix},$$

and we denote by $\widetilde{N}_k^{(n)}$ the vector numerator of $e_k(S_n)$, by $D_k^{(n)}$ its denominator, which is also the denominator of $\beta_k^{(n)}$, and by $N_k^{(n)}$ the numerator of $\beta_k^{(n)}$.

Let us first apply Sylvester's identity to the numerator $D = N_k^{(n)}$ of $\beta_k^{(n)}$. The determinant C is the number corresponding to the determinant obtained by deleting the first and the second rows and columns of D. On the right-hand side of Sylvester's identity, $NW = N_{k-1}^{(n)}$. We move the last row $\langle y_k, \Delta S_n \rangle \cdots \langle y_k, \Delta S_{n+k} \rangle$ of the second determinant of the first product (that is, SE) to the first place. We remark that this determinant is nothing else than $(-1)^{k-1} \langle y_k, N_{k-1}^{(n+1)} \rangle$. The same operation is conducted on the second determinant of the second product (that is, SW) in Sylvester's identity. We have $SW = (-1)^{k-1} \langle y_k, N_{k-1}^{(n)} \rangle$. Thus, since $NE = N_{k-1}^{(n+1)}$, we obtain

$$C N_k^{(n)} = N_{k-1}^{(n)} (-1)^{k-1} \langle y_k, N_{k-1}^{(n+1)} \rangle - N_{k-1}^{(n+1)} (-1)^{k-1} \langle y_k, N_{k-1}^{(n)} \rangle.$$

We perform the same manipulations on $\widetilde{N}_k^{(n)}$ and $D_k^{(n)}$ and then express $e_k(S_n)$ and $\beta_k^{(n)}$ as the ratios of determinants given above. The determinant C disappears in these ratios. Dividing both sides of these two identities by the product $D_{k-1}^{(n)} D_{k-1}^{(n+1)}$ and suppressing the signs $(-1)^{k-1}$, which are in common everywhere, we obtain the $S\beta$-algorithm. ∎

Notice that

$$\beta_k^{(n)} = a_0 \Delta S_n + \cdots + a_k \Delta S_{n+k},$$

where the coefficients a_i are the same as those of the linear combination giving $e_k(S_n)$, and that from its determinantal expression, $\langle y_i, \beta_k^{(n)} \rangle = 0$ for $i = 1, \ldots, k$. The vector $\beta_k^{(n)}$ can also be expressed as a Schur's complement:

$$\beta_k^{(n)} = \Delta S_n - [\Delta^2 S_n, \ldots, \Delta^2 S_{n+k-1}](Y^T[\Delta^2 S_n, \ldots, \Delta^2 S_{n+k-1}])^{-1} Y^T \Delta S_n,$$

where $Y = [y_1, \ldots, y_k]$.

The *Minimal Polynomial Extrapolation* method (MPE) [157] and the *Reduced Rank Extrapolation* method (RRE) [218, 363, 446] can be obtained directly from the expression of the MMPE by taking $y_i = \Delta S_{n+i-1}$ for the first one, and $y_i = \Delta^2 S_{n+i-1}$ for the second. Notice that the two STEAs and the MMPE can be used not only for vector sequences, but also for matrix and tensor ones, which is not the case with the MPE and RRE.

All these methods, which generalize the Shanks transformation, received much attention. They can be used either for the acceleration of sequences, or as fixed point methods. The corresponding literature about their convergence, their implementation, and their applications can easily be found. A partial list of references is [137, 215, 352, 354, 355, 543, 562, 580, 583, 589, 591, 601, 633, 634, 635]. Their scaling properties, which are important in physical applications, were studied in [449]. For a complete history of their discovery, and testimonies by some of their authors, see [135]. See [447] for a complete set of papers dealing with these topics.

Let us also mention an important link with numerical linear algebra. All transformations with a kernel of the form (6.1) furnish the exact solution of a system of linear equations (see [257] for a review[5]). Indeed, consider iterations of the form $S_{n+1} = MS_n + b, n = 0, 1, \ldots$, where $I - M$ is nonsingular. Let S be the vector that satisfies $S = MS + b$. Thus it follows that $S_{n+1} - S = M(S_n - S)$, that is, $S_n - S = M^n(S_0 - S)$ for all n. Let $P_k(\xi) = a_0 + a_1\xi + \cdots + a_k\xi^k$ be the minimal polynomial of the matrix M for the vector $S_0 - S$. Its degree k is less than or equal to the dimension p of the system. We have, for all n, $M^n P_k(M)(S_0 - S) = P_k(M)M^n(S_0 - S) = a_0(S_n - S) + \cdots + a_k(S_{n+k} - S) = 0$, which shows that the sequence (S_n) belongs to the kernel of the transformation. Thus for all n, $e_k(S_n) = S$, which means that all these transformations are direct methods for solving a system of linear equations. Obviously, for storage reasons when k is large, this property cannot be used in practice. Refinements of this result can be found in [85, 581].

6.5 Formal Orthogonal Polynomials

We previously saw that formal orthogonal polynomials were discussed by several researchers; see Sect. 3.3.

The concept of formal orthogonality was taken up again in [89] and subsequent papers of this author, and developed in [94]. All possible recurrence relationships between adjacent families of formal orthogonal polynomials were obtained. They

---

[5] See Sect. 7.4 for the testimony of Walter Gander (b. 1944).

lead to recurrences allowing one to compute recursively any sequence of Padé approximants in the Padé table; see [94] for a Fortran program.

Additional properties were also proved; let us discuss them. Usually, the proof of the Christoffel–Darboux identity (Theorem 2.20 in Sect. 2.4.2) makes use of a three-term recurrence relationship. A direct proof of this identity was given in [101]. It included the case of formal orthogonal polynomials. In the same paper it was also demonstrated that if a family of polynomials satisfies the Christoffel–Darboux identity, it also satisfies a three-term recurrence relationship, and, thus, it is a family of formal orthogonal polynomials with respect to a linear functional whose moments can be computed one by one. Let us give this result.

Theorem 6.2 *Let $\{P_k\}$ be a family of polynomials such that for all $k \geq 0$,*

- *P_k has the exact degree k,*

- *$\gamma_k[P_{k+1}(x)P_k(t) - P_{k+1}(t)P_k(x)] = (x - t)\sum_{i=0}^{k} a_i P_i(x)P_i(t)$, where $\gamma_k \neq 0$, and the a_i are coefficients independent of k.*

Then these polynomials satisfy a three-term recurrence relationship of the form (2.18), which means that by the Shohat–Favard theorem (Theorem 2.19), they form a family of FOPs.

Proof We have

$$\gamma_k[P_{k+1}(x)P_k(t) - P_{k+1}(t)P_k(x)] = (x - t)a_k P_k(x)P_k(t) + (x - t)\sum_{i=0}^{k-1} a_i P_i(x)P_i(t)$$

$$= (x - t)a_k P_k(x)P_k(t) + (x - t)\gamma_{k-1}[P_k(x)P_{k-1}(t) - P_k(t)P_{k-1}(x)].$$

Thus for all x and t,

$$P_k(t)[\gamma_k P_{k+1}(x) - a_k x P_k(x) + \gamma_{k-1}P_{k-1}(x)]$$
$$= P_k(x)[\gamma_k P_{k+1}(t) - a_k t P_k(t) + \gamma_{k-1}P_{k-1}(t)].$$

Therefore, for all x,

$$\frac{\gamma_k P_{k+1}(x) - a_k x P_k(x) + \gamma_{k-1}P_{k-1}(x)}{P_k(x)} = b_k,$$

where b_k is a constant, that is,

$$\gamma_k P_{k+1}(x) = (a_k x + b_k)P_k(x) - \gamma_{k-1}P_{k-1}(x), \tag{6.5}$$

which shows that the polynomials satisfy a three-term recurrence relation of the form (2.18). Hence by the Shohat–Favard theorem (Theorem 2.19), they constitute a family of FOPs.

Let us now give expressions for γ_k, γ_{k-1}, a_k, and b_k. Let t_k be the coefficient of x^k in P_k. Then identifying the coefficients of x^{k+1} on both sides of (6.5) yields

$\gamma_k t_{k+1} = a_k t_k$. Multiply (6.5) by x^{k-1} and apply the linear functional c. We get

$$\gamma_k c(x^{k-1} P_{k+1}) = a_k c(x^k P_k) + b_k c(x^{k-1} P_k) - \gamma_{k-1} c(x^{k-1} P_{k-1}).$$

But by the orthogonality condition, $c(x^{k-1} P_{k+1}) = c(x^{k-1} P_k) = 0$, and thus

$$a_k c(x^k P_k) = \gamma_{k-1} c(x^{k-1} P_{k-1}).$$

If we set $h_k = c(P_k^2)$, then $h_k = t_k c(x^k P_k)$, and we have, since for all k, $t_k \neq 0$,

$$\frac{a_k h_k}{t_k} = \gamma_{k-1} \frac{h_{k-1}}{t_{k-1}}.$$

Setting $C_{k+1} = \gamma_{k-1}/\gamma_k$, we get

$$C_{k+1} = \frac{a_k h_k t_{k-1}}{t_k h_{k-1} \gamma_k} = \frac{t_{k+1}}{t_k} \frac{h_k t_{k-1}}{t_k h_{k-1}} = \frac{t_{k-1} t_{k+1}}{t_k^2} \frac{h_k}{h_{k-1}}.$$

Multiplying (6.5) by P_k and applying c leads to

$$\gamma_k c(P_k P_{k+1}) = a_k c(x P_k^2) + b_k c(P_k^2) - \gamma_{k-1} c(P_k P_{k-1}).$$

Therefore, by the orthogonality conditions $c(P_k P_{k+1}) = c(P_k P_{k-1}) = 0$, we have $a_k \alpha_k = -b_k h_k$, where $\alpha_k = c(x P_k^2)$. Setting $B_{k+1} = b_k/\gamma_k$, we get

$$B_{k+1} = -\frac{\alpha_k a_k}{h_k \gamma_k} = -\frac{\alpha_k}{h_k} \frac{t_{k+1}}{t_k}.$$

Setting finally $A_{k+1} = a_k/\gamma_k$, we obtain the recurrence relation (2.18) and the expressions for the coefficients given after Theorem 2.16.

Moreover,

$$A_{k+1} = \frac{a_k}{\gamma_k} = \frac{t_{k+1}}{t_k}, \qquad C_{k+1} = \frac{\gamma_{k-1}}{\gamma_k} = \frac{A_{k+1} h_k}{A_k h_{k-1}}.$$

Thus

$$C_{k+1} = \frac{a_k}{\gamma_k} \frac{h_k}{h_{k-1}} \frac{\gamma_{k-1}}{a_{k-1}} = \frac{a_k h_k}{a_{k-1} h_{k-1}} C_{k+1}.$$

Therefore, there exists $\gamma \neq 0$ such that for all k, $a_k h_k = \gamma$, that is, $a_k = \gamma h_k^{-1}$. Then

$$\gamma_k = a_k \frac{t_k}{t_{k+1}} = \gamma \frac{t_k}{h_k t_{k+1}},$$

and the relation of the theorem becomes exactly that of Theorem 2.20 after canceling γ from both sides. ∎

This result is quite important, since it shows that the three-term recurrence relationship (relation (2.18) of Theorem 2.16) and the Christoffel–Darboux identity

(Theorem 2.20) are two equivalent characterizations of a family of FOPs, thanks to the Shohat–Favard theorem (Theorem 2.19).

All these results were generalized to the adjacent families of formal orthogonal polynomials $P_k^{(n)}$ and their associated polynomials $Q_k^{(n)}$ [104]. They were obtained under the assumption that all the formal orthogonal polynomials $P_k^{(n)}$ involved exist and have the exact degree k corresponding to their localization in a table where the polynomials with the same lower index are in a column, and those with the same upper index in a descending diagonal. This assumption is equivalent to the nonnullity of all the Hankel determinants involved (we recall that in this case, the linear functional c is called *definite*).

The reciprocals of these theoretical results also exist. Moreover, we have the following additional result.

Theorem 6.3 *For all k and n, the polynomials $P_k^{(n)}, P_k^{(n+1)}, P_{k-1}^{(n+1)}, P_k^{(n-1)}$, and $P_{k+1}^{(n-1)}$ have no common zero. Moreover, $P_k^{(n)}(0) \neq 0$.*

Let us now discuss the case in which some Hankel determinants vanish (that is, the case in which the functional c is *indefinite*). A preliminary study of this problem is due to Householder in 1971 [338]. The linear independence of the columns of Hankel determinants was studied in depth by André Draux in his doctoral thesis [206]. When some Hankel determinants $H_k^{(n)}$ are null, the first problem is to find the rank of the corresponding Hankel matrix. The nullity of certain Hankel determinants implies the nonexistence of certain formal orthogonal polynomials, or the fact that they have degree strictly lower than their index k. Then if several adjacent Hankel determinants vanish, the locations and the sizes of the corresponding regions of the table (described above) containing them have to be determined. These regions are, in fact, square blocks similar to those encountered in the Padé table (see Sect. 2.3.1). Thus it is first necessary to understand whether the recurrence relation used hits such a block on its north or west side. After that, it is necessary to find the degree of the next polynomial satisfying all the requirements, that is, the size of the block. And finally, new recurrences between existing formal orthogonal polynomials of maximal degree need to be established. These relations are used to compute recursively sequences of Padé approximants in the presence of blocks of identical approximants in the Padé table. The same relations are those that allow one to cure breakdowns and near-breakdowns in Lanczos-type algorithms, as will be explained in the next section. The corresponding Fortran programs for computing these polynomials in the general case are given in [210].

Let us give an idea how to treat such a situation. Up to now, the kth polynomial of the family (assuming without lost of generality that it was monic) was denoted by P_k; it existed and had exact degree k. A name will now be given only to the existing polynomials of maximal degree. Therefore, the kth polynomial of the family will be denoted by P_{n_k}, and it will have degree $n_k \geq k$ exactly. Thus it satisfies the orthogonality conditions $c(x^i P_{n_k}(x)) = 0$ for $i = 0, \ldots, n_k - 1$. We now have to determine the next existing polynomial $P_{n_{k+1}}$ of the family. Its degree $n_{k+1} > n_k$ is the smallest integer such that $c(x^i P_{n_k}(x)) = 0$ for $i = n_k, \ldots, n_{k+1} - 2$, and $c(x^{n_{k+1}-1} P_{n_k}(x)) \neq 0$. It satisfies the orthogonality conditions $c(x^i P_{n_{k+1}}(x)) = 0$ for $i = 0, \ldots, n_{k+1} - 1$, and it is recursively computed by a relation of the form

$$P_{n_{k+1}}(x) = A_{k+1}(x)P_{n_k}(x) - C_{k+1}P_{n_{k-1}}(x), \quad k = 0, 1, \ldots,$$

with $n_{-1} = -1, P_{-1}(x) = 0, n_0 = 0, P_0(x) = 1, C_1 = 0$, and where now A_{k+1} is a monic polynomial of degree $m_k = n_{k+1} - n_k$. Setting $A_{k+1}(x) = a_0 + a_1 x + \cdots + a_{m_k-1} x^{m_k-1} + x^{m_k}$ and taking the orthogonality conditions into account leads to the following system, whose solution gives C_{k+1} and the a_i (for the sake of simplicity, we omit the variable x):

$$C_{k+1} = c(x^{n_k+m_k-1} P_{n_k})/c(x^{n_k-1} P_{n_{k-1}}),$$
$$a_{m_k-1}c(x^{n_k+m_k-1} P_{n_k}) + c(x^{n_k+m_k} P_{n_k}) = C_{k+1}c(x^{n_k} P_{n_{k-1}}),$$
$$\vdots$$
$$a_0 c(x^{n_k+m_k-1} P_{n_k}) + \cdots + a_{m_k-1}c(x^{n_k+2m_k-2} P_{n_k}) + c(x^{n_k+2m_k-1} P_{n_k}),$$
$$= C_{k+1}c(x^{n_k+m_k-1} P_{n_{k-1}}).$$

The treatment of near-breakdown is quite similar after replacing the constant C_{k+1} by a polynomial.

Formal orthogonality is related to the *umbral calculus*, a symbolic method first introduced by John Blissard (1803–1875) in 1861. In the 1930s and 1940s, Eric Temple Bell (1883–1960) attempted to set the umbral calculus on a rigorous basis. In the 1970s, Gian-Carlo Rota (1932–1999) and Steven Roman developed the umbral calculus by means of linear functionals on the vector space of polynomials [534].

Formal orthogonality also reveals to be essential in Lanczos's and Krylov's subspace methods for the iterative solution of systems of linear equations (for an introduction, see [140]), for the computation of matrix functions, and for the solution of other problems in linear and nonlinear algebra [276], etc.

The theoretical framework for the study of linear forms was given by Pascal Maroni (b. 1933) [436, 437], who also introduced the notion of formal $1/d$-orthogonality [435]. This kind of orthogonality leads to polynomials satisfying a recurrence relation of order greater than 2. It has applications in vector Padé approximation [658]. Formal orthogonality is a particular case of the concept of biorthogonality described in [105].

Orthogonal rational functions, which generalized orthogonal polynomials, are studied in [152].

6.6 Lanczos's Method

In 1950, Cornelius Lanczos proposed an algorithm for transforming a matrix into a similar tridiagonal matrix [386]. Thanks to the Cayley–Hamilton theorem, this method can also be used for solving a system of linear equations, and Lanczos turned it into an iterative method that furnishes the exact solution in a finite number of steps not greater than the dimension of the system [387]. Lanczos's method was the subject of many investigations, and a historical account of it was given in [277]. See [448] for an up-to-date and full presentation of the subject. Many algorithms for the implementation of Lanczos's method have been obtained, among them the famous *conjugate gradient algorithm* of Magnus Hestenes (1906–1992) and Eduard Stiefel [327] when the matrix is symmetric positive definite, and the *biconjugate gradient algorithm* due to Roger Fletcher (1939–2016) in the general case [243].

Since the problem to be solved belongs to the field of linear algebra, most of the papers on the subject use techniques related to it. However, as shown in [94], for example, the problem is also strongly connected to the theory of formal orthogonal polynomials. Such a connection was, in fact, known to Lanczos (see [387]), but although several papers emphasize the important role of orthogonal polynomials in the theory, it was mostly forgotten and replaced by pure linear algebraic techniques.

Let us introduce it again by such techniques. Consider a system of p linear equations $Ax = b$ in p unknowns. Lanczos's method consists, starting from an arbitrary x_0, in constructing the sequence of vectors (x_k) entirely defined by the two conditions

$$x_k - x_0 \in K_k(A, r_0),$$
$$r_k = b - Ax_k \perp K_k(A^*, y),$$

where $K_k(A, u) = \text{span}(u, Au, \ldots, A^{k-1}u)$, and y is a nonzero vector; K_k is called a *Krylov subspace*. The first condition can be written

$$x_k - x_0 = -a_1 r_0 - a_2 A r_0 - \cdots - a_k A^{k-1} r_0.$$

Multiplying this relation by A and adding and subtracting b gives

$$r_k = r_0 + a_1 A r_0 + a_2 A^2 r_0 + \cdots + a_k A^k r_0.$$

The second condition above, the orthogonality condition, leads to

$$(A^{*^i} y, r_k) = 0 \quad \text{for} \quad i = 0, \ldots, k - 1.$$

Replacing r_k by its expression above, the following system of linear equations is obtained:

$$a_1(y, A^{i+1} r_0) + a_2(y, A^{i+2} r_0) + \cdots + a_k(y, A^{i+k}) r_0 = -(y, A^i r_0), \quad i = 0, \ldots, k - 1.$$

It completely defines x_k, and thus r_k, if the Hankel determinant

$$H_k^{(1)} = \begin{vmatrix} c_1 & c_2 & \cdots & c_k \\ c_2 & c_3 & \cdots & c_{k+1} \\ \vdots & \vdots & & \vdots \\ c_k & c_{k+1} & \cdots & c_{2k-1} \end{vmatrix}$$

is different from zero.

Let us now explain the approach with formal orthogonal polynomials. Setting $P_k(\xi) = a_0 + a_1\xi + \cdots + a_k\xi^k$, we have $r_k = P_k(A)r_0$. Let us now set $c_i = (y, A^i r_0)$ for $i = 0, 1, \ldots$ and define the linear functional c acting on polynomials by $c(\xi^i) = c_i$. The preceding system for the coefficients a_i can be written as

$$c(\xi^i P_k(\xi)) = 0 \quad \text{for} \quad i = 0, \ldots, k-1.$$

These conditions mean that P_k is the polynomial of degree at most k belonging to the family of formal orthogonal polynomials with respect to the linear functional c. These polynomials satisfy $P_k(0) = 1$, and it follows that

$$P_k(\xi) = \begin{vmatrix} 1 & \xi & \xi^2 & \cdots & \xi^k \\ c_0 & c_1 & c_2 & \cdots & c_k \\ \vdots & \vdots & \vdots & & \vdots \\ c_{k-1} & c_k & c_{k+1} & \cdots & c_{2k-1} \end{vmatrix} \Big/ H_k^{(1)}.$$

Thus

$$r_k = \begin{vmatrix} r_0 & Ar_0 & A^2 r_0 & \cdots & A^k r_0 \\ c_0 & c_1 & c_2 & \cdots & c_k \\ \vdots & \vdots & \vdots & & \vdots \\ c_{k-1} & c_k & c_{k+1} & \cdots & c_{2k-1} \end{vmatrix} \Big/ H_k^{(1)}, \quad x_k = \begin{vmatrix} x_0 & -r_0 & -Ar_0 & \cdots & -A^{k-1} r_0 \\ c_0 & c_1 & c_2 & \cdots & c_k \\ \vdots & \vdots & \vdots & & \vdots \\ c_{k-1} & c_k & c_{k+1} & \cdots & c_{2k-1} \end{vmatrix} \Big/ H_k^{(1)}.$$

Let us mention that the recurrence relations between adjacent families of formal orthogonal polynomials (see Sect. 2.4.1) allow one to recover all known recursive algorithms for the implementation of Lanczos's method and to propose new ones [140]. The theory of formal orthogonal polynomials also leads to methods for curing breakdowns and near-breakdowns in these methods, to obtain recursive algorithms for the implementation of the CGS of Peter Sonneveld (b. 1976) [604] and the Bi-CGStab of Hendrik Albertus van der Vorst (b. 1944) [656], and to treat their breakdowns and near-breakdowns. These methods are based on the recurrence relations found by Draux [206] that allow one to jump over orthogonal polynomials that do not exist or that exist and are badly computed. There are too many references by the authors of this book and Hassane Sadok (b. 1957) to quote here. They can easily be found. See also [298] for a review.

Now let us set $A = I - M$, consider the iterates $S_{n+1} = MS_n + b$ for $n = 0, 1, \ldots$, and define the residuals as $r_n = b - AS_n$. It can be proved by induction that for all

n, $\Delta S_n = (-1)^{i-1} A^{i-1} r_n$ for $i = 1, 2, \ldots$. Notice that we also have $\Delta^i r_n = (-1)^i A^i r_n$ for $i = 0, 1, \ldots$. Using these relations in the determinantal formula given above for x_k, it was proved, after some manipulations of the rows (additions of columns and multiplication of the even ones by -1), that Lanczos's method and the first topological Shanks transformation (see formula (6.4)) produce identical sequences [94, pp. 184–190], namely that

$$x_k = e_k(S_0), \quad k = 0, 1, \ldots.$$

Obviously, this is only a theoretical result, since it is much too costly to use the STEA for implementing Lanczos's method. Let us recall that in [756], Wynn tried to find, without success, a connection between the vector ε-algorithm and the conjugate gradient algorithm when A is symmetric positive definite.

6.7 Solution of Nonlinear Equations

Let us consider the problem of finding a fixed point $x = F(x)$ of F, or equivalently, a zero x of a nonlinear equation $f(x) = 0$, where $F, f : \mathbb{R} \longmapsto \mathbb{R}$. Newton's method $x_{n+1} = x_n - f(x_n)/f'(x_n)$ is asymptotically quadratically convergent under some assumptions, but it requires the mathematical expression for $f'(x)$. The fixed point iterative method due to Johan Frederik Steffensen in 1933 [615] was already explained in Sect. 3.1. It consists in the iterates

$$x_{n+1} = x_n - \frac{(F(x_n) - x_n)^2}{F(F(x_n)) - 2F(x_n) + x_n}, \quad n = 0, 1, \ldots,$$

from a given x_0. Obviously, this method is based on the Aitken process. Indeed, setting $u_0 = x_n$, computing two basic iterates $u_1 = F(u_0)$ and $u_2 = F(u_1)$, and applying the Aitken process to these three iterates produces x_{n+1}. Then the process is repeated with $u_0 = x_{n+1}$. Steffensen did not quote Aitken in his paper, and his discovery seems to have been obtained independently. The convergence of this method is also quadratic, under assumptions quite similar to Newton's, but it does not need the derivative of F. Setting $f(x) = F(x) - x$, we see that these iterations can be written in a compact way as

$$x_{n+1} = x_n - \frac{f(x_n)}{f(x_n + f(x_n)) - f(x_n)} f(x_n),$$

which shows that $f'(x_n)$ has been approximated by $[f(x_n + f(x_n)) - f(x_n)]/f(x_n)$ in Newton's method. An extension of Steffensen's method for accelerating iterations with superlinear convergence was proposed by Alexander Markowicz Ostrowski [478]. Another extrapolation method without memory was proposed by Richard F. King in 1980 [369]. The literature proposing methods consisting in replacing $f'(x_n)$

in Newton's method is vast. Let us mention only [14, 367, 474]; it is pointless to give more references here.

The complex mapping $z \longmapsto F(z)$ can be considered a dynamical system. In [343], Arieh Iserles (b. 1947) replaced it by $z \longmapsto F_n(z)$, where F_n is an appropriately constructed quotient of Hankel determinants, and he studied the local behavior of such a procedure in the vicinity of a fixed point $\hat{z}$. Obviously, Steffensen's method enters into this framework but also the ratio of determinants defining the Shanks transformation, that is, the scalar ε-algorithm. In a second paper [344], he determined the basin of attraction (whose complement is the Julia set) of $\hat{z}$. The question treated here was: *How does the attractivity type of $\hat{z}$ change when F is replaced by F_n?*

Steffensen's method was extended to a real Hilbert space by Claude Lemaréchal (b. 1944) [406] in 1971. It consists in the two-step iterations

$$x_{n+1} = x_n - \frac{(F(x_n) - x_n, F(F(x_n)) - 2F(x_n) + x_n)}{(F(F(x_n)) - 2F(x_n) + x_n, F(F(x_n)) - 2F(x_n) + x_n)}(F(x_n) - x_n),$$

where $(\cdot, \cdot)$ denotes the usual inner product. It converges under some monotonicity conditions. This method reduces to Steffensen's in $\mathbb{R}$. A three-step method in the same style was proposed in [434], extended in [116, 117], modified again in [532] by Christophe Roland and Ravi Varadhan, and applied to the acceleration of the *Expectation–Maximization* algorithm (EM) [198], which is widely used for the computation of maximum likelihood estimates in a variety of incomplete-data problems. Then the same authors "squared" their method, a procedure that simply consists in applying it twice within each cycle of a first-order extrapolation method, and produced the so-called SQUAREM methods [672]. In particular, they study the case in which the first-order extrapolation method is the MPE or the RRE (see Sect. 6.4 and below). This type of method is very effective, as shown, for example, in a tomography problem [533]. A new class of acceleration schemes containing the SQUAREM methods as a particular case was proposed in [54], and a restarting procedure was also examined. An improvement in the method for circumventing the problem of stagnation was given in [55] and compared with the ε-algorithm in [56].

A first extension of Steffensen's method to a system of nonlinear equations $x = F(x)$ was given in 1952 by Rudolf Ludwig (1910–1969), of Braunschweig University, in a short paper [424]. He treated each component separately, and, neglecting second order terms, he obtained a difference equation similar to that defining the kernel of the Shanks transformation. A new estimate of each component is then given as a ratio of determinants close to the ratio (2.13) for the Shanks transformation e_k.

Peter Henrici proposed another extension of Steffensen's method to systems of nonlinear equations in his book [319, pp. 115ff.], published in 1964. For finding x such that $x = F(x)$, he started from an arbitrary vector $x_0 = u_0$, and first computed the vectors (u_i) by $u_{i+1} = F(u_i)$ for $i = 0, \ldots, p$, where p is the dimension of the system. Then his method consists in the iterations

$$x_{n+1} = x_n - \Delta U_0(\Delta^2 U_0)^{-1}\Delta u_0,$$

where $U_0 = [u_0, \ldots, u_{p-1}]$, and where Δ is the usual forward difference operator applied to each column of U_0. Then the process is restarted from $u_0 = x_{n+1}$. The quadratic convergence of this method was proved in [474, pp. 373ff.] under certain assumptions. The study was taken again by Tatsuo Noda in 1981 (in Japanese) and later (in English) in [469] and subsequent papers, and his results were extended by Yves Nievergelt in 1991 [468]. Henrici's method can be recursively implemented by the H-algorithm [139], which is a particular case of the MMPE in which the auxiliary vectors y_i are those of the canonical basis of $\mathbb{R}^p$.

The next generalization to systems was obtained independently by several authors within a quite short period of time [79, 264, 595]. All these authors used the vector ε-algorithm of Wynn [732] (see Sect. 5.1.3).

Starting from a given $x_0 = u_0$, they computed $u_1, \ldots, u_{2p}$ by the basic iterations $u_{i+1} = F(u_i)$ for $i = 0, \ldots, 2p - 1$. Then applying the vector ε-algorithm to them, they got $x_{n+1} = \varepsilon_{2p}^{(0)}$ and restarted the basic iterations from $u_0 = x_{n+1}$. The proofs of the quadratic convergence of the procedure given by all these authors under some assumptions had a gap. However, the numerical results show that the property holds. Wynn discussed this method in [800]. The vector ε-algorithm can be replaced by the first topological ε-algorithm, or the first simplified topological ε-algorithm (see Sect. 6.4), for which a correct proof of the quadratic convergence was given by Hervé Le Ferrand in 1992 [399]. The same procedure can be applied with the MPE, the MMPE, or the RRE, which need fewer basic iterations but lack a recursive algorithm for their implementation (except the MMPE whose implementation is costly). Their quadratic convergence was studied by Khalide Jbilou and Hassane Sadok [353], and by Sidi [591].

A related method for systems of nonlinear equations is due to Donald Gordon Marcus Anderson (b. 1937) [15]. It is called *Anderson acceleration*, or *Anderson mixing*, or *Direct Inversion in the Iterative Subspace* (DIIS). This method recently received much attention [225, 231, 465, 503, 531, 562, 643, 678]. It is discussed in [16] and [137]. See [226] and [590] for a review of these results and [135] for a historical account of these procedures.

6.8 Padé Approximants

In this section, we review applications of Padé approximants and some of their generalizations. Let us mention only some of these applications. Padé approximants are widely used in numerical analysis, critical phenomena, theoretical physics, control theory for model reduction, and partial realization; see, for example, [110, 153, 286]. They also have applications in mechanics [156]. Padé approximants and Thiele's interpolation appeared in Laplace transform inversion [415, 416]; see also [176]. For a robust algorithm that circumvents, as much as possible, the instability problems due to the ill-conditioning and the Froissart doublets in the computation of

Padé approximants, see [47, 278]. They can also be written in a barycentric form with free parameters whose choice allows one to improve their robustness and their stability [130] (on this topic, see also [58]). The reason why a barycentric representation is preferable for rational approximation is explained in [236]. Rational functions presenting simultaneously an approximation-through-order and an interpolation property were defined in [128]. They are useful for removing spurious poles.

New convergence results for Padé approximants were obtained by Doron Shaul Lubinsky [420]. Generalizations of Montessus's theorem to scalar and vector rational interpolation were given by Edward Barry Saff (b. 1944) [544, 545]. See their other papers related to Padé approximation on their respective homepages. All contributions to convergence theory cannot be mentioned here.

Let us now discuss generalizations of Padé approximants. In Sect. 2.4.1, we explained what Padé-type approximants are. They were in introduced in [93]. They are rational functions with an arbitrary denominator whose numerator is defined so that the power series expansion of the approximant in ascending powers of the variable coincides with the series as far as possible. The main difference with a Padé approximant is that the denominator of a Padé-type approximant can be arbitrarily chosen, thus giving some degrees of freedom that can be exploited for convergence or approximation reasons [158, 220, 274, 365, 428, 429, 505]. It needs the knowledge of fewer coefficients of the series, but its degree of approximation is also lower.

Many other generalizations of Padé approximants have been studied. Let us mention them without entering into the details, which can be found in [142], where the original references are given.

A generalization to series whose coefficients are functions of a second variable is given in [292]. In partial Padé approximants, a part of the numerator (that is, some zeros) and a part of the denominator (that is, some poles) can be arbitrarily chosen, and their respective remaining parts are obtained by imposing that the development of the approximant coincides with that of the series as far as possible. The series expansion of a multipoint Padé approximant simultaneously agrees with the Taylor's series of a function at several points. Newton–Padé approximants were defined for Newton series [174], Fourier–Padé approximants for Fourier series [168], Chebyshev–Padé for series in Chebyshev polynomials [485, 486], and for Legendre series [242]. For an application of Fourier–Padé approximants to the solution of nonlinear partial differential equations, see [689]. A generalization of Padé approximants for series in a family of orthogonal polynomials was given in [332]. The numerator is a linear combinations of these polynomials, and the denominator is expressed as a linear combination of a second family of orthogonal polynomials. Another approach was considered in [258]. Padé approximants for Laurent series have also been defined; they are treated in [151].[6]

---

[6] See Sect. 7.1 for the testimony of Adhemar François Bultheel, b. 1948, one of the authors of this reference.

In Cauchy-type approximation, the denominator is a power series and the numerator is a polynomial whose coefficients are taken to obtain the maximal degree of approximation. Padé approximants for series of functions can also be constructed, in particular for series in Chebyshev polynomials.

Padé–Hermite approximants were already described in Sect. 2.3.1. Let us recall them. Given two formal power series f_1 and f_2, they consist in finding two polynomials P_1 and P_2 such that the series expansion of $f_1(x)P_1(x) + f_2(x)P_2(x)$ has as many vanishing first terms as possible. Obviously, the concept can be generalized to more than two series. Quadratic Padé approximation consists in finding the polynomials P, Q, and R such that the series expansion of $f^2 P + f Q + R$ satisfies as many as possible accuracy–through–order conditions [566]. Multivariate Padé approximants concern series in several variables. As is often the case in going from dimension one to dimension two, several generalizations are possible. In this case, each researcher who worked on this problem has his own one. Obviously, all the preceding generalizations can also be combined together, thus leading to more generalizations. Rational Hermite interpolation was discussed in [175, 687].

The notion of A-acceptability, due to Germund Dahlquist (1925–2005) [189], is an important property of numerical methods for the integration of differential equations. It is related to rational approximations of e^z. An approximant is A-acceptable if it is analytic in the complex left half-plane and if its modulus there does not exceed one. A vast literature exists on this problem. The A-acceptability of Padé approximants to the exponential function was studied by Byron L. Ehle [219] (see also [342] and [182]).

We now consider a formal power series $f(t) = \sum_{i=0}^{\infty} c_i t^i$, where the c_i are vectors in $\mathbb{C}^d$. Thus $f(t) \in \mathbb{C}^d$. The problem is to construct Padé approximants of f. Obviously, the usual Padé approximants corresponding to each component of f can be constructed separately. However, two better ways for constructing approximants of f were studied: simultaneous approximants by Marcelis Gerrit de Bruin [146], and vector approximants due to Jeannette van Iseghem [658, 660]. In both cases, a rational approximant $(P_1/Q, \ldots, P_d/Q)$ is obtained, and the scalar Padé approximants are recovered when $d = 1$. In the simultaneous case, the degrees of the numerators P_i have to satisfy some constraints, while in the vector case, they all have the same degree. Although the ideas are rather similar, the approximants differ, except when $\deg(Q) = nd$ and $\deg(P_i) = n$, for $i = 1, \ldots, d$.

The vector Padé approximants are simpler than the simultaneous ones. They are given by

$$[m + k - 1/m]_f(t) = \frac{\begin{vmatrix} t^m \sum_{i=0}^{k-1} c_i t^i & \cdots & \sum_{i=0}^{m+k-1} c_i t^i \\ c_k & \cdots & c_{m+k} \\ \vdots & & \vdots \\ c_{k+n-1} & \cdots & c_{m+k+n-1} \\ c_{k+n}^{(j)} & \cdots & c_{m+k+n}^{(j)} \end{vmatrix}}{\begin{vmatrix} t^m & \cdots & 1 \\ c_k & \cdots & c_{m+k} \\ \vdots & & \vdots \\ c_{k+n-1} & \cdots & c_{m+k+n-1} \\ c_{k+n}^{(j)} & \cdots & c_{m+k+n}^{(j)} \end{vmatrix}}, \quad j = 0, \ldots, d-1,$$

with $m = nd + j$, $0 \le j < d$. Only the elements of the first row of the numerator are vectors, and are manipulated as such. The other rows are formed by scalars as follows: each vector c_i has to be replaced by its d scalar components. Thus a row $c_i, \ldots, c_{m+i}$ of vectors corresponds to the d scalar rows containing their components. Similarly, $c_i^{(j)}$ is put for the first j components of the vector c_i. Thus, the rows of the numerator from the second one represent $nd + j = m$ scalars located in its $m + 1$ columns. The approximant $[m + k - 1/m]_f$ is the vector obtained as the linear combination of the $m + 1$ vectors of the first row of the numerator divided by the scalar determinant in the denominator. These determinants have dimension $m + 1$. The numerator is a polynomial of degree $m + k - 1$ with vector coefficients, and the denominator is a scalar polynomial of degree m. The order of approximation is $m + k + n - 1$ at least, and $m + k + n$ for the first j components. These approximants are optimal in the sense that it is impossible to improve simultaneously the order of approximation of all components. A cross-rule, extending that of Wynn, holds between these approximants [659]. Their study was based on a theory of formal *vector orthogonal polynomials* (also called of dimension d): they satisfy a recurrence relationship of order $d + 1$ whose coefficients can be computed by an extension of the qd-algorithm [661]. They also satisfy a Shohat–Favard-type theorem. This notion of d orthogonality can be related to that of $1/d$ orthogonality due to Pascal Maroni [435]. These approximants can be recursively computed by the *Recursive Projection Algorithm* (RPA) or by the *Compact Recursive Projection Algorithm* (CRPA) [96]. More details can be found in [142, 143]. Other generalizations of Padé approximants to vector series can be obtained from the vector or topological extensions of the ε-algorithm.

Matrix Padé approximants of the exponential function were considered in 1961 by Richard Steven Varga (b. 1928) for the numerical solution of self-adjoint parabolic partial differential equations [673]. Let us mention that their computation presents difficulties related to numerical stability [454, 455]. In 1971, Jorma J. Rissanen (1932–2020) published a paper [528] (submitted 13 August 1971) in which he defined matrix polynomials as follows. He first considered two polynomials $P(x) = \sum_{i=0}^m P_i x^i$, $P_i \in \mathbb{R}^{s \times t}$, and $Q(x) = \sum_{i=0}^n Q_i x^i$, $Q_i \in \mathbb{R}^{s \times s}$, and if the determinant $q(x)$ of the matrix $Q(x)$ does not vanish for all values of x, he considered the rational form

$$C(x) = [Q(x)]^{-1}P(x) = T(x)/q(x) = \sum_{i=0}^{\infty} C_i x^i,$$

where, as he said, $T(x)$ is a matrix polynomial *determined* by P and Q. Then let $F(x) = \sum_{i=0}^{\infty} A_i x^i$, $A_i \in \mathbb{R}^{s \times t}$, be a matrix series. If $A_0, \ldots, A_N$ determine a polynomial P of degree m with matrix coefficients, and a monic polynomial q of degree n with scalar coefficients such that $C_i = A_i$ for $i = 0, \ldots, N$, and there is no common factor other than ± 1 in q and all the elements of P, then Rissanen defined the pair (P, q) as the Padé approximant $[m/n]$ of F. Writing that $F(x)q(x) - P(x) = O(x^{N+n})$ with $N \geq \max(m, n)$ and identifying the coefficients, he obtained

$$\sum_{i=0}^{m} q_i A_{r-i} = P_r, \quad r = 0, \ldots, m, \quad \text{and} \quad \sum_{i=0}^{n} q_i A_{r-i} = 0, \quad r = m+1, \ldots, N.$$

Any solution $q_0, \ldots, q_n$ of the second system allows one to compute directly the matrices P_i via the first one. His definition extends that of Wynn in [720]. Then Rissanen explained the computational procedure in detail, and gave an algorithm with its flow chart and numerical examples.

Padé approximants for series whose coefficients are square matrices have been used by physicists for a long time, because they have many applications in theoretical physics, in the partial realization problem in system theory, and in statistics. They can be easily introduced via Padé-type approximants, as done by André Draux [207]. We will follow his presentation. Let A be a noncommutative algebra on a commutative field $\mathbb{K}$ of characteristic 0 and with a unit element I. Let f be a formal power series with coefficients $c_i \in A$. Let us define two linear functionals ^Rc and ^Lc on the space of polynomials with coefficients in A by

$$^Rc(ax^i) = ac_i, \quad ^Lc(ax^i) = c_i a, \quad a \in A.$$

Let v_k be an arbitrary polynomial of degree k with coefficients in A and such that the coefficient of x^k is invertible. We set

$$\widetilde{v}_k(t) = t^t v_k(t^{-1}),$$
$$^L w_{k-1}(t) = {}^Lc((v_k(x) - v_k(t))(x - t)^{-1}), \qquad {}^L\widetilde{w}_{k-1}(t) = t^{k-1}\, {}^L w_{k-1}(t^{-1}),$$
$$^R w_{k-1}(t) = {}^Rc((v_k(x) - v_k(t))(x - t)^{-1}), \qquad {}^R\widetilde{w}_{k-1}(t) = t^{k-1}\, {}^R w_{k-1}(t^{-1}).$$

The left and right matrix Padé-type approximants are then defined by

$$^L(k-1/k)_f(t) = {}^L\widetilde{w}_{k-1}(t)(\widetilde{v}_k(t))^{-1} = f(t) + O(t^k),$$
$$^R(k-1/k)_f(t) = (\widetilde{v}_k(t))^{-1}\, {}^R\widetilde{w}_{k-1}(t) = f(t) + O(t^k).$$

Let us now choose the polynomial v_k that satisfies the conditions

$$^Lc(x^i v_k(x)) = 0 \quad \text{or} \quad {}^Rc(x^i v_k(x)) = 0, \quad i = 0, \ldots, k-1.$$

The polynomials $\{v_k\}$ form the family (up to a normalization condition) of matrix formal orthogonal polynomials with respect to ^Lc or ^Rc, and the matrix Padé-type approximants become the matrix Padé approximants denoted by square brackets, and one has

$$^L[k-1/k]_f(t) = \;^R[k-1/k]_f(t) = f(t) + O(t^{2k}).$$

These matrix Padé approximants, which are uniquely defined, satisfy the same algebraic properties as the scalar Padé approximants. The block structure of the Padé table (which no longer consists of square blocks) has been studied, and recursive algorithms for the computation of any sequence of Padé approximants have been obtained via the connection with matrix formal orthogonal polynomials. A cross rule, similar to that of Wynn, also holds, and an ε-algorithm as well [208].

6.9 Continued Fractions

Converging factors for continued fractions were the topic of several papers. In [98], general conditions on the choice of the converging factors in order to obtain a modified continued fraction converging faster than the initial one are given. A device for controlling the error is studied. It was known that limit periodic continued fractions converge linearly. The reciprocal of this result was proved in [121], and thus the only linearly converging continued fractions are the limit periodic ones. For a survey on these topics and the corresponding references, see [417].

Another type of continued fractions that have been studied are branched continued fractions as defined in [66]. They are used for the approximation of functions in two variables [187] and their interpolation by an extension of Thiele's formula for functions [593]. See [65] for a review of this topic, which is still active. A general recent reference on the use of continued fractions for computing special functions is the book [186].

Wynn published many papers on noncommutative continued fractions [746, 751, 757, 765, 767]. After him, several researchers worked on the same subject. The first reference after his works on this topic seems to be the dissertation of Pia Pflüger, a student of Peter Henrici, at ETH Zürich in 1966, on matrix continued fractions [495] (see also [322]). The topic was taken up again in 1967 by Wyman Glen Fair and Yudell Leo Luke (1918–1983) [230], who, by means of a sequence of linear fractional transformations, obtained a formal continued fraction representation of the solution to the matrix Riccati equation. See also the Ph.D. thesis [227] of Wyman Glen Fair, defended at the University of Kansas in 1969, and his papers [228, 229].

Research on this topic is ongoing. Thomas Lee Hayden in 1974 [314] considered that *the proper generalization of continued fractions to complex Banach spaces is the limit of holomorphic maps generated by composition of linear fractional maps.* Wynn is not quoted. In 1976, using the definition of Fair [228], Nicolae Negoescu (1931–2004) extended the convergence theorem of Julius Worpitzky [701] to noncommutative continued fractions [463]. An extension of the convergence theorem of Pringsheim [512] to the noncommutative case was proved in 1981 by Hans Denk and Max Riederle [199]. They gave 7 references about mathematical applications of noncommutative continued fractions, 15 on general papers concerning them (3 of them by Wynn [746, 751, 765], and 3 by P.I. Bodnarchuk dating to 1969–1970), and 10 on nonstandard continued fractions (with [767]). Convergence theorems for matrix continued fractions were given by David A. Field in 1984 [235]. He quotes [746, 751]. In 2000, several convergence criteria for noncommutative continued fractions whose elements are matrices were given in [517]. Let us also mention the habilitation of Hendrik Baumann, defended in 2017, which contains many references [44]. Adam Doliwa [203], in 2019, studied double-sided continued fractions whose coefficients are noncommuting symbols. In his abstract, he wrote:

> We start with presenting the analogs of the standard results from the theory of continued fractions, including their (right and left) simple fractions decomposition, the Euler–Minding summation formulas, and the relations between nominators [*i.e., numerators*] and denominators of the simple fraction decompositions. We also transfer to the noncommutative double-sided setting the standard description of the continued fractions in terms of 2×2 matrices presenting also a weak version of the Serret theorem [*a result on the approximation of two irrational numbers*].

Wynn's paper [746] is mentioned.

6.10 Varia

In the *general presentation* of this book, we mentioned that the topics covered in this book had, and still have, many applications in numerical analysis and in applied mathematics without giving the corresponding references, which can quite easily be found on the internet. They concern convergence acceleration, summation of divergent series, treatment of the Gibbs phenomenon, numerical solution of integral equations, numerical linear algebra, estimation of the errors in the solution of linear systems, Tikhonov regularization, the PageRank and the multi PageRank problems, tensor eigenvalues, treatment of breakdown and near-breakdown in Lanczos-type methods, model reduction, solution of nonlinear equations, and computation of matrix functions.

As a bonus, let us now highlight the fact that convergence acceleration methods, Padé approximants, orthogonal polynomials, and continued fractions have an impact in domains quite far away from those considered before, such as theoretical physics, number-theoretic problems, theoretical computer science, combinatorics, statistics, integrable systems, and general relativity.

6.10.1 Theoretical Physics

Let us now explain the interest of theoretical physicists in nonlinear convergence acceleration methods, Borel resummation, and Padé approximants (see [159]).

Perturbation theory is an important tool for describing real quantum systems, as it turns out to be very difficult to find exact solutions to the Schrödinger equation for Hamiltonians of even moderate complexity. The Hamiltonians to which exact solutions are known, such as the hydrogen atom, the quantum harmonic oscillator, and the particle in a box, are too simple to adequately describe most systems. Using perturbation theory, we can use the known solutions of these simple Hamiltonians to generate solutions for more complicated systems. The idea is to start with a simple system for which a mathematical solution is known, and add an additional "perturbing" Hamiltonian representing a weak disturbance to the system. If the disturbance is not too large, the various physical quantities associated with the perturbed system can be expressed as "corrections" to those of the simple system. For example, by adding a perturbative electric potential to the quantum-mechanical model of the hydrogen atom, tiny shifts in the spectral lines of hydrogen caused by the presence of an electric field can be calculated.

The expressions produced by perturbation theory are not exact, but they can lead to accurate results as long as the expansion parameter, say α, is very small. Typically, the results are expressed in terms of finite power series in α that seem to converge to the exact values when summed to higher order. After a certain order however, the results become increasingly worse, since the series are usually divergent (being asymptotic series). The earliest use of perturbation theory was the problem of the orbit of the Moon. Under nonrelativistic gravity, the orbit of the Moon is exactly an ellipse as described by Kepler when there are only two gravitating bodies (say, the Earth and the Moon). However, when there are three or more objects (the Earth, the Moon, the Sun, the other planets, ...), Kepler's theory is not entirely satisfactory when the other gravitational interactions are introduced using the formalism of general relativity. The simplified problem consisting of the Earth and the Moon is solved as a first approximation, and then "perturbed" to include the gravitational attractions of the Sun and other objects. This is the famous n-body problem in celestial mechanics [201].

When developing quantum electrodynamics in the 1930s, Max Born, Werner Heisenberg, Pascual Jordan, and Paul Dirac discovered that in perturbative corrections many integrals were divergent. Renormalization was first developed in quantum electrodynamics (QED) to make sense of infinite integrals in perturbation theory.

One way of describing the perturbation theory corrections' divergences was discovered in 1947–1949 by Hans Kramers, Hans Bethe, Julian Schwinger, Richard Feynman, and Shin'ichirō Tomonaga, and systematized by Freeman John Dyson (1923–2020) who wrote in 1952 [217]:

> An argument is presented which leads tentatively to the conclusion that all the power-series expansions currently in use in quantum electrodynamics are divergent after the renormalization of mass and charge. The divergence in no way restricts the accuracy of practical

calculations that can be made with the theory, but raises important questions of principle
concerning the nature of the physical concepts upon which the theory is built.

However, his arguments were slightly naive, not entirely convincing, and im-
precise. The correct study of the divergence of perturbative series in quantum field
theories was initiated by the Russian physicist Lev Nikolaevich Lipatov (1940–2017)
[412], and systematically developed by the group of theoretical physics at Saclay
under the leadership of Jean Zinn-Justin[7] (b. 1943). At the beginning of the sixties,
theoretical physicists began to use Padé approximants. One of their first applications
was [28], where the authors treated the scattering length of a hard core potential,
and the calculation of the binding energy of a Fermi gas of hard spheres. Many other
examples are given by Marcel Froissart [253] and Jean-Louis Basdevant (b. 1939)
[36, 37].

As Froissart explained, in 1967, Daniel Bessis (b. 1933) and Modesto Pusterla
(b. 1932) used Padé approximants to sum the perturbation series in the theory of
renormalized fields [59]. It is a series, probably asymptotic, whose terms become
rapidly quite complicated, which makes precise information on the properties of
the general term difficult to obtain. In order to obtain a meaningful finite answer
from such series, two techniques have been used: Padé approximation and the Borel
transform [71]. However, each of these techniques presents some drawbacks. Padé
approximants can not converge at all, or they can tend to a wrong limit. Borel
transform provides, in principle, a correct answer even in most cases where Padé
approximants fail, but its practical implementation is quite difficult, since it is based
on conformal mapping in the complex plane.

The drawbacks of both methods were avoided by combining them in the so-called
Borel–Padé method [419]. We will explain it without entering into the technical
details. Let us consider the asymptotic series $f(z) = \sum_{n=0}^{\infty} c_n z^n$, and assume that
the coefficients have a factorial growth $c_n \sim (-\alpha)^n n!$ for large n. Its Borel sum is
the series $B(z) = \sum_{n=0}^{\infty} c_n z^n / n!$. It has a nonzero radius of convergence. Indeed, the
factorial growth of the expansion coefficients has been removed, which leads to a
function B that is analytic in some neighborhood of the origin in the complex plane.
The Borel sum converges for $|z| < 1/\alpha$. The function B is called *Borel summable* if
the series can be summed and analytically continued over the whole positive axis.

The Borel transformation restores the factorial in each term of the expansion.
The integral $F(z) = z^{-1} \int_0^{\infty} e^{-t/z} B(z) \, dt$ is absolutely convergent in a domain of the
complex plane, and $F(z) = f(z)$. Thus the integral $F(z)$ provides a formal sum for
the asymptotic series $f(z)$, and the Taylor expansion of F around the origin coincides
with the series f [571]. The main drawback of this method is that usually B is not
known, since in practice, only a finite set of numerically computed coefficients c_n is
available; see [708, Secs. 42.5–7]. The summation of series by Borel transformation
was introduced in 1976 by Jean-Jacques Loeffel (1932–2013) [413]. Since the series
B cannot be used in practice, the idea was to replace $B(z)$ in the definition of F by
$[n + k/k]_B(z)$ with $n \geq -1$, thus giving rise to the Borel–Padé approximants. Many
papers on this method have been published over the years.

---

[7] See his testimony in Sect. 7.12.

As explained in [419, 439], the combination of the Borel summability method with the first confluent form of the ε-algorithm of Wynn for summing the Born series of scattering integral equations is equivalent, in a limiting case, to the construction of the sequence of $[n/n-1]$ Padé approximants to the same series. Very good results were thus obtained in practice, but physicists were then faced with another problem: the proof of the convergence of the method. This is why the Padé approximants involved in the Borel–Padé method were replaced by Padé-type approximants [440]. The convergence of the method was proved for a suitable choice of the generating polynomials of the Padé-type approximants. These works on renormalization opened the way to the standard model of particle physics.

Jean Zinn-Justin wrote [707]:

> In strong interactions, the perturbative expansion has been used for a long time, only in order to understand the analytic properties of the scattering amplitudes, or to estimate Born terms. The reason for this was that the coupling constants were found to be relatively large, and therefore in many cases the perturbation series was useless.

> Moreover, it was not in general possible to find a good non relativistic approximation to the scattering problem, in order to compute first order approximations of the mass spectra as in quantum electrodynamics, so that the problem of finding bound states or resonances, starting from an interaction lagrangian, was open.

> A way to solve these different problems is to find a method of summation for divergent series.

> Among these methods, the Padé approximation seems to have very pleasant mathematical and physical properties.

Loewner's theorem concerns the theory of monotone matrix functions, i.e., functions f such that $B - A$ is positive definite implies $f(B) - f(A)$ is also positive definite for all pairs of self-adjoint matrices. In 1934, Charles Loewner (1893–1968) proved that f on (a, b) is a monotone matrix function on all such $n \times n$ pairs (for all n) if and only if f is real analytic on (a, b) and has an analytic continuation to the upper half-plane with a positive imaginary part there [414]. As explained by Eugene Paul Wigner (1902–1995), Nobel Prize in Physics in 1963, and John von Neumann (1903–1957) in [696], this theorem plays a role in the investigation of the analytic nature of the collision matrix and the derivative matrix R involved in the quantum-mechanical wave function of a collision system. These authors studied the properties of R by developing it into a continued fraction. Loewner's theorem is the central theme of the book of Barry Simon (b. 1946) [594], where various proofs using continued fractions, Padé approximants, and multipoint Padé approximants are given. Another reference on monotone matrix functions is the book by William Francis Donoghue, Jr. (1921–2002) [205].

6.10.2 Number Theory

Padé approximants and continued fractions are instrumental in the prof of irrationality of certain numbers. Recall the proof by Lambert of the irrationality of π. A general criterion for irrationality is due to Axel Thue (1863–1922) in 1909 [638]. Let us give a sample of papers dealing with this problem. Other references can easily be found on the internet.

A new proof of the irrationality of $\zeta(2)$ and $\zeta(3)$ (the Riemann zeta function is defined by $\zeta(s) = \sum_{n=1}^{\infty} 1/n^s$) was given by Marc Prévost [507] who, using Padé approximants, gave a different proof of the identities obtained by Roger Apéry (1916–1994) [20]. Another example of their use is given in [508]. The number $\zeta(3)$ is also considered in [605] by Padé–Hermite approximation.

More recently, Walter van Assche (b. 1958) [654] gave:

> [. . .] a method for proving that a given real number is irrational. It amounts to constructing explicit rational approximants to the real number which are better than possible in case the number is rational. The rational approximants are obtained by evaluating a Hermite-Padé rational approximant to explicit functions at a convenient (integer) point.

6.10.3 Computer Science and Control Theory

At the end of the 1960s, the theory of automata and languages was a very active domain, with Marcel-Paul Schützenberger (1929–1996) as one of the world's leading experts. He developed a theory of noncommutative rational and algebraic series that extended and deepened the theory of formal context-free languages. The introduction of linear algebra made it possible to quantify ambiguity in algebraic languages and to obtain algebraic proofs. The theory of rational series in one variable extends remarkably to rational series in several variables [558]. He was followed in this direction by Michel Fliess (b. 1945), who worked on the algebraic and arithmetic aspects of noncommutative formal power series, together with their applications in theoretical computer science, mostly formal language theory. As explained by Christophe Reutenauer (b. 1953) in [519], his results on Hankel matrices were influential:

> This work of Fliess has had much influence, especially in Control Theory (bilinear systems). Let me recall first the classical case. Consider a linearly recurrent sequence (a_n) and form the infinite matrix $(a_{i+j})_{i,j\geq0}$, called the Hankel matrix. Then it is well-known that this matrix has finite rank (the rank of an infinite matrix is defined as in the finite case). Conversely, if the rank of this matrix is finite, then the sequence is linearly recurrent; and this rank is equal to the length of the shortest linear recursion satisfied by the sequence. Fliess [244] [*our bibliography*] extends all this to noncommutative series.

Control theory is an interdisciplinary domain [110, 154]. It concerns the behavior of a *dynamical system* represented by linear differential or difference equations. The system receives an input (the *control variables*), transforms it following some laws, and an output is measured. The problem is, by a procedure called *feedback*, to regulate the input in order to obtain a desired output. Such a system can also have inaccessible variables, the *state variables*. There are continuous and discrete systems. They can be simultaneously described by the equations

$$\mathcal{D}x(t) = Ax(t) + Bu(t), \tag{6.6}$$

$$y(t) = Cx(t), \tag{6.7}$$

where the variable t is the time, $\mathcal{D}$ is the differential operator $\mathcal{D} = d/dt$ in the continuous case, and the forward difference operator defined by $\mathcal{D}x(t) = x(t+1)$ in the discrete one. The vector $x \in \mathbb{R}^n$ is the *state vector*, $u \in \mathbb{R}^m$ is the *input*, $y \in \mathbb{R}^p$ is the *output*, with $A \in \mathbb{R}^{n \times n}, B \in \mathbb{R}^{n \times m}, C \in \mathbb{R}^{p \times n}$. The integer n is the dimension of the system.

The problem can be treated from two different approaches: the *state-space* approach, which consists in studying the system (6.6) and (6.7) in the time domain, and the *frequency-domain* approach, which makes use of the Laplace transform for a continuous system and the z-transform for a discrete one (see Sect. 5.8). Their interest lies in the fact that the dynamical behavior of a complicated differential or difference equation can be studied using only algebraic operations. Setting $s = \omega\sqrt{-1}$, where ω is the frequency, the Laplace transform moves the problem from the time domain to the frequency domain.

Let g be a function of t. Its Laplace transform $\widetilde{g}$ is defined by

$$\widetilde{g}(s) = \int_0^\infty g(t)e^{-st}\, dt.$$

Assuming for simplicity (which does not restrict the generality) that $x(0) = 0$, the Laplace transform of (6.6) gives $s\widetilde{x}(s) = A\widetilde{x}(s) + B\widetilde{u}(s)$, that is, $\widetilde{x}(s) = (sI - A)^{-1}B\widetilde{u}(s)$. Then the Laplace transform of (6.7) leads to $\widetilde{y}(s) = C\widetilde{x}(s) = C(sI - A)^{-1}B\widetilde{u}(s) = G(s)\widetilde{u}(s)$. The matrix $G(s) = C(sI-A)^{-1}B$ is called the *transfer function* of the system. Each element g_{ij} of G represents the transfer function between the ith input and the jth output, all other inputs being set equal to zero. They are rational functions with a numerator of degree $n-1$ at most, and a common denominator of degree n. Then y is obtained by inverting the Laplace transform $\widetilde{y}$.

We have

$$G(s) = \frac{1}{s}\sum_{i=0}^\infty c_i s^{-i} = -\sum_{i=0}^\infty \widetilde{c}_i s^i,$$

with $c_i = CA^i B \in \mathbb{R}^{p \times m}$ and $\widetilde{c}_i = CA^{-i-1}B$ if A is nonsingular. We also have

$$G(s) = (G_0 s^{n-1} + G_1 s^{n-2} + \cdots + G_{n-1})/P_n(s),$$

with $G_0 = CB$, $G_i = CB_i B$ for $i = 1, \ldots, n-1$, and where $P_n(s) = \det(sI - A) = s^n + a_1 s^{n-1} + \cdots + a_n$ is the characteristic polynomial of A, and the matrices B_i are computed by the Leverrier–Faddeev–Souriau method [607]:

$$B_1 = A + a_1 I, \qquad a_1 = -\text{tr}(A),$$
$$B_i = AB_{i-1} + a_i I, \qquad a_i = -\text{tr}(AB_{i-1})/i, \quad i = 2, \ldots, n-1.$$

Notice that moreover, $B_n = 0$ and $A^{-1} = -B_{n-1}/a_n$ with $a_n = \det A$.

Realization theory consists in determining a system having a given transfer function G. In practice, G is known from experimental results. The solution of the problem depends on the information available on G, which can be its algebraic expression, its power series expansion around some points, or only its first Taylor coefficients, or the values of G at selected points. Complete realization consists in determining the matrices A, B, and C to obtain a given function G. If only a finite number N of the coefficients $\widetilde{c}_i$ are known, *partial realization* consists in finding matrices $\widetilde{A}$, $\widetilde{B}$, and $\widetilde{C}$ of the same dimensions as A, B, and C such that

$$H(s) = \widetilde{C}(sI - \widetilde{A})^{-1}\widetilde{B} = -\sum_{i=0}^{\infty} h_i s^i$$

satisfies $h_i = \widetilde{C}\widetilde{A}^{-i-1}\widetilde{B} = \widetilde{c}_i$ for $i = 0, \ldots, N - 1$. Obviously, Padé and Padé-type approximants can be used for the solution of this problem. Padé-type approximants are particularly attractive since an important characteristic of a transfer function is its stability, which is satisfied if all the eigenvalues of A, that is, the zeros of $P_n(s) = \det(sI - A)$, have strictly negative real parts.

Another way of solving the realization problem is *model reduction*, which consists in looking for a system of dimension k smaller than n. In other words, the transfer function G, which has a denominator of degree n, is approximated by another rational function with a denominator of degree k. One way of solving this problem is to use Lanczos's *biorthonormalization method* [386], which constructs the matrices $V_k = [v_1, \ldots, v_k]$ and $W_k = [w_1, \ldots, w_k]$ such that $W_k^T V_k = I$ with $J_k = W_k^T A V_k$ tridiagonal. As proved in [291], in the case $m = p = 1$, if $v_1 = B$ and $w_1 = C^T$, then the choice $\widehat{A} = J_k$, $\widehat{B} = W_k^T B$ and $\widehat{C} = CV_k$ leads to

$$CA^{i-1}B = \widehat{C}\widehat{A}^{i-1}\widehat{B}, \quad i = 1, \ldots, 2k.$$

It follows that

$$G_k(s) = CV_k(sI - W_k A V_k)^{-1}W_k B = [k - 1/k]_G(s).$$

This approach for obtaining a reduced model is called *Padé via Lanczos* (PVL) [232]. A drawback is that the stability property of the transfer function is not always preserved. For that reason, partial Padé approximants, where a part of the numerator and a part of the denominator can be chosen, are preferred [24] (see Sect. 6.8).

In general, it is not possible to approximate nonlinear systems by linear ones, as explained in [326]. The input–output behavior of a dynamical analytic system has to be coded by a noncommutative formal power series in several variables. Padé-type and Padé approximants to such series were constructed in [325] and related to continued fractions.

6.10.4 Combinatorics

Let us begin by recalling some definitions.

A symmetric function in n variables with values in an abelian group is a function that does not change for any permutation of its variables. A particular case concerns symmetric polynomials. The elementary symmetric polynomials in n variables $x_1, \ldots, x_n$, denoted by $e_k(x_1, \ldots, x_n)$, for $k = 0, \ldots, n$, are defined by

$$e_0(x_1, \ldots, x_n) = 1,$$
$$e_1(x_1, \ldots, x_n) = \sum_{1 \le i \le n} x_i,$$
$$e_2(x_1, \ldots, x_n) = \sum_{1 \le i \le j \le n} x_i x_j,$$

and so on until $e_n(x_1, \ldots, x_n) = x_1 x_2 \cdots x_n$. In number theory and combinatorics, a partition of a positive integer n, also called an integer partition, is a way of writing n as a sum of positive integers. Two sums that differ only in the order of their summands are considered the same partition. Schur polynomials, named after Issai Schur, are certain symmetric polynomials, indexed by partitions, that generalize the elementary symmetric polynomials.

Let $x = (x_1, \ldots, x_n)$ be a set of variables and let Λ be the algebra of symmetric functions in x. Bases for this algebra are indexed by partitions $\lambda = (\lambda_1, \ldots, \lambda_n)$, i.e., λ is a weakly decreasing sequence of nonnegative integers called parts. Associated with any partition is an alternant, which is the determinant $a_\lambda = \det(x_i^{\lambda_j})$. In particular, for the partition $\lambda = (n - 1, n - 2, \ldots, 0)$ one has the Vandermonde determinant. In his thesis [559], Schur defined the functions that bear his name as $s_\lambda = a_{\lambda+\delta}/a_\delta$, where addition of partitions is componentwise. It is clear from this equation that s_λ is a symmetric homogeneous polynomial of degree $\sum_i \lambda_i$. Schur functions are a special basis for the algebra of symmetric functions Λ.

In the abstract of his book [392], Alain Lascoux (1944–2013) wrote:

> The theory of symmetric functions is an old topic in mathematics which is used as an algebraic tool in many classical fields. With λ-rings, one can regard symmetric functions as operators on polynomials and reduce the theory to just a handful of fundamental formulas. One of the main goals of the book is to describe the technique of λ-rings. The main applications of this technique to the theory of symmetric functions are related to the Euclid algorithm and its occurrence in division, continued fractions, Padé approximants, and orthogonal polynomials.

Obviously, it is impossible to enter into the mathematical details but let us only mention that the cross rule of Wynn for Padé approximants [754] (see Sect. 5.1.1.1) is recovered by using the vocabulary of symmetric functions. Continued fractions, reciprocal differences, and orthogonal polynomials are also expressed in the same language.

In [393], the authors give, among others, the following result:

Given a linear functional, it is known how to write an orthogonal basis in terms of moments [...] the following proposition is a rewriting, in terms of Schur functions, of the classical expressions of orthogonal polynomials in terms of Hankel determinants involving moments.

Lascoux had an impressive culture on determinants, but he treated them by means of symmetric functions and Schur functions (which are given by determinants in complete or elementary symmetric functions); see [504] for an analysis of his works. In [390], he declared in his abstract:

Hankel determinants can he viewed as special Schur symmetric functions. This provides, without computations of determinants or matrices, most of the algebra of Hankel determinants and explains their links with other classical fields.

and he added (translation from French):

We prefer to detail the properties in question from the symmetric functions. This makes the links with theories such as the theory of elimination, Euclidean division, continued fractions, the resultant and the Bezoutian, recurring sequences, the rational interpolation (Padé approximants) and orthogonal polynomials, the localization of the zeros of the polynomials. For our part, we do them derive from their common roots in the theory of symmetric functions.

Lascoux gave a combinatorial description of powers of continued fractions in [391]. Noncommutative continued fractions, noncommutative Padé approximants, and noncommutative orthogonal polynomials are treated in [265] by means of quasi-determinants and quasi-Schur functions; see also [377].

Another topic studied by combinatorists concerns orthogonal polynomials. In combinatorial models, some finite structures have been introduced (permutations, trees, ...), and combinatorial interpretations of the coefficients of their recurrence relationship have been given. The combinatorial study of identities on orthogonal polynomials was reviewed by Dominique Foata (b. 1934) in [245]. In [675], Xavier Gérard Viennot considered formal orthogonal polynomials and introduced weighted paths in order to give combinatorial (i.e., with bijections) proofs of their classical properties, for example, the equivalence between the orthogonality and the three–term recurrence. Classical results are shown to be a consequence of the construction of bijections and correspondence between finite structures such as paths and tilings. Viennot also gave a combinatorial interpretation of the qd-algorithm that is derived from the geometry of the paths, without involving the usual determinant manipulations [676].

A major contributor to the combinatorial interpretation of continued fractions was Philippe Flajolet (1948–2011), a member of the French Academy of Sciences. A conference, *Philippe Flajolet and Analytic Combinatorics: Conference in the memory of Philippe Flajolet*, was held at Paris-Jussieu, 14–16 December 2011. After this conference, a project for a seven-volume *Collected Works of Philippe Flajolet* was started. Unfortunately, the project was not finished. In Volume V, in a draft of the introduction to Chapter 3, on continued fractions, Viennot wrote about him:

Analytic continued fractions theory was put at the combinatorial level with a beautiful interpretation in terms of certain weighted paths (the so-called Motzkin paths). This interpretation is described in the famous seminal paper [239] [*our bibliography*]. Combining this interpretation with some combinatorial bijections between classical combinatorial objects and these weighted paths leads Philippe Flajolet to a fireworks of combinatorial proofs of many classical developments of power series into continued fraction in connection with special functions, together with the explicit computation of integrated costs for some classical data structures such as lists, priority queue, dictionary, each corresponding to classical orthogonal polynomials (Hermite, Laguerre, Charlier, . . .).

In his paper [239], Flajolet explained:

In this paper we present a geometrical interpretation of continued fractions together with some of its enumerative consequences. The basis is the equivalence between the characteristic series of positive labelled paths in the plane and the universal continued fraction of the Jacobi type. The equivalence can be asserted in the strong form of an equality in the set of formal series in non-commutative variables [. . .]

Paths we wish to consider here are positive paths in the x–y plane, which consist of only three types of steps: rises, levels and falls. More precisely we start with three step vectors $a = (1, 1)$, $b = (1, -1)$, $c = (1, 0)$ called respectively *rise* vector, *fall* vector and *level* vector; to each word $u = u_1 u_2 \cdots u_n$ on the alphabet $\{a, b, c\}$ is associated a sequence of points $M_0 M_1 \cdots M_n$ where $M_0 = O = (0, 0)$ and for each j s.t. $1 \leq j \leq n, OM_j = OM_{j-1} + u_j$ [. . .]

The connections between powers series and continued fractions were described in terms of matrix equations by Stieltjes [623]. Flajolet showed that the elements of the Stieltjes matrix have simple combinatorial interpretations. Enumerative properties of continued fractions and their convergents are then studied. Flajolet came back to continued fractions in the coauthored paper [241].

A combinatorial interpretation of Padé approximants was given in his doctoral thesis, under Xavier Gérard Viennot, by Emmanuel Roblet [530], who explained what he did:

We develop combinatorial models for three families of rational approximants: Padé approximants, two-point Padé approximants and vector Padé approximants. Our method consists in writing the coefficients of formal power series as sums of valuations of some labelled paths on the square lattice. The geometry of these paths provides a uniform framework for computing Hankel determinants, investigating the structure of the Padé table and studying the three classes of continued fractions linked to these approximants, respectively named P-fractions, T-fractions, Łukasiewicz multicontinued fractions. Our models also provide new algorithms for the computation of these approximants [. . .]

On Motzkin's paths, Łukasiewicz words, series reversion as explained by Lagrange [381], continued fractions, Padé approximants, and Hankel determinants, see [23].

Philippe Flajolet was the founder of analytic combinatorics, a new branch of combinatorics. It consists in expressing a problem in combinatorial terms and then passing to an analytic representation of combinatorial objects via generating functions (this first step is qualified by Flajolet as a symbolic method). The tools of complex analysis (computation of residue, saddle point, localization of the singularities) make it possible then to obtain results on the behavior of the generating series (in particular on its coefficients). This approach can be seen as a cousin of the analytic theory of numbers. His book with Robert Sedgewick (b. 1946) in this domain [240] received the 2019 Steele Prize for Mathematical Exposition. Analytical combinatorics finds a natural breeding ground for applications in the average analysis of algorithms, because it is often based on the study of some key parameters of the discrete structures of computer science (trees, graphs, words, permutations ...). Flajolet and Sedgewick have thus studied the effectiveness of numerous variants of sorting algorithms, the appearance of motifs in text or in trees [564].

6.10.5 Statistics

In the statistical analysis of time series, autoregressive-moving-average (ARMA) models provide a parsimonious description of a (weakly) stationary stochastic process in terms of two polynomials, one for the autoregression (AR) and the second for the moving average (MA). Given a time series of data X_t, the ARMA model is a tool for understanding and, perhaps, predicting future values in this series. The AR part involves regressing the variable on its own lagged (i.e., past) values. The MA part involves modeling the error term as a linear combination of error terms occurring contemporaneously and at various times in the past. The model is usually referred to as the ARMA(p, q) model, where p is the order of the AR part and q is the order of the MA part. The problem is to determine the values of p and q. A second-order stationary process X admits an ARMA(p, q) minimal representation if and only if the sequence X_t satisfies a linear difference equation of minimal order p from the initial rank $q - p + 1$. In his doctoral thesis [52], Alain Berlinet (b. 1952) applied the scalar ε-algorithm for that purpose, and he proved that X admits an ARMA(p, q) minimal representation if and only if $\varepsilon_{2p}^{(n)} = 0$ for all $n \geq q - p + 1$, and $\varepsilon_{2p}^{(q-p)} \neq 0$. The proof is based on the kernel of the Shanks transformation and on the nullity or not of the Hankel determinants appearing in its numerator.

Padé approximants, that is, the ε-algorithm, were also used for predicting the behavior of relevant economic variables in data series with a certain degree of certainty [279].

6.10.6 Integrable Systems

A dynamical system is called *completely integrable* if its motion at each time t can be expressed analytically in terms of the initial values of the variables. This is, for example, the case of the simplified differential equation of the pendulum, which is a *continuous* integrable system, and of the corresponding discretized equation, which is a *discrete* integrable system. In spite of their nonlinearity, integrable systems possess quite a remarkable regularity, exhibiting smooth behavior in all regions of the phase space. In contrast, a nonintegrable system often exhibits chaotic motion. The first nonintegrable system was studied in the mid 1950s by the Nobel Prize winner Enrico Fermi (1901–1954) and his collaborators. It represents the time evolution of a group of particles coupled together by nonlinear springs and constrained to move in only one direction. The system consists of a set of coupled differential equations and it has 64 degrees of freedom (so the phase space has dimension 128). All Hamiltonian systems with one degree of freedom are completely integrable. Hamiltonian systems with more degrees of freedom that are completely integrable form a restricted, but extremely important, class.

Integrable systems are closely related to convergence acceleration methods. Indeed, in [749], Wynn studied partial differential equations associated with certain nonlinear algorithms. Conversely, theoretical physicists have considered two-variable difference equations as the discretization of partial differential equations. As explained in [484]:

> Integration schemes for ordinary or partial differential equations can be greatly improved if they conserve exactly the quantities that are conserved by the initial (continuous) systems. The literature on this topic is important and new schemes are regularly produced incorporating these complete or partial integrability features.

These authors then studied what happens in the Aitken Δ^2 process when the denominator vanishes. In that case, the transformed sequence

> [...] diverges and this divergence propagates indefinitely with the iterations of the mapping. This is, clearly, an undesirable feature for a numerical scheme. The next question is, thus, whether one can derive accelerator schemes for which the (spontaneously appearing) singularities are confined to a few iteration steps.

Since singularity confinement was proposed as an integrability criterion for discrete systems, this is an important problem. It is clearly related to the papers of Wynn [716] and [743]. Then the authors turn to the ε-algorithm, which

> [...] is expected to be integrable. And in fact it is! The reason is simple: this algorithm, considered as a two-variable difference equation, is precisely the (discrete) lattice potential KdV equation.

The ϱ-algorithm is also studied. Finally, a new convergence acceleration algorithm based on a modified Korteweg–de Vries lattice equation is derived that shows that the connection works in both directions.

Japanese and Chinese researchers, sometimes with occidental colleagues, devoted quite a few papers to the link between both domains. A first example was given in [315], where the lattice Boussinesq equation gave rise to a new acceleration algorithm.[8] A q-difference lattice Boussinesq equation and its confluent form were also proposed, and their molecule solutions were constructed. It can be used as a numerical convergence acceleration algorithm for computing approximations of the limit of a function when its argument tends to infinity [628]. Similarly, the

---

[8] See the testimony of Xing-Biao Hu in Sect. 7.5, and that of E.J. Weniger in Sect. 7.11.

η-algorithm and a generalized version of the ϱ-algorithm are considered as integrable discretizations, respectively, of the discrete and the cylindrical KdV equations [461]. And the purpose of the paper [462]

> [...] is to establish a relation between the ε-algorithm and the Toda molecule equation, to give a physical reason for convergence acceleration from the equation, and to estimate quantitatively the error caused by the algorithm by means of the solution for the equation.

A q-difference version of the ε-algorithm was proposed in [316]. It could be considered the q-difference version of the modified Toda molecule equation. In a similar way as the quotient–difference algorithm is related to the discrete-time Toda equation, a new quotient–quotient–difference algorithm was presented in [608] for a higher-order analogue of the discrete-time Toda equation. In the same spirit, a new algorithm for computing Padé approximations of given power series was derived in [453] from the discrete relativistic Toda molecule equation.

Hirota's bilinear method [331] is quite useful in the solution of nonlinear differential and difference equations like those described above. This method was invented by the Japanese theoretical physicist Ryogo Hirota (1932–2015). When the mathematical expressions of quantities computed by a recursive algorithm like, for example, the $\varepsilon_k^{(n)}$ given by the ε-algorithm have to be derived, Hirota's method consists in writing them in the form of a ratio $\varepsilon_k^{(n)} = G_k^{(n)}/F_k^{(n)}$, plugging them into the rule of the algorithm for obtaining difference equations linking the $G_k^{(n)}$ and the $F_k^{(n)}$, and solving them. In [134], we showed how this method can lead to a novel proof that the ε-algorithm of Wynn implements the Shanks transformation, and reciprocally, that the quantities it computes are expressed as ratios of Hankel determinants as given by Shanks. New identities between Hankel determinants and the quantities involved in Hirota's method were obtained, and they form the basis of our proof. Then the same bunch of results was shown to hold also for the confluent form of the ε-algorithm. Hirota's method was also used in [119] for deriving a multistep extension of the scalar ε-algorithm. The confluent form of this algorithm was studied in [120]. Cross rules and nonabelian lattice equations for the discrete and confluent nonscalar ε-algorithms were given in [113]. Let us mention that the first confluent form of the ε-algorithm is related to the Lotka–Volterra equation. Indeed, recall that its rule is $\varepsilon_{k+1}(t) = \varepsilon_{k-1}(t) + [\varepsilon_k'(t)]^{-1}$. Setting $M_k(t) = \varepsilon_k'(t)$ and $N_k(t) = M_k(t)M_{k+1}(t)$ leads to the Lotka–Volterra equation $N_k'(t) = N_k(t)[N_{k-1}(t) - N_{k+1}(t)]$. Similar results hold for the confluent form of the multistep ε-algorithm.

Hirota's method and determinantal identities were also used in [627] to derive a new sequence transformation from the molecule solution of an extended discrete Lotka–Volterra equation. Its convergence and stability properties were studied, and the transformation was shown to be effective on some kinds of linearly convergent sequences and factorially convergent sequences.

6.10.7 General Relativity

It seems strange that continued fractions could have something to do with general relativity. However, it is the case. The *Mixmaster universe* is a solution of Einstein's field equations whose aim is a better understanding of the dynamics of the early universe. The hope is to show that the early universe underwent an oscillatory, chaotic epoch. The problem arises from the difficulty in explaining the observed homogeneity of causally disconnected regions of space in the absence of a mechanism that sets the same initial conditions everywhere.

This problem was considered in [432], where the authors present

[. . .] an extension of the classical Gauss-Kuzmin theorem about the distribution of continued fractions, which in particular allows one to take into account some congruence properties of successive convergents. This result has an application to the Mixmaster Universe model in general relativity. We then study some averages involving modular symbols and show that Dirichlet series related to modular forms of weight 2 can be obtained by integrating certain functions on real axis defined in terms of continued fractions.

A link to the noncommutative geometry developed by Alain Connes (b. 1947) is explained [178].

Chapter 7
Testimonies

We contacted several researchers who worked on the topics treated in this book or related to them, and asked them for their testimony about the influence played by the main contributors mentioned, in particular Peter Wynn, and how they developed their ideas in their own work. We left them completely free to write what they wanted. They allowed us to reproduce their contributions here, and they are all published below as they were received, without any modification. These testimonies reflect their own views on the domain, and provide an account of their own work. As the reader will see, each testimony has its own flavor, and all are quite different. The last section of this chapter contains the testimonies of the authors of this book, in which they express their personal feelings.

Thus, this chapter highlights the links between the various topics and the various actors encountered in the preceding chapters, thus providing complementary and personal views on the matters treated in this book, together with additional explanations and results. It also allows the reader to have an idea on the latest developments.

No family names and mathematical topics contained in these testimonies were put into the index. The bibliographic citations made by the researchers have been inserted at the end of each corresponding testimony (and not in the general bibliography) and the numbers refer to them. After each reference, the pages where they are quoted are mentioned in italics.

7.1 Adhemar Bultheel

Adhemar Bultheel[1] (b. 1948) is emeritus professor at the *Katholieke Universiteit te Leuven*, Leuven, Belgium, since 2009. He was president of the Belgian Mathematical Society in 2002–2005. He is a specialist in rational approximation, orthogonal functions, and linear algebra. He is currently the author (or coauthor) of several books, around 200 research papers, and more than 85 papers published in proceedings or book chapters.

---

[1] Homepage: https://people.cs.kuleuven.be/~adhemar.bultheel/.

© Springer Nature Switzerland AG 2020
C. Brezinski, M. Redivo-Zaglia, *Extrapolation and Rational Approximation*,
https://doi.org/10.1007/978-3-030-58418-4_7

Testimony

(Not So Much) About Wynn

Because I was supervising the practical sessions associated with courses given by Professor Ludo Buyst about numerical analysis and another one about Padé approximation and continued fractions in the 1970s, I became interested in the computation of Padé approximation for my PhD, which was originally supervised by him. Ludo Buyst was the first to introduce these courses at the University of Leuven, and he later founded the Computer Science Department. I was not the first one to work on these topics, because Luc Wuytack worked on rational approximation and continued fractions before me. He had left Leuven and had a position in Antwerp when I started.

I met the Δ^2 technique and related methods for the acceleration of convergence, but I never did explicit research on these methods, although they were always in the background. I became aware of more general approaches only through the work of Brezinski and by being a jury member in PhD defenses about the topic from some of his students.

For my PhD I was working on all the recursive algorithms to wander through the Padé table and about matrix generalizations that had a linear algebra counterpart dealing with (block) Hankel matrices (for diagonal paths) or Toeplitz matrices (for horizontal paths). Especially in the continued fraction interpretation of these algorithms, QD and epsilon-like algorithms popped up.

When my supervisor was switched to Patrick Dewilde (who shortly there after left Leuven for the TU Delft), I was directed toward more analytic problems that originated from electrical engineering and systems science (closely connected to transmission lines, scattering, and information and coding theory).

The Hankel-type problems were related to system realization (which is essentially Padé approximation at infinity) and model reduction. The recurrences relate to recurrence for orthogonal polynomials on (part of) the real line and corresponding moment problems. They also relate to Krylov-type methods in linear algebra, but that came later in the European project Rolls (1993–1996) and in work with my student Marc Van Barel.

The Toeplitz-type recurrences were more related to filtering and signal processing, the spectral factorization problem, and orthogonal polynomials on the unit circle. My PhD contribution on this topic was to give an interpretation to a multipoint version of these algorithms, which meant to provide orthogonal rational functions instead of orthogonal polynomials. I did the matrix version in both cases (Hankel and Toeplitz).

My PhD was written in Dutch with the title "Rekursieve Rationale Benaderingen" (Recursive Rational Approximations), which I defended in 1979 at KU Leuven.

During that period I of course came in contact with many other people while attending conferences on Padé and rational approximation, continued fractions, systems theory, signal processing, orthogonal polynomials, and linear algebra. I do not think I ever met Wynn at one of these conferences (at least I do not remember).

The only text I ever produced that had explicitly ϵ algorithm in the title was [2]. It was only written up as a report [1],[2] and the qd-algorithm is also the subject of [3].

Chapter 7 of [4] is entitled "Rhombus algorithms" and was inspired by Rutishauser's qd-algorithm [7], Gragg's $\pi\zeta$-algorithm [5], and by John McCabe [6], and of course Wynn.

References

1. A. Bultheel. Epsilon and qd algorithms for matrix- and 2D-Padé approximation. Technical Report TW57, Department of Computer Science, K.U. Leuven, April 1982.
   (Cited on page 219.)
2. A. Bultheel. Epsilon and qd algorithms for the matrix Padé and 2-D Padé problem. Presented at Nato Advanced Study Intitutes programme: Computational aspects of complex analysis, Braunlage, DE, July 26–August 6, 1982.
   (Cited on page 219.)
3. A. Bultheel. Quotient-difference relations in connection with AR filtering. In E. Lueder and E. Gleissner, editors, *Proc. ECCTD'83, Stuttgart*, pages 395–399, Berlin, 1983. VDE Verlag.
   (Cited on page 219.)
4. A. Bultheel. *Laurent series and their Padé approximations*, volume OT-27 of *Oper. Theory: Adv. Appl.* Birkhäuser Verlag, Basel–Boston, 1987.
   (Cited on page 219.)
5. W.B. Gragg. The Padé table and its relation to certain algorithms of numerical analysis. *SIAM Rev.*, 14:1–62, 1972.
   (Cited on page 219.)
6. J.H. McCabe. A formal extension of the Padé table to include two-point Padé quotients. *J. Inst. Math. Appl.*, 15:363–372, 1975.
   (Cited on page 219.)
7. H. Rutishauser. Der Quotienten-Differenzen-Algorithmus. *Zeit. für Angewandte Math.*, 5:233–251, 1954.
   (Cited on page 219.)

7.2 Annie Cuyt

Annie Cuyt[3] (b. 1956) is a professor at the University of Antwerp in Belgium. She is a member of the *Royal Flemish Academy of Belgium for Science and the Arts*. She is an expert in rational and Padé approximation, continued fractions, the qd-algorithm, and related topics. She has published over 130 research papers and several books.

---

[2] Available as http://nalag.cs.kuleuven.be/papers/ade/TW/TW57.pdf.

[3] Homepage https://www.uantwerpen.be/en/staff/annie-cuyt/.

Testimony

From Hankel, Aitken, and Wynn to de Prony, Rutishauser, and Henrici

Abstract

What do the epsilon-algorithm, Padé approximants, Gaussian quadrature rules, exponential analysis, and tensor decomposition have in common? Hankel matrices! While Padé approximants belong to rational approximation theory, Gaussian quadrature rules are high-order numerical integration rules derived from orthogonal polynomial families, exponential analysis is a parametric spectral analysis tool, and tensor decomposition is situated in mathematical data analysis.

Between all these applications, computational tools (the qd-algorithm, polynomial interpolation, generalized eigenvalue solvers, ...) and mathematical theorems (convergence results for Padé approximants, Froissart doublet behaviour, computational complexity analysis, ...) can now be exchanged, because it is possible to rewrite one problem statement in the form of another. This opens up a whole new world to explore.

The formally orthogonal Hadamard polynomials introduced in Sect. 7.2.1 are related to Padé denominator polynomials in Sect. 7.2.2. The Hadamard polynomial zeroes form the nodes of Gaussian quadrature rules in Sect. 7.2.3, and in Sect. 7.2.4 the Hadamard polynomial equals the so-called Prony polynomial of exponential analysis. The zeroes are also the nodes of the Vandermonde structured factor matrices in Sect. 7.2.5 in the decomposition of a higher-order tensor with Hankel structured slices.

All of the above have been generalized to higher dimensions, with the preservation of several connections, as briefly discussed in Sect. 7.2.6.

Introduction

Research in numerical approximation theory introduces one to many great mathematicians, such as the already mentioned Hankel, Sylvester, Schweins, Aitken, Wynn, Rutishauser, de Prony, Padé, Henrici, Brezinski, to name just a few. They are the giants on whose shoulders we stand.

The rich world of determinant identities leads to Wynn's epsilon-algorithm and gives rise to several generalizations when moving to more general concepts than complex numbers, such as vectors, multivariate functions, matrices, tensors, and nonlinear operators, etc.

Less well–known but equally important are the connections between several seemingly disjoint topics. For instance, the connection between sparse interpolation studied in the computer algebra community and Padé approximation was already pointed out in [23] by connecting to the method of de Prony [18], which is at the basis of the parametric spectral analysis method called exponential analysis. The connection with formal orthogonal polynomials, Padé-type approximation, and Gaussian quadrature rules are further elaborated by Brezinski in [4]. More recently, we completed the circle by relating all of the above to tensor decomposition methods [11] and establishing several multivariate generalizations, thereby preserving as many connections as possible.

Let us now discuss the different problem statements one by one, pointing out the relationships as we go along. Only at the end, we shall briefly touch upon their multivariate versions.

7.2.1 Classical Orthogonal Polynomials

We denote by $\mathbb{C}[z]$ the linear space of polynomials in the variable z with complex coefficients and define the linear functional $\lambda : \mathbb{C}[z] \to \mathbb{C}$ that associates the number e_i with the monomial t^i, so

$$\lambda(t^i) = e_i, \qquad i = 0, 1, \dots .$$

Formally, we can write

$$\sum_{i=0}^{\infty} e_i z^i = \lambda \left(\frac{1}{1 - tz} \right).$$

The sequence of polynomials $V_m(z), m = 0, 1, \dots$, given by

$$V_m(z) = \sum_{j=0}^{m} b_{m-j} z^j = b_m + b_{m-1} z + \dots + b_0 z^m, \qquad b_0 \neq 0,$$

and satisfying the conditions

$$\lambda(t^i V_m(t)) = 0, \qquad i = 0, \dots, m - 1, \tag{7.1}$$

is a sequence of polynomials orthogonal with respect to the linear functional λ [4, p. 40], since (7.1) implies that

$$\lambda(V_k(t) V_m(t)) = 0, \qquad k \neq m.$$

Let us further define the Hankel determinants

$$H_m^{(s)} = \begin{vmatrix} e_s & \cdots & e_{s+m-1} \\ \vdots & & \vdots \\ e_{s+m-1} & \cdots & e_{s+2m-2} \end{vmatrix}, \qquad H_0^{(s)} = 1, \qquad s = 0, 1, \dots .$$

The linear functional λ is called definite if

$$H_m^{(0)} \neq 0, \qquad m = 0, 1, \ldots.$$

From (7.1), the linear system

$$\sum_{j=0}^{m} e_{i+j} b_{m-j} = 0, \qquad i = 0, \ldots, m-1, \tag{7.2}$$

is obtained directly, which allows one to compute the orthogonal polynomial $V_m(z)$ up to a normalization. If λ is definite, then the linear system of Eqs. (7.2) has maximal rank. With $b_0 = 1$, the monic orthogonal polynomial $V_m(z)$ is then given by

$$V_m(z) = \frac{1}{H_m^{(0)}} \begin{vmatrix} e_0 & e_1 & \ldots & e_m \\ \vdots & \vdots & & \vdots \\ e_{m-1} & e_m & \ldots & e_{2m-1} \\ 1 & z & \ldots & z^m \end{vmatrix}, \qquad V_0(z) = 1.$$

In [19] the orthogonal polynomial $V_m(z)$ is called the formally orthogonal Hadamard polynomial.

7.2.2 Connection with Padé Approximation

Let at first $V_m(z)$ be general and not necessarily satisfy (7.1). For a given $V_m(z)$ we define the associated polynomials $W_{m-1}(z)$ of degree $m-1$ by [4, p. 10]

$$W_{m-1}(z) := \lambda \left(\frac{V_m(z) - V_m(t)}{z - t} \right)$$
$$= \lambda[b_{m-1} + b_{m-2}(z+t) + \cdots + b_0(z^{m-1} + z^{m-2}t + \cdots + zt^{m-2} + t^{m-1})].$$

We also define the reverse polynomials

$$\widetilde{V}_m(z) := z^m V_m(1/z),$$
$$\widetilde{W}_{m-1}(z) := z^{m-1} W_{m-1}(1/z),$$

which jointly satisfy [4, p. 11]

$$\sum_{i=0}^{\infty} e_i z^i - \frac{\widetilde{W}_{m-1}(z)}{\widetilde{V}_m(z)} = \sum_{i=m}^{\infty} d_i z^i. \tag{7.3}$$

The rational function $\widetilde{W}_{m-1}(z)/\widetilde{V}_m(z)$ is called a Padé-type approximant of degree $m-1$ in the numerator and m in the denominator to the formal power series

$$F(z) := \sum_{i=0}^{\infty} e_i z^i. \tag{7.4}$$

Now let $V_m(z)$ satisfy the orthogonality conditions (7.1). Then (7.3) improves to [4, pp. 34–35]

$$\sum_{i=0}^{\infty} e_i z^i - \frac{\widetilde{W}_{m-1}(z)}{\widetilde{V}_m(z)} = \sum_{i=2m}^{\infty} d_i z^i. \tag{7.5}$$

The rational function $\widetilde{W}_{m-1}(z)/\widetilde{V}_m(z)$ is then called a Padé approximant of degree $m-1$ in the numerator and m in the denominator to the series $F(z)$ and often denoted by $[m-1/m]_F(z)$. So the reverse of the Padé denominator

$$\widetilde{V}_m(z) = \sum_{j=0}^{m} b_{m-j} z^{m-j},$$

which can be computed from the linear system (7.2), is the orthogonal polynomial $V_m(z)$. After its coefficients b_j have been computed, the Padé numerator

$$\widetilde{W}_m(z) = \sum_{j=0}^{m-1} a_j z^j$$

can be obtained from the linear system

$$\sum_{j=0}^{m} e_{i-j} b_j = a_i, \qquad i = 0, \dots, m-1,$$

where $e_i := 0$ for $i < 0$.

7.2.3 Connection with Gaussian Quadrature

In the sequel of this section we consider z a parameter and t the variable. Also, the numbers e_i are classical moments, given on the standard interval $[-1, 1]$ for the weight function $w(z)$,

$$e_i := \int_{-1}^{1} w(t) t^i \, dt, \qquad \int_{-1}^{1} w(t) \, dt > 0.$$

The formal power series $F(z)$ defined in (7.4) then equals

$$F(z) = \lambda \left(\frac{1}{1 - tz} \right) = \int_{-1}^{1} w(t) \frac{1}{1 - tz} \, dt.$$

Let $z_i^{(m)}, i = 1, \ldots, m$, denote the zeroes of $V_m(z)$, so with $b_0 = 1$ we have

$$V_m(z) = (z - z_1^{(m)}) \cdots (z - z_m^{(m)}).$$

Consider the Hermite interpolating polynomial $p_{m-1}(t; z)$ for $1/(1 - tz)$ through the interpolation points $z_1^{(m)}, \ldots, z_m^{(m)}$, so

$$
\begin{aligned}
p_{m-1}(t; z) &= \sum_{i=1}^{m} \left(\frac{1}{1 - z_i^{(m)} z} \right) \frac{V_m(t)}{(t - z_i^{(m)}) V_m'(z_i^{(m)})} \\
&= \sum_{i=1}^{m} \left(\frac{1}{1 - z_i^{(m)} z} \right) \frac{V_m(t) - V_m(z_i^{(m)})}{(t - z_i^{(m)}) V_m'(z_i^{(m)})}.
\end{aligned}
$$

Then the approximation

$$
\lambda \left(\frac{1}{1 - tz} \right) \approx \lambda \left(p_{m-1}(t; z) \right)
$$

$$
= \sum_{i=1}^{m} \lambda \left(\frac{V_m(t) - V_m(z_i^{(m)})}{(t - z_i^{(m)}) V_m'(z_i^{(m)})} \right) \frac{1}{1 - z_i^{(m)} z}
$$

is a quadrature rule for the integration of $1/(1 - tz)$, with nodes $z_1^{(m)}, \ldots, z_m^{(m)}$ and weights

$$
\begin{aligned}
A_i^{(m)} &= \lambda \left(\frac{V_m(t) - V_m(z_i^{(m)})}{(t - z_i^{(m)}) V_m'(z_i^{(m)})} \right) \\
&= \frac{W_{m-1}(z_i^{(m)})}{V_m'(z_i^{(m)})}.
\end{aligned}
$$

In other words,

$$
\lambda \left(\frac{1}{1 - tz} \right) \approx \sum_{i=1}^{m} A_i^{(m)} \frac{1}{1 - z_i^{(m)} z}.
$$

When the $z_i^{(m)}, i = 1, \ldots, m$, are all distinct, then the quadrature rule is a Gaussian rule guaranteeing correctness up to an including polynomial of degree $2m - 1$: for $q(z) \in \mathbb{C}[z]$ of degree $2m - 1$, we have

$$
\lambda \left(q(t) \right) = \int_{-1}^{1} q(t) \, dt = \sum_{i=1}^{m} A_i^{(m)} q(z_i^{(m)}).
$$

So the nodes and weights of a formal m-point Gaussian quadrature rule are closely connected with the orthogonal and associated polynomials $V_m(z)$ and $W_{m-1}(z)$.

7.2.4 Connection with Exponential Analysis

Let us consider a different situation, in which the e_i follow a very structured rule,

$$e_i = \lambda(t_i) = \sum_{j=1}^{m} \alpha_j \Phi_j^i, \qquad \alpha_j, \Phi_j \in \mathbb{C}. \tag{7.6}$$

Here the $\Phi_j, j = 1, \ldots, m$, are (for simplicity) assumed to be mutually distinct and given by $\Phi_j = \exp(2\pi\phi_j/M)$ with $\phi_j, j = 1, \ldots, m \in \mathbb{C}$ and M satisfying $|\Im(\phi_j)| < M/2$ to avoid periodicity problems [14, 22].

Because of the nature of the e_i, the Hankel matrices

$$\mathcal{H}_m^{(s)} = \begin{pmatrix} e_s & \cdots & e_{s+m-1} \\ \vdots & & \vdots \\ e_{s+m-1} & \cdots & e_{s+2m-2} \end{pmatrix}, \qquad \mathcal{H}_0^{(s)} = 1, \qquad s = 0, 1, \ldots,$$

can be factorized as [13]

$$\mathcal{H}_m^{(s)} = \mathcal{V}_m \mathcal{D}_\alpha \mathcal{D}_\Phi^s \mathcal{V}_m^T, \tag{7.7}$$

where $\mathcal{V}_m$ is the Vandermonde matrix

$$\mathcal{V}_m = \left(\Phi_j^{i-1}\right)_{i,j=1}^{m}$$

and $\mathcal{D}_\alpha$ and $\mathcal{D}_\Phi$ are $m \times m$ diagonal matrices filled with the vectors $(\alpha_1, \ldots, \alpha_m)$ and $(\Phi_1, \ldots, \Phi_m)$. In addition, the power series $F(z)$ given by (7.4) reduces to the rational function

$$F(z) = \sum_{i=0}^{\infty} \lambda(t^i) z^i$$
$$= \sum_{i=0}^{\infty} \left(\sum_{j=1}^{m} \alpha_j \Phi_j^i\right) z^i = \sum_{j=1}^{m} \frac{\alpha_j}{1 - \Phi_j z}.$$

Because of the consistency property for Padé approximants [4, p. 36], the denominator polynomials of $F(z)$ and $[m-1/m]_F(z)$ must equal

$$(1 - \Phi_1 z) \cdots (1 - \Phi_m z),$$

which in turn equals $\bar{V}_m(z)$. The zeroes $z_i^{(m)}, i = 1, \ldots, m$, of $V_m(z)$, which are the Gaussian quadrature nodes in the previous section, are in this exponentially structured special case given by $z_i^{(m)} = \Phi_i$. In the literature, the polynomial $V_m(z)$ is called the Prony polynomial [20]: its zeroes Φ_i are the atoms in the exponential sum (7.6).

In case the ϕ_j and α_j are given, then the e_i can be computed. In the inverse case, where the e_i are somehow known or measured, the ϕ_j and α_j can be extracted from the $e_0, \ldots, e_{2m-1}$ as follows. Because of (7.7), the Φ_j are the generalized eigenvalues [21] of

$$\mathcal{H}_m^{(1)} v_j = \Phi_j \mathcal{H}_m^{(0)} v_j,$$

where the $v_j, j = 1, \ldots, m$, are the right generalized eigenvectors. The constraint $|\Im(\phi_j)| < M/2$ now allows one to extract the complex value ϕ_j unambiguously from the generalized eigenvalue Φ_j. With the Φ_j known, the linear coefficients α_j are obtained from the Vandermonde system of interpolation conditions

$$\sum_{j=1}^{m} \alpha_j \Phi_j^i = e_i, \qquad i = s, \ldots, s+m-1, \qquad 0 \le s \le m.$$

The extraction of the nonlinear parameters ϕ_j and the linear coefficients α_j from the equidistant samples

$$e_i = \sum_{j=1}^{m} \alpha_j \exp(2\pi(i/M)\phi_j), \qquad i = 0, \ldots, 2m-1,$$

is a frequent problem statement in signal processing.

7.2.5 Connection with Tensor Decomposition

With the e_i given by (7.6), we fill an order-n tensor $T \in \mathbb{C}^{m_1 \times \cdots \times m_n}$, where

$$2 \le m_k \le m, \qquad 1 \le k \le n, \qquad 3 \le n \le 2m-1,$$

$$\sum_{k=1}^{n} m_k = 2m + n - 1,$$

and

$$T_{i_1, \ldots, i_n} := e_{i_1 + \cdots + i_n - n}, \qquad 1 \le i_k \le m_k. \tag{7.8}$$

The tensor of smallest order $n = 3$ is of size $m \times m \times 2$, and the one of largest order $n = 2m - 1$ is symmetric and of size $2 \times \cdots \times 2$. For the sequel we generalize the definition of the square Hankel matrix above to cover rectangular Hankel structured matrices

$$\mathcal{H}_{m_1,m_2}^{(s)} = \begin{pmatrix} e_s & e_{s+1} & \cdots & e_{s+m_2-1} \\ e_{s+1} & e_{s+2} & \cdots & e_{s+m_2} \\ \vdots & \vdots & \ddots & \vdots \\ e_{s+m_1-1} & e_{s+m_1} & \cdots & e_{s+m_1+m_2-2} \end{pmatrix}.$$

The tensor slices $T_{\cdot,\cdot,i_3,\ldots,i_n}$ of our tensor T then equal

$$T_{\cdot,\cdot,i_3,\ldots,i_n} = \mathcal{H}_{m_1,m_2}^{(i_3+\cdots+i_n-n+2)}$$

and so are Hankel structured. The tensor T decomposes as

$$
T = \sum_{j=1}^{m} \alpha_j \begin{pmatrix} 1 \\ \Phi_j \\ \vdots \\ \Phi_j^{m_1-1} \end{pmatrix} \circ \cdots \circ \begin{pmatrix} 1 \\ \Phi_j \\ \vdots \\ \Phi_j^{m_n-1} \end{pmatrix}, \tag{7.9}
$$

where $\circ$ denotes the outer product and the Φ_j are mutually distinct. The decomposition (7.9) is easily verified by checking the element at position $(i_1, \ldots, i_n)$ on the left-hand side and the right-hand side of (7.9). The factor matrices are the rectangular Vandermonde structured matrices

$$
\mathcal{V}_{m_k,m} = \left(\Phi_j^{i-1} \right)_{i=1,j=1}^{m_k,m}, \qquad 1 \le k \le n.
$$

Because of the Vandermonde structure of the factor matrices with $m_k \le m, k = 1, \ldots, n$, their Kruskal rank equals m_k for all k. Since $m_1 + \cdots + m_n = 2m+n-1$, we find that the sum of the Kruskal ranks of the n factor matrices of the rank-m tensor T is bounded below by $2m + n - 1$. Hence the Kruskal condition is satisfied, and the uniqueness of the decomposition is guaranteed.

7.2.6 Multidimensional Generalizations

The concept of the formally orthogonal polynomial $V_m(z)$ is generalized in [10], for different radial weight functions, to so-called spherical orthogonal polynomials. The latter differ from several other definitions of multivariate orthogonal polynomials, in that they preserve the connections laid out here in Sects. 7.2.2 and 7.2.3. At the heart are again Hankel matrices and determinants, but now parameterized [9, 17].

Homogeneous multivariate Padé approximants, as defined in [6, 7], can be computed using the epsilon-algorithm [5] and can also be obtained from the spherical orthogonal polynomials in a way similar to that described here in Sect. 7.2.2 [1, 2]. The homogeneous definition satisfies a very strong projection property, in the sense that this multivariate Padé approximant reduces to the univariate Padé approximant on every one-dimensional subspace.

A whole lot of Gaussian cubature rules on the disk can be united in a single approach when developing the rules from these spherical orthogonal polynomials [3]. What's more, the nodes and weights of such Gaussian cubature rules on the disk can be obtained as the solution of a multivariate Prony-like system of interpolation conditions [3]. And this brings us to the next connection.

The result that an m-term exponential analysis problem of the form

$$
\sum_{j=1}^{m} \alpha_j \exp(2\pi(i/M)\phi_j) = e_i, \qquad i = 0, \ldots, 2m-1,
$$

can be solved uniquely for the $\alpha_j, \phi_j, j = 1, \ldots, m$, from only $2m$ samples e_i was only recently generalized to the d-dimensional setting [15] in its full flavour. The equations

$$\sum_{j=1}^{m} \alpha_j \exp(\langle x_i, \phi_j \rangle) = e_i, \qquad \phi_j \in \mathbb{C}^d, \qquad x_i \in \mathbb{R}^d,$$

can be solved for the vectors ϕ_j and the coefficients α_j from a mere $(d+1)m$ samples collected at vectors x_i. This number of samples is also the theoretical minimal number [15]. While the $2m$ samples in the one-dimensional problem are collected uniformly, the $(d+1)m$ samples in the d-dimensional problem are located on d parallel lines constructed from a basis in $\mathbb{R}^d$.

This multivariate Prony generalization is in turn closely connected to both multivariate and multidimensional generalizations of the Padé approximant concept and to various tensor decompositions [11, 16]. Among other things, we also mention an algorithm to locate zeroes of the homogeneous Padé denominator [8], which, as we know from Sect. 7.2.4, are related to the atoms in the exponential analysis problem. The detailed interpretation of the output of this algorithm is through the convergence result for homogeneous Padé approximants given in [12].

References

1. B. Benouahmane. Approximants de Padé "homogènes" et polynômes orthogonaux à deux variables. *Rend. Mat. (7)*, 11:673–689, 1991.
   (Cited on page 227.)
2. B. Benouahmane and A. Cuyt. Multivariate orthogonal polynomials, homogeneous Padé approximants and Gaussian cubature. *Numer. Algorithms*, 24:1–15, 2000.
   (Cited on page 227.)
3. B. Benouahmane, A. Cuyt, and I. Yaman. Near-minimal cubature formulae on the disk. *IMA J. Numer. Anal.*, 39(1):297–314, 2019.
   (Cited on page 227.)
4. C. Brezinski. *Padé type approximation and general orthogonal polynomials.* ISNM 50, Birkhäuser Verlag, Basel, 1980.
   (Cited on pages 221, 222, 223, and 225.)
5. A. Cuyt. The epsilon-algorithm and multivariate Padé approximants. *Numer. Math.*, 40:39–46, 1982.
   (Cited on page 227.)
6. A. Cuyt. Multivariate Padé approximants. *J. Math. Anal. Appl.*, 96:283–293, 1983.
   (Cited on page 227.)
7. A. Cuyt. Multivariate Padé approximants revisited. *BIT*, 26:71–79, 1986.
   (Cited on page 227.)
8. A. Cuyt. On the convergence of the multivariate "homogeneous" qd-algorithm. *BIT*, 34:535–545, 1994.
   (Cited on page 228.)
9. A. Cuyt, B. Benouahmane, Hamsapriye, and I. Yaman. Symbolic-numeric Gaussian cubature rules. *Appl. Numer. Math.*, 61:929–945, 2011.
   (Cited on page 227.)
10. A. Cuyt, B. Benouahmane, and B. Verdonk. Spherical orthogonal polynomials and symbolic-numeric Gaussian cubature formulas. In M. Bubak et al., editors, *LNCS 3037*, pages 557–560, Berlin, 2004. Springer-Verlag.
    (Cited on page 227.)

11. A. Cuyt, F. Knaepkens, and W. Lee. From exponential analysis to Padé approximation and tensor decomposition, in one and more dimensions. In V.P. Gerdt et al., editors, *LNCS11077*, pages 116–130, 2018. Proceedings CASC 2018, Lille (France).
   (Cited on pages 221 and 228.)
12. A. Cuyt and D. Lubinsky. A de Montessus theorem for multivariate homogeneous Padé approximants. *Ann. Numer. Math.*, 4:217–228, 1997.
   (Cited on page 228.)
13. A. Cuyt and W. Lee. Sparse interpolation and rational approximation. volume 661 of *Contemporary Mathematics*, pages 229–242, Providence, RI, 2016. American Mathematical Society.
   (Cited on page 225.)
14. A. Cuyt and W. Lee. How to get high resolution results from sparse and coarsely sampled data. *Appl. Comput. Harmon. Anal.*, 48(3):1066–1087, 2020.
   (Cited on page 225.)
15. A. Cuyt and W. Lee. Multivariate exponential analysis from the minimal number of samples. *Adv. Comput. Math.*, 44:987–1002, 2018.
   (Cited on page 228.)
16. A. Cuyt, W. Lee, and X. Yang. On tensor decomposition, sparse interpolation and Padé approximation. *Jaén J. Approx.*, 8(1):33–58, 2016.
   (Cited on page 228.)
17. A. Cuyt, I. Yaman, A. Ibrahimoglu, and B. Benouahmane. Radial orthogonality and Lebesgue constants on the disk. *Numer. Algorithms*, 61:291–313, 2012.
   (Cited on page 227.)
18. R. de Prony. Essai expérimental et analytique sur les lois de la dilatabilité des fluides élastiques et sur celles de la force expansive de la vapeur de l'eau et de la vapeur de l'alkool, à différentes températures. *J. Ec. Poly.*, 1:24–76, 1795.
   (Cited on page 220.)
19. P. Henrici. *Applied and computational complex analysis I.* John Wiley & Sons, New York, 1974.
   (Cited on page 222.)
20. F.B. Hildebrand. *Introduction to numerical analysis.* Mc Graw Hill, New York, 1956.
   (Cited on page 225.)
21. Y. Hua and T.K. Sarkar. Matrix pencil method for estimating parameters of exponentially damped/undamped sinusoids in noise. *IEEE Trans. Acoust., Speech, Signal Process.*, 38:814–824, 1990.
   (Cited on page 226.)
22. H. Nyquist. Certain topics in telegraph transmission theory. *Trans. Am. Inst. Electr. Eng.*, 47(2):617–644, April 1928.
   (Cited on page 225.)
23. L. Weiss and R.N. McDonough. Prony's method, Z-transforms, and Padé approximation. *SIAM Rev.*, 5:145–149, 1963.
   (Cited on page 220.)

7.3 André Draux

André Draux (b. 1943) is emeritus professor at the *Institut National des Sciences Appliquées* in Rouen, France. In December 1981, he defended a *Doctorat d'État es Sciences Mathématiques* at the University of Lille. Peter Henrici was the president of his defense committee. In this work he gave all the recurrence relations between formal orthogonal polynomials when some of them do not exist. These relations are instrumental in the computation of the nonnormal Padé table and in the treatment of breakdown and near-breakdown in Krylov subspace methods for solving linear systems.

Testimony

The aim of this presentation is to explain how the theory of formal orthogonal polynomials was introduced. At the end of the seventies I began a study of exponential-type approximants [4] that approach a formal power series $f(x)$ as far as possible:

$$f(x) = \sum_{i=0}^{\infty} \frac{c_i}{i!}, \quad c_i \in \mathbb{R}, \forall i \in \mathbb{N}.$$

In a first stage, these approximants had a simple form

$$F_k(x) = \sum_{j=0}^{k} \alpha_j e^{m_j x},$$

and I wanted to find the α_j and m_j such that $f(x) - F_k(x) = O(x^{2k+2})$.

In a second stage, because this problem of approximation could have a link with the solution of a homogeneous linear differential equation with constant coefficients, I changed the form of these approximants in order to obtain a more general problem. This form became

Problem 1

$$F_p(x) = \sum_{j=0}^{\ell_p} S_j(x) e^{m_{j,p} x}$$

with $S_j(x) \in \mathcal{P}$, the vector space of real polynomials in one variable, $\deg S_j(x) = q_{j,p} - 1$, $\forall j \in \mathbb{N}$, $0 \le j \le \ell_p$, and $\sum_{j=0}^{\ell_p} q_{j,p} = p$. We looked for the following families:

$$S_j(x) = \sum_{i=i_j}^{i_{j+1}-1} k_{i,p} x^{i-i_j}, \quad j \in \mathbb{N}, 0 \le j \le \ell_p,$$

with $i_0 = 0$ and $i_j = \sum_{s=0}^{j-1} q_{s,p}$ for $j \in \mathbb{N}$, $1 \le j \le \ell_p + 1$. The unknowns are the $m_{j,p}$ and the $k_{i,p}$. Using the expansion of $F_p(x)$ as a formal power series, we wanted to obtain the following result:

$$f(x) - F_p(x) = O(x^r) \quad \text{with } r \ge 2p.$$

In this case, the unknowns $m_{j,p}$ and $k_{i,p}$ satisfied the following system

$$M_{2p} K_p = C_{2p}, \tag{7.10}$$

where M_{2p} is a $2p \times p$ matrix depending on the $m_{j,p}$, K_p is a vector of $\mathbb{R}^p$ with entries $k_{0,p}, \ldots, k_{p-1,p}$, and C_{2p} is a vector of $\mathbb{R}^{2p}$ with entries $c_0, \ldots, c_{2p-1}$.

The homogeneous linear differential equation of order p with constant coefficients that is linked with Problem 1 is

Problem 2

$$\sum_{i=0}^{p} \lambda_{i,p} y^{(p-i)}(x) = 0, \tag{7.11}$$

where $y^{(p-i)}(x)$ is the derivative of order $p - i$ of $y(x)$. We looked for the solutions $F_p(x)$ of this Eq. (7.11) such that if the expansion of $F_p(x)$ as a formal power series is used, we had $f(x) - F_p(x) = O(x^r)$ with $r \geq 2p$. The $\lambda_{i,p}$ were obtained as the solution of a linear system

$$H_p^{(0)} \Lambda_p = -\widehat{C}_p, \tag{7.12}$$

where $H_p^{(0)}$ is a Hankel matrix $H_p^{(0)} = (c_{i+j})_{i,j=0}^{p-1}$, c_{i+j} is in row i and column j, Λ_p is a vector of $\mathbb{R}^p$ with entries $\lambda_{p,p}, \ldots, \lambda_{1,p}$, $\widehat{C}_p$ is a vector of $\mathbb{R}^p$ with entries $c_p, \ldots, c_{2p-1}$, and $\lambda_{0,p} = 1$.

Let M_p be the matrix obtained by keeping the first p rows and the first p columns of the matrix M_{2p}. So we had a new system

$$M_p K_p = C_p. \tag{7.13}$$

We proved that the system (7.10) is equivalent to the systems (7.12) and (7.13). Therefore Problems 1 and 2 are identical, and to obtain the solution of Problem 1 it was necessary to obtain general results about the solution of the system (7.12). It was the first step of a general study of the solutions of (7.12) when $H_p^{(0)}$ is regular or when it is singular.

When $H_p^{(0)}$ is regular, the result is well known (see, for example, the books of Chihara [3] and Brezinski [1]). But when $H_p^{(0)}$ is singular, the results were absolutely new. This study is linked with the consistency of linear systems. Moreover, when $H_p^{(0)}$ is regular, the polynomial $P_p(x) = \sum_{i=0}^{p} \lambda_{i,p} x^{p-i}$ is orthogonal with respect to a linear functional c whose moments are $c(x^i) = c_i, \forall i \in \mathbb{N}$.

The new problem was: "what happens when $H_p^{(0)}$ is singular"? We found that in this case, if $H_{p_1}^{(0)}$ and $H_{p_2}^{(0)}$ are regular ($p_1 < p_2$) and $H_p^{(0)}$ singular $\forall p$ such that $p_1 + 1 \leq p \leq p_2 - 1$, then the system (7.12) has an infinity of solutions for $p_1 + 1 \leq p \leq p_2 - 1 - [\frac{p_2-p_1+1}{2}]$ and no solution for $p_2 - [\frac{p_2-p_1+1}{2}] \leq p \leq p_2 - 1$. Here $[\cdot]$ is the integer part of the content between the square brackets. Moreover, when $p_1 + 1 \leq p \leq p_2 - 1 - [\frac{p_2-p_1+1}{2}]$, the polynomial $P_p(x)$, solution of (7.12), is such that $P_p(x) = P_{p_1}(x) w_{p-p_1}(x)$, where $w_{p-p_1}(x)$ is an arbitrary polynomial of degree $p - p_1$ and $P_{p_1}(x)$ is the unique solution of the linear system $H_{p_1}^{(0)} \Lambda_{p_1} = -\widehat{C}_{p_1}$; $P_{p_1}(x)$ was called regular orthogonal polynomial when $H_{p_1}^{(0)}$ is regular, and $P_p(x)$ satisfies orthogonality conditions

$$c(P_p(x)x^j) = 0 \text{ for } j = 0, \ldots, p_1 + p_2 - p - 2,$$
$$\neq 0 \text{ for } j = p_1 + p_2 - p - 1.$$

These polynomials were called singular orthogonal polynomials. All the obtained polynomials were called formal orthogonal polynomials. When $p_2 - [\frac{p_2-p_1+1}{2}] \leq p \leq p_2 - 1$, $P_p(x)$ is a quasi-orthogonal polynomial, since $c(P_p(x)x^j) = 0$ for $j = 0, \ldots, p_1 + p_2 - p - 2 < p - 1$.

It is well known that the orthogonal polynomials $P_p(x)$ satisfy a three-term recurrence relation when $H_p^{(0)}$ is regular $\forall p \in \mathbb{N}$ (see [3]). The new fact was that three consecutive regular orthogonal polynomials also satisfied a three-term recurrence relation. If p_1, p_2, p_3 with $p_1 < p_2 < p_3$ are the degrees of three consecutive regular orthogonal polynomials, we have

$$P_{p_3}(x) = \omega_{p_3-p_2}(x)P_{p_2}(x) - \beta_{p_3}P_{p_1}(x),$$

where $\omega_{p_3-p_2}(x)$ is a polynomial of degree $p_3 - p_2$ and β_{p_3} is a nonzero constant. The consequence of this property is that the set of all the consecutive regular orthogonal polynomials is exactly the same set obtained using the Euclidean algorithm. Therefore two consecutive regular orthogonal polynomials have no common zeros.

From this starting point I used the book of Claude Brezinski [1] to define the sequence of properties to prove. All these results are also described in [5].

- Associated polynomials $Q_p(t)$ and their properties,
  $Q_p(t) = c\left(\frac{P_p(x)-P_p(t)}{x-t}\right)$.
- System of adjacent orthogonal polynomials and their properties. They are obtained using the adjacent formal functionals $c^{(j)}$ defined from their moments $c^{(j)}(x^i) = c_{i+j}$, $\forall i \in \mathbb{N}$, $\forall j \in \mathbb{Z}$ with the convention that $c_k = 0$ if $k < 0$.
- Table P of all the adjacent orthogonal polynomials and properties of the square blocks in which the Hankel matrices are singular.
- Detection of all the singular matrices M. If M_k is a regular matrix, then the orthogonal polynomial P_k exists and is unique when the leading coefficient (i.e., the coefficient of x^k) is fixed. If $c(x^{k+j-1}P_k(x)) = 0$, $\forall j = 1, \ldots, s$ and $c(x^{k+s}P_k(x)) \neq 0$, then the matrices $M_{k+1}, \ldots, M_{k+s}$ are singular and M_{k+s+1} is regular.
- Generalization of the qd-algorithm when the table P contains square blocks.
- Obtention of all the three-term recurrence relations along rows, columns, and antidiagonals, and all types of staircase relations.
- Padé approximants in this general case.

After that I also studied other domains linked with these formal orthogonal polynomials:

- Gaussian quadratures.
- Two-point Padé approximants.
- Approximants of series of functions.
- Generalization of Favard's theorem.

More recently I published a paper [6] about the quasi-orthogonal polynomials when some Hankel matrices are singular. A linear functional c is said to be quasi-definite if all the Hankel matrices $H_p^{(0)}$ are regular for all $p \in \mathbb{N}$. In the case of a quasi-definite linear functional c, Chihara [2] proved in 1957 that three consecutive quasi-orthogonal polynomials satisfy a three-term recurrence relation with polynomial coefficients. In [6] we extended this property in the case of a non-quasi-definite linear functional c.

References

1. C. Brezinski, Padé-Type Approximation and General Orthogonal Polynomials, ISNM vol. 50, Birkhäuser, Basel, 1980.
   (Cited on pages 231 and 232.)
2. T.S. Chihara, On quasi-orthogonal polynomials, Proc. Amer. Math. Soc. 8 (1957) 765–767.
   (Cited on page 233.)
3. T.S. Chihara, An Introduction to Orthogonal Polynomials, Gordon and Breach, New York, 1978.
   (Cited on pages 231 and 232.)
4. A. Draux, Approximants de type exponentiel. Polynômes orthogonaux, Publication ANO 27, Lille University, october 1980.
   (Cited on page 230.)
5. A. Draux, Polynômes Orthogonaux Formels. Applications, LNM 974, Springer-Verlag, Berlin, 1983.
   (Cited on page 232.)
6. A. Draux, On quasi-orthogonal polynomials of order r, Integral Transforms and Special Functions 27 (2016) 747–765.
   (Cited on page 233.)

7.4 Walter Gander

Walter Gander[4] (b. 1944) is emeritus professor at the *Eidgenössische Technische Hochschule* in Zürich, Switzerland, since 2009. After his diploma in mathematics at ETH Zürich in 1968, he became assistant of Heinz Rutishauser. He obtained his Ph.D. in mathematics in 1973 under the supervision of Peter Henrici. He has authored 7 books and over 70 papers in scientific computing and numerical linear algebra.

Testimony

I think I learned from two sources for the first time about the epsilon-algorithm. The first was a lecture series by Heinz Rutishauser about "Limitierungsverfahren",

---

[4] Homepage http://people.inf.ethz.ch/gander/.

in which he discussed acceleration methods for slowly convergent series. The second time was in studying the literature for my PhD thesis of 1973 with the title "Numerische Implementationen des Rombergschen Extrapolationsverfahrens, mit Anwendung auf die Summation unendlicher Reihen." I studied the methods discussed in the book Sauer-Szabo: Mathematische Hilfsmittel des Ingenieurs. Teil III, H. Interpolation und genäherte Quadratur, von R. Bulirsch und H. Rutishauser, Springer 1968. The epsilon algorithm is described in that book.

Later in 1986 I spent a summer at Bell Labs with Gene Golub, and we worked on a problem solving linear equations by extrapolation. Eventually this work was published:

Walter Gander, Gene H. Golub and Dominik Gruntz, Solving Linear Equations by Extrapolation in: Supercomputing, NATO ASI Series F: Computer and Systems Sciences, No 62, Janusz S. Kowalik (Ed.), pp. 279–295, Springer-Verlag Berlin, 1989.

About this paper you [*authors' note: that is, C.B.*] wrote a comment to Gene!

Fri, 19 Jan 90
dear Gene, many thanks for your very interesting report on "Solving linear equations by extrapolation". I would like to give you some precisions. The TEA gives exactly the same iterates as the conjugate or bi-conjugate gradient as proved in my book "Pade-type approximation and general orthogonal polynomials". Your system (31) shows that the 4 transformations considered can be implemented by my H-algorithm (H for Henrici) or by solving the system by a bordering method. We now know particular rules for avoiding (at least partially) numerical instability, see my paper on the Schur complement (I am sending it to you). There are also some related papers which can be interesting for you. In particular when cycling, as explained page 16 of your NA-89-11 report, with m=size of the system, one obtain, under some assumptions, a quadratic method for solving systems of nonlinear equations. I think that you shall attend the meeting in Copper Mountain next April. Thus we shall discuss there. I am also interested by the reports NA-89-01 to NA-89-9. Best regards and a very happy new year! Claude Brezinski.

In my lectures at ETH, I gave lectures about the epsilon-algorithms. I also invited Avram Sidi to visit ETH. Finally, the vector-epsilon-algorithms were also described in our book:

Walter Gander, Martin J. Gander, Felix Kwok, Scientific Computing, an Introduction Using Maple and Matlab, Springer, 2014.

7.5 Xing-Biao Hu

Xing-Biao Hu[5] (b. 1962) is a researcher at the Academy of Mathematics and Systems Science of the Chinese Academy of Sciences. He also holds a professorship at the University of the Chinese Academy of Sciences. His current research focuses on the interdisciplinary study of integrable systems including convergence acceleration algorithms via discrete integrable systems, links between formal orthogonal polynomials and integrable systems, integrable numerical schemes of soliton equations, integrable combinatorics, and numerical study of N-periodic wave solutions to integrable physical models.

Testimony

Wynn's ε Algorithm, Its Generalizations and Their Links to Discrete Integrable Systems

My early research interest was in Hirota's bilinear method and soliton equations. After attending the SIDE II workshop organized by Profs. Peter A. Clarkson and Frank Nijhoff in Canterbury, UK, in 1996, discrete integrable systems became my favorite research field. Influenced by works of Profs. Papageorgiou, Grammaticos, and Ramani [14], Nagai and Satsuma [12], and Nagai, Tokihiro, and Satsuma [13] on the connections between the ε, η, and ϱ algorithms and discrete potential KdV, discrete KdV, and cylindrical KdV equations, I began to collect and learn material on convergence acceleration methods. Thanks to Prof. Claude Brezinski for his review article [1], which helped me to gradually understand the development of convergence acceleration in the twentieth century, and also introduced me to many important references in this field, including Prof. Wynn's papers [16, 17, 18, 19, 20, 21, 22, 23].

In 2008, inspired by the work [11] of Prof. Yoshimasa Nakamura and his students on successfully designing the discrete Lotka–Volterra equation as an efficient algorithm for computing the singular values of a matrix, the following question naturally appeared as how to design new convergence acceleration algorithms from fully discrete integrable systems. For this purpose, on 15 Dec. 2008 I wrote to Prof. Claude Brezinski for the first time to invite him to visit my institute for a possible collaboration. To my happiness, Prof. Claude Brezinski soon replied and promised to visit my institute after October 2009.

It happened that together with Prof. Zaijiu Shang, of the Academy of Mathematics and Systems Science (AMSS), and Prof. Shufang Xu, from Peking University, I co-organized a long-term program called "Study on Structures and Algorithms for Dynamical Systems" supported by the Morningside Centre of AMSS in 2009. We had a weekly seminar given by experts and students. As a part of the activity of the

---

[5] Homepage: http://www.cc.ac.cn/staff/hxb.html.

program, my students Juan Hu, Yi He, and Jian-Qing Sun et al. took turns speaking about the book by C. Brezinski and M. Redivo-Zaglia [5]. During the activities of the program, we invited some experts, including Prof. Moody Chu, of North Carolina State University, to give a series of lectures based on [7].

It was my pleasure to attend the Luminy09 conference "Approximation and extrapolation of convergent and divergent sequences and series" organized by C. Brezinski, M. Redivo-Zaglia, and E.J. Weniger during the period 28 Sept.–2 Oct. 2009. This was the first time I attended a professional workshop on convergence acceleration and related topics. It provided with me a great opportunity to communicate with leading experts in the field and learn about recent developments and progress in this field. During 1–15 Nov. 2009, Prof. Brezinski visited my institute and gave four lectures on convergence acceleration, formal orthogonal polynomials, Padé approximations, continued fractions, etc., and had many discussions with my students Yi He and Jian-Qing Sun and myself. As a product of these discussions, we finished two joint papers [3, 4].

During 8–22 Dec. 2009, Prof. E.J. Weniger visited my institute for 2 weeks. He gave four lectures on sequence transformations, etc. During his visit, Yi He, Jian-Qing Sun, and I collaborated with Prof. Weniger to work on how to design a lattice Boussinesq equation as a new convergence acceleration algorithm. Based on the observation that discrete potential KdV and lattice Boussinesq equations are members of the lattice Gelfand–Dikii hierarchy, finally we successfully achieved our goal and found two-step Shanks transformations [10]. It is the guidance and collaborations of Profs. Brezinski and Weniger that enabled us to quickly cross the frontier of the field of convergence acceleration.

In 2011, Yi He, Jian-Qing Sun, and I together with C. Brezinski and M. Redivo-Zaglia, successfully generalized Wynn's ε-algorithm to the multistep ε-algorithm for computing the multistep Shanks transformation [2]. Furthermore, Jian-Qing Sun, Xiang-Ke Chang, Yi He, and I established a connection between an extended multistep ε-algorithm and the discrete-time generalized Lotka–Volterra system [15]. Later on, motivated by the work on the Brezinski–Durbin–Redivo-Zaglia transformation [6], X.K. Chang, Y. He, Shi-Hao Li, and I generalized this transformation to higher-order sequence transformations via the algebraic tool of Pfaffians, and we designed a discrete integrable system as a numerical algorithm for computing higher-order Brezinski–Durbin–Redivo-Zaglia transformations [8].

Finally, it should be mentioned that during 10–14 Oct. 2011, I was invited by C. Brezinski, M. Redivo-Zaglia, and E.J. Weniger to attend the special session "Extrapolation and Convergence Acceleration" of SC2011 in S. Margherita di Pula, Sardinia, Italy, where I had fruitful discussions with Prof. E.J. Weniger, eventually producing the joint paper [9] through collaboration with X.-K. Chang, Y. He, and J.-Q. Sun. In the paper we construct new sequence transformations based on Wynn's ε and ϱ algorithms. The recurrences of the new algorithms include the recurrences of Wynn's ε and ϱ algorithms and of Osada's generalized ϱ algorithm as special cases.

References

1. Brezinski, C. Convergence acceleration during the 20th century. Numerical analysis 2000, Vol. II: Interpolation and extrapolation. J. Comput. Appl. Math., 122 (2000), no. 1–2, 1–21. *(Cited on page 235.)*

2. Brezinski, Claude; He, Yi; Hu, Xing-Biao; Redivo-Zaglia, Michela; Sun, Jian-Qing. Multistep ε-algorithm, Shanks' transformation, and the Lotka–Volterra system by Hirota's method. Math. Comp. 81 (2012), no. 279, 1527–1549. *(Cited on page 236.)*

3. Brezinski, Claude; He, Yi; Hu, Xing-Biao; Sun, Jian-Qing. A generalization of the G-transformation and the related algorithms. Appl. Numer. Math. 60 (2010), no. 12, 1221–1230. *(Cited on page 236.)*

4. Brezinski, Claude; He, Yi; Hu, Xing-Biao; Sun, Jian-Qing. Cross rules of some extrapolation algorithms. Inverse Problems 26 (2010), no. 9, 095013, 22 pp. *(Cited on page 236.)*

5. Brezinski, Claude; Redivo-Zaglia, Michela. Extrapolation methods. Theory and practice. With 1 floppy disk (5.25 inch). Studies in Computational Mathematics, 2. North-Holland Publishing Co., Amsterdam, 1991. x+464 pp *(Cited on page 236.)*

6. Brezinski, Claude; Redivo-Zaglia, Michela. Generalizations of Aitken's process for accelerating the convergence of sequences. Comput. Appl. Math. 26 (2007), no. 2, 171–189. *(Cited on page 236.)*

7. Chu, Moody T. Linear algebra algorithms as dynamical systems. Acta Numer. 17 (2008), 1–86. *(Cited on page 236.)*

8. X.K. Chang, Y. He, X.B. Hu, S.H. Li. A new integrable convergence acceleration algorithm for computing Brezinski–Durbin–Redivo-Zaglia's sequence transformation via Pfaffians. Numer. Algorithms 78 (2018), no. 1, 87–106. *(Cited on page 236.)*

9. X.K. Chang, Y. He, X.B. Hu, J.Q. Sun, E.J. Weniger. Construction of new generalizations of Wynn's epsilon and rho algorithm by solving finite difference equations in the transformation order. Numer. Algorithms 83 (2020), no. 2, 593–627. *(Cited on page 236.)*

10. He, Yi; Hu, Xing-Biao; Sun, Jian-Qing; Weniger, Ernst Joachim. Convergence acceleration algorithm via an equation related to the lattice Boussinesq equation. SIAM J. Sci. Comput. 33 (2011), no. 3, 1234–1245. *(Cited on page 236.)*

11. Nakamura Yoshimasa. Functionality of Integrable Systems. Kyoritsu Shuppan Co., Tokyo, Japan 2006 *(Cited on page 235.)*

12. Nagai, A.; Satsuma, J. Discrete soliton equations and convergence acceleration algorithms. Phys. Lett. A 209 (1995), no. 5–6, 305–312. *(Cited on page 235.)*

13. Nagai, Atsushi; Tokihiro, Tetsuji; Satsuma, Junkichi. The Toda molecule equation and the ε-algorithm. Math. Comp. 67 (1998), no. 224, 1565–1575. *(Cited on page 235.)*

14. Papageorgiou, V.; Grammaticos, B.; Ramani, A. Integrable lattices and convergence acceleration algorithms. Phys. Lett. A 179 (1993), no. 2, 111–115. *(Cited on page 235.)*

15. Sun, Jian-Qing; Chang, Xiang-Ke; He, Yi; Hu, Xing-Biao. An extended multistep Shanks transformation and convergence acceleration algorithm with their convergence and stability analysis. Numer. Math. 125 (2013), no. 4, 785–809. *(Cited on page 236.)*

16. Wynn, P. On a device for computing the $e_m(S_n)$ tranformation. Math. Tables Aids Comput. 10 (1956), 91–96.
    (Cited on page 235.)
17. Wynn, P. On a procrustean technique for the numerical transformation of slowly convergent sequences and series. Proc. Cambridge Philos. Soc. 52 (1956), 663–671.
    (Cited on page 235.)
18. Wynn, P. Confluent forms of certain non-linear algorithms. Arch. Math. 11 (1960), 223–236.
    (Cited on page 235.)
19. Wynn, P. Acceleration techniques for iterated vector and matrix problems. Math. Comp. 16 (1962), 301–322.
    (Cited on page 235.)
20. Wynn, P. Singular rules for certain non-linear algorithms. BIT 3 (1963) 175–195
    (Cited on page 235.)
21. Wynn, P. Partial differential equations associated with certain non-linear algorithms. ZAMP 15 (1964) 273–289
    (Cited on page 235.)
22. Wynn, P. Upon systems of recursions which obtain among the quotients of the Padé table. Numer. Math. 8 (1966), 264–269.
    (Cited on page 235.)
23. Wynn, P. On the convergence and stability of the epsilon algorithm. SIAM J. Numer. Anal. 3 (1966), no. 1, 91–122.
    (Cited on page 235.)

7.6 William B. Jones

William Branham Jones is emeritus professor of Mathematics at the Department of Mathematics of the University of Colorado. He obtained a Ph.D. in mathematics from Vanderbilt University, Nashville, Tennessee, USA, in 1963 with the help of Professor Wolfgang J. Thron of the University of Colorado at Boulder (UCB), who suggested a suitable thesis project in the analytic theory of continued fractions. This was the beginning of a lifetime collaboration between them, culminating in the two important publications [359, 360]. Then, Jones accepted a 1-year visiting assistant professorship in the UCB Mathematics Department, and in 1964, he obtained an assistant professorship in the Engineering Mathematics Department at UCB. Two years later, the two mathematics departments merged, and he remained on that faculty until his retirement in December 1996. Bill Jones is a recognized specialist in continued fractions, with around 110 papers on this topic. See [295] for his full biography.

Testimony

The advent of high-speed digital computers in the mid-1950s created a rebirth of interest and activity in applied and computational mathematics, including approximation theory, nonlinear approximation, Padé approximants, and continued fractions. Dr. Peter Wynn was among the first mathematicians to work in these areas in the computer age. In addition to his own extensive creative research, he gave inspiration, guidance and mentoring to a number of young mathematicians at the beginning of their careers. It is therefore appropriate that we acknowledge here the extensive, valuable contributions to mathematics made by Peter Wynn. It is also appropriate to express appreciation to my long-term friend and colleague Claude Brezinski for making the present volume a reality. [*Bill Jones has never met M.R.-Z., the second author of this book*].

7.7 Pierre-Jean Laurent

Pierre-Jean Laurent (b. 1937) proved important results on Richardson's and Romberg's extrapolation methods. These methods were considered by Wynn in [711] and [793]. P.-J. Laurent defended his doctoral thesis on extrapolation methods at the University of Grenoble, France, under the supervision of Jean Kuntzmann (1912–1992), a pioneer in numerical analysis in France.[6] He spent his career in Grenoble as a professor. After his thesis, P.-J. Laurent became interested by approximation and optimisation (he wrote a book on these topics) and CAGD. He was a member of the doctoral committee of C.B. with Noël Gastinel (1925–1984), Francis Ceschino, and Jean Kuntzmann.

Testimony

From the Romberg Algorithm to Richardson Extrapolation

In the first part of my thesis (1960–1961), I was particularly interested in Monte–Carlo methods. These methods seemed quite attractive and original. The probabilistic analogies implemented for solving a deterministic problem were very astute and of various kinds. A great number of studies focused on the generation of random numbers following a given density law; see for instance the elegant and very efficient Von Neumann method for the exponential law. But one property was common to all Monte–Carlo methods: they were all very greedy in computer time! Remember that for increasing the accuracy by 2, you need to multiply the number of random tests by 4. They were rightly nicknamed the "dispair" methods, to be implemented only where classical methods could not be considered. This was often the case for problems of high dimensions. Therefore, many studies aimed at reducing the variance of the estimators. That was precisely the point I was working on. In my thesis I recalled and developed some of the procedures: stratified sampling, weighted sampling, controlled variable method, compensated variable method, and so on. However, the method for which I contributed substantially was very close to, not to say identical with, the Richardson principle. It was "Richardson, random version" and I called it "systematic sampling on a modified function." When computing a simple integral,

---

[6] See https://www.idref.fr/030565677.

it consisted in observing (using the Euler–Maclaurin development and the Bernoulli polynomials) that the variance of the estimator admitted an even development in powers of $1/n$. The idea was to construct a new estimator by conveniently combining several values of n, not just two values (C.R. Acad. Sci. 253, July 1961, 610–612). Thus, one obtains a new estimator whose variance acts in $1/n^{2r}$, $r > 1$. The method was tested on several problems, in particular on Fredholm integral equations, and it yielded spectacular results.

A turn happened end of 1961 (or beginning of 1962) when F.L. Bauer gave a talk in Grenoble on the Romberg method. His talk was very concrete and calculatory: starting from the trapeze method with step h, he showed that combining it with the same method of step $h/2$ and coefficients $-1/3, 4/3$, one obtained the Simpson method. Smiles... Then combining two Simpsons with coefficients $-1/15, 16/15$, one obtained Newton–Cotes. Surprise! And proceeding likewise, one obtained a new, still unknown formula. Fascination!!! At the end of his talk I pointed out that the same coefficients ($-1/3$ and $4/3$) were found in a totally different context, namely in reducing the variance for the Monte–Carlo method. F.L. Bauer listened and agreed, without the slightest comment. I cannot say whether he was aware that this was a general principle. Nor do I know whether other studies of that time had made the connection with the Richardson extrapolation. For example, H. Rutishauser, who published his paper in Num. Math. in 1963 on an extension of the Romberg principle, did not mention Richardson. I can only tell how it was for myself. Things moved a lot during the week following F.L. Bauer's talk! As it had worked for the variance development and for the computation of integrals, it seemed evident to me that it should also work for any other problem, provided there was a development in powers of h. The only difference was that at this time, I did not choose systematically a geometric progression of rate $1/2$ and that I did not build the famous triangular array of Romberg. I did consider the rate $1/2$ for two steps, but not necessarily for the sequence.

Another point ought to be mentioned: J. Kuntzmann had worked a lot on the integration of differential equations, in particular using Runge–Kutta methods. He published a number of papers on the subject and even a book, and he largely insisted on the subject in the classes he taught. A particular procedure was in current use: the double-step method for estimating the error. If one knows that the error is in h^2, the error evaluation for the solution obtained with step h is $|S(h) - S(2h)|/3$. These two approximate solutions were not used to obtain the more precise result $(4\,S(h) - S(2h))/3$, but the principle is the same. I vaguely remember that J. Kuntzmann expressly mentioned the h^2-extrapolation according to Richardson, referring to his first paper published in 1911 (not the one of 1927 where he developed his method). The only, though important, difference was that one does not systematically pursue with steps in geometric ratio $1/2$. For integration methods in h^4, we of course made use of the error estimation $|16S(h) - S(2h)|/15$.

Now advancing rapidly, I experimentally applied the Richardson principle to other problems and especially to an elliptic partial differential equation, the heat problem, for instance. In this particular case, as h tends to 0, the computation cost of the solution $S(h)$ explodes far more quickly than in the case of a simple integral. Thus it was natural to consider step sequences converging more slowly to 0. Numerical applications showed that it was not reasonable either to choose the slowest convergence, i.e., by inverse of successive integers, since then it is the sum of the coefficients that explodes, which proves disastrous for the stability by propagating the errors. Hence, a theoretical question arises rather naturally: what is the best progression to ensure convergence as well as stability? Of course, the geometric progression was a first-line candidate.

In September 1962, my wife and I went to the United States for 2 months, she to MIT for mechanical translation and I to Harvard, distant by only a few miles. The exceptional ambience of Harvard's sumptuous library may very well have been conducive for finding the condition $h_i/h_{i+1} \geq \alpha$ for $\alpha > 1$, called the α-condition, for convergence as well as stability. What surprised me most was to find a condition that was necessary as well as sufficient. I sent the proof – handwritten on airmail stationary—to J. Kuntzmann. Convergence problems in approximation theory having been intensively investigated in all directions, J. Kuntzmann solicited the opinion of J. Favard, professor at the École Polytechnique in Paris and a great expert in this field, so as to be sure of the originality of this result. I was hardly back in France when J. Favard amiably invited me to his Grenoble home. He straightaway told me, first of all, that my result was original and very nice indeed, but that one could (should) eliminate half of my proof by applying the Banach–Steinhaus theorem!!! My first proof (of which I lost track) had indeed been much longer than the one you can find in my thesis and my papers (C.R. Acad. Sci. 256, février 1963, 1435–1437), since it implicitly contained this fundamental theorem of functional analysis. Forging ahead, J. Favard bluntly announced that he would be happy to preside over my thesis examining board, a signal favour! Perhaps so as to remedy to my insufficiencies in functional analysis, he suggested for my second subject, the closed graph (or continuous inverse) theorem, of which I made thorough use later on in spline-function theory.

7.8 David Levin

David Levin[7] is a professor at Tel-Aviv University, Israel. After working on convergence acceleration methods (he found quite powerful sequence transformations bearing his name), he moved to splines, subdivision, moving least-squares, multivariate approximation, quadrature methods, CAGD, and computer graphics.

---

[7] Homepage https://english.tau.ac.il/profile/levin.

Testimony

In 1970, I began to work on my M.Sc. degree in geophysics, and I was looking for a subject for a thesis. I found a nice problem, involving waves near a 3D corner, and started working on it. I had many good ideas, but none of them worked.

Then, Professor Ivor Longman, of the Geophysics Department handed, me a copy of the 1955 paper by Daniel Shanks on the e_m-algorithm. It is a very nice paper, and I still keep the same copy I got in 1970. I also learned about Wynn's ρ-algorithm.

In view of these two algorithms, I was challenged to find a method that would work for both oscillatory and nonoscillatory slowly convergent series. I derived three transformations, and I looked for names for these transformations. The combination "tuv" means goodness in Hebrew, so I chose these letters to denote the different transformations. Perhaps I should have called these transformations "Levin-type" transformations. . .

The t-u-v-transformations were the subject of my M.Sc. dissertation and the subject of my first presentation at my first conference, at Canterbury 1972. The paper was published in 1973, and in 1974, while working on my Ph.D., I got a letter from Richard Bellman, inviting me to write a book on convergence acceleration.

I could not accept the offer, simply since I did not feel mature enough for such a project.

In my Ph.D. thesis I presented explicit expressions for multivariate Padé approximations of general orders, and I also extended the idea in the t-u-v-transformations to deal with series of orthogonal polynomials.

After my Ph.D., in 1975, I began working with Avram Sidi. Our paper on the D- and d-transformations, which we derived, is still, in my view, the most elegant and complete study of convergence acceleration of practical infinite series and integrals appearing in applied mathematics. After this work with Sidi I left the subject of convergence acceleration. Sidi has continued the work, studying the convergence theory and the computational aspects of the D- and d- transformations.

Even today, I receive requests to review papers on convergence acceleration. Most of these papers suggest minor changes in the "Levin-type" transformations, which I find disappointing. It seems that no one tries to deal with really challenging series such as Sidi and I considered in our joint paper.

In 1976–1977 I did my postdoc at Brunel University, England. There I worked on numerical conformal mapping with Nick Papamichael and on CAGD with John Gregory. I have also worked on collocation approximation for integral equations and for PDE.

In 1978 I joined Tel Aviv University, and I began my collaboration with Nira Dyn. First, we worked on approximation by radial basis functions. Then, after coming up with the four-point subdivision scheme, for more than 30 years we are still working on the analysis of subdivision schemes and related topics.

I have also worked on other subjects, among which my main contributions are:

1. Set-valued approximation using the signed-distance function,
2. A method for the numerical integration of rapidly oscillatory functions,

3. My paper on "The approximation power of moving least-squares",
4. Moving least-squares projection for the reconstruction of surfaces,
5. Approximation of functions with singularities,
   and recently
6. Extension of functions,
7. Fixed-point theory for sequences of maps.

7.9 Naoki Osada

Naoki Osada[8] is a Japanese mathematician and a professor at the Tokyo Woman's Christian University. His research relates to various extrapolation methods, in particular the ϱ-algorithm, and their historical roots.

Testimony

Around 1983, I noticed by numerical experiment that

$$\frac{M_{4n} - M_{2n}}{M_{2n} - M_n} \approx \frac{M_{2n} - I}{M_n - I},$$

where M_n is the n panels midpoint rule for a certain improper integral I. I was writing a program to compute an improper integral based on this property, and the computational result of the program was good. After examining a textbook on numerical analysis, the method I used was the Aitken Δ^2 process. After this I began studying extrapolation.

In 1986, I proved the acceleration theorem for the ρ-algorithm of Peter Wynn, and succeeded in generalizing the ρ-algorithm. I summarized these results in a paper, but it was too long, so I divided it into two. I submitted the paper concerning the acceleration theorem to Math. Comput., concerning the generalization to SIAM J. Numer. Anal.. The former paper was rejected, but the latter paper [2] was accepted. On account of submitting to SIAM, Claude Brezinski invited me to the international congress on extrapolation and rational approximation held at CIRM in Luminy, France, in September 1989. I attended this congress, so I became acquainted with many researchers in this field.

I did not submit the former paper to another journal, but I recorded the paper in my doctoral thesis [3] in 1993. Brezinski, who read my doctoral thesis, advised me to submit the result. Thus I rewrote it and submitted the paper. This paper [4] was published in Numerishe Mathematik and was reviewed by Wynn in Zbl. Math.

In 1996 I received a letter and his offprints from Wynn. I sent a letter and my offprints to Wynn. Then I received a long letter from him. At that time I was burdened by responsibilities at the university, so a long time passed without my replying. Not

[8] A list of papers can be found at http://www.lab.twcu.ac.jp/~osada/index-e.html.

only did I lose the opportunity to learn a lot from him, I regret that I had been impolite to him.

Since I learned from Brezinski's paper [1] that it was Takakazu Seki who discovered the Aitken Δ^2 process, when examining a book on the history of Japanese mathematics, I found out that the Richardson extrapolation was discovered by Seki's disciple Katahiro Takebe. I wrote about Seki and Takebe's results in the introduction of my doctoral thesis. I disseminated their results in lectures in Japan.

Until I gave a lecture on the acceleration method of Seki in 2007, I had not studied it. I began to study the history of mathematics, especially Seki's and Newton's mathematics, on that occasion. I wrote about Seki's and Newton's extrapolation methods in [5] and [6], respectively.

References

1. C. Brezinski, Introduction and historical survey, in Jean-Paul Delahaye, Sequence Transformations, Springer, 1988.
   (Cited on page 244.)
2. N. Osada, A convergence acceleration method for some logarithmically convergent sequences, SIAM journal on numerical analysis 27 (1) (1990), 178–189.
   (Cited on page 243.)
3. N. Osada, Acceleration methods for slowly convergent sequences and their applications, Doctoral thesis, Nagoya University, 1993.
   (Cited on page 243.)
4. N. Osada, An acceleration theorem for the ρ-algorithm, Numerische Mathematik 73 (4) (1996), 521–531.
   (Cited on page 243.)
5. Naoki Osada, The early history of convergence acceleration methods, Numerical Algorithms 60 (2012), 205–221.
   (Cited on page 244.)
6. N. Osada, Isaac Newton's "Of Quadrature by Ordinates", Archive for history of exact sciences 67 (4) (2013), 457–476.
   (Cited on page 244.)

7.10 Avram Sidi

Avram Sidi[9] (b. 1947) is professor emeritus at Technion—Israel Institute of Technology, Haifa, Israel. His research interests include extrapolation methods, Padé approximation, rational interpolation and approximation, quadrature, Krylov subspace methods, etc. He has published two books on extrapolation, one on scalar sequences and one on vector sequences, and about 150 papers.

---

[9] Homepage http://www.cs.technion.ac.il/~asidi/.

Testimony

My Research Related to Peter Wynn's Work

7.10.1 Work Related to the Scalar Epsilon Algorithm: Part 1

7.10.1.1 Review of the ϵ-Algorithm

One of Wynn's most famous and interesting contributions is his ϵ-*algorithm*, published in [69], for implementing the transformation of Shanks [27]. Basically, the Shanks transformation is defined via the linear equations

$$A_r = A_n^{(j)} + \sum_{i=1}^{n} \beta_i a_{r+i}, \quad r = j, j+1, \ldots, j+n; \quad j, n \geq 0, \tag{7.14}$$

where the A_r and the a_r are related as in

$$A_k = \sum_{i=0}^{k} a_i, \quad k = 0, 1, \ldots, \tag{7.15}$$

$A_n^{(j)}$ and the β_i are the unknowns, and $A_n^{(j)}$ is the quantity we would like to determine. The ϵ-algorithm solves these equations for the $A_n^{(j)}$ recursively in j and n:

1. Set $\epsilon_{-1}^{(j)} = 0$ and $\epsilon_0^{(j)} = A_j$, $j = 0, 1, \ldots$.
2. Compute the $\epsilon_k^{(j)}$ by the recurrence

$$\epsilon_{k+1}^{(j)} = \epsilon_{k-1}^{(j+1)} + \frac{1}{\epsilon_k^{(j+1)} - \epsilon_k^{(j)}}, \quad j, k = 0, 1, \ldots.$$

Wynn showed that

$$\epsilon_{2n}^{(j)} = e_n(A_j) \quad \text{and} \quad \epsilon_{2n+1}^{(j)} = \frac{1}{e_n(\Delta A_j)} \quad \text{for all } j \text{ and } n; \quad \Delta A_j = A_{j+1} - A_j. \tag{7.16}$$

The Shanks transformation and the ϵ-algorithm are treated in great detail in Sidi [46, Chapter 16].

7.10.1.2 The FS/qd Algorithm

In case the $A_n^{(j)}$ are defined via the equations

$$A_r = A_n^{(j)} + \sum_{i=1}^{n} \beta_i u_{r+i}, \quad r = j, j+1, \ldots, j+n; \quad j, n \geq 0, \tag{7.17}$$

instead of those in (7.14), and the A_r and the u_r are *not* necessarily related in any way—unlike the A_r and the a_r, which are related as in (7.15)—$A_n^{(j)}$ and the β_i are the unknowns, and $A_n^{(j)}$ is the quantity we would like to determine. The $A_n^{(j)}$ can be computed recursively by the FS/qd algorithm of Sidi [44], [46, Chapter 21], as follows:

1. For $j = 0, 1, \ldots$, set

$$e_0^{(j)} = 0, \quad q_1^{(j)} = \frac{u_{j+1}}{u_j}, \quad M_0^{(j)} = \frac{A_j}{u_j}, \quad N_0^{(j)} = \frac{1}{u_j}.$$

2. For $j = 0, 1, \ldots$, and $n = 1, 2, \ldots$, compute recursively

$$e_n^{(j)} = q_n^{(j+1)} - q_n^{(j)} + e_{n-1}^{(j+1)}, \quad q_{n+1}^{(j)} = \frac{e_n^{(j+1)}}{e_n^{(j)}} q_n^{(j+1)},$$

$$M_n^{(j)} = \frac{M_{n-1}^{(j+1)} - M_{n-1}^{(j)}}{e_n^{(j)}}, \quad N_n^{(j)} = \frac{N_{n-1}^{(j+1)} - N_{n-1}^{(j)}}{e_n^{(j)}}.$$

3. For $j, n = 0, 1, \ldots$, set

$$A_n^{(j)} = \frac{M_n^{(j)}}{N_n^{(j)}}.$$

Concerning the FS/qd algorithm, the following remarks are noteworthy:

1. The problem in (7.17) arises in relation to the higher-order G-transformation of Gray, Atchison, and McWilliams [15].
2. Clearly, the FS/qd algorithm can also be used to implement the Shanks transformation for the sequence $\{A_m\}$, by letting $u_m = A_m - A_{m-1}$ in computing the quantities $q_1^{(j)}$, $M_0^{(j)}$, and $N_0^{(j)}$, and is as efficient as the ϵ-algorithm in this case.
3. The FS/qd algorithm combines in a clever way the FS-algorithm of Ford and Sidi [6] and the qd-algorithm of Rutishauser [25, 26], and turns out to be very efficient computationally.
   Given the linear systems

$$A_r = A_n^{(j)} + \sum_{i=1}^{n} \beta_i g_i(r), \quad r = j, j+1, \ldots, j+n; \quad j, n \geq 0, \tag{7.18}$$

where the unknowns are $A_n^{(j)}$ and $\beta_1, \ldots, \beta_n$, the FS algorithm is an efficient procedure that computes the $A_n^{(j)}$ recursively. Note that the $g_i(m)$ do not need to have a specific structure in this case.

7.10.2 Work Related to the Scalar Epsilon Algorithm: Part 2

7.10.2.1 Epsilon Algorithm on Dirichlet-Type Series

One of the first results related to the Shanks transformation and treated by Wynn [71] concerns the analysis of $\epsilon_{2n}^{(j)} = A_n^{(j)}$ in the case

$$A_m \sim A + \sum_{k=1}^{\infty} \alpha_k \lambda_k^m \quad \text{as } m \to \infty, \tag{7.19}$$

when the λ_k satisfy

$$|\lambda_1| > |\lambda_2| > |\lambda_3| > \cdots, \quad \lim_{k \to \infty} \lambda_k = 0; \quad \lambda_k \text{ all positive or all negative.} \tag{7.20}$$

By relaxing this condition on the λ_k, in [41], Sidi extended Wynn's result as follows:

Theorem 7.1 *Assume*

$$\lambda_k \neq 1 \text{ and distinct for all } k; \quad |\lambda_1| \geq |\lambda_2| \geq |\lambda_3| \geq \cdots \geq \cdots, \quad \lim_{k \to \infty} \lambda_k = 0. \tag{7.21}$$

Assume, in addition, that

$$|\lambda_n| > |\lambda_{n+1}| = \cdots = |\lambda_{n+r}| > |\lambda_{n+r+1}| \geq \cdots. \tag{7.22}$$

Then

$$\epsilon_{2n}^{(j)} - A = \sum_{p=n+1}^{n+r} \alpha_p \left(\prod_{i=1}^{n} \frac{\lambda_p - \lambda_i}{1 - \lambda_i} \right)^2 \lambda_p^j + o(\lambda_{n+1}^j) \quad \text{as } j \to \infty,$$

$$= O(\lambda_{n+1}^j) \quad \text{as } j \to \infty, \tag{7.23}$$

whether $\{A_m\}$ *converges or not.*

7.10.2.2 Epsilon Algorithm on General Dirichlet-Type Series

In [41], Sidi proved two extensions of Theorem 7.1 by generalizing (7.19). The first extension is Theorem 7.2 below. The technique used in the proof of this theorem is a refined version of that developed in Sidi and Bridger [61].

Theorem 7.2 *Assume*

$$A_m \sim A + \sum_{k=1}^{\infty} P_k(m) \lambda_k^m \quad \text{as } m \to \infty, \tag{7.24}$$

where

$\lambda_k \neq 1$ *and distinct for all* k; $\quad |\lambda_1| \geq |\lambda_2| \geq |\lambda_3| \geq \cdots$; $\quad \lim_{k \to \infty} \lambda_k = 0$, $\quad$ (7.25)

and for each k, $P_k(m)$ *is a polynomial in* m *of degree exactly* $p_k \geq 0$ *and with leading coefficient* $e_k \neq 0$:

$$|\lambda_1| \geq \cdots \geq |\lambda_t| > |\lambda_{t+1}| = \cdots = |\lambda_{t+r}| > |\lambda_{t+r+1}|, \quad (7.26)$$

and order the λ_k *such that*

$$\bar{p} \equiv p_{t+1} = \cdots = p_{t+\mu} > p_{t+\mu+1} \geq \cdots \geq p_{t+r}. \quad (7.27)$$

Set

$$\omega_k = p_k + 1, \quad k = 1, 2, \ldots, \quad (7.28)$$

and let

$$n = \sum_{k=1}^{t} \omega_k. \quad (7.29)$$

Then

$$\epsilon_{2n}^{(j)} - A = j^{\bar{p}} \sum_{s=t+1}^{t+\mu} e_s \left[\prod_{i=1}^{t} \left(\frac{\lambda_s - \lambda_i}{1 - \lambda_i} \right)^{2\omega_i} \right] \lambda_s^j + o(j^{\bar{p}} \lambda_{t+1}^j) \ as \ j \to \infty,$$

$$= O(j^{\bar{p}} \lambda_{t+1}^j) \ as \ j \to \infty, \quad (7.30)$$

whether $\{A_m\}$ *converges or not.*

The second extension is Theorem 7.3, which deals with the remaining cases in which (7.29) in Theorem 7.2 is *not* satisfied, in the sense that $n \neq \sum_{k=1}^{t} \omega_k$.

Theorem 7.3 *Assume that in Theorem 7.2, (7.29) is replaced by*

$$\sum_{k=1}^{t} \omega_k < n < \sum_{k=1}^{t+r} \omega_k \quad (7.31)$$

and let

$$\tau = n - \sum_{k=1}^{t} \omega_k. \quad (7.32)$$

This time, however, also allow $t = 0$ *and define* $\sum_{k=1}^{0} \omega_k = 0$. *Denote by* $\mathrm{IP}(\tau)$ *the nonlinear integer programming problem*

$$\text{maximize } g(\sigma); \ g(\sigma) = \sum_{k=t+1}^{t+r} (\omega_k \sigma_k - \sigma_k^2)$$

$$\text{subject to } \sum_{k=t+1}^{t+r} \sigma_k = \tau \ and \ 0 \leq \sigma_k \leq \omega_k, \ t+1 \leq k \leq t+r, \quad (7.33)$$

and denote by $G(\tau)$ the (optimal) value of $g(\sigma)$ at the solution to IP(τ).

Then, provided IP(τ) *has a unique solution for* σ_k, $k = t + 1, \ldots, t + r$, $\epsilon_{2n}^{(j)}$ *satisfies*

$$\epsilon_{2n}^{(j)} - A = O(j^{G(\tau+1)-G(\tau)} \lambda_{t+1}^j) \quad \text{as } j \to \infty, \tag{7.34}$$

whether $\{A_m\}$ *converges or not.* [IP($\tau + 1$) *is not required to have a unique solution.*]

An algorithm for solving the problem IP(τ) in an efficient way and for determining whether it has a unique solution or not has been given in Kaminski and Sidi [17].

7.10.3 Work Related to the Scalar Epsilon Algorithm: Part 3

7.10.3.1 Epsilon Algorithm on *Linear Sequences*

In [71], Wynn also considers the case in which the sequence $\{A_m\}$ is such that

$$A_m \sim A + (-1)^m \sum_{i=0}^{\infty} \alpha_i m^{-1-i} \quad \text{as } m \to \infty. \tag{7.35}$$

Garibotti and Grinstein [8] generalize (7.35) considerably and prove the following result:

Theorem 7.4 *Let* $\{A_m\}$ *be such that*

$$A_m \sim A + \zeta^m \sum_{i=0}^{\infty} \alpha_i m^{\gamma-i} \quad \text{as } m \to \infty; \quad \zeta \neq 1, \quad \alpha_0 \neq 0. \tag{7.36}$$

1. Provided $\gamma \neq 0, 1, \ldots, n - 1$, *one has*

$$\epsilon_{2n}^{(j)} - A \sim (-1)^n \alpha_0 \frac{n! \, [\gamma]_n}{(\zeta - 1)^{2n}} \zeta^{j+2n} j^{\gamma-2n} \quad \text{as } j \to \infty, \tag{7.37}$$

where $[\gamma]_n = \gamma(\gamma - 1) \cdots (\gamma - n + 1)$ *when* $n > 0$ *and* $[\gamma]_0 = 1$, *as before.*

2. If γ *is an integer,* $0 \leq \gamma \leq n - 1$, *and* $\alpha_{\gamma+1} \neq 0$, *then*

$$\epsilon_{2n}^{(j)} - A \sim \alpha_{\gamma+1} \frac{(n - \gamma - 1)!(n + \gamma + 1)!}{(\zeta - 1)^{2n}} \zeta^{j+2n} j^{-2n-1} \quad \text{as } j \to \infty. \tag{7.38}$$

These results are valid whether $\{A_m\}$ *converges or not.*

Sequences $\{A_m\}$ satisfying (7.36) are called *linear sequences*.

7.10.3.2 Epsilon Algorithm on General Linear Sequences

In Sidi [54], the following theorem that generalizes part 1 in Theorem 7.4 was proved:

Theorem 7.5 *Let the sequence* $\{A_m\}$ *be such that*

$$A_m \sim A + \sum_{k=1}^{p} \zeta_k^m \sum_{i=0}^{\infty} \beta_{ki} m^{\gamma_k - i} \quad as \ m \to \infty; \quad \beta_{k0} \neq 0, \quad k = 1, \ldots, p, \quad (7.39)$$

where $p > 1$ *is arbitrary and*

$$|\zeta_1| = \cdots = |\zeta_p| = \theta; \quad \zeta_k \neq 1 \ distinct, \quad \gamma_k \neq 0, 1, 2, \ldots, arbitrary, \quad (7.40)$$

with the γ_k *ordered as in*

$$\mathfrak{R}\gamma_1 \geq \mathfrak{R}\gamma_2 \geq \cdots \geq \mathfrak{R}\gamma_p. \quad (7.41)$$

Assume that the positive integer n *in* $\epsilon_{2n}^{(j)}$ *is such that the integer programming problem*

$$\max_{s_1, \ldots, s_p} g(s_1, \ldots, s_p); \quad g(s_1, \ldots, s_p) = \sum_{k=1}^{p} \left[s_k(\mathfrak{R}\gamma_k) - s_k(s_k - 1) \right]$$

$$\text{(7.42)}$$

$$\textit{subject to} \quad s_1 \geq 0, \ldots, s_p \geq 0 \quad \textit{and} \quad \sum_{k=1}^{p} s_k = n; \quad s_k \ integers,$$

which we denote by $I\mathcal{P}_n$, *has a unique solution for integer* s_k, *which we shall denote by* $(s_1', \ldots, s_p')$. *Let us also denote* $g(s_1', \ldots, s_p')$ *by* σ_n. *Then* $\epsilon_{2n}^{(j)}$ *exist for all large* j, *and satisfy*

$$\epsilon_{2n}^{(j)} - A = O\left(\theta^j j^{\sigma_{n+1} - \sigma_n}\right) \quad as \ j \to \infty. \quad (7.43)$$

Sequences $\{A_m\}$ satisfying (7.39) are called *general linear sequences*.

Remarks

1. In [54], it is also shown that the Shanks transformation/epsilon algorithm is stable numerically and its stability property is quantified.
2. Examples of sequences $\{A_m\}$ that satisfy the conditions of Theorem 7.5 are sequences of partial sums of classical Fourier series and generalized Fourier series such as series of orthogonal polynomials (e.g., Legendre series), series of special functions (e.g., Bessel function series), and more. See Sidi [38].
3. The problem $I\mathcal{P}_n$ is analyzed in detail and an algorithm for it is given in [54, Appendix].
4. In case $I\mathcal{P}_n$ does not have a unique solution, we cannot conclude that the $\epsilon_{2n}^{(j)}$ exist for all large j, and therefore, we cannot conclude anything about the convergence of the sequence $\{\epsilon_{2n}^{(j)}\}_{j=0}^{\infty}$.

5. Fortunately, as proved in [54, Appendix], there exist infinitely many positive integers n for which $\mathcal{IP}_n$ has a unique solution, which means that Theorem 7.5 applies to an infinite number of the columns of the epsilon table, and this is significant.

For example, in case $\Re\gamma_1 = \cdots = \Re\gamma_p \equiv \beta$, $\mathcal{IP}_n$ has a unique solution (only) for $n = pv$, for each $v \in \{1, 2, \ldots\}$. In such a case, $s'_k = v$, $k = 1, \ldots, p$, and we have

$$\epsilon_{2pv}^{(j)} - A = O(\theta^j \, j^{\beta - 2v}) \quad \text{as } j \to \infty, \quad v = 1, 2, \ldots.$$

Comparing this with

$$\epsilon_0^{(j)} - A = A_j - A = O(\theta^j \, j^\beta) \quad \text{as } j \to \infty,$$

it becomes clear that acceleration of convergence takes place. Numerical examples show that $\epsilon_{2(pv+i)}^{(j)} - A$, $i = 1, \ldots, p - 1$, for which the problems $\mathcal{IP}_{pv+i}$ do not have unique solutions, are all approximately of the order of $\epsilon_{2pv}^{(j)} - A$. This suggests that (7.43) is optimal.

6. Finally, Theorem 7.5 continues to apply if instead of (7.39), we have the general case of

$$A_m \sim A + \sum_{k=1}^{q} \zeta_k^m \sum_{i=0}^{\infty} \beta_{ki} m^{\gamma_k - i} \quad \text{as } m \to \infty; \quad \beta_{k0} \neq 0, \quad k = 1, \ldots, q,$$

with

$$|\zeta_1| = \cdots = |\zeta_p| > |\zeta_{p+1}| \geq \cdots \geq |\zeta_q|; \quad \zeta_k \neq 1.$$

7.10.3.3 The Generalized Richardson Extrapolation Process GREP on *General Linear Sequences*

The convergence of the generalized linear sequences $\{A_m\}$ can be accelerated very efficiently also using the Generalized Richardson Extrapolation Process GREP$^{(p)}$ of Sidi [29]. We now let $\{A_m\}$ be such that

$$A_m \sim A + \sum_{k=1}^{p} \zeta_k^m \sum_{i=0}^{\infty} \beta_{ki} m^{\gamma_k - i} \quad \text{as } m \to \infty; \quad \beta_{k0} \neq 0, \quad k = 1, \ldots, p, \quad (7.44)$$

$$\zeta_k \text{ arbitrary}, \quad \zeta_k \neq 1, \quad k = 1, \ldots, p. \quad (7.45)$$

For these sequences, GREP$^{(p)}$ produces approximations $A_n^{(p,j)}$ with $n \equiv (n_1, \ldots, n_p)$ ($n_1, \ldots, n_p$ are arbitrary positive integers) to A that are defined via the linear systems

$$A_l = A_n^{(p,j)} + \sum_{k=1}^{p} \zeta_k^l \sum_{i=0}^{n_k - 1} \bar{\beta}_{ki} (\alpha + l)^{\gamma_k - i}, \quad j \leq l \leq j + \sum_{k=1}^{p} n_k, \quad (7.46)$$

where α is some arbitrary fixed positive scalar and $\bar{\beta}_{ki}$ are additional unknowns.

The following theorem was proved in Sidi [51]:

Theorem 7.6 *Let the sequence* $\{A_m\}$ *be as in* (7.44)–(7.45). *Then, for every* n,

$$A_n^{(p,j)} - A = \sum_{k=1}^{m} O(\zeta_k^j \, j^{\gamma_k - 2n_k}) \quad as \; j \to \infty.$$

When compared with $A_j - A = \sum_{k=1}^{m} O(\zeta_k^j \, j^{\gamma_k})$ as $j \to \infty$, this result shows that GREP$^{(p)}$ is a true convergence acceleration method for the sequences considered. In addition, it is shown in [51] that GREP$^{(p)}$ is stable for the case being studied, and its stability property is quantified.

7.10.4 Work Related to the Scalar Epsilon Algorithm: Part 4

7.10.4.1 Complex Series Approach for Fourier Series

The Shanks transformation (hence ϵ-algorithm) is a very effective method for summing real Fourier series

$$F_c(x) = \sum_{n=0}^{\infty} a_n \cos nx, \quad F_s(x) = \sum_{n=0}^{\infty} b_n \sin nx.$$

In [72], Wynn proposes to apply the epsilon algorithm to the complex series

$$G(x) = \sum_{n=0}^{\infty} c_n e^{inx},$$

instead of applying it directly to $F_c(x)$ or $F_s(x)$. If $\epsilon_{2n}^{(j)}$ are the approximations to $G(x)$, then $\Re\epsilon_{2n}^{(j)}$ and $\Im\epsilon_{2n}^{(j)}$ are approximations to $F_c(x)$ and $F_s(x)$, respectively. This way of applying the ϵ-algorithm requires *half* the number of terms needed when applying it to the real series $F_c(x)$ and $F_s(x)$ to achieve comparable accuracy.

This complex series approach of Wynn was generalized by Sidi [40] as follows: We first note that $\cos nx$ and $\sin nx$ are the solutions of the linear homogeneous ODE $u'' + n^2 u = 0$; if we call one of them the solution of the first kind, then the other is the solution of the second kind. We can now consider other series $\sum_{n=0}^{\infty} a_n \phi_n(x)$ when $\phi_n(x)$ are solutions of a second-order linear homogeneous ODE that arises from Sturm–Liouville problems, by introducing the (properly normalized) second solution (function of the second kind) $\psi_n(x)$. Such series are commonly known as *generalized Fourier series*.

7.10.4.2 Complex Series Approach for Generalized Fourier Series

More generally, assume we want to accelerate the convergence of an infinite series $F(x)$ given in the form

$$F(x) := \sum_{n=0}^{\infty} [b_n \phi_n(x) + c_n \psi_n(x)], \tag{7.47}$$

where the b_n and c_n are known and the functions $\phi_n(x)$ and $\psi_n(x)$ satisfy

$$\rho_n^{\pm}(x) \equiv \phi_n(x) \pm \mathrm{i}\psi_n(x) = e^{\pm \mathrm{i}n\omega x} g_n^{\pm}(x), \tag{7.48}$$

ω being some fixed real positive constant, and

$$g_n^{\pm}(x) \sim n^\epsilon \sum_{j=0}^{\infty} \delta_j^{\pm}(x) n^{-j} \text{ as } n \to \infty, \tag{7.49}$$

for some fixed ϵ that can be complex in general. Then

$$\phi_n(x) = \frac{1}{2} \left[\rho_n^+(x) + \rho_n^-(x) \right] \text{ and } \psi_n(x) = \frac{1}{2\mathrm{i}} \left[\rho_n^+(x) - \rho_n^-(x) \right]. \tag{7.50}$$

The simplest and most widely treated members of the series above are the classical Fourier series, with $\phi_n(x) = \cos nx$ and $\psi_n(x) = \sin nx$. Other examples are:

1. $\phi_n(x) = \cos \lambda_n x$ and $\psi_n(x) = \sin \lambda_n x$, $\lambda_n \sim n \sum_{j=0}^{\infty} \alpha_j n^{-j}$ as $n \to \infty$, $\alpha_0 \neq 0$.
2. $\phi_n(x) = T_n(x)$ and $\psi_n(x) = U_n(x)$. (Chebyshev polynomials of the first and second kinds).
3. $\phi_n(x) = P_n(x)$ and $\psi_n(x) = -\frac{2}{\pi} Q_n(x)$. (Legendre polynomials and associated Legendre functions of the second kind.)
4. $\phi_n(x) = J_n(x)$ and $\psi_n(x) = Y_n(x)$. (Bessel functions of the first and second kinds.)

The approach proposed in [40] for accelerating the convergence of the series $F(x)$, *assuming that the coefficients b_n and c_n are known*, is as follows:

1. Define the series $B^{\pm}(x)$ and $C^{\pm}(x)$ by

$$B^{\pm}(x) := \sum_{n=1}^{\infty} b_n \rho_n^{\pm}(x) \text{ and } C^{\pm}(x) := \sum_{n=1}^{\infty} c_n \rho_n^{\pm}(x), \tag{7.51}$$

and observe that

$$F_\phi(x) := \sum_{n=1}^{\infty} b_n \phi_n(x) = \frac{1}{2} \left[B^+(x) + B^-(x) \right] \text{ and}$$

$$F_\psi(x) := \sum_{n=1}^{\infty} c_n \psi_n(x) = \frac{1}{2\mathrm{i}} \left[C^+(x) - C^-(x) \right], \tag{7.52}$$

and that

$$F(x) = F_\phi(x) + F_\psi(x). \tag{7.53}$$

2. Apply the transformation of Shanks (equivalently, the ϵ algorithm) or the Levin–Sidi d-transformation [21] to the series $B^\pm(x)$ and $C^\pm(x)$ and then invoke (7.52) and (7.53). The number of the b_n and c_n needed to achieve a desired level of accuracy for the series $B^\pm(x)$ and $C^\pm(x)$ is practically *half* of that needed for the series $F(x)$. In addition, numerical computations seem to show that to achieve a certain level accuracy, the d-transformation uses fewer terms than the ϵ-algorithm.

In connection with this approach, we note that when the functions $\phi_n(x)$ and $\psi_n(x)$ and the coefficients b_n and c_n are all *real*, it is enough to treat the two complex series $B^+(x)$ and $C^+(x)$ as $B^-(x) = \overline{B^+(x)}$ and $C^-(x) = \overline{C^+(x)}$, so that $F_\phi(x) = \Re B^+(x)$ and $F_\psi(x) = \Im C^+(x)$ in such cases.

For more on this subject, see also [46, Chapter 13].

7.10.5 Work Related to the Scalar Epsilon Algorithm: Part 5

7.10.5.1 The Levin–Sidi $d^{(m)}$-Transformation

The $d^{(m)}$-transformation was developed in [21] for accelerating the convergence of a large class of sequences that includes, among many others, the linear and general linear sequences discussed above, on which the epsilon algorithm is effective. It was developed following a rigorous treatment of the asymptotic expansions of the tails of infinite series. In some special cases, the approximations obtained from the $d^{(m)}$-transformation are actually those obtained from the Shanks transformation.

We begin with the following definition of the sequence class $\mathbf{b}^{(m)}$:

The infinite sequence $\{a_n\}_{n=1}^\infty$ belongs to the family $\mathbf{b}^{(m)}$ if the a_n satisfy a homogeneous linear difference equation of the form

$$a_n = \sum_{k=1}^m p_k(n) \Delta^k a_n,$$

where $\Delta^0 a_n = a_n$, $\Delta a_n = a_{n+1} - a_n$, and $\Delta^r a_n = \Delta(\Delta^{r-1} a_n)$, $r = 1, 2, \ldots$, and

$$p_k(n) \sim \sum_{s=0}^\infty p_{ks} n^{i_k - s} \quad \text{as } n \to \infty; \quad i_k \leq k \quad \text{integers}.$$

It is shown in [21] that provided $\{a_n\}_{n=1}^\infty \in \mathbf{b}^{(m)}$ and provided the series $\sum_{k=1}^\infty a_k$ converges, S being its sum, one has

$$A_{n-1} \sim S + \sum_{k=0}^{m-1} n^{\rho_k}(\Delta^k a_n) \sum_{i=0}^{\infty} \frac{\beta_{ki}}{n^i} \quad \text{as } n \to \infty,$$

where

$$A_r = \sum_{i=1}^{r} a_i, \quad r = 1, 2, \ldots; \quad A_0 = 0,$$

and the ρ_k are integers satisfying

$$\rho_k \leq \left[\max_{k+1 \leq s \leq m} (i_s - s) \right] + k + 1 \leq k + 1, \quad k = 0, 1, \ldots, m - 1.$$

This asymptotic expansion forms the basis of the $d^{(m)}$-transformation.

The $d^{(m)}$-transformation, in its most user-friendly form, produces approximations $d_n^{(m,j)}$ to S, the sum of the series, that are defined via the linear systems of equations

$$A_{R_l} = d_n^{(m,j)} + \sum_{k=1}^{m} R_l^k (\Delta^{k-1} a_{R_l}) \sum_{i=0}^{n_k-1} \frac{\bar{\beta}_{ki}}{(R_l + \alpha)^i}, \quad j \leq l \leq j + N; \quad N = \sum_{k=1}^{m} n_k,$$

where

$$A_r = \sum_{i=1}^{r} a_i, \quad r = 1, 2, \ldots; \quad A_0 = 0,$$

$\bar{\beta}_{ki}$ are auxiliary unknowns, and $n \equiv (n_1, \ldots, n_m)$, n_k being nonnegative integers. In addition, $\alpha \geq 0$ is a scalar and the R_l are integers, all at our disposal, and

$$1 \leq R_0 < R_1 < R_2 < \cdots .$$

(The "default" choice is $R_l = l + 1, l = 0, 1, \ldots$.)

1. It can be shown that if $\{a_n\} \in \mathbf{b}^{(m)}$, then $\{a_n\} \in \mathbf{b}^{(m')}$ with $m' > m$. This implies that

$$\mathbf{b}^{(1)} \subset \mathbf{b}^{(2)} \subset \mathbf{b}^{(3)} \subset \cdots ;$$

thus, unlike most other sequence transformations, the d-transformations have an ever increasing scope, that is, they accelerate the convergence of more classes of infinite series.

2. Note that the sequences $\{a_n\}$ associated with the linear sequences we have discussed in Theorem 7.4 [in (7.36)] are in $\mathbf{b}^{(1)}$. The sequences $\{a_n\}$ associated with the general linear sequences we have discussed in Theorem 7.5 [in (7.44)–(7.45)] are in $\mathbf{b}^{(p)}$. In addition, the only nonlinear convergence acceleration methods that are applicable to general linear sequences are the ϵ-algorithm and the $d^{(m)}$-transformation with suitable m.

3. Numerical experiments show that the best approximations to the sum of $\sum_{n=1}^{\infty} a_n$ are produced by the sequences $\{d_{(\nu,\nu,\ldots,\nu)}^{(m,j)}\}_{\nu=0}^{\infty}$, which can be computed very efficiently via the W-algorithm of Sidi [32] for $m = 1$ and via the $W^{(m)}$-algorithm

of Ford and Sidi [6] for all $m \geq 1$. (We describe the W-algorithm later as it is developed for implementing the $\tilde{d}^{(m)}$-transformation.)

4. The convergence and stability properties of the $d^{(1)}$-transformation have been studied in detail in several publications of Sidi (see [40, 42, 43, 45], and [46, Chapters 8 and 9]).

5. The t-transformation and u-transformation of Levin [20] are actually $d^{(1)}$-transformations with the choice $R_l = l + 1$. They have been analyzed in Sidi [28, 30].

6. Finally, when applied to sequences of partial sums of infinite power series $\sum_{n=0}^{\infty} c_n z^n$, the $d^{(m)}$-transformation produces rational approximations of the sums of these series. These approximations possess Padé-like properties too. See Sidi and Levin [63].

7.10.5.2 The Sidi $\tilde{d}^{(m)}$-Transformation

In [46, Chapter 6], Sidi discusses a new class of infinite series $\sum_{n=1}^{\infty} a_n$, where the asymptotic expansions of a_n as $n \to \infty$ involve fractional powers of n^{-1}. Thus, we say that $\{a_n\} \in \tilde{\mathbf{b}}^{(m)}$, $m \in \{1, 2, \ldots\}$, if a_n satisfies a first-order linear homogeneous difference equation of the form

$$a_n = p(n) \Delta a_n, \quad p(n) \sim \sum_{i=0}^{\infty} p_i \, n^{q/m - i/m} \quad \text{as } n \to \infty, \quad \text{integer } q \leq m,$$

the general solution of this difference equation being of the form

$$a_n = [\Gamma(n)]^{s/m} e^{Q(n)} w(n), \quad Q(n) = \sum_{i=0}^{m-1} \theta_i \, n^{1-i/m}, \quad w(n) \sim \sum_{i=0}^{\infty} w_i \, n^{\gamma - i/m} \text{ as } n \to \infty.$$

Clearly, if $m = 1$, then $\tilde{\mathbf{b}}^{(1)} = \mathbf{b}^{(1)}$. When $m \geq 2$, the classes $\tilde{\mathbf{b}}^{(m)}$ are much richer, since they give rise to different types of behaviour, mathematically and numerically.

It is shown in [46, Chapter 6] that when $\sum_{k=1}^{\infty} a_k$ converges, S being its sum, one has

$$A_{n-1} \sim S + n^{q/m} a_n \sum_{i=0}^{\infty} \beta_i n^{-i/m} \quad \text{as } n \to \infty.$$

This asymptotic expansion forms the basis of the $\tilde{d}^{(m)}$-transformation, which accelerates the convergence of the infinite series $\sum_{n=1}^{\infty} a_n$, with $\{a_n\} \in \tilde{\mathbf{b}}^{(m)}$.

In its most user-friendly form, the $\tilde{d}^{(m)}$-transformation produces approximations $\tilde{d}_n^{(m,j)}$ to S, the sum of the series $\sum_{n=1}^{\infty} a_n$, that are defined via the linear systems of equations

$$A_{R_l} = \tilde{d}_n^{(m,j)} + R_l \, a_{R_l} \sum_{i=0}^{n} \frac{\bar{\beta}_i}{(R_l + \alpha)^{i/m}}, \quad j \leq l \leq j + n.$$

The R_l are as described in our treatment of the $d^{(m)}$-transformation. In addition, choices of the R_l via APS and GPS that we described there are very useful in the application of the $\tilde{d}^{(m)}$-transformation. For additional information and numerical examples, see Sidi [56].

7.10.5.3 The W-Algorithm for $\tilde{d}^{(m)}$-Transformation

The approximations $\tilde{d}_n^{(m,j)}$, with arbitrary m, can be computed recursively by the W-algorithm of Sidi [32]; see also [46, Chapter 7]. Here are the steps of this algorithm, with $A_r = \sum_{i=1}^r a_i$ and $\omega_r = r a_r$:

1. For $j = 0, 1, \ldots,$

$$M_0^{(j)} = \frac{A_{R_j}}{\omega_{R_j}}, \quad N_0^{(j)} = \frac{1}{\omega_{R_j}}, \quad K_0^{(j)} = (-1)^j |M_0^{(j)}|.$$

2. For $j = 0, 1, \ldots,$ and $n = 1, 2 \ldots,$ compute

$$M_n^{(j)} = \frac{M_{n-1}^{(j+1)} - M_{n-1}^{(j)}}{R_{j+n}^{-1/m} - R_j^{-1/m}}, \quad N_n^{(j)} = \frac{N_{n-1}^{(j+1)} - N_{n-1}^{(j)}}{R_{j+n}^{-1/m} - R_j^{-1/m}}, \quad K_n^{(j)} = \frac{K_{n-1}^{(j+1)} - K_{n-1}^{(j)}}{R_{j+n}^{-1/m} - R_j^{-1/m}}.$$

3. For $j = 0, 1, \ldots,$ and $n = 1, 2 \ldots,$ compute

$$A_n^{(j)} = \frac{M_n^{(j)}}{N_n^{(j)}} \equiv \tilde{d}_n^{(m,j)}, \quad \Lambda_n^{(j)} = \left| \frac{K_n^{(j)}}{N_n^{(j)}} \right|.$$

Note that the quantities $\Lambda_n^{(j)}$ provide a very accurate measure of the stability issue in the following sense: Assume that all computations are carried out in floating-point arithmetic with roundoff unit u and that $(\Lambda_n^{(j)}/\bar{A}_n^{(j)}) u = O(10^{-p})$, where $\bar{A}_n^{(j)}$ is the computed $A_n^{(j)}$. Then, as an approximation to S, $\bar{A}_n^{(j)}$ possesses p correct significant digits close to convergence. Note that the computation of the $\Lambda_n^{(j)}$ involves no additional cost.

With $m = 1$, the W-algorithm implements also the $d^{(1)}$-transformation, since the $\tilde{d}^{(1)}$- and $d^{(1)}$-transformations are identical.

7.10.5.4 Treatment of Numerical Stability Issues

When applying sequence transformations, we are often confronted with the issue of numerical instability, which limits the accuracy of the approximations obtained in floating-point arithmetic. Since the integers R_l in the $d^{(m)}$- and $\tilde{d}^{(m)}$-transformations are at our disposal, we can achieve numerical stability by a judicious choice of the integers R_l. Depending on the nature of the sequence $\{a_n\}$, we have found that the following two choices are very useful:

1. *Arithmetic Progression Sampling (APS)*:

$$R_l = \lfloor \kappa(l+1) \rfloor, \quad l = 0, 1, \ldots, \quad \kappa \geq 1.$$

When κ is an integer, the sequence $\{R_l\}$ is a true arithmetic progression, with $R_{l+1} - R_l = \kappa, l = 0, 1, \ldots$. Otherwise, it is an approximate arithmetic progression, with $\left|(R_{l+1} - R_l) - \kappa\right| < 1, l = 0, 1, \ldots$.

2. *Geometric Progression Sampling (GPS)*:

$$R_0 = 1; \quad R_l = \max\{\lfloor \sigma R_{l-1} \rfloor, (l+1)\}, \quad l = 1, 2, \ldots, \quad \sigma > 1.$$

If σ is an integer ≥ 2, then $\{R_l\}$ is a true geometric progression, with $R_{l+1}/R_l = \sigma, l = 0, 1, \ldots$. Otherwise, it is an asymptotic geometric progression, with $\lim_{l \to \infty}(R_{l+1}/R_l) = \sigma$.[10] This enables us to generate a sequence of integers R_l that grow like σ^l, moderately, by choosing $\sigma \in (1, 2)$.

Choosing $\kappa = 1, 2, \ldots$, APS can be used to stabilize, in certain cases, some of the other known sequence transformations, the ϵ-algorithm being one of them. GPS is unique to the $d^{(m)}$- and $\tilde{d}^{(m)}$-transformations, however.

For more details, see [46, Chapter 10]. For a detailed survey of numerical stability issues in convergence acceleration, see Sidi [53]. For many numerical examples involving the application of the $\tilde{d}^{(m)}$-transformation with APS and GPS, see Sidi [56].

7.10.6 Work Related to the Scalar Epsilon Algorithm: Part 6

7.10.6.1 Connection of the Shanks Transformation with the Padé Table

As was shown by Shanks [27], the approximations $\epsilon_{2n}^{(j)}$ obtained by applying the Shanks transformation to an infinite power series are Padé approximants from this series. Specifically, if the Shanks transformation is being applied to the sequence $A_m(z) = \sum_{k=0}^{m} c_k z^k$, $m = 0, 1, \ldots$, then $\epsilon_{2n}^{(j)}(z) = f_{j+n,n}(z)$, where $f_{m,n}(z)$ stands for the $[m/n]$ Padé approximant from the power series $f(z) = \sum_{k=0}^{\infty} c_k z^k$. If it exists, $f_{m,n}(z)$ is the rational function $P_{m,n}(z)/Q_{m,n}(z)$ with $\deg P_{m,n} \leq m$ and $\deg Q_{m,n} \leq n$, and it satisfies

$$f(z) - f_{m,n}(z) = O(z^{m+n+1}) \quad \text{as } z \to 0.$$

---

[10] It can be shown that

$$R_l = l + 1, \quad l = 0, 1, \ldots, L-1; \quad R_l = \lfloor \sigma R_{l-1} \rfloor, \quad l = L, L+1, \ldots; \quad L = \left\lceil \frac{2}{\sigma - 1} \right\rceil.$$

7.10.6.2 Theorem of de Montessus and Its Extensions and Refinements

Theorem of de Montessus

A most important topic concerning the Padé table is that of convergence of the row sequences $\{f_{m,n}(z)\}_{m=0}^{\infty}$, with fixed $n = 0, 1, \ldots$. The best known result in this connection is the theorem of de Montessus concerning the case in which $f(z)$ is meromorphic in the complex plane:

Theorem 7.7 *Let $f(z)$ be analytic at $z = 0$ and meromorphic in the disk $K = \{z : |z| < R\}$ and let it have n poles in K counting multiplicities. Then, the row sequence $\{f_{m,n}(z)\}_{m=0}^{\infty}$ converges to $f(z)$ uniformly in any compact subset of K excluding the poles of $f(z)$, such that*

$$\limsup_{m \to \infty} |f(z) - f_{m,n}(z)|^{1/m} \le |z/R|. \tag{7.54}$$

Extension of the Theorem of de Montessus

In case the singularities of $f(z)$ on the boundary of the disk K, namely on $\partial K = \{z : |z| = R\}$, are all poles, Theorem 7.7 can be improved substantially. This improvement, presented originally in Sidi [35, Theorem 3.3], is quantitative in nature, and we state it next.

Theorem 7.8 *Let $f(z)$ be analytic at $z = 0$ and meromorphic in the disk K and on its boundary ∂K. Let $z_1, \ldots, z_t$ be the poles of $f(z)$ in K and let $\omega_1, \ldots, \omega_t$ be their respective multiplicities. Define $Q(z) = \prod_{j=1}^{t}(1 - z/z_j)^{\omega_j}$ and $n = \sum_{j=1}^{t} \omega_j$. Similarly, let $\hat{z}_1, \ldots, \hat{z}_r$ be the poles of $f(z)$ on ∂K, and let $\hat{\omega}_1, \ldots, \hat{\omega}_r$ be their respective multiplicities. Thus, for each $j \in \{1, \ldots, r\}$, the Laurent expansion of $f(z)$ about $z = \hat{z}_j$ is given by*

$$f(z) = \sum_{i=1}^{\hat{\omega}_j} \frac{\hat{a}_{ji}}{(1 - z/\hat{z}_j)^i} + \Theta_j(z); \quad \hat{a}_{j\hat{\omega}_j} \ne 0, \quad \Theta_j(z) \text{ analytic at } \hat{z}_j. \tag{7.55}$$

Let us now order the $\hat{z}_j$ on ∂K such that

$$\hat{\omega}_1 = \hat{\omega}_2 = \cdots = \hat{\omega}_\mu > \hat{\omega}_{\mu+1} \ge \cdots \ge \hat{\omega}_r, \tag{7.56}$$

and set $\bar{p} = \hat{\omega}_1 - 1$. Then there holds

$$f(z) - f_{m,n}(z) = \frac{m^{\bar{p}}}{\bar{p}!} \sum_{j=1}^{\mu} \frac{\hat{a}_{j\hat{\omega}_j}}{1 - z/\hat{z}_j} \left[\frac{Q(\hat{z}_j)}{Q(z)} \right]^2 \left(\frac{z}{\hat{z}_j} \right)^{m+n+1} + o(m^{\bar{p}} |z/R|^m)$$

$$= O(m^{\bar{p}} |z/R|^m) \quad \text{as } m \to \infty, \tag{7.57}$$

uniformly in any compact subset of $K\backslash\{z_1, \ldots, z_t\}$. This result is best possible as $m \rightarrow \infty$.

Extension of the Theorem of de Montessus to Intermediate Rows of the Padé Table

Theorem 7.8 is not valid when $q < n < q + \sum_{j=1}^r \hat{\omega}_j$, where q is the number of poles z_j of $f(z)$ in K counted according to their multiplicities ω_j, that is, $q = \sum_{j=1}^t \omega_j$. Rows of the Padé table for which n takes on such values are called *intermediate rows*. Note that intermediate rows may appear not only when $n > q$; when $f(z)$ has multiple poles and/or a number of poles with equal modulus in K, they appear with $n < q$ as well. Thus, intermediate rows are at least as common as those treated by de Montessus's theorem.

The convergence problem of intermediate rows was treated partially (for some special cases) in a series of papers by Wilson [66, 67, 68]. The complete solution for the general case was given by Sidi [35, Theorem 6.1]. The following convergence result pertaining to the convergence of intermediate rows in the most general case is part of Sidi [35, Theorem 6.1], and it gives a surprisingly simple condition sufficient for the convergence of the *whole* sequence $\{f_{m,n}(z)\}_{m=0}^{\infty}$. This condition involves the nonlinear integer programming problem IP(τ) we discussed earlier.

Theorem 7.9 *Let $f(z)$ be precisely as in Theorem 7.8 and let $n = q + \tau$ with $0 < \tau < \sum_{j=1}^r \hat{\omega}_j$. Denote by* IP($\tau$) *the nonlinear integer programming problem*

$$\text{maximize } g(\sigma); \quad g(\sigma) = \sum_{k=1}^r (\hat{\omega}_k \sigma_k - \sigma_k^2)$$

$$\text{subject to } \sum_{k=1}^r \sigma_k = \tau \text{ and } 0 \le \sigma_k \le \hat{\omega}_k, \ 1 \le k \le r. \tag{7.58}$$

Then $\{f_{m,n}(z)\}_{m=0}^{\infty}$ converges uniformly to $f(z)$ in any compact subset of $K\backslash\{z_1, \ldots, z_t\}$, provided IP($\tau$) *has a unique solution for $\sigma_1, \ldots, \sigma_r$. If we denote by $G(\tau)$ and $G(\tau+1)$ the (optimal) values of $g(\sigma)$ at the solutions to* IP(τ) *and* IP($\tau + 1$), *respectively, then there holds*

$$f(z) - f_{m,n}(z) = O(m^{G(\tau+1)-G(\tau)}|z/R|^m) \text{ as } m \rightarrow \infty. \tag{7.59}$$

[Note that IP($\tau + 1$) *need not have a unique solution.]*

Generalized König Theorem

The following result concerning the poles of Padé approximants is known as the *generalized König Theorem*.

Theorem 7.10 *Let $f(z)$ be as in Theorem 7.7 and denote its poles by $\xi_1, \ldots, \xi_n$. Here, the ξ_i are not necessarily distinct and are ordered such that $|\xi_1| \leq \cdots \leq |\xi_n|$. Define $Q(z) = \prod_{j=1}^{n}(1 - z/\xi_j)$. Let $Q_{m,n}(z)$ be the denominator of $f_{m,n}(z)$, normalized such that $Q_{m,n}(0) = 1$. Then*

$$\limsup_{m \to \infty} |Q_{m,n}(z) - Q(z)|^{1/m} \leq |\xi_n/R| \ as \ m \to \infty. \tag{7.60}$$

The special case in which $n = 1$ was proved originally by König [19]. The general case follows from a closely related theorem of Hadamard [16], and was proved by Golomb [9], and more recently, by Gragg and Householder [11].

If the poles of $f(z)$ in K are as in Theorem 7.8, then the result in (7.60) can be refined, as shown in Sidi [35], and it reads

$$Q_{m,n}(z) - Q(z) = O(m^{\alpha}|\xi_n/R|^m) \ as \ m \to \infty, \ \alpha \geq 0 \ \text{some integer.} \tag{7.61}$$

Of course, what Theorem 7.10 implies is that for all large m, $f_{m,n}(z)$ has precisely n poles that tend to the poles $\xi_1, \ldots, \xi_n$ of $f(z)$ in K. If we let the poles of $f(z)$ and their multiplicities and $\bar{p}$ be as in Theorem 7.8, then for each $j \in \{1, \ldots, t\}$, $f_{m,n}(z)$ has precisely ω_j poles $z_{jl}(m)$, $l = 1, \ldots, \omega_j$, that tend to z_j. Also, the p_jth derivative of $Q_{m,n}(z)$, the denominator of $f_{m,n}(z)$, has a zero $z_j'(m)$ that tends to z_j. More specifically, we have the following quantitative results, whose proofs are given in Sidi [35, Theorem 3.1].

Theorem 7.11

(i)   *In Theorem 7.7,*

$$\limsup_{m \to \infty} \left| z_{jl}(m) - z_j \right|^{1/m} = \left| z_j/R \right|^{1/\omega_j},$$

$$\limsup_{m \to \infty} \left| \frac{1}{\omega_j} \sum_{l=1}^{\omega_j} z_{jl}(m) - z_j \right|^{1/m} = \left| z_j/R \right|,$$

$$\limsup_{m \to \infty} \left| z_j'(m) - z_j \right|^{1/m} = \left| z_j/R \right|. \tag{7.62}$$

(ii)   *In Theorem 7.8, we obtain the following more refined results:*

$$z_{jl}(m) - z_j = O([m^{\bar{p}}|z_j/R|^m]^{1/\omega_j}) \ as \ m \to \infty,$$

$$\frac{1}{\omega_j} \sum_{l=1}^{\omega_j} z_{jl}(m) - z_j = O(m^{\bar{p}}|z_j/R|^m) \ as \ m \to \infty,$$

$$z_j'(m) - z_j = O(m^{\bar{p}}|z_j/R|^m) \ as \ m \to \infty. \tag{7.63}$$

The first of the results in (7.62) and (7.63) were given earlier by Gončar [10]. The version of (7.63) that is given in [35] is actually more refined in that it provides the first term of the asymptotic expansion of $z_{jl}(m) - z_j$:

$$z_{jl}(m) \sim z_j + E_{jl}(m)(m^{\bar{p}}|z_j/R|^m)^{1/\omega_j} \quad \text{as } m \to \infty, \tag{7.64}$$

where $\{E_{jl}(m)\}_{m=0}^{\infty}$ is some bounded sequence with a subsequence that has a nonzero limit.

7.10.6.3 Two-Point Padé Approximants

We defined a Padé approximant to be a rational function whose Maclaurin expansion agrees with a given power series $\sum_{k=0}^{\infty} c_k z^k$ as far as possible. This idea can be generalized as follows:

Definition 7.1 *Let the formal power series* $F_r(z) := \sum_{k=0}^{\infty} c_{rk}(z - z_r)^k$, $r = 1, \ldots, q$, *with* z_r *distinct, be given. Define* $\boldsymbol{\mu} = (\mu_1, \ldots, \mu_q)$. *Then the rational function* $\hat{f}_{\boldsymbol{\mu},n}(z)$ *with degrees of numerator and denominator at most* m *and* n *respectively, such that* $m + n + 1 = \sum_{r=1}^{q}(\mu_r + 1)$, *is the* q-point Padé approximant *of type* $(\mu_1, \ldots, \mu_q)$ *from* $\{F_r(z)\}_{r=1}^{q}$ *if it satisfies*

$$\hat{f}_{\boldsymbol{\mu},n}(z) - F_r(z) = O((z - z_r)^{\mu_r+1}) \quad \text{as } z \to z_r, \ r = 1, \ldots, q. \tag{7.65}$$

The case that has received significant attention is that of $q = 2$. This case can be standardized by sending the points z_1 and z_2 to 0 and ∞ by a Möbius transformation. In its "symmetric" form introduced in Sidi [31], this case of two-point Padé approximants can thus be formulated as follows:

Definition 7.2 *Let the function* $f(z)$ *satisfy*

$$f(z) \sim \frac{c_0}{2} + c_1 z + c_2 z^2 + \cdots \quad \text{as } z \to 0,$$

$$f(z) \sim -\left(\frac{c_0}{2} + \frac{c_{-1}}{z} + \frac{c_{-2}}{z^2} + \cdots\right) \quad \text{as } z \to \infty; \ c_0 \neq 0. \tag{7.66}$$

For any two integers i *and* j *such that* $i + j$ *is even, we define the* two-point Padé approximant $\hat{f}_{i,j}(z)$ *to be the rational function*

$$\hat{f}_{i,j}(z) = \frac{P(z)}{Q(z)} = \frac{\sum_{k=0}^{m} \alpha_k z^k}{\sum_{k=0}^{m} \beta_k z^k}, \quad \beta_0 = 1, \ m = (i + j)/2, \tag{7.67}$$

that satisfies

$$\hat{f}_{i,j}(z) - f(z) = O(z^i, z^{-j-1}) \equiv \begin{cases} O(z^i) & \text{as } z \to 0, \\ O(z^{-j-1}) & \text{as } z \to \infty, \end{cases} \tag{7.68}$$

provided $\hat{f}_{i,j}(z)$ *exists.*

Note that in case $f(0) + f(\infty) \neq 0$, the asymptotic expansions of $f(z)$ do not have the symmetric form of (7.66). The symmetric form can be achieved simply by adding a constant to $f(z)$. For example, if $\phi(z) \sim \sum_{i=0}^{\infty} \gamma_i z^i$ as $z \to 0$ and $\phi(z) \sim \sum_{i=0}^{\infty} \delta_i / z^i$ as $z \to \infty$, and $\gamma_0 + \delta_0 \neq 0$, then $f(z) = \phi(z) - (\gamma_0 + \delta_0)/2$ has asymptotic expansions as in (7.66), with $c_0 = \gamma_0 - \delta_0$ and $c_i = \gamma_i$ and $c_{-i} = -\delta_i$, $i = 1, 2, \ldots$.

The following results and their proofs can be found in Sidi [31, Theorems 1, 2, 3].

Theorem 7.12 *When $\hat{f}_{i,j}(z)$ exists, it is unique.*

Theorem 7.13 *Let $g(z) = 1/f(z)$ in the sense that $g(z)$ has asymptotic expansions as $z \to 0$ and $z \to \infty$ obtained by inverting those of $f(z)$ in (7.66) appropriately. If $\hat{g}_{i,j}(z)$ is the two-point Padé approximant from $g(z)$ precisely as in Definition 7.2, then $\hat{g}_{i,j}(z) = 1/\hat{f}_{i,j}(z)$.*

Theorem 7.14

1. The β_k can be obtained by solving the linear equations

$$\sum_{s=0}^{m} c_{r-s} \beta_s = 0, \ r = i - m, i - m + 1, \ldots, i - 1.$$

With the β_k determined, $f_{i,j}(z)$ can be computed via

$$f_{i,j}(z) = \frac{\sum_{s=0}^{m} \beta_s z^s S_{m-s-1}(z)}{\sum_{s=0}^{m} \beta_s z^s} \quad \text{if } i \geq m,$$

and

$$f_{i,j}(z) = \frac{\sum_{s=0}^{m} \beta_s z^s S_{-s-1}(z)}{\sum_{s=0}^{m} \beta_s z^s} \quad \text{if } i \geq m.$$

Here

$$S_0(z) = \frac{c_0}{2}, \ S_k(z) = S_{k-1}(z) + c_k z^k, \ k = \pm 1, \pm 2, \ldots .$$

[Note that $S_k(z) = c_0/2 + \sum_{i=1}^{k} c_i z^i$ and $S_{-k}(z) = -c_0/2 - \sum_{i=1}^{k-1} c_{-i} z^{-i}$ for $k = 1, 2, \ldots$.]

2. The approximant $\hat{f}_{i,j}(z)$ has the determinant representation

$$\hat{f}_{i,j}(z) = \frac{P_{i,j}(z)}{Q_{i,j}(z)},$$

$$Q_{i,j}(z) = \begin{vmatrix} 1 & z & \cdots & z^m \\ c_{i-1} & c_{i-2} & \cdots & c_{i-m-1} \\ c_{i-2} & c_{i-3} & \cdots & c_{i-m-2} \\ \vdots & \vdots & & \vdots \\ c_{i-m} & c_{i-m-1} & \cdots & c_{-j} \end{vmatrix},$$

and $P_{i,j}(z)$ is obtained from $Q_{i,j}(z)$ by replacing the first row of the latter by the row vector

$$\left(S_{m-1}(z), z S_{m-2}(z), \ldots, z^m S_{-1}(z) \right) \quad \text{if } i \geq m,$$

and by the row vector

$$\left(S_{-1}(z), z S_{-2}(z), \ldots, z^m S_{-m-1}(z)\right) \quad \text{if } i \leq m.$$

Note that the β_k can also be computed *recursively*, as proved in [31, Theorems 11, 12].

Finally, two-point Padé approximants are convergents of certain continued fractions that are known as M-fractions and T-fractions. (See [31, Theorems 7–10] and McCabe and Murphy [22].) For example, the approximants $\hat{f}_{r,r}(z), \hat{f}_{r+1,r-1}(z), r = 1, 2, \ldots$, are consecutive convergents of a continued fraction of the form

$$c + \frac{dz}{1 + ez +} \frac{\lambda_1}{\mu_1 +} \frac{w_1 z}{1} \frac{\lambda_2}{+\mu_2 +} \frac{w_2 z}{1} + \ldots ; \quad \lambda_i + \mu_i = 1 \text{ for all } i,$$

with $\hat{f}_{1,1}(z) = c + dz/(1 + ez)$, etc.

7.10.7 Work Related to the Vector Epsilon Algorithm: Part 1

7.10.7.1 Vector Epsilon Algorithm (VEA) and Vector Padé Approximants

Consider the problem of solving a system of linear or nonlinear algebraic equations $\Psi(x) = 0$, where $\Psi : \mathbb{C} \to \mathbb{C}^N$, where N is large. Normally, one rewrites this system in the form $x = f(x)$ and computes a sequence of approximations $\{x_m\}_{m=0}^\infty$ to the solution via the fixed-point iterations

$$x_{m+1} = f(x_m), \quad m = 0, 1, \ldots, ; \quad \text{with some } x_0.$$

In most problems of practical interest, these sequences converge very slowly or even diverge. Their convergence can be accelerated using suitable extrapolation methods designed specifically for vectors. Vector extrapolation methods, with their theory and their applications, are studied in detail in the recent book [59] by Sidi.

In [70], Wynn generalized his scalar ϵ-algorithm and developed the Vector Epsilon Algorithm (VEA), the very first vector extrapolation method, for accelerating the convergence of vector sequences $\{x_m\} \in \mathbb{C}^N$. This very elegant algorithm reads (in form) the same as the scalar epsilon algorithm; that is,

1. Set $\epsilon_{-1}^{(j)} = 0$ and $\epsilon_0^{(j)} = x_j, j = 0, 1, \ldots$.
2. Compute the $\epsilon_k^{(j)}$ by the recurrence

$$\epsilon_{k+1}^{(j)} = \epsilon_{k-1}^{(j+1)} + \frac{1}{\epsilon_k^{(j+1)} - \epsilon_k^{(j)}}, \quad j, k = 0, 1, \ldots .$$

Here $1/z$ is the Samelson inverse of the vector z defined as

$$\frac{1}{z} = z^{-1} = \frac{\overline{z}}{z^*z}.$$

The $\epsilon_{2k}^{(j)}$ are approximations to the limit (or antilimit) of the sequence $\{x_m\}$. Much has been written on VEA.

VEA was analyzed extensively in the papers by Graves-Morris [12, 13], and Graves-Morris and Jenkins [14]. As shown in these papers, VEA is closely related to vector-valued Padé approximants from vector-valued power series $\sum_{i=0}^{\infty} u_i z^i$.

7.10.7.2 Other Vector Extrapolation Methods

Following the development of VEA, other extrapolation methods to accelerate the convergence of vector sequences $\{x_m\}$ were invented. These are:

1. The Topological Epsilon Algorithm (TEA) of Brezinski [1], which is an interesting variant of VEA.
   Actually, Brezinski gives two different versions of TEA, namely, TEA1 and TEA2; each of these versions is based on a different interpretation of the vector inverse $1/z$. TEA, along with the scalar and vector extrapolation methods, was also discussed in detail in Brezinski [2].
2. The Minimal Polynomial Extrapolation (MPE) of Cabay and Jackson [4].
3. The Reduced Rank Extrapolation (RRE) of Kaniel and Stein [18], Eddy [5], and Mešina [23].
   (Note that the versions of [18] and [5] are essentially identical. The version of [23] is different. The mathematical equivalence of the approaches of [5] and [23] was proved by Smith et al. [65].)
4. The Modified Minimal Polynomial Extrapolation (MMPE) of Brezinski [1], Pugachev [24], and Sidi et al. [62].
5. The SVD-based Minimal Polynomial Extrapolation (SVD-MPE) of Sidi [55].

Given the vector sequence $\{x_m\}$, each of these methods generates a two-dimensional array of approximations (to the limit or antilimit of $\{x_m\}$) $s_{n,k}$, which are of the form

$$s_{n,k} = \sum_{j=0}^{k} \gamma_j \, x_{n+j+\nu}, \quad \text{such that} \quad \sum_{j=0}^{k} \gamma_j = 1, \quad \text{for some integer } \nu \geq 0.$$

Here the γ_j are scalars that depend nonlinearly on the vectors x_m that are used in constructing $s_{n,k}$.

Choosing n and k, $(k < N)$, and letting

$$u_m = x_{m+1} - x_m, \quad w_m = x_{m+2} - 2x_{m+1} + x_m, \quad m = 0, 1, \ldots,$$

the γ_j are determined (theoretically) as follows:

1. *TEA:*

- Choose a nonzero vector g, and determine $c_0, c_1, \ldots, c_{k-1}$ as the solution to the linear system

$$\sum_{j=0}^{k-1} c_j(g, u_{n+i+j}) = -(g, u_{n+i+k}), \quad i = 0, 1, \ldots, k-1.$$

- Set $c_k = 1$ and $\alpha = \sum_{i=0}^{k} c_i$ and set $\gamma_j = c_j/\alpha, j = 0, 1, \ldots, k$.

2. *MPE:*

- Determine $c_0, c_1, \ldots, c_{k-1}$ as the solution to

$$\min_{c_0, \ldots, c_{k-1}} \left\| \sum_{j=0}^{k-1} c_j u_{n+j} + u_{n+k} \right\|_2.$$

- Set $c_k = 1$ and $\alpha = \sum_{i=0}^{k} c_i$ and set $\gamma_j = c_j/\alpha, j = 0, 1, \ldots, k$.

3. *RRE:*

- Determine $\gamma_0, \gamma_1, \ldots, \gamma_k$ as the solution to

$$\min_{\gamma_0, \ldots, \gamma_k} \left\| \sum_{j=0}^{k} \gamma_j u_{n+j} \right\|_2 \quad \text{subject to} \quad \sum_{j=0}^{k} \gamma_j = 1.$$

4. *MMPE:*

- Choose linearly independent vectors $g_1, \ldots, g_k$, and determine $c_0, c_1, \ldots, c_{k-1}$ as the solution to the linear system

$$\sum_{j=0}^{k-1} c_j(g_i, u_{n+j}) = -(g_i, u_{n+k}), \quad i = 1, \ldots, k.$$

- Set $c_k = 1$ and $\alpha = \sum_{i=0}^{k} c_i$ and set $\gamma_j = c_j/\alpha, j = 0, 1, \ldots, k$.

5. *SVD-MPE:*

- Determine $c = [c_0, c_1, \ldots, c_k]^T$ as the solution to

$$\min_{c_0, \ldots, c_k} \left\| \sum_{j=0}^{k} c_j u_{n+j} \right\|_2 \quad \text{subject to} \quad \|c\|_2 = 1.$$

The solution is the right singular value of the matrix $U = [u_n|u_{n+1}|\cdots|u_{n+k}]$.
- Set $\alpha = \sum_{i=0}^{k} c_i$ and $\gamma_j = c_j/\alpha, j = 0, 1, \ldots, k$.
The vector c is the right singular vector corresponding to the smallest singular value of the matrix U. Thus, if we order the singular values of U such that

$\sigma_0 \geq \sigma_1 \geq \cdots \geq \sigma_k$, and denote the respective right singular vectors by $\boldsymbol{h}_0, \boldsymbol{h}_1, \ldots, \boldsymbol{h}_k$, then we have $\boldsymbol{c} = \boldsymbol{h}_k$.

The vectors $\boldsymbol{s}_{n,k}$ generated by these methods have determinant representations, analogous to those for the Shanks transformation and VEA, that have the following unified form:

$$
\boldsymbol{s}_{n,k} = \frac{\begin{vmatrix} \boldsymbol{x}_n & \boldsymbol{x}_{n+1} & \cdots & \boldsymbol{x}_{n+k} \\ u_{0,0} & u_{0,1} & \cdots & u_{0,k} \\ u_{1,0} & u_{1,1} & \cdots & u_{1,k} \\ \vdots & \vdots & & \vdots \\ u_{k-1,0} & u_{k-1,1} & \cdots & u_{k-1,k} \end{vmatrix}}{\begin{vmatrix} 1 & 1 & \cdots & 1 \\ u_{0,0} & u_{0,1} & \cdots & u_{0,k} \\ u_{1,0} & u_{1,1} & \cdots & u_{1,k} \\ \vdots & \vdots & & \vdots \\ u_{k-1,0} & u_{k-1,1} & \cdots & u_{k-1,k} \end{vmatrix}}, \quad u_{i,j} = \begin{cases} (\boldsymbol{g}, \boldsymbol{u}_{n+i+j}) & \text{for TEA,} \\ (\boldsymbol{u}_{n+i}, \boldsymbol{u}_{n+j}) & \text{for MPE,} \\ (\boldsymbol{w}_{n+i}, \boldsymbol{u}_{n+j}) & \text{for RRE,} \\ (\boldsymbol{g}_{i+1}, \boldsymbol{u}_{n+j}) & \text{for MMPE,} \\ (\boldsymbol{h}_i, \boldsymbol{u}_{n+j}) & \text{for SVD-MPE.} \end{cases}
$$

The determinant representations are due to Brezinski [1] for TEA, Sidi [33] for MPE and RRE, Brezinski [1] and Sidi et al. [62] for MMPE, and Sidi [55] for SVD-MPE.

The determinant representations have proved to be very useful in analyzing the convergence behavior of $\boldsymbol{s}_{n,k}$ as $n \to \infty$, k being held fixed, when the sequence $\{\boldsymbol{x}_m\}$ is generated by a linear fixed-point iterative method; see [33, 36, 61], and [62]. The study of errors in $\boldsymbol{s}_{n,k}$ with fixed n and k is given in [64]. In a recent paper by Sidi [60], the convergence of RRE is addressed for the general case in which the sequence $\{\boldsymbol{x}_m\}$ is generated from a nonlinear system by a fixed-point iteration method.

The determinant representations have been very useful also in proving that there are some very interesting recurrence relations among the different $\boldsymbol{s}_{n,k}$ from each of the methods MPE, RRE, MMPE, and TEA. See Brezinski [3] and Ford and Sidi [7]. Concerning MMPE and TEA (each one separately), we have

$$
\boldsymbol{s}_{n,k+1} = \alpha_{nk} \boldsymbol{s}_{n,k} + \beta_{nk} \boldsymbol{s}_{n+1,k}, \quad \alpha_{nk} + \beta_{nk} = 1.
$$

Concerning MPE and RRE (each one separately) we have

$$
\boldsymbol{s}_{n,k+1} = \alpha_{nk} \boldsymbol{s}_{n,k} + \beta_{nk} \boldsymbol{s}_{n+1,k-1} + \gamma_{nk} \boldsymbol{s}_{n+1,k}, \quad \alpha_{nk} + \beta_{nk} + \gamma_{nk} = 1.
$$

Recently, Sidi [58] proved two general results that show that MPE and RRE are related in more than one way. The first of these results concerns the stagnation of RRE when MPE fails; it reads as follows:

$$
\boldsymbol{s}_{n,k}^{RRE} = \boldsymbol{s}_{n,k-1}^{RRE} \quad \Leftrightarrow \quad \boldsymbol{s}_{n,k}^{MPE} \text{ does not exist.}
$$

The second result concerns the general case in which $s_{n,k}^{MPE}$ exists; part of it reads

$$\mu_k s_{n,k}^{RRE} = \mu_{k-1} s_{n,k-1}^{RRE} + \nu_k s_{n,k}^{MPE}, \quad \mu_k = \mu_{k-1} + \nu_k,$$

where μ_k, μ_{k-1}, ν_k are positive scalars depending only on $s_{n,k}^{RRE}, s_{n,k-1}^{RRE}, s_{n,k}^{MPE}$, respectively.

All the methods discussed here are Krylov subspace methods when applied to sequences $\{x_m\}$ generated by linear fixed-point iterative methods in solving linear systems. In particular, TEA is equivalent to the method of Lanczos (see Brezinski [1]), MPE and RRE are, respectively, the full orthogonalization method (FOM) and the generalized residual method (GMR) (see Sidi [34]).

7.10.8 Work Related to the Vector Epsilon Algorithm: Part 2

7.10.8.1 Vector-Valued Rational Approximations from the Vector Epsilon Algorithm

As we mentioned earlier, when applied to vector-valued power series $\sum_{i=0}^{\infty} u_i z^i$, VEA produces rational approximations to the sums of these series. This subject was analyzed extensively in the papers by Graves-Morris [12, 13], and Graves-Morris and Jenkins [14]. As shown in these papers, VEA is closely related to vector-valued Padé approximants from the series considered.

7.10.8.2 Vector-Valued Rational Approximations from Other Vector Extrapolation Methods

It turns out that when the vector extrapolation methods MPE, MMPE, and TEA are applied to the sequence $\{x_m(z)\}_{m=0}^{\infty}$ of partial sums $x_m(z) = \sum_{i=0}^{m-1} u_i z^i$, $m = 0, 1, \ldots$, they generate approximations $s_{n,k}(z)$ that are vector-valued rational functions. We call these approximation procedures SMPE, SMMPE, and STEA, respectively. Of these, STEA is due to Brezinski [1], while SMPE and SMMPE were developed by Sidi [37].

Ultimately, all the approximations $s_{n,k}(z)$ generated by these methods are of the form

$$s_{n,k}(z) = \frac{\sum_{j=0}^{k} c_j z^{k-j} x_{n+j}(z)}{\sum_{j=0}^{k} c_j z^{k-j}} \equiv \frac{p_{n,k}(z)}{q_{n,k}(z)},$$

where $p_{n,k}(z)$ is a vector-valued polynomial of degree at most $n + k - 1$ and $q_{n,k}(z)$ is a scalar-valued polynomial of degree k.

Here the scalars c_j are determined in different ways by each method. In addition, these approximations have determinant representations, analogous to those for the Shanks transformation and VEA, that have the following unified form:

$$
s_{n,k}(z) = \frac{\begin{vmatrix} z^k \boldsymbol{x}_n(z) & z^{k-1}\boldsymbol{x}_{n+1}(z) & \cdots & z^0 \boldsymbol{x}_{n+k}(z) \\ u_{0,0} & u_{0,1} & \cdots & u_{0,k} \\ u_{1,0} & u_{1,1} & \cdots & u_{1,k} \\ \vdots & \vdots & & \vdots \\ u_{k-1,0} & u_{k-1,1} & \cdots & u_{k-1,k} \end{vmatrix}}{\begin{vmatrix} z^k & z^{k-1} & \cdots & z^0 \\ u_{0,0} & u_{0,1} & \cdots & u_{0,k} \\ u_{1,0} & u_{1,1} & \cdots & u_{1,k} \\ \vdots & \vdots & & \vdots \\ u_{k-1,0} & u_{k-1,1} & \cdots & u_{k-1,k} \end{vmatrix}}, \qquad u_{i,j} = \begin{cases} (\boldsymbol{u}_{n+i}, \boldsymbol{u}_{n+j}) & \text{for MPE,} \\ (\boldsymbol{g}_{i+1}, \boldsymbol{u}_{n+j}) & \text{for MMPE,} \\ (\boldsymbol{g}, \boldsymbol{u}_{n+i+j}) & \text{for TEA,} \end{cases}
$$

The determinant representations are due to Brezinski [1] for TEA, Sidi [33] for MPE and RRE, Brezinski [1] and Sidi et al. [62] for MMPE, and Sidi [55] for SVD-MPE.

The determinant representations have proved to be very useful in analyzing the behavior of $s_{n,k}(z)$ as $n \to \infty$, when the sequence $\{\boldsymbol{x}_m(z)\}$ is the sequence of partial sums of the Maclaurin series of $\boldsymbol{f}(z)$ when $\boldsymbol{f}(z)$ is analytic at $z = 0$ and meromorphic in a disk $K = \{z : |z| < R\}$. The convergence theory is of the Montessus type, analogous to what we reviewed earlier for Padé approximants.

It was shown in Sidi [39] that when applied to the series $\sum_{i=0}^{\infty} \boldsymbol{u}_i z^i$, $\boldsymbol{u}_i = A^i \boldsymbol{b}$, i=0,1,..., A being a square matrix, the approximations $s_{n,k}(z)$ generated by these methods are related very closely to the Krylov subspace methods for the matrix eigenvalue problem for the matrix A. Specifically, the reciprocals of the poles of $s_{n,k}(z)$ are the Ritz values of A, and the corresponding residues are the corresponding Ritz vectors.

7.10.8.3 Vector-Valued Rational Interpolation

A nice feature of the vector-valued rational approximations from power series is that they possess a Padé-like property, namely, if $\boldsymbol{f}(z) = \sum_{i=0}^{\infty} \boldsymbol{u}_i z^i$, then

$$
\boldsymbol{f}(z) - s_{n,k}(z) = O(z^{n+k}) \quad \text{as } z \to 0.
$$

That is, $s_{n,k}(z)$ interpolates $\boldsymbol{f}(z)$ at $z = 0$ in the sense of Hermite $n + k$ times.

In [47], Sidi extended this idea to arbitrary rational interpolation in the complex plane. Thus, given the not necessarily distinct points of interpolation $\xi_1, \xi_2, \ldots$, we want to design a vector-valued rational interpolant $\boldsymbol{r}_{p,k}(z)$, of the form

$$
\boldsymbol{r}_{p,k}(z) = \frac{\boldsymbol{u}_{p,k}(z)}{v_{p,k}(z)} = \frac{\sum_{j=0}^{k} c_j \psi_{1,j}(z) \boldsymbol{f}_{j+1,p}(z)}{\sum_{j=0}^{k} c_j \psi_{1,j}(z)}, \quad c_k \neq 0,
$$

where

$$
\psi_{r,s}(z) = \prod_{i=r}^{s} (z - \xi_i), \quad s \geq r \geq 1; \quad \psi_{r,r-1}(z) = 1,
$$

and $f_{m,m'}(z)$ is the polynomial of interpolation to $f(z)$ (in the sense of Hermite) at the points $\xi_m, \xi_{m+1}, \ldots, \xi_{m'}$. As a result, $\boldsymbol{u}_{p,k}(z)$ is a vector-valued polynomial of degree at most $p - 1$, $v_{p,k}(z)$ is a scalar-valued polynomial of degree k, and $c_0, c_1, \ldots, c_k$ are scalars to be determined. Using the known theory of polynomial interpolation, it can easily be shown that with arbitrary c_i,

$$\boldsymbol{r}_{p,k}(\xi_i) = f(\xi_i), \quad i = 1, \ldots, p.$$

The c_i are determined in different manners for three different methods, which we denote by IMPE, IMMPE, and ITEA. We do not go into detail here. We will only mention that these interpolants have determinant representations that can be unified as follows:

$$\boldsymbol{r}_{p,k}(z) = \frac{\begin{vmatrix} \psi_{1,0}(z)f_{1,p}(z) & \psi_{1,1}(z)f_{2,p}(z) & \cdots & \psi_{1,k}(z)f_{k+1,p}(z) \\ u_{1,0} & u_{1,1} & \cdots & u_{1,k} \\ u_{2,0} & u_{2,1} & \cdots & u_{2,k} \\ \vdots & \vdots & & \vdots \\ u_{k,0} & u_{k,1} & \cdots & u_{k,k} \end{vmatrix}}{\begin{vmatrix} \psi_{1,0}(z) & \psi_{1,1}(z) & \cdots & \psi_{1,k}(z) \\ u_{1,0} & u_{1,1} & \cdots & u_{1,k} \\ u_{2,0} & u_{2,1} & \cdots & u_{2,k} \\ \vdots & \vdots & & \vdots \\ u_{k,0} & u_{k,1} & \cdots & u_{k,k} \end{vmatrix}},$$

where

$$u_{i,j} = \begin{cases} (\boldsymbol{d}_{i,p+1}, \boldsymbol{d}_{j+1,p+1}) & \text{for IMPE,} \\ (\boldsymbol{g}_i, \boldsymbol{d}_{j+1,p+1}) & \text{for IMMPE,} \\ (\boldsymbol{g}, \boldsymbol{d}_{j+1,p+i}) & \text{for ITEA,} \end{cases}$$

with $\boldsymbol{d}_{r,s} = f[\xi_r, \xi_{r+1}, \ldots, \xi_s]$ being the divided difference of $f(z)$ over the set of points $\{\xi_r, \xi_{r+1}, \ldots, \xi_s\}$. Here the vectors $\boldsymbol{g}_1, \ldots, \boldsymbol{g}_k$ are linearly independent, and $\boldsymbol{g} \neq \boldsymbol{0}$.

These interpolants possess several desirable properties:

- As $\xi_i \to 0$, $i = 1, 2, \ldots$, $\boldsymbol{r}_{n+k,k}(z)$ from IMMPE and ITEA approach $\boldsymbol{s}_{n,k}(z)$ from SMMPE and STEA, respectively. The limit of $\boldsymbol{r}_{n+k,k}(z)$ from IMPE is slightly different from that of $\boldsymbol{s}_{n,k}(z)$ from SMPE.
- $\boldsymbol{r}_{p,k}(z)$ is a symmetric function of the $\xi, \xi_1, \ldots, \xi_p$. That is, the order in which the ξ_i are introduced does not change $\boldsymbol{r}_{p,k}(z)$.
- If $f(z)$ is a vector-valued rational function, then it is reproduced by each of the interpolation procedures IMPE, IMMPE, and ITEA.
- All three procedures have a Montessus-type convergence theory that is analogous to that relevant to Padé approximants.

For all these, see Sidi [48, 49, 50, 52], and [57].

References

1. C. Brezinski. Généralisations de la transformation de Shanks, de la table de Padé, et de l'ε-algorithme. *Calcolo*, 12:317–360, 1975.
   (Cited on pages 265, 267, 268, and 269.)
2. C. Brezinski. *Accélération de la Convergence en Analyse Numérique*. Number 584 in Lecture Notes in Mathematics. Springer-Verlag, Berlin, 1977.
   (Cited on page 265.)
3. C. Brezinski. Recursive interpolation, extrapolation and projection. *J. Comp. Appl. Math.*, 9:369–376, 1983.
   (Cited on page 267.)
4. S. Cabay and L.W. Jackson. A polynomial extrapolation method for finding limits and antilimits of vector sequences. *SIAM J. Numer. Anal.*, 13:734–752, 1976.
   (Cited on page 265.)
5. R.P. Eddy. Extrapolating to the limit of a vector sequence. In P.C.C. Wang, editor, *Information Linkage Between Applied Mathematics and Industry*, pages 387–396, New York, 1979. Academic Press.
   (Cited on page 265.)
6. W.F. Ford and A. Sidi. An algorithm for a generalization of the Richardson extrapolation process. *SIAM J. Numer. Anal.*, 24:1212–1232, 1987.
   (Cited on pages 246 and 256.)
7. W.F. Ford and A. Sidi. Recursive algorithms for vector extrapolation methods. *Appl. Numer. Math.*, 4:477–489, 1988. Originally appeared as Technical Report No. 400, Computer Science Dept., Technion–Israel Institute of Technology, (1986).
   (Cited on page 267.)
8. C.R. Garibotti and F.F. Grinstein. Recent results relevant to the evaluation of infinite series. *J. Comp. Appl. Math.*, 9:193–200, 1983.
   (Cited on page 249.)
9. M. Golomb. Zeros and poles of functions defined by power series. *Bull. Amer. Math. Soc.*, 49:581–592, 1943.
   (Cited on page 261.)
10. A.A. Gončar. Poles of rows of the Padé table and meromorphic continuation of functions. *Math. USSR-Sbornik*, 43:527–546, 1982.
    (Cited on page 261.)
11. W.B. Gragg and A.S. Householder. On a theorem of König. *Numer. Math.*, 8:465–468, 1966.
    (Cited on page 261.)
12. P.R. Graves-Morris. Vector valued rational interpolants I. *Numer. Math.*, 42:331–348, 1983.
    (Cited on pages 265 and 268.)
13. P.R. Graves-Morris. Vector valued rational interpolants II. *IMA J. Numer. Anal.*, 4:209–224, 1984.
    (Cited on pages 265 and 268.)
14. P.R. Graves-Morris and C.D. Jenkins. Vector valued rational interpolants III. *Constr. Approx.*, 2:263–289, 1986.
    (Cited on pages 265 and 268.)
15. H.L. Gray, T.A. Atchison, and G.V. McWilliams. Higher order G-transformations. *SIAM J. Numer. Anal.*, 8:365–381, 1971.
    (Cited on page 246.)
16. J. Hadamard. Essai sur l'étude des fonctions données par leur développement de Taylor. *J. Math. Pures Appl. (4)*, 8:101–186, 1892.
    (Cited on page 261.)
17. M. Kaminski and A. Sidi. Solution of an integer programming problem related to convergence of rows of Padé approximants. *Appl. Numer. Math.*, 8:217–223, 1991.
    (Cited on page 249.)

18. S. Kaniel and J. Stein. Least-square acceleration of iterative methods for linear equations. *J. Optimization Theory Appl.*, 14:431–437, 1974.
    (Cited on page 265.)
19. J. König. Über eine Eigenschaft der Potenzreihen. *Math. Ann.*, 23:447–449, 1884.
    (Cited on page 261.)
20. D. Levin. Development of non-linear transformations for improving convergence of sequences. *Intern. J. Computer Math.*, B3:371–388, 1973.
    (Cited on page 256.)
21. D. Levin and A. Sidi. Two new classes of nonlinear transformations for accelerating the convergence of infinite integrals and series. *Appl. Math. Comp.*, 9:175–215, 1981. Originally appeared as a Tel Aviv University preprint in 1975.
    (Cited on page 254.)
22. J.H. McCabe and J.A. Murphy. Continued fractions which correspond to power series at two points. *J. Inst. Maths. Applics.*, 17:233–247, 1976.
    (Cited on page 264.)
23. M. Mešina. Convergence acceleration for the iterative solution of the equations $X = AX + f$. *Comput. Methods Appl. Mech. Engrg.*, 10:165–173, 1977.
    (Cited on page 265.)
24. B.P. Pugachev. Acceleration of the convergence of iterative processes and a method of solving systems of nonlinear equations. *U.S.S.R. Comput. Math. Math. Phys.*, 17:199–207, 1978.
    (Cited on page 265.)
25. H. Rutishauser. Anwendungen des Quotienten-Differenzen-Algorithmus. *Z. Angew. Math. Phys.*, 5:496–508, 1954.
    (Cited on page 246.)
26. H. Rutishauser. Der Quotienten-Differenzen-Algorithmus. *Z. Angew. Math. Phys.*, 5:233–251, 1954.
    (Cited on page 246.)
27. D. Shanks. Nonlinear transformations of divergent and slowly convergent sequences. *J. Math. and Phys.*, 34:1–42, 1955.
    (Cited on pages 245 and 258.)
28. A. Sidi. Convergence properties of some nonlinear sequence transformations. *Math. Comp.*, 33:315–326, 1979.
    (Cited on page 256.)
29. A. Sidi. Some properties of a generalization of the Richardson extrapolation process. *J. Inst. Maths. Applics.*, 24:327–346, 1979.
    (Cited on page 251.)
30. A. Sidi. Analysis of convergence of the T-transformation for power series. *Math. Comp.*, 35:833–850, 1980.
    (Cited on page 256.)
31. A. Sidi. Some aspects of two-point Padé approximants. *J. Comp. Appl. Math.*, 6:9–17, 1980.
    (Cited on pages 262, 263, and 264.)
32. A. Sidi. An algorithm for a special case of a generalization of the Richardson extrapolation process. *Numer. Math.*, 38:299–307, 1982.
    (Cited on pages 255 and 257.)
33. A. Sidi. Convergence and stability properties of minimal polynomial and reduced rank extrapolation algorithms. *SIAM J. Numer. Anal.*, 23:197–209, 1986. Originally appeared as NASA TM-83443 (1983).
    (Cited on pages 267 and 269.)
34. A. Sidi. Extrapolation vs. projection methods for linear systems of equations. *J. Comp. Appl. Math.*, 22:71–88, 1988.
    (Cited on page 268.)
35. A. Sidi. Quantitative and constructive aspects of the generalized Koenig's and de Montessus's theorems for Padé approximants. *J. Comp. Appl. Math.*, 29:257–291, 1990.
    (Cited on pages 259, 260, and 261.)

36. A. Sidi. Convergence of intermediate rows of minimal polynomial and reduced rank extrapolation tables. *Numer. Algorithms*, 6:229–244, 1994.
   (Cited on page 267.)

37. A. Sidi. Rational approximations from power series of vector-valued meromorphic functions. *J. Approx. Theory*, 77:89–111, 1994.
   (Cited on page 268.)

38. A. Sidi. Acceleration of convergence of (generalized) Fourier series by the d-transformation. *Annals Numer. Math.*, 2:381–406, 1995.
   (Cited on page 250.)

39. A. Sidi. Application of vector-valued rational approximation to the matrix eigenvalue problem and connections with Krylov subspace methods. *SIAM J. Matrix Anal. Appl.*, 16:1341–1369, 1995.
   (Cited on page 269.)

40. A. Sidi. Convergence analysis for a generalized Richardson extrapolation process with an application to the $d^{(1)}$-transformation on convergent and divergent logarithmic sequences. *Math. Comp.*, 64:1627–1657, 1995.
   (Cited on pages 252, 253, and 256.)

41. A. Sidi. Extension and completion of Wynn's theory on convergence of columns of the epsilon table. *J. Approx. Theory*, 86:21–40, 1996.
   (Cited on page 247.)

42. A. Sidi. Further convergence and stability results for the generalized Richardson extrapolation process GREP$^{(1)}$ with an application to the $D^{(1)}$-transformation for infinite integrals. *J. Comp. Appl. Math.*, 112:269–290, 1999.
   (Cited on page 256.)

43. A. Sidi. The generalized Richardson extrapolation process GREP$^{(1)}$ and computation of derivatives of limits of sequences with applications to the $d^{(1)}$-transformation. *J. Comp. Appl. Math.*, 122:251–273, 2000.
   (Cited on page 256.)

44. A. Sidi. A new algorithm for the higher order G-transformation. Preprint, Computer Science Dept., Technion–Israel Institute of Technology, 2000.
   (Cited on page 246.)

45. A. Sidi. New convergence results on the generalized Richardson extrapolation process GREP$^{(1)}$ for logarithmic sequences. *Math. Comp.*, 71:1569–1596, 2002.
   (Cited on page 256.)

46. A. Sidi. *Practical Extrapolation Methods: Theory and Applications*. Number 10 in Cambridge Monographs on Applied and Computational Mathematics. Cambridge University Press, Cambridge, 2003.
   (Cited on pages 245, 246, 254, 256, 257, and 258.)

47. A. Sidi. A new approach to vector-valued rational interpolation. *J. Approx. Theory*, 130:177–187, 2004.
   (Cited on page 269.)

48. A. Sidi. Algebraic properties of some new vector-valued rational interpolants. *J. Approx. Theory*, 141:142–161, 2006.
   (Cited on page 270.)

49. A. Sidi. A de Montessus type convergence study for a vector-valued rational interpolation procedure. *Israel J. Math.*, 163:189–215, 2008.
   (Cited on page 270.)

50. A. Sidi. A de Montessus type convergence study of a least-squares vector-valued rational interpolation procedure. *J. Approx. Theory*, 155:75–96, 2008.
   (Cited on page 270.)

51. A. Sidi. Asymptotic analysis of a generalized Richardson extrapolation process on linear sequences. *Math. Comp.*, 79:1681–1695, 2010.
   (Cited on page 252.)

52. A. Sidi. A de Montessus type convergence study of a least-squares vector-valued rational interpolation procedure II. *Comput. Methods Funct. Theory*, 10:223–247, 2010. *(Cited on page 270.)*

53. A. Sidi. Survey of numerical stability issues in convergence acceleration. *Appl. Numer. Math.*, 60:1395–1410, 2010. *(Cited on page 258.)*

54. A. Sidi. Acceleration of convergence of general linear sequences by the Shanks transformation. *Numer. Math.*, 119:725–764, 2011. *(Cited on pages 250 and 251.)*

55. A. Sidi. SVD-MPE: An SVD-based vector extrapolation method of polynomial type. *Appl. Math.*, 7:1260–1278, 2016. Special issue on Applied Iterative Methods. *(Cited on pages 265, 267, and 269.)*

56. A. Sidi. Acceleration of convergence of some infinite sequences $\{A_n\}$ whose asymptotic expansions involve fractional powers of n. *arXiv e-prints*, 2017. arXiv:1703.06495 [math.NA]. *(Cited on pages 257 and 258.)*

57. A. Sidi. A de Montessus type convergence study for a vector-valued rational interpolation procedure of epsilon class. *Jaen J. Approx.*, 9:85–104, 2017. *(Cited on page 270.)*

58. A. Sidi. Minimal polynomial and reduced rank extrapolation methods are related. *Adv. Comput. Math.*, 43:151–170, 2017. *(Cited on page 267.)*

59. A. Sidi. *Vector Extrapolation Methods with Applications.* Number 17 in SIAM Series on Computational Science and Engineering. SIAM, Philadelphia, 2017. *(Cited on page 264.)*

60. A. Sidi. A convergence study for reduced rank extrapolation on nonlinear systems. *Numer. Algorithms*, 2020. First online 20 August 2019. *(Cited on page 267.)*

61. A. Sidi and J. Bridger. Convergence and stability analyses for some vector extrapolation methods in the presence of defective iteration matrices. *J. Comp. Appl. Math.*, 22:35–61, 1988. *(Cited on pages 247 and 267.)*

62. A. Sidi, W.F. Ford, and D.A. Smith. Acceleration of convergence of vector sequences. *SIAM J. Numer. Anal.*, 23:178–196, 1986. Originally appeared as NASA TP-2193 (1983). *(Cited on pages 265, 267, and 269.)*

63. A. Sidi and D. Levin. Rational approximations from the d-transformation. *IMA J. Numer. Anal.*, 2:153–167, 1982. *(Cited on page 256.)*

64. A. Sidi and Y. Shapira. Upper bounds for convergence rates of vector extrapolation methods on linear systems with initial iterations. Technical Report 701, Computer Science Dept., Technion–Israel Institute of Technology, 1991. Appeared also as NASA Technical memorandum 105608, ICOMP-92-09 (1992). *(Cited on page 267.)*

65. D.A. Smith, W.F. Ford, and A. Sidi. Extrapolation methods for vector sequences. *SIAM Rev.*, 29:199–233, 1987. Erratum: *SIAM Rev.,* 30:623–624, 1988. *(Cited on page 265.)*

66. R. Wilson. Divergent continued fractions and polar singularities. *Proc. London Math. Soc.*, 26:159–168, 1927. *(Cited on page 260.)*

67. R. Wilson. Divergent continued fractions and polar singularities II. Boundary pole multiple. *Proc. London Math. Soc.*, 27:497–512, 1928. *(Cited on page 260.)*

68. R. Wilson. Divergent continued fractions and polar singularities III. Several boundary poles. *Proc. London Math. Soc.*, 28:128–144, 1928. *(Cited on page 260.)*

69. P. Wynn. On a device for computing the $e_m(S_n)$ transformation. *Mathematical Tables and Other Aids to Computation*, 10:91–96, 1956. *(Cited on page 245.)*

70. P. Wynn. Acceleration techniques for iterated vector and matrix problems. *Math. Comp.*, 16:301–322, 1962. *(Cited on page 264.)*

71. P. Wynn. On the convergence and stability of the epsilon algorithm. *SIAM J. Numer. Anal.*, 3:91–122, 1966. *(Cited on pages 247 and 249.)*

72. P. Wynn. Transformations to accelerate the convergence of Fourier series. In *Gertrude Blanche Anniversary Volume*, pages 339–379, Wright Patterson Air Force Base, 1967. *(Cited on page 252.)*

7.11 Ernst Joachim Weniger

Ernst Joachim Weniger is a theoretical chemist and physicist who was affiliated with the University of Regensburg in Germany. For his research, he became interested in convergence acceleration and extrapolation methods, in particular for divergent series. He introduced many new ideas in these domains. He has written about 80 papers on these topics, and their application to various problems in science.

Testimony

In this letter [*to C.B. and M.R.-Z.*] I want to describe how I got involved in the convergence acceleration and summation business.

My current research interests obviously depend strongly on my scientific past (shifts in interest are usually not completely discontinuous). Therefore, I will first give a short review of my earlier work with an emphasis on those things that influenced my later work.

In my diploma thesis [129] as well as in my PhD thesis [130], both of which I did under the supervision of Otto Steinborn at the Institut für Physikalische und Theoretische Chemie of the University of Regensburg, I worked on molecular multicenter integrals of exponentially decaying functions. This is an essentially mathematical research topic that has played a major role in the development of molecular electronic structure theory. Numerous authors have worked on this topic, but until now the mathematical and computational problems occurring in this context have not yet been solved in a completely satisfactory way. So in some sense, I also failed. A review of the extensive older literature can be found in articles by Browne [35], Dalgarno [49], Harris and Michels [63], and Huzinaga [69]. More recent developments with a special emphasis on my own work are discussed in my article [155].

I continued to work on multicenter integrals also after my PhD thesis, and a substantial part of my publications can be related more or less strongly to the problem of evaluating multicenter integrals [61, 67, 68, 113, 114, 115, 116, 117, 129, 130, 131, 144, 149, 152, 162, 163, 164, 165, 166, 167, 168, 169, 170, 171, 172, 173].

The exponentially decaying functions that I used in my diploma and PhD thesis and also later on were the so-called reduced Bessel functions [112, Eqs. (3.1) and (3.2)]

$$\hat{k}_\nu(z) = (2/\pi)^{1/2} z^\nu K_\nu(z), \tag{7.69}$$

where $K_\nu(z)$ is a modified Bessel function of the second kind, and their anisotropic generalizations, the so-called B functions [54, Eq. (2.14)]:

$$B_{n,\ell}^m(\beta, r) = \frac{1}{2^{n+\ell}(n+\ell)!} \, \hat{k}_{n-1/2}(\beta r) \, \mathcal{Y}_\ell^m(\beta r). \tag{7.70}$$

Here, $\beta > 0$, $n \in \mathbb{Z}$, and $\mathcal{Y}_\ell^m$ is a regular solid harmonic. The historical development of the B functions is discussed in [155].

The most important analytical result of my PhD thesis is undoubtedly the observation that a B function possesses an extremely compact Fourier transform [130, Eq. (7.1-6) on p. 160]:

$$\bar{B}_{n,\ell}^m(\beta, p) = (2\pi)^{-3/2} \int e^{-ip \cdot r} \, B_{n,\ell}^m(\beta, r) \, d^3 r = (2/\pi)^{1/2} \frac{\beta^{2n+\ell-1}}{[\beta^2 + p^2]^{n+\ell+1}} \, \mathcal{Y}_\ell^m(-ip). \tag{7.71}$$

This is the most consequential and also the most often cited result of my PhD thesis [130, Eq. (7.1-6) on p. 160]. Later, the Fourier transform (7.71) was published in [163, Eq. (3.7)]. Independently and almost simultaneously, (7.71) was also derived by Niukkanen [87, Eqs. (57)–(58)].[11]

The exceptionally simple Fourier transform (7.71) explains why multicenter integrals of B functions are often significantly simpler than the corresponding integrals of other exponentially decaying functions such as Slater-type functions (see for example the articles by Safouhi, and Safouhi and Weniger [99, 100, 155] or the PhD thesis by Slevinsky [109] and references therein).

Other frequently cited articles that originated from the work for my PhD thesis are the Computer Physics Communications article [162], which describes FORTRAN programs for the recursive calculation of a string of so-called Gaunt coefficients (integrals of the product of three spherical harmonics over the surface of the unit sphere in $\mathbb{R}^3$)

$$\langle \ell_3 m_3 | \ell_2 m_2 | \ell_1 m_1 \rangle = \int \left[Y_{\ell_3}^{m_3}(\Omega) \right]^* Y_{\ell_2}^{m_2}(\Omega) Y_{\ell_1}^{m_1}(\Omega) \, d\Omega, \tag{7.72}$$

and the Physical Review A article [164], which discusses the numerical evaluation of convolution integrals of B functions.

---

[11] A. V. Niukkanen, of the Academy of Sciences of the USSR in Moscow, had for a while the strange habit of publishing articles that were very similar to some of my articles. It was a strange experience for me that there was somebody in Moscow whose brain apparently functioned very much like my own.

Although I am now much more interested in other things, my initial work on multicenter integrals undeniably shaped me as a researcher and strongly influenced also my later research interests. Therefore, I shall try to give you a very condensed introduction to this highly interdisciplinary field of research with a special emphasis on my own activities.

Probably the simplest example of such a molecular multicenter integral is the so-called *overlap integral*,

$$S(f, g, A, B) = \int_{\mathbb{R}^3} \left[f(r - A) \right]^* g(r - B) \, \mathrm{d}^3 r, \qquad (7.73)$$

which is closely related to the in mathematics more common *convolution integral*. Here, $f, g : \mathbb{R}^3 \to \mathbb{C}$ are suitable functions[12] – in electronic structure theory usually called atomic orbitals – which are centered at two nuclei with coordinates A and B, respectively. Integration extends over the whole three-dimensional space $\mathbb{R}^3$.

The most complicated molecular multicenter integrals, which can occur in the context of a molecular electronic structure calculation on the ab initio *level* (a so-called Hartree–Fock–Roothaan calculation), are the six-dimensional 4-center integrals

$$\begin{aligned} &I(f, g, u, v; A, B, C, D) \\ &= \int \left[f(r - A) \right]^* g(r - B) \frac{1}{|r - r'|} \left[u(r' - C) \right]^* v(r' - D) \, \mathrm{d}^3 r \, \mathrm{d}^3 r', \end{aligned}$$
$$(7.74)$$

which describe the electrostatic or Coulomb interaction of two nonclassical electron densities $\left[f(r - A) \right]^* g(r - B)$ and $\left[u(r' - C) \right]^* v(r' - D)$.

The evaluation of multicenter integrals is difficult because of different centers occurring in the integrand. This effectively prevents the straightforward separation of the three- and six-dimensional integrals into products of simpler integrals. These problems are at least conceptually related to the problem of separating the variables of a partial differential equation.

In the case of multicenter integrals of the physically well motivated exponentially decaying functions, a separation of the integration variables is particularly difficult.[13] Therefore, sophisticated mathematical techniques such as addition theorems or Fourier transformation have to be used in order to accomplish something useful. Unfortunately, these techniques—although in principle successful—lead to fairly complicated expressions whose efficient and reliable evaluation is far from trivial. Typical expressions for these multicenter integrals are either multiple (nested) infi-

---

[12] Simple examples are $f(r) = \exp(-\alpha r)$ and $g(r) = \exp(-\beta r)$ with $\alpha, \beta > 0$ and $r = |r|$.

[13] Because of the problems with the evaluation of the notoriously difficult multicenter integrals of exponentially decaying functions, molecular electronic structure calculations are now normally done with the help of Gaussian basis functions. These functions have many highly undesirable features, and they also lead to slow convergence of approximation schemes because of their nonphysical nature. They have only one, albeit decisive advantage: their multicenter integrals can be evaluated comparatively easily.

nite series expansions involving a large number of special functions or complicated multidimensional integral representations—many of them highly oscillatory—which have to be evaluated by numerical quadrature.

During the work for my PhD thesis and also later, I had to derive new explicit expressions for multicenter integrals, usually in terms of special functions.[14] These expressions then had to be programmed.[15] Consequently, I always pursued a dual approach, since I worked both analytically and numerically. The ultimate goal had always been to find a way of determining the numerical values of molecular multicenter integrals both efficiently and reliably, which normally required first some nontrivial analytical work.

From a technical and methodological point of view, the most important mathematical tool of my PhD thesis was special function theory. Consequently, I first had to learn a lot about the mathematical properties of special functions and also about their efficient and reliable evaluation. For me, the second part turned out to be more difficult than it may seem at first sight. At that time, special functions were still considered to be predominantly analytical tools, whereas only relatively little work had been done on the numerical evaluation of special functions.

My personal situation was aggravated by the fact that during my PhD thesis, the University of Regensburg was served by a Telefunken TR440 mainframe. This computer was the result of an ill-conceived and ultimately unsuccessful government-sponsored research funding with the intention of helping to build up a national (German) computer industry. In any case, the TR440 mainframe was a stillborn brain child of this well-meaning but intellectually defective policy. Ultimately, it was commercially unsuccessful.[16] At that time, there were either IBM-style mainframes having a 32-bit architecture, or for predominantly computational purposes CDCs and Crays having a 64-bit architecture. Instead, the TR440 had a 48-bit architecture, which was probably meant to be some compromise between the 32- and 64-bit worlds. Obviously, this was a compromise, but unfortunately, it was a very bad compromise. It implied that on a TR440 one could use neither IBM nor CDC programs, which at that time were at least partly written in machine code. Thus, the TR440 had almost no software for special functions, which meant that I had to program everything myself. This was good for thoroughly learning special function theory, but very bad for producing new results within a reasonably short amount of time.

In the meantime, this has of course changed, most likely because of the impact of computer algebra systems like Maple and Mathematica, which offer routines that allow the evaluation of numerous special functions with an in principle arbitrary precision. The recent books by Cuyt et al. [46], Gil et al. [56], and the NIST Handbook [91], and the reviews by Gill et al. [57] and Temme [120] provide convincing

---

[14] These extensive analytical manipulations involving special functions were responsible for the strong nineteenth century flavor of the nonnumerical part of my PhD thesis and also of some of my later analytical work on multicenter integrals.

[15] Initially, I used exclusively FORTRAN 66 and later FORTRAN 77. Now I routinely also use the computer algebra system Maple for numerical calculations.

[16] Consequently, computer centers of universities were strongly urged to buy these superfluous machines.

evidence that a lot of work is currently being done by numerical mathematicians on the efficient and reliable evaluation of special functions.

As a consequence of the dual approach that I had to pursue during my PhD thesis, I ultimately developed a pronounced interest in special functions—both from an analytical and a numerical perspective—as well as a keen interest in sophisticated numerical techniques. Ultimately, this turned out to be a good basis for my later interest in sequence transformations.

Also after my PhD thesis, special functions have played a major role in my research. In numerous articles, special functions were simply very important technical tools. But there are also several articles dealing directly with special functions. Example are the articles [71, 74, 133, 141, 147, 150, 153]. My more recent articles [27, 154, 155, 156, 157, 158, 170] also treat topics related to special function theory.

Molecular multicenter integrals can be viewed to be some kind of super-complicated special functions. Like special functions, they can often be expressed in terms of a large variety of (usually complicated) series expansions, which can also be used for their evaluation. Often, it is also possible to derive alternative integral representations, usually in terms of nonphysical integration variables, which have to be evaluated by means of numerical quadrature.[17]

Series expansions for multicenter integrals had been my principal numerical tool in my PhD thesis and also later. During my PhD thesis, I lacked completely a sufficiently detailed knowledge of more sophisticated numerical tools for the evaluation of infinite series (this had not been part of my mathematical training). Therefore, I had to evaluate my infinite series expansions in a very pedestrian way by adding up their terms sequentially. Needless to say, I was frequently not at all satisfied with the observed convergence rates of my series expansions (see for example [164, Table II]).

In view of my frequent convergence problems, it would have been a natural idea to try to speed up convergence with the help of nonlinear sequence transformations. Even at that time, this had not been a new idea. To the best of my knowledge, this approach was first pursued in 1967 by Petersson and McKoy [92], who had used several different convergence accelerators for the evaluation of series expansions for multicenter integrals. Unfortunately, I was at that time completely ignorant of the more powerful nonlinear, but also nonregular, transformations and also about Padé approximants, which—as I found out later—often accomplish spectacular improvements of convergence. Of course, I was also ignorant of the article by Petersson and McKoy [92].[18] During my PhD thesis, I knew only the classical theory of linear series transformations as described in the book by Knopp [77]. However, these linear transformations turned out to be useless in the case of my challenging series expansions. Therefore, I ignored series and sequence transformations completely in my PhD thesis.

---

[17] I have only very little practical experience with numerical quadrature. There is only one article [169] dealing with Gauss quadrature. However, Hassan Safouhi usually evaluates multicenter integrals via integral representations combined with numerical quadrature.

[18] If you want to find something in the literature, you first have to know what you should be looking for. Moreover, at that time I did not have the help of Google and the internet.

Let me summarize: at the end of my PhD thesis, I was fully aware of the limitations of the conventional approach of adding up the terms of a series successively, but I simply did not know a better (numerical) alternative.

Shortly after the completion of my PhD thesis, Jiří Čížek, a 1968 emigré from the J. Heyrovsky Institute in Prague,[19] visited Regensburg to give a talk about his current work, which involved a lot of divergent series. But divergent series were already then of considerable interest for me. I had already done some preliminary reading, but I had never worked either *with* or *on* divergent series. In my PhD work on infinite series expansions for multicenter integrals, I always made sure that my series did not diverge.

In any case, I apparently asked Jiří Čížek during his talk some questions that were not completely stupid, because afterward I was asked whether I would be interested in doing postdoctoral work at the Department of Applied Mathematics of the University of Waterloo.

My work with Jiří Čížek in Waterloo in 1983 turned out to be very consequential, because I learned quite a few new things that I could never have learned in Regensburg and that strongly influenced my later research interests.

In Waterloo, I worked on distributive expansions of a plane wave $\exp(\pm i\boldsymbol{p} \cdot \boldsymbol{r})$, which converge weakly with respect to the norm of either the Hilbert space $L^2(\mathbb{R}^3)$ or the Sobolev space $W_2^{(1)}(\mathbb{R}^3)$ [131]. This work was inspired by an article by Shibuya and Wulfam [103], who expanded a plane wave in terms of the four-dimensional hyperspherical harmonics that had been introduced by Fock [55] in his classic treatment of the so-called *accidental* degeneracy of the hydrogen atom.

However, I did not understand the mathematical basis of the article by Shibuya and Wulfman [103] at all, and I still think that my lack of understanding was not my fault. In quantum physics, it is quite common to speak quite a lot about Hilbert spaces without necessarily understanding the underlying mathematical theory. It is usually tacitly assumed that bound-state wave functions are square integrable with respect to an integration over the whole $\mathbb{R}^3$ (see for example [29]). Thus, if we want to expand $\exp(\pm i\boldsymbol{p} \cdot \boldsymbol{r})$ in terms of a complete and orthonormal function set, the underlying Hilbert space must be based on an inner product involving an integration over the whole $\mathbb{R}^3$. The most natural Hilbert space for functions $f: \mathbb{R}^3 \to \mathbb{C}$ would be the Hilbert space

$$L^2(\mathbb{R}^3) = \left\{ f: \mathbb{R}^3 \to \mathbb{C} \;\middle|\; \int |f(\boldsymbol{r})|^2 \, \mathrm{d}^3 \boldsymbol{r} < \infty \right\} \tag{7.75}$$

of square integrable functions. However, the plane wave $\exp(\pm i\boldsymbol{p} \cdot \boldsymbol{r})$ obviously does not belong to this or to related Hilbert spaces. Thus, the expansion introduced by Shibuya and Wulfman [103] in terms of hyperspherical harmonics cannot be an expansion that converges in the mean with respect to the norm of a suitable Hilbert space. Instead, it must be a divergent orthogonal expansion. This follows at once from the *Riesz–Fischer theorem* (see for example [60, Theorem 7.43 on p.

---

[19] Further biographical details on Jiří Čížek and a review of his research can be found in the article [121].

191]). However, questions of convergence or divergence never bothered Shibuya and Wulfman [103]. They are typical theoretical physicists who optimistically believe that all limiting processes converge.[20]

But if the expansion of Shibuya and Wulfman [103] is a divergent orthogonal expansion, one is confronted with the nontrivial question in what sense this expansion is to be interpreted and under what conditions it can safely be applied. For me, this question was important, because I intended to actually apply the expansion of Shibuya and Wulfman [103] or other related expansions in Fourier integrals,[21] which played a major role in my PhD thesis.

It took me quite a while to find an answer. The only reasonable interpretation that I could find was that the expansion of Shibuya and Wulfman [103] is a *weakly convergent* expansion, i.e., essentially a distribution or generalized function, which is divergent when used alone, but which gives meaningful and thus convergent results when used in suitable functionals.

In my opinion, the essential features of weak convergence in contrast to strong convergence can be explained most easily by considering the Euclidean vector space $\mathbb{C}^\infty$ of *infinite* row or column vectors $u = (u_1, u_2, \dots)$ with complex coefficients u_n, which is a comparatively simple model for more complicated function spaces. By equipping the vector space $\mathbb{C}^\infty$ with the inner product $(u|v) = \sum_{n=1}^\infty [u_n]^* v_n$, we obtain the corresponding Hilbert space $\ell^2 \subset \mathbb{C}^\infty$ with norm $\|u\| = \sqrt{(u|u)} = \sum_{n=1}^\infty |u_n|^2$. The condition $\|u\| < \infty$, which defines the Hilbert space ℓ^2, can be satisfied only if the coefficients $u_n \in \mathbb{C}$ of u decay sufficiently fast as $n \to \infty$. A sufficient asymptotic condition inspired by the Dirichlet series for the Riemann zeta function is $|u_n|^2 \sim n^{-\alpha-1}$ with $\alpha > 0$ as $n \to \infty$.

Let us now assume that some vector $w \in \mathbb{C}^\infty$ cannot be normalized, i.e., $\|w\|^2 = (w|w) = \sum_{n=1}^\infty |w_n|^2 < \infty$ does not hold. Thus, $w \notin \ell^2$, but this does not imply that all inner products (w, u) with normalizable $u \in \ell^2$ do not exist. If the coefficients u_n of u decay sufficiently rapidly as $n \to \infty$, then the insufficiently slow decay of the coefficients w_n of w can be compensated, and the infinite series $(w|u) = \sum_{n=1}^\infty [w_n]^* u_n$ converges to a finite result, although the series $(w|w) = \sum_{n=1}^\infty |w_n|^2$ does not.

Obviously, $|(w|u)| < \infty$ cannot hold for arbitrary $u \in \ell^2$, but only for the elements of a suitably restricted *proper* subspace $\mathcal{W} \subset \ell^2 \subset \mathbb{C}^\infty$ of normalizable vectors such that $\omega \in \mathcal{W}$ implies $|(w|\omega)| < \infty$. Accordingly, my interpretation of weak convergence resembles at least conceptually the theory of rigged Hilbert spaces or Gelfand triplets (see for example [13] and [50] and references therein). A very readable account of rigged Hilbert spaces from the perspective of quantum

---

[20] Like many other physicists before and after them, Shibuya and Wulfman [103] ignored the obvious, but nevertheless very consequential, fact that Hilbert space theory applies only to Hilbert space elements.

[21] Traditionally, Fourier integrals are evaluated with the help of the so-called Rayleigh expansion for the plane wave containing spherical Bessel functions and spherical harmonics. However, spherical Bessel functions are according to experience always troublesome, both in integrals and numerically. Thus, avoiding them is definitely desirable, no matter whether we have to evaluate such a Hankel-type integral numerically or analytically.

mechanics and their relationship with Dirac's bra and ket formalism can be found in the book by Ballentine [3, Chapter 1.4].

In principle, it would be desirable to specify for a given w the whole set $\mathcal{W}$, but this may not be so easy. In practice, it may be sufficient to specify only a sufficiently large subset of $\mathcal{W}$. Let us for instance assume that the coefficients w_n of a vector $w \notin \ell^2$ are all finite and satisfy $|w_n| \sim n^\beta$ with $\beta \geq -1/2$ as $n \to \infty$. Thus, w cannot be normalized. If, however, the coefficients ω_n of a vector $\omega \in \ell^2$ are all finite and satisfy $|\omega_n| \sim n^\alpha$ with $\alpha < -\beta - 1$ as $n \to \infty$, then the infinite series $(w|\omega) = \sum_{n=1}^{\infty} [w_n]^* \omega_n$ converges and the inner product $(w|\omega)$ makes sense.

However, the condition $|\omega_n| \sim n^\alpha$ with $\alpha < -\beta - 1$ as $n \to \infty$ does not suffice to specify the whole set $\mathcal{W}$ of vectors ω with $|(w|\omega)| < \infty$. It can happen that the series $\sum_{n=1}^{\infty} [w_n]^* \omega_n$ diverges but is summable.[22] If summability techniques are included in our arsenal of mathematical techniques, we could even discard the otherwise essential requirement $\omega \in \ell^2$ and try to make inner products $(w|\omega)$ with both $w, \omega \notin \ell^2$ mathematically meaningful. A (very) condensed review of the classical summability methods associated with the names Cesàro, Abel, and Riesz can be found in the book by Zayed [177, Chapter 1.11.1]. More detailed treatments of linear summability methods can be found in specialized monographs such as the books by Boos [14], Hardy [62], Knopp [77], and Powell and Shah [94].

In any case, it should be clear that the theory of weakly convergent expansions is still far from complete. In the case of Fourier transformation of functions belonging to $L^2(\mathbb{R}^3)$, which I treated in my Waterloo article [131], this was comparatively easy, simply because $L^2(\mathbb{R}^3)$ is invariant under Fourier transformation (see for example [95, Theorem IX.6 on p. 10]). However, in the case of other functions, understanding the subtleties turned out to be much more difficult. For example, because of its physical relevance, I would very much like to construct weakly convergent expansions for the Coulomb potential $1/|r - r'|$. Unfortunately, the Coulomb potential is a difficult beast, and so far, I have not been successful.

I remarked before that at that time (and also later), Jiří Čížek had been very much interested in divergent series. So, my second research topic in Waterloo was also related to divergent series. I looked at the summation of divergent series, as they for instance occur in quantum-mechanical perturbation theory or in special function theory, by means of continued fractions[23] and Padé approximants.

In Waterloo, I mainly studied continued fractions for the asymptotic expansion of the modified Bessel function of the second kind [128, pp. 202–203],

$$K_\nu(z) \sim [\pi/(2z)]^{1/2} e^{-z} \, {}_2F_0\big(1/2 + \nu, 1/2 - \nu; -1/(2z)\big), \qquad |z| \to \infty, \quad (7.76)$$

but also continued fractions for some quantum-mechanical perturbation expansions.

Jiří Čížek had conjectured that the elements of Rutishauser's QD scheme [96, 97, 98] for the asymptotic series (7.76) possess some simple large-order asymptotics

---

[22] Divergent but summable series also played a major role in my later article [154], in which I investigated the analyticity of Laguerre series.

[23] My main sources of information on the theory of continued fractions was initially the second volume of Henrici [66] and later also the monograph by Jones and Thron [75].

that could be deciphered numerically. Unfortunately, the results of my numerical investigations were inconclusive, and I produced nothing worth publishing.[24]

So, you see that in Waterloo in 1983, I was really dealing quite a lot with divergent series. In this way, I definitely learned quite a bit on divergent series.

In spite of my failure to produce anything useful, I returned to Regensburg at the beginning of 1984 with the firm conviction that rational approximants could be extremely useful numerical tools and that I should try to use them also in different contexts.

There was something else that I learned in 1983 in Waterloo. Late in 1983, shortly before the end of my stay, Keith Geddes presented an early version of Maple to the public. At this occasion, I could try out Maple on a Unix workstation. However, the available memory on this Unix workstation was—as it was customary at that time—ridiculously small and definitely no match for my ambitions caused by my naïve enthusiasm. Therefore, my jobs were usually terminated because of insufficient memory long before I could achieve anything useful. Nevertheless, this short episode convinced me that computer algebra had a bright future. Unfortunately, it took quite a few years before this bright future materialized. When I returned to Regensburg, there were neither Unix workstations nor a Maple license. Years later, when things had changed for the better,[25] I used Maple so much[26] that my articles [43, 131, 132, 133, 142, 148, 149, 174, 175] were cited in the Mathematica books by Trott [122, 123], although I never used Mathematica.

When I returned to Regensburg, I had to make a decision about the direction of my future research. Roughly at that time, I also became aware of the book by Wimp [176], which was the first modern monograph on sequence transformations written in English. From this book, I profited quite a lot. By studying the references in Wimp's book, I also learned that the real competence in the business of sequence transformations could be found on this side of the Atlantic. However, reading in French was not easy for me. I had a so-called classical education, which means that in high school I studied Latin for nine years, classical Greek for 6 years, and English for 3 years. In addition, I did not learn much about mathematics and the sciences, which was not necessarily the best preparation for a scientific career in the computer age.

In Waterloo, I had mainly used continued fractions. However, I was in some sense frightened by the highly developed and very extensive mathematical theory of continued fractions. I did not see how I *alone* could get new results within a reasonably short amount of time.[27] In my opinion, Padé approximants and sequence transformations were in some sense newer and less mature research topics than continued fractions. Consequently, I was optimistic that I would not have to learn

---

[24] However, the summation of the asymptotic series (7.76) for $K_\nu(z)$ with the help of sequence transformations was later studied in the article [159].

[25] After some time, Maple had become available on DOS/Windows PCs.

[26] I even have an article [41, 42] mentioning Maple explicitly in its title, and in 1996, when I was a visiting professor in Waterloo, I collaborated with the Symbolic Computation Group which is essentially the mathematical brain behind the Maple Company.

[27] In Regensburg, I did not know anybody who could have helped me with continued fractions.

so much in order to produce something new. In addition, Padé approximants and sequence transformations were in my opinion better suited for numerical work and therefore offered better chances of accomplishing something useful within a limited amount of time. I still believe that it was the right decision to focus on sequence transformations and not on continued fractions.

When I returned to Regensburg, I first tried to apply sequence transformations for the evaluation of molecular integrals. As a kind of preliminary study, we first published some kind of propaganda article [172]. More series work was done in the two articles [173] and [61], which actually should be seen as a single (long) article. In the first one, we developed new explicit expressions for certain multicenter integrals, and in the second one we studied the evaluation of these fairly complicated expressions with the help of sequence transformations. In the case of linear convergence, we always used the epsilon algorithm, and in the case of logarithmic convergence, we always used Levin's u transformation. Compared to untransformed infinite series representations, our approach based on sequence transformations yielded substantial improvements in convergence.

During that period, I predominantly tried to get to know practically useful sequence transformations. I profited a lot from the at that time very new articles by Smith and Ford [110, 111] that compared the numerical performance of virtually all transformations known at that time.[28] These articles also aroused my interest in the different variants of Levin's transformation, simply because of their observed versatility and power.

In addition to multicenter integrals, we also tried to apply sequence transformations to infinite series representations for special functions and auxiliary functions. In [170], we studied the auxiliary function

$$F_m(z) = \int_0^1 u^{2m} e^{-zu^2}\, du, \qquad m \in \mathbb{N}_0, \quad z \in \mathbb{R}_+, \tag{7.77}$$

which corresponds to an incomplete gamma function and which plays a major role in the evaluation of multicenter integrals of Gaussian functions.[29] Our article [170] mentions in Eq. (9) on p. 344 for the first time a variant of the so-called S transformation (compare also Ref. [8] on p. 346). Our article [170] appeared in the proceedings of a NATO Advanced Research Workshop *Numerical Determination of the Electronic Structure of Atoms, Diatomic and Polyatomic Molecules*, which took place in April 1988 in Versailles. This was later of some importance when Sidi called the S transformation in his book [105] the *Sidi transformation*.

In the articles by Smith and Ford [110, 111], one finds some simple rules about how to apply sequence transformations efficiently. For example, [110] emphasized frequently that the variants of Levin's transformation are particularly effective in the case of strictly alternating series. In order to check this recommendation, I looked for suitable examples in the literature. A not so very obvious example is the following *strictly alternating* power series for the digamma function [2, Eq. (6.3.14)]:

---

[28] Do not forget that at that time I was predominantly interested in practical applications.

[29] An article by Mathar [82] on this function appeared in Numerical Algorithms.

$$\psi(1+z) = \gamma + z \sum_{v=0}^{\infty} \zeta(v+2)(-z)^v . \qquad (7.78)$$

To my utter dismay, the usual variants of Levin's transformation performed quite poorly in the case of the power series (7.78). I was quite a bit frustrated, because I did not understand my bad results at that time.[30]

However, my failure in accelerating the convergence of the alternating power series (7.78) effectively with the help of Levin's transformation as well as some similar other experiences taught me that I should invest more time and effort to understand better the theoretical properties of nonlinear sequence transformations. Apparently, the simple recommendations of Smith and Ford [110, 111] were too simple to be completely reliable.

More or less as a by-product of these studies, I was able to find several new transformations. The majority of them can be found in my long *Computer Physics Reports* [132], which is also my most often cited publication.

My most important contribution to the theory of sequence transformations in my Computer Physics Reports is most likely my treatment of Levin-type transformations [132, Sections 7–9]. In [132, Section 3.2] I had introduced a comparatively simple construction principle for sequence transformations based on annihilation operators. The starting point is the following model sequence:[31]

$$s_n = s + \omega_n z_n^{(k)}, \qquad k, n \in \mathbb{N}_0 . \qquad (7.79)$$

This ansatz tacitly assumes that the products $\omega_n z_n^{(k)}$ provide sufficiently accurate approximations to the remainders $r_n = s_n - s$ of the sequence $\{s_n\}_{n=0}^{\infty}$ to be transformed.

The key quantities in (7.79) are the so-called *correction terms* $z_n^{(k)}$. The superscript k characterizes their complexity. In practice, k denotes the number of unspecified parameters occurring *linearly* in $z_n^{(k)}$. Since the ω_n are assumed to be known, the approach based on (7.79) conceptually boils down to the determination of the unspecified parameters in $z_n^{(k)}$ and the subsequent elimination of $\omega_n z_n^{(k)}$ from s_n. Often, this approach leads to clearly better results than the construction and elimination of other approximations to r_n.

Let us now assume that *linear* operators $\hat{T}_k$ can be found that annihilate the correction terms $z_n^{(k)}$ according to $\hat{T}_k(z_n^{(k)}) = 0$ for fixed k and for all $n \in \mathbb{N}_0$. We apply this annihilation operator $\hat{T}_k$ to the ratio $[s_n - s]/\omega_n = z_n^{(k)}$. Since $\hat{T}_k$ annihilates $z_n^{(k)}$ and is by assumption also linear, the resulting sequence transformation

---

[30] I understood my disappointing numerical results only much later, when I prepared my article [150], in which I studied the truncation error of the power series (7.78) in a very detailed way. My disappointing results also showed that one should be extremely careful with statistical arguments. Let us assume that the claim of Smith and Ford [110] is true in 95% of all cases. This does not help at all if the problem under consideration belongs to the remaining 5%.

[31] Here I do not describe the original version in [132, Section 3.2], but a slightly upgraded version described in [27, Section 4].

$$\mathcal{T}_k(s_n, \omega_n) = \frac{\hat{T}_k(s_n/\omega_n)}{\hat{T}_k(1/\omega_n)} = s \tag{7.80}$$

is exact for the model sequence (7.79) (see for example [132, Eq. (3.2-11)] or [27, Eq. (4.2)]).

With the help of this construction principle it is trivially simple to (re)derive the Levin-type transformations discussed in [132, Sections 7–9]. This is all that I originally had intended to achieve with my construction principle.

However, you [*Weniger means C.B., and coauthors M.R.-Z, and Ana Matos*] showed in the articles [31, 32] and [33] that my construction principle is actually much more general than I had originally anticipated, and that the majority of the currently known sequence transformations can be derived in this way. Obviously, this was a pleasant surprise for me.

As a special case of the general sequence transformation (7.80), I obtained the following expression [132, Eqs. (7.1-6) and (7.1-7)] for Levin's general sequence transformation [79]:

$$\mathcal{L}_k^{(n)}(\beta, s_n, \omega_n) = \frac{\Delta^k[(\beta + n)^{k-1} s_n/\omega_n]}{\Delta^k[(\beta + n)^{k-1}/\omega_n]} \tag{7.81}$$

$$= \frac{\displaystyle\sum_{j=0}^{k} (-1)^j \binom{k}{j} \frac{(\beta + n + j)^{k-1}}{(\beta + n + k)^{k-1}} \frac{s_{n+j}}{\omega_{n+j}}}{\displaystyle\sum_{j=0}^{k} (-1)^j \binom{k}{j} \frac{(\beta + n + j)^{k-1}}{(\beta + n + k)^{k-1}} \frac{1}{\omega_{n+j}}}, \qquad k, n \in \mathbb{N}_0. \tag{7.82}$$

Here $\beta > 0$ is a *positive* scaling parameter, and $\{\omega_n\}_{n=0}^{\infty}$ is a sequence of remainder estimates that are assumed to be explicitly known.

Avram Sidi apparently never liked my modification of Levin's general sequence transformation,[32] but David Levin recently used exactly my modified notation (7.81) and (7.82) [1, Eq. (2.4)].

In practice, one normally uses Levin's transformation in combination with Levin's simple remainder estimates $\omega_n = \Delta s_{n-1}$, $\omega_n = (\beta + n)\Delta s_{n-1}$, and $\omega_n = -[\Delta s_{n-1}][\Delta s_n]/[\Delta^2 s_{n-1}]$, which lead to the so-called t, u, and v variants of Levin's transformation [79, Eqs. (28), (58), and (67)], or the remainder estimate $\omega_n = \Delta s_n$ suggested by Smith and Ford [110, Eq. (2.5)], which leads to what I called Levin's d transformation.

Levin's transformation is undoubtedly the most important Levin-type transformation. Its u variant [132, Eq. (7.3-5)] is used internally in the computer algebra system Maple as a numerical tool to overcome convergence problems (see for example [45, pp. 51 and 125] and [65, p. 258]).

Superficially, Levin's t transformation seems to be superior to Levin's d transformation, since $t_k^{(n)}(\beta, s_n)$ needs only the input of $s_n, s_{n+1}, \ldots, s_{n+k}$, whereas

---

[32] I am wondering whether I have ever done anything upon which Avram Sidi does not look with utmost contempt.

$d_k^{(n)}(\beta, s_n)$ requires the input of $s_n, s_{n+1}, \ldots, s_{n+k+1}$. Accordingly, $t_k^{(n)}(\beta, s_n)$ has been used almost exclusively in the literature, whereas $d_k^{(n)}(\beta, s_n)$ has largely been neglected by other authors.

I never liked the idea behind the t transformation. The best *simple* estimate for the truncation error of a *convergent* power series is the first term neglected in the partial sum [77, p. 259]. The first term neglected is also an estimate of the truncation error of a factorially divergent hypergeometric series $_2F_0(a, b, -z)$ with $z > 0$, if its parameters a and b satisfy $a + n, b + n > 0$ for some $n \in \mathbb{N}_0$ [37, Theorem 5.12-5]. These examples suggest using the remainder estimate $\omega_n = \Delta s_n$ of [110, Eq. (2.5)] for convergent or divergent power series.

When I wrote my Computer Physics Report, my dislike of $\omega_n = \Delta s_{n-1}$ had been largely instinctive in nature. However, in [151, Eq. (4.46)] I explicitly showed that Levin's $t_k^{(n)}(\beta, s_n)$ transformation is nothing but the $d_k^{(n-1)}(\beta+1, s_{n-1})$ transformation in disguise. This follows at once from the simple relationships

$$\frac{s_n}{\Delta s_{n-1}} = \frac{s_{n-1}}{\Delta s_{n-1}} + 1 \tag{7.83}$$

and

$$\Delta^k \mathcal{P}_{k-1}(n) \frac{s_n}{\Delta s_{n-1}} = \Delta^k \mathcal{P}_{k-1}(n) \frac{s_{n-1}}{\Delta s_{n-1}}, \tag{7.84}$$

where $\mathcal{P}_{k-1}(n)$ is an essentially arbitrary polynomial of degree $k - 1$ in n. In [151, p. 1227], I remarked:

Numerical cancellation increases the risk of losing accuracy. Therefore, it is probably wiser not to use the t-type initial conditions $u_n = s_n/\Delta s_{n-1} \cdots$, but instead the d-type initial conditions $u_n = s_{n-1}/\Delta s_{n-1}$.

My derivation of Levin's transformation via (7.79) is based on the fact that the correction term

$$z_n^{(k)} = \sum_{j=0}^{k-1} \frac{c_j^{(k)}}{(\beta + n)^j}, \qquad k \in \mathbb{N}, \quad n \in \mathbb{N}_0, \quad \beta > 0, \tag{7.85}$$

is annihilated by the weighted difference operator $\hat{T}_k = \Delta(\beta + n)^{k-1}$, yielding (7.81) and (7.82). From my PhD thesis, I had been familiar with hypergeometric series containing Pochhammer symbols $(a)_m = \Gamma(a + m)/\Gamma(a)$. Consequently, it was clear to me that the alternative correction term

$$z_n^{(k)} = \sum_{j=0}^{k-1} \frac{c_j^{(k)}}{(\beta + n)_j}, \qquad k \in \mathbb{N}, \quad n \in \mathbb{N}_0, \quad \beta > 0, \tag{7.86}$$

is annihilated by the alternative weighted difference operator $\hat{T}_k = \Delta(\beta + n)_{k-1}$, yielding the sequence transformation [132, Eqs. (8.2-6) and (8.2-7)]

$$\mathcal{S}_k^{(n)}(\beta, s_n, \omega_n) = \frac{\Delta^k \left[(\beta + n)_{k-1} s_n / \omega_n \right]}{\Delta^k \left[(\beta + n)_{k-1} / \omega_n \right]} \tag{7.87}$$

$$= \frac{\displaystyle\sum_{j=0}^{k} (-1)^j \binom{k}{j} \frac{(\beta + n + j)_{k-1}}{(\beta + n + k)_{k-1}} \frac{s_{n+j}}{\omega_{n+j}}}{\displaystyle\sum_{j=0}^{k} (-1)^j \binom{k}{j} \frac{(\beta + n + j)_{k-1}}{(\beta + n + k)_{k-1}} \frac{1}{\omega_{n+j}}}, \qquad k, n \in \mathbb{N}_0. \tag{7.88}$$

The ratio (7.88) was originally derived by Sidi [104] for the construction of explicit expressions for Padé approximants of some special hypergeometric series. However, Sidi's article [104] provides no evidence that he intended to use this ratio as a sequence transformation. Moreover, I am not aware of any article by Sidi in which the properties of the sequence transformation $\mathcal{S}_k^{(n)}(\beta, s_n, \omega_n)$ were discussed or where it was applied.

Later, $\mathcal{S}_k^{(n)}(\beta, s_n, \omega_n)$ was used in the master's thesis of Shelef [102] (in Hebrew) for the numerical inversion of Laplace transforms, but it seems that this master's thesis was not published elsewhere.[33] The first refereed and generally accessible article in which an application of (7.88) as a sequence transformation was described is [170], where its u variant $y_k^{(n)}(\beta, s_n) = \mathcal{S}_k^{(n)}(\beta, s_n, (\beta + n)\Delta s_{n-1})$ was employed for the evaluation of the auxiliary function (7.77) used in molecular electronic structure calculations [170, Eq. (9)]. The mathematical properties of $\mathcal{S}_k^{(n)}(\beta, s_n, \omega_n)$ as a sequence transformation and in particular its connection with factorial series, which turned out to be of considerable importance later, were developed independently of Sidi and Shelef[34] in [132, Sections 8 and 13] (compare also [170, Ref. [8]]).

In his book, Sidi called $\mathcal{S}_k^{(n)}(\beta, s_n, \omega_n)$ the *Sidi S transformation* [105, Chapters 6.3.3 and 19.3]. It is, however, now common to call either $\mathcal{S}_k^{(n)}(\beta, s_n, \omega_n)$ or its variant $\delta_k^{(n)}(\beta, s_n) = \mathcal{S}_k^{(n)}(\beta, s_n, \Delta s_n)$ the *Weniger transformation* (see for example the articles by Borghi, Cvetič and Yu, Gil, Segura, and Temme, Li, Zhang and Tian, Temme [15, 18, 19, 47, 57, 80, 120] or the book by Gil et al. [56, Eq. (9.53) on p. 287]). This terminology is also used in the NIST Handbook of Mathematical Functions [91, §3.9(v) Levin's and Weniger's Transformations].[35]

$\mathcal{S}_k^{(n)}(\beta, s_n, \omega_n)$ and in particular its delta variant $\delta_k^{(n)}(\beta, s_n) = \mathcal{S}_k^{(n)}(\beta, s_n, \Delta s_n)$ were later used with considerable success for the evaluation of special functions [70, 71, 74, 132, 133, 135, 136, 141, 147, 154, 159], for the summation of divergent perturbation expansions [36, 40, 41, 42, 44, 72, 73, 133, 135, 136, 138, 139, 140,

---

[33] Before the internet, a master's thesis was essentially an internal document, but not a generally accessible publication, simply because it was not publicly available and it was usually not possible to become aware of it (except by a personal communication).

[34] Apparently, factorial series are completely ignored in Sidi's book [105], although I had written quite a lot about them and their usefulness for the $\mathcal{S}$ transformation [132]. It really seems that Sidi likes to ignore me and my contributions.

[35] I have the impression that after the publication of the NIST Handbook Sidi stopped trying to push others aggressively to call the $\mathcal{S}$ transformation the Sidi transformation.

142, 143, 151, 174, 175], and for the prediction of unknown perturbation series coefficients [9, 72, 73, 143]. More recently, the delta transformation was employed with considerable success in optics for the study of nonparaxial free-space propagation of optical wavefields [26, 28, 48, 80, 81] and for the numerical evaluation of diffraction catastrophes [15, 16, 17, 18, 19, 20, 22, 23, 24] (see also the detailed review by Borghi [25] and references therein).

My derivation of (7.86) and (7.88) was of course a triviality, but when I first derived $S_k^{(n)}(\beta, s_n, \omega_n)$ in 1986, I had no idea what the truncated factorial series in (7.86) really was.[36] In the contemporary mathematical literature as well as in my personal mathematical training, factorial series play a very minor role. However, I soon arrived at the conclusion that this current lack of interest is not completely justified. In the case of discrete data—like the partial sums of an infinite series—factorial series have some distinct advantages over their direct competitors, the inverse power series.

Factorial series had been employed already in Stirling's classic book *Methodus Differentialis* [118], which appeared in [119]. Some years ago, a new annotated translation of Sterling's book was published by Tweddle [124], who remarked that Stirling was not the inventor of factorial series. Apparently, Stirling became aware of factorial series through the work of the French mathematician Nicole [124, p. 174] (see also [125]). However, Stirling used factorial series quite extensively in his classic *Methodus Differentialis* [118, 119] and thus did a lot to popularize them.

The late nineteenth and the early twentieth century was in some sense the *golden era* of factorial series. Its theory was then fully developed by a variety of authors. Fairly complete surveys of the older literature as well as thorough treatments of their properties can be found in older books on finite differences such as [83, 84, 86], and [88, 89, 90]. Factorial series are also discussed in the books by Knopp [77] and Nielsen [85] on infinite series.

In the contemporary mathematical literature, factorial series are largely neglected and are used by a few specialists only. Some more recent references can be found in [156], and for even more recent references, see [21, 25, 27, 51, 52, 76, 126]. In any case, I am quite happy that I could contribute to a modest renaissance of factorial series.

After the publication of my Computer Physics Report [132], I also treated other theoretical aspects of sequence transformations. Inspired by Aitken's iterated Δ^2 process, I treated iterated sequence transformations in my article [134]. In my article [135], I constructed a sequence transformation that interpolates—depending on the value some parameter—between Levin's transformation and the S transformation, which had—in spite of their apparent similarity—behaved quite differently in the summation of some quantum-mechanical perturbation expansions (anharmonic oscillators) [40, 175]. In my article [136], I argued that the power of nonlinear and nonregular sequence transformations is more due to their nonregularity than their nonlinearity. In my article [147], I studied the impact of what I called *irregular*

---

[36] There was a long detour until I finally found out that the classic book by Nielsen [86], which had already been waiting on my bookshelf for several years without being touched, was actually a very useful reference on factorial series.

input data on the performance of sequence transformations. In our article [64], we constructed a convergence acceleration algorithm via an equation related to the lattice Boussinesq equation.

I have always been interested in practical applications of convergence acceleration and summation methods. For example, in our articles [38, 160, 161], we applied convergence acceleration techniques for the extrapolation of quantum chemical crystal orbital and cluster electronic structure calculations for oligomers to their infinite chain limits of stereoregular *quasi*-one-dimensional organic polymers.

In view of my interest in divergent series, it is not surprising that divergent series in general and divergent perturbation expansions from quantum physics in particular have played a major role in my research. Dyson [53] had already argued that perturbation expansions in quantum electrodynamics must diverge factorially, and ever since the seminal articles by Bender and Wu [10, 11] on the perturbation expansions of anharmonic oscillators it has been clear that quantum-mechanical perturbation theory produces almost by default factorially divergent perturbation expansions. A good source on divergent perturbation expansions in quantum mechanics and in higher field theories is the book edited by Le Guillou and Zinn-Justin [78], in which many of the relevant articles are reprinted.

Therefore, it was for me a fairly natural step to take a closer look at the summation of strongly divergent quantum-mechanical perturbation expansions, which I did in my articles [9, 40, 41, 43, 72, 73, 133, 135, 137, 138, 139, 140, 142, 143, 147, 174, 175].

Divergent perturbation expansions are power series expansions for a physical property $\mathcal{P}$ in a physically motivated coupling constant β. Normally, these expansions diverge factorially or even hyper-factorially for every nonzero coupling constant $\beta > 0$. Standard summation techniques including sequence transformations produce reasonably good summation results only for sufficiently small coupling constants β, but the physically relevant range of the coupling constant is usually the whole positive semi-axis $\beta \in [0, \infty)$. If we nevertheless want to compute $\mathcal{P}(\beta)$ for larger values of β, straightforward summation will not suffice.[37] Instead, we must combine summation techniques with more sophisticated analytical techniques, which in physics are usually called *renormalization* techniques.[38]

In my articles [40, 41, 43, 107, 108, 140, 142], we applied a renormalization technique introduced by Čížek and Vrscay [39] and worked out by Vinette and Čížek [127] with considerable success for the summation of the divergent perturbation expansions of the so-called anharmonic oscillators (numerical techniques alone would not suffice).

What are probably my most spectacular results were presented in [140, Tables II–IV] and [142, Tables I and II]. There, I computed the coefficients of the renormalized strong coupling and the strong coupling expansions of the quartic, sextic, and octic anharmonic oscillators. These coefficients, which allow a very convenient com-

---

[37] This is a much more challenging problem than the determination of the large-z behavior of a function $f(z)$ from a *convergent* power series about $z = 0$.

[38] In practice, renormalization techniques are essentially (very) skillful variable transformations $\beta \in [0, \infty) \longmapsto \kappa \in [a, b]$ with a, b finite, which make it possible to incorporate at least some information about the large-β behavior of $\mathcal{P}(\beta)$ into the new expansion in κ.

putation of sufficiently accurate approximations to the ground state energies,[39] were obtained by summing the corresponding weak coupling perturbation expansions with the help of $\delta_k^{(n)}(\beta, s_n) = S_k^{(n)}(\beta, s_n, \Delta s_n)$. In my opinion, the coefficients in [140, Tables II–IV] and [142, Tables I and II] demonstrate convincingly how far you can get if you combine powerful numerical techniques with intelligent renormalization techniques.

Closely related to the summation of divergent power series by Padé approximants and other rational approximants is *Padé prediction* of unknown series coefficients and its generalization to other sequence transformations.[40] It seems that this idea was first formulated by Gilewicz [58], and further developed in articles by Sidi and Levin [106] and Brezinski [30]. In physics, these ideas were first used in the article by Samuel et al. [101]. I applied prediction methods in my articles [72, 73, 143].

In these articles, we were satisfied to obtain approximations to some unknown series coefficients. But in [9] we were more ambitious. In quantum theory, it has traditionally been emphasized that physical observables are to be represented by Hermitian operators, because only then is it guaranteed that observables represented by operators are real. However, in recent years it was found that this assumption is not completely true. So-called non-Hermitian but $\mathcal{PT}$-symmetric operators were found whose eigenvalues were also real (see for example [6, 7, 12]).

In the case of perturbation expansions of Hermitian operators, it is frequently possible to prove rigorously that these expansions are Stieltjes series, which for example guarantees that certain sequences of Padé approximants to such a series converge. This is obviously a good thing. In the case of perturbation expansions of non-Hermitian but $\mathcal{PT}$-symmetric operators, such things were not known at that time. Thus, it was also unclear whether certain subsequences in their Padé tables would converge.

So in [9] we used some Padé prediction techniques formulated in [145, 146] to find numerical evidence that the factorially divergent perturbation expansion of a certain non-Hermitian $\mathcal{PT}$-symmetric anharmonic oscillator, whose perturbation series coefficients had been computed in [8], is a Stieltjes series. This was not known at that time. In the meantime, however, our numerical conjecture was proved rigorously by Grecchi et al. [59].

Rigorous convergence proofs have been and still are a major problem for Padé approximants and sequence transformations (in particular in the case of large transformation orders). In our article [27], we analyzed the Padé and the delta summation of the factorially divergent Euler series

$$\mathcal{E}(z) \sim \sum_{m=0}^{\infty} (-1)^m \, m! \, z^m = {}_2F_0(1, 1; -z), \qquad z \to 0, \qquad (7.89)$$

---

[39] If these coefficients are available, a programmable pocket calculator should suffice for the computation of the corresponding energies for all $\beta \in [0, \infty)$.

[40] One must not forget that the computation of perturbation series coefficients is in some cases extremely difficult. Therefore, it is a priori desirable if other approaches to the computation of approximations to the perturbation series coefficients are available.

whose summation and interpretation had already been studied by Euler (see for example the books by Bromwich [34, pp. 323–324] and Hardy [62, pp. 26–29] or the articles by Barbeau [4] and Barbeau and Leah [5]), respectively, in the case of large transformation orders $k \to \infty$.

If $|\arg(z)| < \pi$, the Euler series is asymptotic as $z \to 0$ in the sense of Poincaré [93] to the so-called *Euler integral*

$$\mathcal{E}(z) = \int_0^\infty \frac{\exp(-t)}{1 + zt}\, dt, \qquad |\arg(z)| < \pi. \tag{7.90}$$

The input data of both our Padé and delta summations were the partial sums

$$\mathcal{E}_n(z) = \sum_{\nu=0}^n (-1)^\nu \nu!\, z^\nu, \qquad n \in \mathbb{N}_0, \tag{7.91}$$

which can, according to

$$\mathcal{E}_n(z) = \mathcal{E}(z) + \mathcal{R}_n(\mathcal{E}; z), \qquad n \in \mathbb{N}_0, \quad |\arg(z)| < \pi, \tag{7.92}$$

be expressed by the Euler integral plus a truncation error integral

$$\mathcal{R}_n(\mathcal{E}; z) = -(-z)^{n+1} \int_0^\infty \frac{t^{n+1} \exp(-t)dt}{1 + zt}, \qquad n \in \mathbb{N}_0, \quad |\arg(z)| < \pi, \tag{7.93}$$

which is also a Stieltjes integral.

Our convergence analysis was very much based on advantageous properties of factorial series in general and on Riccardo Borghi's factorial series representation for the remainder $\mathcal{R}_n(\mathcal{E}; z)$ of the Euler series in particular [21, Eq. (52)]:

$$\frac{\mathcal{R}_n(\mathcal{E}; z)}{(-1)^{n+1}(n + 1)!z^{n+1}} = -\sum_{k=0}^\infty \frac{L_k^{(-1)}(1/z)}{z} \frac{k!}{(n + 1)_{k+1}}. \tag{7.94}$$

We are working to generalize our approach in [27] to other infinite series for special functions whose truncation errors can also be expressed as a suitable factorial series.

References

1. Abdalkhani, J. and Levin, D. (2015), On the choice of β in the u-transformation for convergence acceleration, *Numer. Algor.* **70**, 205–213.
   (Cited on page 286.)
2. Abramowitz, M. and Stegun, I.A. (editors) (1972), *Handbook of Mathematical Functions* (National Bureau of Standards, Washington, D. C.).
   (Cited on page 284.)
3. Ballentine, L.E. (1998), *Quantum Mechanics: A Modern Development* (World Scientific, Singapore).
   (Cited on page 282.)

4. Barbeau, E.J. (1979), Euler subdues a very obstreperous series, *Amer. Math. Monthly* **86**, 356–372.
   (Cited on page 292.)
5. Barbeau, E.J. and Leah, P.J. (1976), Euler's 1760 paper on divergent series, *Hist. Math.* **3**, 141–160.
   (Cited on page 292.)
6. Bender, C.M. (2007), Making sense of non-Hermitian Hamiltonians, *Rep. Prog. Phys.* **70**, 947–1018.
   (Cited on page 291.)
7. Bender, C.M. and Boettcher, S. (1998), Real spectra in non-Hermitian Hamiltonians having $\mathcal{PT}$ symmetry, *Phys. Rev. Lett.* **80**, 5243–5246.
   (Cited on page 291.)
8. Bender, C.M. and Dunne, G.V. (1999), Large-order perturbation theory for a non-Hermitian $\mathcal{PT}$-symmetric Hamiltonian, *J. Math. Phys.* **40**, 4616–4621.
   (Cited on page 291.)
9. Bender, C.M. and Weniger, E.J. (2001), Numerical evidence that the perturbation expansion for a non-Hermitian $\mathcal{PT}$-symmetric Hamiltonian is Stieltjes, *J. Math. Phys.* **42**, 2167–2183.
   (Cited on pages 289, 290, and 291.)
10. Bender, C.M. and Wu, T.T. (1969), Anharmonic oscillator, *Phys. Rev.* **184**, 1231–1260.
    (Cited on page 290.)
11. Bender, C.M. and Wu, T.T. (1971), Large-order behavior of perturbation theory, *Phys. Rev. Lett.* **27**, 461–465.
    (Cited on page 290.)
12. Bender, C.M., Boettcher, S., and Meisinger, P.N. (1999), $\mathcal{PT}$-symmetric quantum mechanics, *J. Math. Phys.* **40**, 2201–2229.
    (Cited on page 291.)
13. Böhm, A. (1978), *The Rigged Hilbert Space and Quantum Mechanics* (Springer-Verlag, Berlin).
    (Cited on page 281.)
14. Boos, J. (2000), *Classical and Modern Methods of Summability* (Oxford U. P., Oxford).
    (Cited on page 282.)
15. Borghi, R. (2007), Evaluation of diffraction catastrophes by using Weniger transformation, *Opt. Lett.* **32**, 226–228.
    (Cited on pages 288 and 289.)
16. Borghi, R. (2008a), Summing Pauli asymptotic series to solve the wedge problem, *J. Opt. Soc. Amer. A* **25**, 211–218.
    (Cited on page 289.)
17. Borghi, R. (2008b), On the numerical evaluation of cuspoid diffraction catastrophes, *J. Opt. Soc. Amer. A* **25**, 1682–1690.
    (Cited on page 289.)
18. Borghi, R. (2008c), Joint use of the Weniger transformation and hyperasymptotics for accurate asymptotic evaluations of a class of saddle-point integrals, *Phys. Rev. E* **78**, 026703-1 - 026703-11.
    (Cited on pages 288 and 289.)
19. Borghi, R. (2009), Joint use of the Weniger transformation and hyperasymptotics for accurate asymptotic evaluations of a class of saddle-point integrals. II. Higher-order transformations, *Phys. Rev. E* **80**, 016704-1 - 016704-15.
    (Cited on pages 288 and 289.)
20. Borghi, R. (2010a), On the numerical evaluation of umbilic diffraction catastrophes, *J. Opt. Soc. Amer. A* **27**, 1661–1670.
    (Cited on page 289.)
21. Borghi, R. (2010b), Asymptotic and factorial expansions of Euler series truncation errors via exponential polynomials, *Appl. Numer. Math.* **60**, 1242–1250.
    (Cited on pages 289 and 292.)

22. Borghi, R. (2011a), Evaluation of cuspoid and umbilic diffraction catastrophes of codimension four, *J. Opt. Soc. Amer. A* **28**, 887–896.
   (Cited on page 289.)
23. Borghi, R. (2011b), Optimizing diffraction catastrophe evaluation, *Opt. Lett.* **36**, 4413–4415.
   (Cited on page 289.)
24. Borghi, R. (2012), Numerical computation of diffraction catastrophes with codimension eight, *Phys. Rev. E* **85**, 046704-1 - 046704-14.
   (Cited on page 289.)
25. Borghi, R. (2016), Computational optics through sequence transformations, in T.D. Visser (editor), *Progress in Optics*, volume 61, chapter 1, 1–70 (Academic Press, Amsterdam).
   (Cited on page 289.)
26. Borghi, R. and Santarsiero, M. (2003), Summing Lax series for nonparaxial beam propagation, *Opt. Lett.* **28**, 774–776.
   (Cited on page 289.)
27. Borghi, R. and Weniger, E.J. (2015), Convergence analysis of the summation of the factorially divergent Euler series by Padé approximants and the delta transformation, *Appl. Numer. Math.* **94**, 149–178.
   (Cited on pages 279, 285, 286, 289, 291, and 292.)
28. Borghi, R., Gori, F., Guattari, G., and Santarsiero, M. (2011), Decoding divergent series in nonparaxial optics, *Opt. Lett.* **36**, 963–965.
   (Cited on page 289.)
29. Born, M. (1955), Statistical interpretation of quantum mechanics, *Science* **122**, 675–679.
   (Cited on page 280.)
30. Brezinski, C. (1985), Prediction properties of some extrapolation methods, *Appl. Numer. Math.* **1**, 457–462.
   (Cited on page 291.)
31. Brezinski, C. and Matos, A.C. (1996), A derivation of extrapolation algorithms based on error estimates, *J. Comput. Appl. Math.* **66**, 5–26.
   (Cited on page 286.)
32. Brezinski, C. and Redivo Zaglia, M. (1994a), A general extrapolation procedure revisited, *Adv. Comput. Math.* **2**, 461–477.
   (Cited on page 286.)
33. Brezinski, C. and Redivo Zaglia, M. (1994b), On the kernel of sequence transformations, *Appl. Numer. Math.* **16**, 239–244.
   (Cited on page 286.)
34. Bromwich, T.J.I. (1991), *An Introduction to the Theory of Infinite Series* (Chelsea, New York), 3rd edition. Originally published by Macmillan (London, 1908 and 1926).
   (Cited on page 292.)
35. Browne, J.C. (1971), Molecular wave functions: Calculation and use in atomic and molecular processes, *Adv. At. Mol. Phys.* **7**, 47–95.
   (Cited on page 275.)
36. Caliceti, E., Meyer-Hermann, M., Ribeca, P., Surzhykov, A., and Jentschura, U.D. (2007), From useful algorithms for slowly convergent series to physical predictions based on divergent perturbative expansions, *Phys. Rep.* **446**, 1–96.
   (Cited on page 288.)
37. Carlson, B.C. (1977), *Special Functions of Applied Mathematics* (Academic Press, New York).
   (Cited on page 287.)
38. Cioslowski, J. and Weniger, E.J. (1993), Bulk properties from finite cluster calculations. VIII. Benchmark calculations on the efficiency of extrapolation methods for the HF and MP2 energies of polyacenes, *J. Comput. Chem.* **14**, 1468–1481.
   (Cited on page 290.)
39. Čížek, J. and Vrscay, E.R. (1986), Lower bounds to ground state eigenvalues of the Schrödinger equation via optimized inner projection: Application to quartic and sextic anharmonic oscillators, *Int. J. Quantum Chem. Symp.* **20**, 65–72.
   (Cited on page 290.)

40. Čížek, J., Vinette, F., and Weniger, E.J. (1991), Examples on the use of symbolic computation in physics and chemistry: Applications of the inner projection technique and of a new summation method for divergent series, *Int. J. Quantum Chem. Symp.* **25**, 209–223.
    (Cited on pages 288, 289, and 290.)

41. Čížek, J., Vinette, F., and Weniger, E.J. (1993a), On the use of the symbolic language Maple in physics and chemistry: Several examples, in R.A. de Groot and J. Nadrchal (editors), *Proceedings of the Fourth International Conference on Computational Physics PHYSICS COMPUTING '92*, 31–44 (World Scientific, Singapore).
    (Cited on pages 283, 288, 290, and 295.)

42. Čížek, J., Vinette, F., and Weniger, E.J. (1993b), On the use of the symbolic language Maple in physics and chemistry: Several examples, *Int. J. Mod. Phys. C* **4**, 257–270. Reprint of [41].
    (Cited on pages 283 and 288.)

43. Čížek, J., Weniger, E.J., Bracken, P., and Špirko, V. (1996), Effective characteristic polynomials and two-point Padé approximants as summation techniques for the strongly divergent perturbation expansions of the ground state energies of anharmonic oscillators, *Phys. Rev. E* **53**, 2925–2939.
    (Cited on pages 283 and 290.)

44. Čížek, J., Zamastil, J., and Skála, L. (2003), New summation technique for rapidly divergent perturbation series. Hydrogen atom in magnetic field, *J. Math. Phys.* **44**, 962–968.
    (Cited on page 288.)

45. Corless, R.M. (2002), *Essential Maple 7: An Introduction for Scientific Programmers* (Springer-Verlag, New York).
    (Cited on page 286.)

46. Cuyt, A., Brevik Petersen, V., Verdonk, B., Waadeland, H., and Jones, W.B. (2008), *Handbook of Continued Fractions for Special Functions* (Springer-Verlag, New York).
    (Cited on page 278.)

47. Cvetič, G. and Yu, J.Y. (2000), Borel-Padé vs Borel-Weniger method: A QED and a QCD example, *Mod. Phys. Lett. A* **15**, 1227–1235.
    (Cited on page 288.)

48. Dai, L., Li, J.X., Zang, W.P., and Tian, J.G. (2011), Vacuum electron acceleration driven by a tightly focused radially polarized Gaussian beam, *Opt. Expr.* **19**, 9303–9308.
    (Cited on page 289.)

49. Dalgarno, A. (1954), Integrals occurring in problems of molecular structure, *Math. Tables Aids Comput.* **8**, 203–212.
    (Cited on page 275.)

50. de la Madrid, R. (2005), The role of the rigged Hilbert space in quantum mechanics, *Eur. J. Phys.* **26**, 287–312.
    (Cited on page 281.)

51. Deeb, A., Hamdouni, A., and Razafindralandy, D. (2016), Comparison between Borel-Padé summation and factorial series, as time integration methods, *Discr. Cont. Dyn. Sys. (Ser. S)* **9**, 393–408.
    (Cited on page 289.)

52. Delabaere, E. and Rasoamanana, J.M. (2007), Sommation effective d'une somme Borel par séries de factorielles, *Annal. l'Inst. Fourier* **57**, 421–456.
    (Cited on page 289.)

53. Dyson, D.J. (1952), Divergence of perturbation theory in quantum electrodynamics, *Phys. Rev.* **85**, 32–33.
    (Cited on page 290.)

54. Filter, E. and Steinborn, E.O. (1978), Extremely compact formulas for molecular one-electron integrals and Coulomb integrals over Slater-type atomic orbitals, *Phys. Rev. A* **18**, 1–11.
    (Cited on page 276.)

55. Fock, V. (1935), Zur Theorie des Wasserstoffatoms, *Z. Physik* **98**, 145–154.
    (Cited on page 280.)

56. Gil, A., Segura, J., and Temme, N.M. (2007), *Numerical Methods for Special Functions* (SIAM, Philadelphia).
    (Cited on pages 278 and 288.)
57. Gil, A., Segura, J., and Temme, N.M. (2011), Basic methods for computing special functions, in T.E. Simos (editor), *Recent Advances in Computational and Applied Mathematics*, 67–121 (Springer-Verlag, Dordrecht).
    (Cited on pages 278 and 288.)
58. Gilewicz, J. (1973), Numerical detection of the best Padé approximant and determination of the Fourier coefficients of the insufficiently sampled functions, in P.R. Graves-Morris (editor), *Padé Approximants and Their Applications*, 99–103 (Academic Press, London).
    (Cited on page 291.)
59. Grecchi, V., Maioli, M., and Martinez, A. (2009), Padé summability of the cubic oscillator, *J. Phys. A* **42**, 425208-1–425208-17.
    (Cited on page 291.)
60. Griffel, D.H. (2002), *Applied Functional Analysis* (Dover, Mineola, NY). Originally published by Ellis Horwood (Chichester, 1981).
    (Cited on page 280.)
61. Grotendorst, J., Weniger, E.J., and Steinborn, E.O. (1986), Efficient evaluation of infinite-series representations for overlap, two-center nuclear attraction, and Coulomb integrals using nonlinear convergence accelerators, *Phys. Rev. A* **33**, 3706–3726.
    (Cited on pages 275 and 284.)
62. Hardy, G.H. (1949), *Divergent Series* (Clarendon Press, Oxford).
    (Cited on pages 282 and 292.)
63. Harris, F.E. and Michels, H.H. (1967), The evaluation of molecular integrals for Slater-type orbitals, *Adv. Chem. Phys.* **13**, 205–266.
    (Cited on page 275.)
64. He, Y., Hu, X.B., Sun, J.Q., and Weniger, E.J. (2011), Convergence acceleration algorithm via an equation related to the lattice Boussinesq equation, *SIAM J. Sci. Comput.* **33**, 1234–1245.
    (Cited on page 290.)
65. Heck, A. (2003), *Introduction to Maple* (Springer-Verlag, New York), 3rd edition.
    (Cited on page 286.)
66. Henrici, P. (1977), *Applied and Computational Complex Analysis II* (Wiley, New York).
    (Cited on page 282.)
67. Homeier, H.H.H., Weniger, E.J., and Steinborn, E.O. (1992a), Simplified derivation of a one-range addition theorem of the Yukawa potential, *Int. J. Quantum Chem.* **44**, 405–411.
    (Cited on page 275.)
68. Homeier, H.H.H., Weniger, E.J., and Steinborn, E.O. (1992b), Programs for the evaluation of overlap integrals with *B* functions, *Comput. Phys. Commun.* **72**, 269–287.
    (Cited on page 275.)
69. Huzinaga, S. (1967), Molecular integrals, *Prog. Theor. Phys. Suppl.* **40**, 52–77.
    (Cited on page 275.)
70. Jentschura, U.D. and Lötstedt, E. (2012), Numerical calculation of Bessel, Hankel and Airy functions, *Comput. Phys. Commun.* **183**, 506–519.
    (Cited on page 288.)
71. Jentschura, U.D., Mohr, P.J., Soff, G., and Weniger, E.J. (1999), Convergence acceleration via combined nonlinear-condensation transformations, *Comput. Phys. Commun.* **116**, 28–54.
    (Cited on pages 279 and 288.)
72. Jentschura, U.D., Becher, J., Weniger, E.J., and Soff, G. (2000a), Resummation of QED perturbation series by sequence transformations and the prediction of perturbative coefficients, *Phys. Rev. Lett.* **85**, 2446–2449.
    (Cited on pages 288, 289, 290, and 291.)
73. Jentschura, U.D., Weniger, E.J., and Soff, G. (2000b), Asymptotic improvement of resummations and perturbative predictions in quantum field theory, *J. Phys. G* **26**, 1545–1568.
    (Cited on pages 288, 289, 290, and 291.)

74. Jentschura, U.D., Gies, H., Valluri, S.R., Lamm, D.R., and Weniger, E.J. (2002), QED effective action revisited, *Can. J. Phys.* **80**, 267–284.
(Cited on pages 279 and 288.)

75. Jones, W.B. and Thron, W.T. (1980), *Continued Fractions* (Addison Wesley, Reading, Mass.).
(Cited on page 282.)

76. Karp, D.B. and Prilepkina, E. (2018), An inverse factorial series for a general gamma ratio and related properties of the Nørlund-Bernoulli polynomials, *J. Math. Sci.* **234**, 680–696.
(Cited on page 289.)

77. Knopp, K. (1964), *Theorie und Anwendung der unendlichen Reihen* (Springer-Verlag, Berlin).
(Cited on pages 279, 282, 287, and 289.)

78. Le Guillou, J.C. and Zinn-Justin, J. (editors) (1990), *Large-Order Behaviour of Perturbation Theory* (North-Holland, Amsterdam).
(Cited on page 290.)

79. Levin, D. (1973), Development of non-linear transformations for improving convergence of sequences, *Int. J. Comput. Math. B* **3**, 371–388.
(Cited on page 286.)

80. Li, J., Zang, W., and Tian, J. (2009a), Simulation of Gaussian laser beams and electron dynamics by Weniger transformation method, *Opt. Expr.* **17**, 4959–4969.
(Cited on pages 288 and 289.)

81. Li, J.X., Zang, W., Li, Y.D., and Tian, J. (2009b), Acceleration of electrons by a tightly focused intense laser beam, *Opt. Expr.* **17**, 11850–11859.
(Cited on page 289.)

82. Mathar, R.J. (2003), Numerical representations of the incomplete Gamma function of complex-valued argument, *Numer. Algor.* **36**, 247–264.
(Cited on page 284.)

83. Meschkowski, H. (1959), *Differenzengleichungen* (Vandenhoek & Rupprecht, Göttingen).
(Cited on page 289.)

84. Milne-Thomson, L.M. (1981), *The Calculus of Finite Differences* (Chelsea, New York). Originally published by Macmillan (London, 1933).
(Cited on page 289.)

85. Nielsen, N. (1909), *Lehrbuch der unendlichen Reihen* (Teubner, Leipzig and Berlin).
(Cited on page 289.)

86. Nielsen, N. (1965), *Die Gammafunktion* (Chelsea, New York). Originally published by Teubner (Leipzig and Berlin, 1906).
(Cited on page 289.)

87. Niukkanen, A.W. (1984), Fourier transforms of atomic orbitals. I. Reduction to four-dimensional harmonics and quadratic transformations, *Int. J. Quantum Chem.* **25**, 941–955.
(Cited on page 276.)

88. Nörlund, N.E. (1926), *Leçons sur les Séries d'Interpolation* (Gautier-Villars, Paris).
(Cited on page 289.)

89. Nörlund, N.E. (1929), *Leçons sur les Équations Linéaires aux Différences Finies* (Gautier-Villars, Paris).
(Cited on page 289.)

90. Nörlund, N.E. (1954), *Vorlesungen über Differenzenrechnung* (Chelsea, New York). Originally published by Springer-Verlag (Berlin, 1924).
(Cited on page 289.)

91. Olver, F.W.J., Lozier, D.W., Boisvert, R.F., and Clark, C.W. (editors) (2010), *NIST Handbook of Mathematical Functions* (Cambridge U. P., Cambridge). Available online under http://dlmf.nist.gov/.
(Cited on pages 278 and 288.)

92. Petersson, G.A. and McKoy, V. (1967), Application of nonlinear transformations to the evaluation of multicenter integrals, *J. Chem. Phys.* **46**, 4362–4368.
(Cited on page 279.)

93. Poincaré, H. (1886), Sur les intégrales irrégulières des équations linéaires, *Acta Math.* **8**, 295–344.
(Cited on page 292.)

94. Powell, R.E. and Shah, S.M. (1988), *Summability Theory and Its Applications* (Prentice-Hall of India, New Delhi).
(Cited on page 282.)

95. Reed, M. and Simon, B. (1975), *Methods of Modern Mathematical Physics II: Fourier Analysis, Self-Adjointness* (Academic Press, New York).
(Cited on page 282.)

96. Rutishauser, H. (1954a), Der Quotienten-Differenzen-Algorithmus, *Zeitschr. Angew. Math. Phys. (ZAMP)* **5**, 233–251.
(Cited on page 282.)

97. Rutishauser, H. (1954b), Anwendungen des Quotienten-Differenzen-Algorithmus, *Zeitschr. Angew. Math. Phys. (ZAMP)* **5**, 496–508.
(Cited on page 282.)

98. Rutishauser, H. (1957), *Der Quotienten-Differenzen-Algorithmus* (Birkhäuser, Basel).
(Cited on page 282.)

99. Safouhi, H. (2010a), Bessel, sine and cosine functions and extrapolation methods for computing molecular multi-center integrals, *Numer. Algor.* **54**, 141–167.
(Cited on page 276.)

100. Safouhi, H. (2010b), Integrals of the paramagnetic contribution in the relativistic calculation of the shielding tensor, *J. Math. Chem.* **48**, 601–616.
(Cited on page 276.)

101. Samuel, M.A., Ellis, J., and Karliner, M. (1995), Comparison of the Padé approximation method to perturbative QCD calculations, *Phys. Rev. Lett.* **74**, 4380–4383.
(Cited on page 291.)

102. Shelef, R. (1987), *New numerical quadrature formulas for Laplace transform inversion by Bromwich's integral*, master's thesis, Technion, Israel Institute of Technology, Haifa. In Hebrew.
(Cited on page 288.)

103. Shibuya, T.I. and Wulfman, C.E. (1965), Molecular orbitals in momentum space, *Proc. Roy. Soc. A* **286**, 376–389.
(Cited on pages 280 and 281.)

104. Sidi, A. (1981), A new method for deriving Padé approximants for some hypergeometric functions, *J. Comput. Appl. Math.* **7**, 37–40.
(Cited on page 288.)

105. Sidi, A. (2003), *Practical Extrapolation Methods* (Cambridge U. P., Cambridge).
(Cited on pages 284 and 288.)

106. Sidi, A. and Levin, D. (1983), Prediction properties of the t-transformation, *SIAM J. Numer. Anal.* **20**, 589–598.
(Cited on page 291.)

107. Skála, L., Čížek, J., Kapsa, V., and Weniger, E.J. (1997), Large-order analysis of the convergent renormalized strong-coupling perturbation theory for the quartic anharmonic oscillator, *Phys. Rev. A* **56**, 4471–4476.
(Cited on page 290.)

108. Skála, L., Čížek, J., Weniger, E.J., and Zamastil, J. (1999), Large-order behavior of the convergent perturbation theory for anharmonic oscillators, *Phys. Rev. A* **59**, 102–106.
(Cited on page 290.)

109. Slevinsky, R.M. (2014), *New techniques in numerical integration: The computation of molecular integrals over exponential-type functions*, Ph.D. thesis, Department of Mathematical and Statistical Sciences, University of Alberta, Edmonton, Alberta.
(Cited on page 276.)

110. Smith, D.A. and Ford, W.F. (1979), Acceleration of linear and logarithmic convergence, *SIAM J. Numer. Anal.* **16**, 223–240.
(Cited on pages 284, 285, 286, and 287.)

111. Smith, D.A. and Ford, W.F. (1982), Numerical comparisons of nonlinear convergence accelerators, *Math. Comput.* **38**, 481–499.
      (Cited on pages 284 and 285.)

112. Steinborn, E.O. and Filter, E. (1975), Translations of fields represented by spherical-harmonic expansions for molecular calculations. III. Translations of reduced Bessel functions, Slater-type s-orbitals, and other functions, *Theor. Chim. Acta* **38**, 273–281.
      (Cited on page 276.)

113. Steinborn, E.O. and Weniger, E.J. (1977), Advantages of reduced Bessel functions as atomic orbitals: An application to H_2^+, *Int. J. Quantum Chem. Symp.* **11**, 509–516.
      (Cited on page 275.)

114. Steinborn, E.O. and Weniger, E.J. (1978), Reduced Bessel functions as atomic orbitals: Some mathematical aspects and an LCAO-MO treatment of HeH^{++}, *Int. J. Quantum Chem. Symp.* **12**, 103–108.
      (Cited on page 275.)

115. Steinborn, E.O. and Weniger, E.J. (1990), Sequence transformations for the efficient evaluation of infinite series representations of some molecular integrals with exponentially decaying basis functions, *J. Mol. Struct. (Theochem)* **210**, 71–78.
      (Cited on page 275.)

116. Steinborn, E.O. and Weniger, E.J. (1992), Nuclear attraction and electron interaction integrals of exponentially decaying functions and the Poisson equation, *Theor. Chim. Acta* **83**, 105–121.
      (Cited on page 275.)

117. Steinborn, E.O., Homeier, H.H.H., and Weniger, E.J. (1992), Recent progress on representations for Coulomb integrals of exponential-type orbitals, *J. Mol. Struct. (Theochem)* **260**, 207–221.
      (Cited on page 275.)

118. Stirling, J. (1730), *Methodus Differentialis sive Tractatus de Summatione et Interpolatione Serierum Infinitarum* (London).
      (Cited on pages 289 and 299.)

119. Stirling, J. (1749), *The Differential Method, or, a Treatise Concerning the Summation and Interpolation of Infinite Series* (London). English translation of [118] by F. Holliday.
      (Cited on page 289.)

120. Temme, N.M. (2007), Numerical aspects of special functions, *Acta Numer.* **16**, 379–478.
      (Cited on pages 278 and 288.)

121. Thakkar, A.J. (2015), The life and work of Jiří Čížek, *AIP Conf. Proc.* **1642**, 138–149.
      (Cited on page 280.)

122. Trott, M. (2006a), *The Mathematica GuideBook for Numerics* (Springer, New York).
      (Cited on page 283.)

123. Trott, M. (2006b), *The Mathematica GuideBook for Symbolics* (Springer, New York).
      (Cited on page 283.)

124. Tweddle, I. (2003), *James Stirling's Methodus Differentialis: An Annotated Translation of Stirling's Text* (Springer-Verlag, London).
      (Cited on page 289.)

125. Tweedie, C. (1917), Nicole's contributions to the foundations of the calculus of finite differences, *Proc. Edinb. Math. Soc.* **36**, 22–39.
      (Cited on page 289.)

126. Varin, V.P. (2018), Factorial transformation for some classical combinatorial sequences, *Comput. Math. Math. Phys.* **58**, 1687–1707.
      (Cited on page 289.)

127. Vinette, F. and Čížek, J. (1991), Upper and lower bounds of the ground state energy of anharmonic oscillators using renormalized inner projection, *J. Math. Phys.* **32**, 3392–3404.
      (Cited on page 290.)

128. Watson, G.N. (1966), *A Treatise on the Theory of Bessel Functions* (Cambridge U. P., Cambridge), 2nd edition.
      (Cited on page 282.)

129. Weniger, E.J. (1977), *Untersuchung der Verwendbarkeit reduzierter Besselfunktionen als Basissatz für ab initio Rechnungen an Molekülen. Vergleichende Rechnungen am Beispiel des* H_2^+, Diplomarbeit, Fachbereich Chemie und Pharmazie, Universität Regensburg. *(Cited on page 275.)*

130. Weniger, E.J. (1982), *Reduzierte Bessel-Funktionen als LCAO-Basissatz: Analytische und numerische Untersuchungen*, Ph.D. thesis, Fachbereich Chemie und Pharmazie, Universität Regensburg. A short abstract of this thesis was published in Zentralblatt für Mathematik **523**, 444 (1984), abstract no. 65015.
    (Cited on pages 275 and 276.)

131. Weniger, E.J. (1985), Weakly convergent expansions of a plane wave and their use in Fourier integrals, *J. Math. Phys.* **26**, 276–291.
    (Cited on pages 275, 280, 282, and 283.)

132. Weniger, E.J. (1989), Nonlinear sequence transformations for the acceleration of convergence and the summation of divergent series, *Comput. Phys. Rep.* **10**, 189–371. Los Alamos Preprint math-ph/0306302 (http://arXiv.org).
    (Cited on pages 283, 285, 286, 287, 288, and 289.)

133. Weniger, E.J. (1990), On the summation of some divergent hypergeometric series and related perturbation expansions, *J. Comput. Appl. Math.* **32**, 291–300.
    (Cited on pages 279, 283, 288, and 290.)

134. Weniger, E.J. (1991), On the derivation of iterated sequence transformations for the acceleration of convergence and the summation of divergent series, *Comput. Phys. Commun.* **64**, 19–45.
    (Cited on page 289.)

135. Weniger, E.J. (1992), Interpolation between sequence transformations, *Numer. Algor.* **3**, 477–486.
    (Cited on pages 288, 289, and 290.)

136. Weniger, E.J. (1994a), On the efficiency of linear but nonregular sequence transformations, in A. Cuyt (editor), *Nonlinear Numerical Methods and Rational Approximation II*, 269–282 (Kluwer, Dordrecht).
    (Cited on pages 288 and 289.)

137. Weniger, E.J. (1994b), *Verallgemeinerte Summationsprozesse als numerische Hilfsmittel für quantenmechanische und quantenchemische Rechnungen*, Habilitation thesis, Fachbereich Chemie und Pharmazie, Universität Regensburg. Los Alamos Preprint math-ph/0306048 (http://arXiv.org).
    (Cited on page 290.)

138. Weniger, E.J. (1996a), Nonlinear sequence transformations: A computational tool for quantum mechanical and quantum chemical calculations, *Int. J. Quantum Chem.* **57**, 265–280.
    (Cited on pages 288 and 290.)

139. Weniger, E.J. (1996b), Erratum: Nonlinear sequence transformations: A computational tool for quantum mechanical and quantum chemical calculations, *Int. J. Quantum Chem.* **58**, 319–321.
    (Cited on pages 288 and 290.)

140. Weniger, E.J. (1996c), A convergent renormalized strong coupling perturbation expansion for the ground state energy of the quartic, sextic, and octic anharmonic oscillator, *Ann. Phys. (NY)* **246**, 133–165.
    (Cited on pages 288, 290, and 291.)

141. Weniger, E.J. (1996d), Computation of the Whittaker function of the second kind by summing its divergent asymptotic series with the help of nonlinear sequence transformations, *Comput. Phys.* **10**, 496–503.
    (Cited on pages 279 and 288.)

142. Weniger, E.J. (1996e), Construction of the strong coupling expansion for the ground state energy of the quartic, sextic and octic anharmonic oscillator via a renormalized strong coupling expansion, *Phys. Rev. Lett.* **77**, 2859–2862.
    (Cited on pages 283, 289, 290, and 291.)

143. Weniger, E.J. (1997), Performance of superconvergent perturbation theory, *Phys. Rev. A* **56**, 5165–5168.
(Cited on pages 289, 290, and 291.)

144. Weniger, E.J. (2000a), Addition theorems as three-dimensional Taylor expansions, *Int. J. Quantum Chem.* **76**, 280–285.
(Cited on page 275.)

145. Weniger, E.J. (2000b), Prediction properties of Aitken's iterated Δ^2 process, of Wynn's epsilon algorithm, and of Brezinski's iterated theta algorithm, *J. Comput. Appl. Math.* **122**, 329–356.
(Cited on pages 291 and 301.)

146. Weniger, E.J. (2000c), Prediction properties of Aitken's iterated Δ^2 process, of Wynn's epsilon algorithm, and of Brezinski's iterated theta algorithm, in C. Brezinski (editor), *Numerical Analysis 2000, Vol. 2: Interpolation and Extrapolation*, 329–356 (Elsevier, Amsterdam). Reprint of [145].
(Cited on page 291.)

147. Weniger, E.J. (2001a), Irregular input data in convergence acceleration and summation processes: General considerations and some special Gaussian hypergeometric series as model problems, *Comput. Phys. Commun.* **133**, 202–228.
(Cited on pages 279, 288, 289, and 290.)

148. Weniger, E.J. (2001b), Nonlinear sequence transformations: Computational tools for the acceleration of convergence and the summation of divergent series. Los Alamos preprint math.CA/0107080 (http://arXiv.org).
(Cited on page 283.)

149. Weniger, E.J. (2002), Addition theorems as three-dimensional Taylor expansions. II. *B* functions and other exponentially decaying functions, *Int. J. Quantum Chem.* **90**, 92–104.
(Cited on pages 275 and 283.)

150. Weniger, E.J. (2003), A rational approximant for the digamma function, *Numer. Algor.* **33**, 499–507.
(Cited on pages 279 and 285.)

151. Weniger, E.J. (2004), Mathematical properties of a new Levin-type sequence transformation introduced by Čížek, Zamastil, and Skála. I. Algebraic theory, *J. Math. Phys.* **45**, 1209–1246.
(Cited on pages 287 and 289.)

152. Weniger, E.J. (2005), The spherical tensor gradient operator, *Collect. Czech. Chem. Commun.* **70**, 1225–1271.
(Cited on page 275.)

153. Weniger, E.J. (2007), Asymptotic approximations to truncation errors of series representations for special functions, in A. Iske and J. Levesley (editors), *Algorithms for Approximation*, 331–348 (Springer-Verlag, Berlin).
(Cited on page 279.)

154. Weniger, E.J. (2008), On the analyticity of Laguerre series, *J. Phys. A* **41**, 425207-1–425207-43.
(Cited on pages 279, 282, and 288.)

155. Weniger, E.J. (2009), The strange history of *B* functions or how theoretical chemists and mathematicians do (not) interact, *Int. J. Quantum Chem.* **109**, 1706–1716.
(Cited on pages 275, 276, and 279.)

156. Weniger, E.J. (2010), Summation of divergent power series by means of factorial series, *Appl. Numer. Math.* **60**, 1429–1441.
(Cited on pages 279 and 289.)

157. Weniger, E.J. (2012), On the mathematical nature of Guseinov's rearranged one-range addition theorems for Slater-type functions, *J. Math. Chem.* **50**, 17–81.
(Cited on page 279.)

158. Weniger, E.J. (2019), Comment on "Fourier transform of hydrogen-type atomic orbitals", Can. J. Phys. vol. 96, 724–726 (2018) by N. Yükçü and S. A. Yükçü, *Can. J. Phys.* **97**, 1349–1360.
(Cited on page 279.)

159. Weniger, E.J. and Čížek, J. (1990), Rational approximations for the modified Bessel function of the second kind, *Comput. Phys. Commun.* **59**, 471–493.
     (Cited on pages 283 and 288.)
160. Weniger, E.J. and Kirtman, B. (2003), Extrapolation methods for improving the convergence of oligomer calculations to the infinite chain limit of *quasi*-onedimensional stereoregular polymers, *Comput. Math. Applic.* **45**, 189–215.
     (Cited on page 290.)
161. Weniger, E.J. and Liegener, C. (1990), Extrapolation of finite cluster and crystal-orbital calculations on trans-polyacetylene, *Int. J. Quantum Chem.* **38**, 55–74.
     (Cited on page 290.)
162. Weniger, E.J. and Steinborn, E.O. (1982), Programs for the coupling of spherical harmonics, *Comput. Phys. Commun.* **25**, 149–157.
     (Cited on pages 275 and 276.)
163. Weniger, E.J. and Steinborn, E.O. (1983a), The Fourier transforms of some exponential-type functions and their relevance to multicenter problems, *J. Chem. Phys.* **78**, 6121–6132.
     (Cited on pages 275 and 276.)
164. Weniger, E.J. and Steinborn, E.O. (1983b), Numerical properties of the convolution theorems of *B* functions, *Phys. Rev. A* **28**, 2026–2041.
     (Cited on pages 275, 276, and 279.)
165. Weniger, E.J. and Steinborn, E.O. (1983c), New representations for the spherical tensor gradient and the spherical delta function, *J. Math. Phys.* **24**, 2553–2563.
     (Cited on page 275.)
166. Weniger, E.J. and Steinborn, E.O. (1984), Comment on "Hydrogenlike orbitals referred to another coordinate", *Phys. Rev. A* **29**, 2268–2271.
     (Cited on page 275.)
167. Weniger, E.J. and Steinborn, E.O. (1985), A simple derivation of the addition theorems of the irregular solid harmonics, the Helmholtz harmonics, and the modified Helmholtz harmonics, *J. Math. Phys.* **26**, 664–670.
     (Cited on page 275.)
168. Weniger, E.J. and Steinborn, E.O. (1987), Comment on "Molecular overlap integrals with exponential-type integrals", *J. Chem. Phys.* **87**, 3709–3711.
     (Cited on page 275.)
169. Weniger, E.J. and Steinborn, E.O. (1988), Overlap integrals of *B* functions. A numerical study of infinite series representations and integral representations, *Theor. Chim. Acta* **73**, 323–336.
     (Cited on pages 275 and 279.)
170. Weniger, E.J. and Steinborn, E.O. (1989a), Nonlinear sequence transformations for the efficient evaluation of auxiliary functions for GTO molecular integrals, in M. Defranceschi and J. Delhalle (editors), *Numerical Determination of the Electronic Structure of Atoms, Diatomic and Polyatomic Molecules*, NATO ASI Series, 341–346 (Kluwer, Dordrecht). Proceedings of the NATO Advanced Research Workshop, Versailles, France, 17–22 April 1988.
     (Cited on pages 275, 279, 284, and 288.)
171. Weniger, E.J. and Steinborn, E.O. (1989b), Addition theorems for *B* functions and other exponentially declining functions, *J. Math. Phys.* **30**, 774–784.
     (Cited on page 275.)
172. Weniger, E.J., Grotendorst, J., and Steinborn, E.O. (1986a), Some applications of nonlinear convergence accelerators, *Int. J. Quantum Chem. Symp.* **19**, 181–191.
     (Cited on pages 275 and 284.)
173. Weniger, E.J., Grotendorst, J., and Steinborn, E.O. (1986b), Unified analytical treatment of overlap, two-center nuclear attraction, and Coulomb integrals of *B* functions via the Fourier transform method, *Phys. Rev. A* **33**, 3688–3705.
     (Cited on pages 275 and 284.)
174. Weniger, E.J., Čížek, J., and Vinette, F. (1991), Very accurate summation for the infinite coupling limit of the perturbation series expansions of anharmonic oscillators, *Phys. Lett. A* **156**, 169–174.
     (Cited on pages 283, 289, and 290.)

175. Weniger, E.J., Čížek, J., and Vinette, F. (1993), The summation of the ordinary and renormalized perturbation series for the ground state energy of the quartic, sextic, and octic anharmonic oscillators using nonlinear sequence transformations, *J. Math. Phys.* **34**, 571–609.
(Cited on pages 283, 289, and 290.)
176. Wimp, J. (1981), *Sequence Transformations and Their Applications* (Academic Press, New York).
(Cited on page 283.)
177. Zayed, A.I. (1996), *Handbook of Function and Generalized Function Transformations* (CRC Press, Boca Raton).
(Cited on page 282.)

7.12 Jean Zinn-Justin

Jean Zinn-Justin (b. 1943) studied at École Polytechnique (1962–1964). Then he completed a doctorate in physics at Saclay under the supervision of Marcel Froissart. He spent his entire career there, becoming the head of the Theoretical Physics Laboratory. His works concern quantum field theory and the renormalization group. In particular, in collaboration with B.W. Lee, he gave the first demonstration of the renormalizability of nonabelian gauge theories in the broken symmetry phase, an important step in the construction of the Standard Models of weak and electromagnetic interactions. He was also interested in the properties of matrices and the summation of divergent series. He was the director of the DAPNIA Laboratory at CERN from 2003 to 2008. He has received many distinctions, and he is a member of the French Academy of Sciences.

Testimony

Theoretical physics has often made use of Padé approximants as a mathematical tool to improve the convergence of series, and in this sense, this is not peculiar. Still, a few modern examples are worth noting, which involve quantum field theory, the theory that describes continuous phase transitions in statistical physics and particle physics.

After the successful predictions of perturbative quantum electrodynamics, due to the smallness of the expansion parameter, the fine structure constant (1/137), particle physicists tried to use perturbation theory in the domain of so-called strong interactions (also called hadronic physics). However, to their disappointment, they discovered that the expansion parameters were too large for the perturbative expansion to be useful. Bessis and Pusterla then proposed to use Padé approximants to sum the series in the simple example of self-interacting pions. The method was generalized by Badevant, Bessis, and Zinn-Justin. The results obtained in this way made much better physical sense, especially from the point of view of unitarity.

However, the limit of the methods was that in the 1960s it was difficult to calculate more than two or three terms of perturbation theory, and convergence could not really be checked. Moreover, later it was realized that the results made sense only in a very limited domain of low energies.

In statistical physics, the Padé–Borel summation method was applied to the calculation of critical exponents (Baker, Green, Nickel), yielding interesting estimates, although later the Borel transformation with conformal mapping, based on an improved knowledge of the properties of perturbation theory, led to more precise estimates (Le Guillou, Zinn-Justin).

References

1. D. Bessis, M. Pusterla, Unitary Padé approximants in strong coupling field theory and application to the calculation of the ρ- and f_0-meson Regge trajectories, Il Nuovo Cimento A, 54 (1968) 243–294.
2. J.L. Basdevant, D. Bessis, J. Zinn-Justin, Padé approximants in strong interactions. Two-body pion and kaon systems, Il Nuovo Cimento, 60A (1969) 185–238.
3. J.L. Basdevant, J. Zinn-Justin, Yang-Mills fields and the $\pi\pi$ interaction, Physical Review D, 3 (1971) 1865–1873.
4. J. Zinn-Justin, Strong interactions dynamics with Padé approximants, Physics Reports, 1 (1971) 55–102.
5. G.A. Baker, Jr., B.G. Nickel, M.S. Green, D.I. Meiron, Ising-model critical indices in three dimensions from the Callan-Symanzik equation, Phys. Rev. Lett., 36 (1976) 1351–1354.

7.13 The Authors

The purpose of our testimonies is not to describe our own work but to point out that a large part of the research we did, alone, together, or in collaboration with others, are, in fact, more or less, issued from the circle of ideas originated by Padé, Richardson, Aitken, Shanks, and Wynn, and often directly influenced by them.

After the personal testimony of each of us, we explain our collaboration together, which runs over 30 years. References are to the General Bibliography.

Claude Brezinski

I carried out my military service in a scientific laboratory of the French Army in 1968–1969.[41] The building next to mine was another division of the Army, the *Centre de Calcul Scientifique de l'Armement*, which, among other duties, had to solve numerical analysis problems for the civil and military engineers working for the Defense Ministery. When freed from my military obligations in June 1969, I went to Dr. Francis Ceschino, the head of the numerical analysis group of this service, to ask him whether it was possible to be employed there and simultaneously to work

---

[41] A part of this testimony is reprinted, with permission, from [115].

for a doctoral thesis, the French *Thèse d'État ès Sciences Mathématiques*, on a topic agreeable to him and a university professor. I was hired, and Ceschino proposed that I work on convergence acceleration methods.

At that time in France, there were essentially two centers for doing research in numerical analysis: the University of Paris, where Jacques-Louis Lions (1928–2001) was working on PDEs, and the University of Grenoble with Noël Gastinel, who was interested in other topics. Gastinel agreed to supervise my thesis on convergence acceleration methods.

My first task was to read the papers on the Shanks transformation and on the ε-algorithm of Wynn. Each paper of Wynn cited another paper of his, I read the new paper and found another citation of him. It seemed to be an endless procedure. Thus, I decided to write a letter to Wynn (at that time the internet did not exist), asking him to send me a complete list of his publications. He immediately answered and provided me with many reprints of his papers. This was the beginning of many epistolary exchanges (from September 1969 to January 1992). The formal tone of our first letters changed to more personal messages after some time. Peter gave me much advice, he helped me in the proofs of several of my results, and discussed with me his own research. Once he wrote to me that the only mathematicians who do not make errors are those who never publish. Without his help, my career would certainly have been different. I defended my doctoral thesis on 26 April 1971 with Jean Kuntzmann as the president of the committee, Francis Ceschino, Pierre-Jean Laurent, and Noël Gastinel as its members. In 1963, Kuntzmann and Ceschino together wrote a book entitled *Problèmes Différentiels de Conditions Initiales* (translated into English in 1966). They had not met since, and I was thinking of a quite friendly reunion. One of them, I don't remember who, said to the other *"Bonjour Monsieur"* who answered *"Bonjour Monsieur"*, and they shook hands. End of the conversation! I hope they spoke more together later in the day!

In 1972, Wynn managed to have me invited to participate in the *International Conference on Padé Approximation and Related Matters*, in Boulder, USA, 19–22 June 1972, organized by William Branham Jones and Wolfgang Joseph Thron (for his biography, see [358]), two specialists in Padé approximation and continued fractions. This was the first international conference I attended. There I met George Baker Jr., Michael Barnsley, Daniel Bessis, Roy Chisholm, Albert Edrei, David Filed, John Gammel, Gene Golub, Bill Gragg, Yudell Luke, John Nuttall, Joe Traub, Joseph Walsh, Luc Wuytack, Jean Zinn-Justin, and many others with whom I stayed in touch over the years. From this partial list of speakers, one can see that many topics were covered: numerical analysis, complex interpolation and approximation, continued fractions, moments, quadrature methods, theoretical physics, number theory, solid state physics, critical phenomena, perturbation theory, differential equations, etc. The proceedings were published in *The Rocky Mountain Journal of Mathematics*, Spring 1974, volume 4, number 2, thus spreading these topics among the community of pure and applied mathematicians, various groups of physicists, and engineers. It was the first international congress on these themes, and it was followed by many others.

At this congress I also met Peter for the first time. After the congress, he invited me to give a seminar at the *Centre de Recherches Mathématiques* of the *Université de Montréal*, where he was working at that time. I stayed there a few days discussing mathematics with him, and having meals in good restaurants. Our regular contact continued for many years. Several of my works are a continuation of his papers and vice versa. He discussed with me the results obtained by some of my students and colleagues. He often gave me lengthy explanations. He also took steps so that I could be invited for a long stay in Canada or the USA (but I could not take advantage of those invitations) and to get funds for my visits. I translated for him three papers he wanted to publish in French in the *Comptes Rendus de l'Académie des Sciences de Paris*. Once, he completely rewrote a paper of mine before its submission to a journal. We met at various occasions in Canada (for the ICM 1974 in Vancouver), USA, and France. In particular, he attended the *Colloque d'Analyse Numérique* at La Grande-Motte, France, 26 May–1 June 1975 (where he gave an invited talk entitled *A numerical method for estimating parameters in mathematical models*, and where I taught him how to play *pétanque*), and a congress on Padé approximants in Toulon and Marseille. The last time we met was at the congress *Rational Approximation with Emphasis on Padé Approximants*, Tampa, USA, 15–17 December 1976, organized by Ed Saff. But we continued to correspond regularly until 1992.

In 1981, Peter decided to move to Mexico. He wanted to get away, as he wrote me, *from the many futilities and stupidities of modern academic life for a year or so*. But, because he was able to have a better life there, he decided to stay in that country. He first lived in Guanajuato, where he was associated to the *Centro de Investigación en Matematicás*. The cultural life of the city was very rich, and he often went to the theater and to concerts. He wrote me that

> [. . .] one can certainly get a great deal of work done here. CIMAT [*Centro de Investigación en Matemáticas*] is a small mathematics research center recently set up by the Mexican government. It has a library but as yet no computer, so calculations have to be done in Mexico City.

Although, as he wrote me, he was working hard at mathematics and had settled many matters that previously had troubled him, he did not publish anything any longer. He was also planning to write one or two books and to come to Europe, *giving circus performances*, as he wrote me. But those projects were not realized. However, he was still writing reviews for *Zentralblatt für Mathematik und ihre Grenzgebiete*. Twice a year he would travel to the USA for official reasons, mainly to the University of Texas at San Antonio and to the University of Texas in Austin. During those trips, he left mathematical documents and unfinished manuscripts in the house of his friends Maria and Manuel Berriozábal. We are in the process of analyzing them with the help of Sandy Norman of the University of Texas at San Antonio. Wynn moved to Zacatecas in 1990. His last paper was published in 1981, and his last letter to me is dated 1992. It had been a great pleasure for me to be a close friend of him for many years, to discuss mathematics and other things with him, and to share nice moments together.

Peter Wynn was a *bon vivant*, enjoying good restaurants, but also a man of culture, *un honnête homme* in the sense of the seventeenth century with, in addition, a very British sense of humor. He never asked for a permanent position, and worked only on contracts with various universities or various governmental administrations. He was not interested in making what is commonly called as a *career*. He wanted to be free, free from a permanent position, free from the administrative obligations of a university professor, free from its too numerous meetings, free from students, free from lectures and organizing and grading exams, free to select his own research topics, free to choose where he wanted to live, and he was never married. He was a free man in all meanings of the word.

Without Peter Wynn, the fields of acceleration (or extrapolation) methods, Padé approximation and continued fractions would certainly not have been so highly developed, as well as their applications to fixed-point methods and other topics in numerical analysis and applied mathematics.

At the beginning of my career, I worked for some years on the ε-algorithm and other extrapolation algorithms. In 1975, I obtained the topological ε-algorithm. Then I worked on Padé approximation, and in 1979, I introduced Padé-type approximants. The E-algorithm, which is the most general recursive algorithm for extrapolation known so far, was designed 1 year later. Due to their connection with Padé approximants, I obtained some results on formal orthogonal polynomials, including a new proof and a reciprocal of the Christoffel—Darboux identity. I also gave some new results on limit periodic continued fractions.

Hassane Sadok and I met Michela Redivo-Zaglia in 1989 in Tenerife at a *NATO Advanced Study Institute* organized by Mariano Gasca, of the University of Zaragoza, Spain. I have been collaborating with Michela Redivo-Zaglia since that time. She is simultaneously a numerical analyst and an accomplished computer scientist.

In my career, I have the honor and the pleasure to know and converse with so many students and researchers that it is impossible to list them here. Many of them have provided me new ideas and stimulated me to continue to work on old topics and to discover new ones. This is the proof of the importance, for the development of research, of maintaining scientific collaborations and not to stay cloistered in one's own office, as some people do.

Michela Redivo-Zaglia

I obtained my master in mathematics at the University of Padua in 1975. During the last year of my studies, I had several courses on the use of computers in mathematics, in particular algorithmic development and programming in Fortran and other languages. I was passionate about these subjects and I was recruited by the Computing Center of the university, the only place having a *mainframe* computer at the university. I deepened my knowledge of languages, gave lectures on programming languages, and wrote programs for researchers. I then obtained a position as a computer scientist.

In 1984, I went to the Department of Electronic and Informatics, and I became head of the computer center. I has the responsibility of installing the first computer laboratories for students. At that time, very few people had an experience with operating systems, and I wrote a book on Unix.

But my work was becoming too much routine, and I wanted to go back to mathematics. Thus, in 1987, I contacted Professor Maria Morandi Cecchi, of the Math Department of my university, and we began to collaborate. Together, in 1989, we attended the *NATO Advanced Study Institute on Computation of Curves and Surfaces* in Tenerife, organized by Mariano Gasca, of the University of Zaragoza. There I met Claude Brezinski and Hassane Sadok with whom I discussed. They were much interested in my competency in computer science, I liked the domain they were working on, and we decided to begin a collaboration. They invited me to the congress *Extrapolation and Rational Approximation* that they were organizing in Luminy, near Marseille.

I was very excited by the talks I heard. They were on topics not well known in the numerical analysis community in Italy, and also abroad, and I was convinced of the potential of these methods for applications. This was the start of my new scientific life, and I realized that I wanted to work on these fields of research. It was at the end of this conference that Claude Brezinski suggested that I participate in his project for a book on extrapolation methods, containing also optimized and easy-to-use software. I accepted immediately, since it was a unique opportunity for deepening this subject, and also to bring my contribution, thanks to my knowledge and skills in regard to computers and programming languages. In this project I was, of course, pushed to read and study carefully the related papers of Wynn and others. It was a real discovery to see the amazing ideas Wynn had for optimizing algorithms and checking their numerical instability, and from then on, I always implemented his procedures, and similar ones, in all algorithms and computer programs, for instance when I recently wrote with Claude the public domain MATLAB toolbox EPSFUN [133]. It is important to point out that in the past, computational costs, in terms of memory and execution time, were a serious and critical problem. With time, the technology reduced this drawback considerably. But nowadays, with big data treatment, large-dimensional vectors and matrices, and also tensors, this criticality returned, making it again very important to optimize and reduce the algorithms we use. And for extrapolation methods, the ideas of Wynn are still current.

I worked also in other fields of numerical analysis, in particular numerical linear algebra, formal orthogonal polynomials, and special functions and always, among the works of other researchers, I investigated, often first, the papers of Peter Wynn. Moreover, I suggested their reading to all my master and Ph.D. students at the University of Padua, and all of them found these works very interesting, useful, and current, even the very old ones published in the fifties and sixties.

I'm very happy to have the possibility to continue to contribute to the popularization and appreciation of extrapolation methods, my preferred domain of research, through the publication of papers and software in this field, and to give related research subjects to my students and postdocs. And I have to thank my mentor Claude Brezinski, whom I also had the privilege of having as supervisor of my doctoral thesis, but also several other colleagues, collaborators, and researchers, in particular Peter Wynn (with regret for having known him only through his works), whose scientific results have been very important for my training.

Both Authors

Although C.B. programmed a lot in Fortran years before meeting M.R.-Z., he was not up-to-date with its new developments, such as indented programs and structured programming. In 1989, he had the idea of writing a book on extrapolation methods containing subroutines corresponding to the extrapolation algorithms he planned to describe for making their use easier, and to disseminate them among scientists. He asked Michela if she was willing to collaborate with him on this project. It ended with the publication, in 1991, of our book *Extrapolation Methods. Theory and Practice*, North-Holland. The Fortran routines provided with the book consist of over 300 pages of statements and are freely downloadable [127]. We also worked together on various aspects of extrapolation methods, Padé approximation, formal orthogonal polynomials, and their applications.

In 1991, we (C.B. and M.R.-Z.) and Hassane Sadok attended together the *Copper Mountain Conference on Iterative Methods*. Many speakers were mentioning that they were unable to cure the problem of *breakdown* in the Lanczos's method for solving systems of linear equations. When such a breakdown occurs, the iterations had to be restarted from a new initialization. We immediately realized that we were able to propose a solution to this problem. Indeed, the Lanczos's method is related to formal orthogonal polynomials. Lanczos himself took this approach in his seminal paper on his method, but since the problem was in the field of linear algebra, this connection had been forgotten by researchers working on it. As explained in this book, André Draux was able to jump over square blocks of identical approximants in the Padé table using extended recurrences for formal orthogonal polynomials. The problem of breakdown was a similar one, and we adapted the work of Draux to it. Using the theory of formal orthogonal polynomials, we were able to find a way to treat the problem of breakdown. Our main algorithm had recursive forward jumps, and we named it the *Method of Recursive Zoom*, or MRZ for short, a joke by the two French authors, and the paper was published in the journal *Numerische Mathematik* with an explanation of it! Let us mention that we recently learned that following a suggestion by Gene Golub, Martin Gutknecht was already looking at this problem at that time but from a different point of view. We applied the same treatment, using formal orthogonal polynomials, to the CGS of Peter Sonneveld and the BiCGStab of Henk van der Vorst, and we also solved the problem called *near-breakdown* in these algorithms, which arises from division by a number close to zero and leads to an important propagation of rounding errors. The approach to the Lanczos's method from formal orthogonal polynomials is so simple that C.B. taught it to his master's students in Lille. We also worked together on transpose-free methods for nonsymmetric linear systems.

We used extrapolation methods for improving the solution of several problems: Tikhonov regularization, the PageRank problem, Fredholm integral equations, the Kaczmarz method, etc. We also always tried to publish the software corresponding to our algorithms and to put it in the public domain so that the interested reader could use our programs (some of them being quite tricky to code).

Our collaboration was always on an equality basis. Each of us brought a significant and indistinguishable part to all our papers. The discovery and the development of the results presented in our joint papers would not have been accomplished without the complete and decisive participation of each of us.

But let us stop here. We have already said too much about us.

Chapter 8
Vitæ

In this chapter, we give an account of the life of the main contributors encountered during our exploration of the country of rational extrapolation and approximation. We followed alphabetical order.

8.1 Alexander Craig Aitken

Alexander (Alec) Craig Aitken was born in Dunedin, New Zealand, on 1 April 1895.[1] He attended Otago's High School from 1908 to 1912, where he was not particularly brilliant. But at the age of 15, he realized that he had a real power in mental calculations,[2] and that his memory was extraordinary. He was able to recite the first 1000 decimals of π, and to multiply two numbers of nine digits in a few seconds. He also knew Virgil's Aeneid by heart [603, pp. 266–275]. He was also very good at several sports and began to play violin. He studied mathematics, French, and Latin at the University of Otago in 1913 and 1914. It seems that the professor of mathematics there, David J. Richards, a "temperamental, eccentric Welshman," was lacking in the power to communicate his knowledge to the students, and Aitken's interest in mathematics waned. Richards was trained as an engineer as well as a mathematician, and he was working as an engineer in Newcastle prior to his appointment to the chair of mathematics at Otago in 1907, where he stayed until 1917.

Aitken volunteered in the Otago infantry during World War I, and he took part in the Gallipoli landing and in the campaign in Egypt as a lieutenant.[3] Then he was commissioned in the north of France, and was wounded in the shoulder and foot during the first battle of the Somme on 27 September 1916. Did he meet Lewis Fry Richardson, who was serving in France in the Friend's Ambulance Unit (a Quaker organization) at this time? After a stay in a London hospital, he was invalided home

---

[1] This biography is reprinted, with permission, from [135].

[2] Look at https://www.youtube.com/watch?v=9Es7wHodd9M.

[3] https://www.aucklandmuseum.com/war-memorial/online-cenotaph/record/C34273.

© Springer Nature Switzerland AG 2020

C. Brezinski, M. Redivo-Zaglia, *Extrapolation and Rational Approximation*,
https://doi.org/10.1007/978-3-030-58418-4_8

Alexander Craig Aitken in 1915.
Courtesy of Auckland War Memorial Museum.

in 1917 and spent 1 year of recovering in Dunedin, where he wrote an account of his experiences published later [8]. He managed to keep his violin with him throughout the war.

Aitken resumed his studies at Otago University and graduated with first class honours in languages, but only with second ones in mathematics. He married Winifred Betts in 1920 and became a school teacher at his old Otago High School. Richards's successor in the chair of mathematics was Robert John Tainsh Bell (1876–1963). Bell graduated from the University of Glasgow in 1898 and was appointed lecturer there 3 years later. He was awarded a D.Sc. in 1911, and was appointed professor of pure and applied mathematics at Otago University in 1919. He was the only staff member in the Mathematics Department, lecturing 5 days a week, each day from 8.00 am to 1.00 pm. He retired in 1948 and died in 1963. When Bell required an assistant, he called on Aitken. He encouraged him to apply for a scholarship to study with Edmund Taylor Whittaker (1873–1956) in Edinburgh. Aitken left New Zealand in 1923. His Ph.D. on the smoothing of data, completed in 1925, was considered so outstanding that he was awarded a D.Sc. for it instead of a Ph.D. The same year, Aitken was appointed as a lecturer at the University of Edinburgh, where he stayed for the rest of his life. But the effort expended for obtaining his degree led him to a first severe nervous breakdown in 1927, and then he was periodically affected by such crises. They were certainly in part due to his fantastic memory, which did not fade with time, and he was always remembering the horrors he saw during the war. In a 1935 paper, he introduced the concept of generalized least squares, along with the now standard vector/matrix notation for linear regression. In 1936, Aitken became a reader in statistics, and he was elected a Fellow of the Royal Society. During World War II he worked in Hut 6 of Bletchley Park, decrypting the ENIGMA code. In 1946, he was appointed to Whittaker's chair in mathematics. In Aitken's obituary, Walter Ledermann [403] wrote:

> He was a most inspiring teacher; his lecturing was superb, and his personality made an indelible impression on his students. If in later life they might not remember the entire mathematical content of his lectures, they never forgot their erudition, humanity and wit.

In 1956, Aitken received the prestigious Gunning Victoria Jubilee Prize of the Royal Society of Edinburgh. In 1964, he was elected to the Royal Society of Literature. He was fond of taking long walks, and he wrote poetry throughout his life. Aitken retired from the chair of mathematics at the University of Edinburgh on 30 September 1965. At a meeting of the senate of the university on 19 January 1966, the following special minute was adopted as a tribute to him:

> His lecturing was brilliant and inspired generations of students, both undergraduate and post-graduate [...]

> Edinburgh has always had a great attraction for him and he has resisted the many offers tempting him to go elsewhere. It has most of what he wanted, a congenial job, hills to walk on, and above all, music, concerts, and musical friends. He is a fine (largely self-taught) violinist and viola-player, and a very knowledgeable musician [...]

> Indeed, he is reliably reported as having said that he spent three-quarters of his time thinking about music. This remark reveals how efficiently he must have used the other quarter! He is moreover a creative artist, though only a few intimate friends have been privileged to know his compositions, occasional poems, and the eloquent calligraphy of his manuscript copies of Bach.

Alexander Craig Aitken died in Edinburgh on 3 November 1967.

Aitken wrote a kind of diary of his life from 1923 to 1943, and added some more notes in 1958. They were found by his daughter and published under the title of *To Catch the Spirit* [9] by Peter C. Fenton, of the University of Otago with a biographical introduction covering the years 1895 to 1923. As written in the preface:

> The title of the book derives from a letter taken up with the proof of a certain theorem, which Aitken signed: "Q.E.D. [*quod erat demonstrandum*] and A.C.A." [*ad captandam animam=to catch the spirit of the thing*].

For details about his work and life, see [694].

8.2 George Allen Baker, Jr.

George Allen Baker Jr.
© The Los Alamos Monitor.

George Allen Baker, Jr., was born on 25 November 1932 in Alton, Illinois. His parents were George Allen and Grace Elizabeth (Cummins) Baker. In 1937, his family moved to Davis, California, where he graduated from Davis High School in 1950. He then attended the California Institute of Technology. As a junior he came in fourth place for the national "Putnam" competition. He graduated in 1954. He went to UC Berkeley for his Ph.D. in physics, which he completed in 1956. He did a year of postdoctoral study at Columbia University in New York City. In the summer of 1957, he went to work at the Los Alamos Scientific Laboratory, and in 1961–1962, he worked at the UC San Diego Scripps Institute. In 1964–1965, he took a sabbatical leave at Kings College in London, England. After his return to Los Alamos, he and his family moved to Bellport, NY, to work at the Brookhaven National Laboratory. In 1971–1972, he took a sabbatical at Cornell University. In 1975, he moved back to work at the Los Alamos Scientific Laboratory. In 1976–1977, he visited the *Centre d'Études Nucléaires* in Saclay, near Paris, France, then returned to Los Alamos. In 1982–1983, he took another sabbatical, half time in France and half time at Princeton University, where he was made a laboratory fellow, a promotion for distinguished research. During his career, he published more than 100 technical papers and three books. He was an expert in the area of Padé approximants and its application to Ising models. In 1970, his first monograph, with John Ledel Gammel (1924–2011), was devoted to theoretical and practical aspects of computing methods for mathematical modeling of nonlinear systems [29]. His book *Essentials of Padé Approximants* [26], published in 1975, played a fundamental role in the dissemination of the topic. It was later followed by his masterly book, written with Peter Russell Graves-Morris [31]. He retired in 1995 from Los Alamos Laboratory.

In his spare time, George loved traveling and music. He visited all the continents except Antarctica. Music was a constant in his life. In high school and college, he played the trombone. Throughout his life he sang in the church choir. He frequently attended the opera and other musical concerts. In his final months, music was his great solace. He was a very sweet man. Baker died at home on 24 July 2018.[4]

8.3 Friedrich Ludwig Bauer

Friedrich Ludwig (Fritz) Bauer was born on 10 June 1924 in the Bavarian city of Regensburg, Germany. He earned his *Abitur* in 1942. From 1943 to 1945, he served as a soldier in the Wehrmacht during World War II. From 1946 to 1950, he studied mathematics, theoretical physics, astronomy, and logic at the Ludwig-Maximilian University in Munich.

After 2 years as a teacher, he returned to the university as a teaching assistant to Professor Friedrich Bopp (1909–1987), a German theoretical physicist who contributed to nuclear physics and quantum field theory. He received his Ph.D. in 1952 on *Group-theoretic investigations of the theory of spin wave equations*. Then he went to the *Technische Hochschule* in Munich, where he served as a teaching assistant to Professor Robert Sauer (1898–1970). He obtained his habilitation in 1954 with a work entitled *On quadratically convergent iteration methods for solving algebraic equations and eigenvalue problems*, and he became *Privatdozent*.

In 1958, he moved to the University of Mainz as an associate professor. In 1962, he returned to the *Technische Hochschule* of Munich as a full professor, a position he held until his retirement in 1989.

Although he was originally a physicist, Bauer soon became fascinated by the use of computers in science, and then by computers in general. He covered almost all the aspects of this area: algorithms, arithmetic operations, propagation of rounding errors, computer architecture, ... He was one of the developers of the programming language Algol and, with Klaus Samelson (recall that Wynn used his inverse of a vector for his vector ε-algorithm), he invented the *stack* in 1955, the conceptually simplest way of saving information in a temporary storage location. Stacks are the way to handle recursive function calls and the dynamic runtime behavior of computer programs. Every modern microprocessor incorporates them. In 1968, Bauer also coined the term *software engineering*.

In 1952, Bauer met Rutishauser, who introduced him to Eduard Stiefel. In *My years with Rutishauser*,[5] he wrote:

---

[4] We are grateful to Doron Lubinsky, who found his obituary published in the Los Alamos Monitor on 8 August 2018. We slightly modified and extended it.

[5] Available at https://www.cs.umd.edu/users/oleary/cggg/bauer.pdf.

Friedrich Ludwig Bauer.
Courtesy of Manfred Broy.

In the first half of the 20th century, in Pure Mathematics at a few occasions, say with the 10th Problem of Hilbert or the word problem of Dehn and other decision problems, a resort was made to the century-old linguistic usage of 'algorithm' in the sense of 'a general, a priori determined procedure giving a solution in a finite number of steps'. While the word was barely used in Applied Mathematics, since it seemed to be connected with trivialities like 'Euclidean algorithm' or 'algorithm of multiplication', a sharpening of the concept had been made in the course of Turing's ideas on computability. Therefore, it was not astonishing that in the vicinity of Stiefel the word 'algorithm' shows up for numerical methods that terminate steadfastly. In fact, Stiefel and Hestenes wrote 1952 in the summary of their joint paper: 'An iterative algorithm is given for solving a system of linear equations'. Consequently, Rutishauser in 1954 called his method for the determination of eigenvalues 'Quotienten-Differenzen-Algorithmus'. Peter Wynn, who later was my assistant at Mainz, introduced in 1956 the 'epsilon-Algorithmus', and in 1958 I introduced the 'g-Algorithmus' and the 'η-Algorithmus', both being continued fraction algorithms, the links between the previously mentioned ones. 'Algorithm' pretty soon was a catch-phrase of numerical and nonnumerical mathematics; ALGOL stood for algorithmic language.

In numerical analysis, Bauer's contributions included matrix algorithms for linear systems and eigenvalue problems, continued fractions, nonlinear sequence transformations, and rounding error analysis. With Alston Scott Householder (1904–1993), he was a pioneer in the use of norms in matrix analysis, thus opening the way to Richard S. Varga in the United States, and Noël Gastinel in France. Bauer also introduced the concept of the field of values subordinate to a norm—a notion that also turned out to be useful in functional analysis. The Bauer–Fike theorem is a standard result in the perturbation theory of the eigenvalue of a complex-valued diagonalizable matrix.

After 1975, Bauer devoted most of his efforts to computer science. He was also interested in its history and, more generally, in the history of mathematics. He received many distinctions. He was the doctoral advisor of 39 students, including Peter Wynn in 1959. He died on 26 March 2015.

Bauer's curriculum vitae can be found in the preface of volume 417 (2006), pages 299–300, of the journal *Linear Algebra and Its Applications*, and on several web sites.

8.4 Thomas Nall Eden Greville

Thomas Nall Eden Greville.
Courtesy of Adi Ben-Israel.

Thomas Nall Eden Greville was born on 27 December 1910 in New York. After a B.A. degree from the University of the South (Sewanee, Tennessee) in 1930, and an M.A. degree in 1932, he received a Ph.D. from the University of Michigan in 1933 with a thesis entitled *Invariance of the Property of Admissibility Under Certain General Types of Transformations*. In the depression years, such a doctorate was more of a liability than an asset. He worked with life insurance companies until he obtained in 1940 a position with the U.S. Bureau of the Census in connection with life tables. He then spent 21 years with the federal government in various departments.

In 1962, he accepted a visiting position with the Mathematics Research Center at the University of Wisconsin-Madison (see [165]), and from 1963 to 1985, he worked there as a mathematics professor. On leave from MRC, he served as an advisor to the National Center for Research Statistics from 1973 to 1976. He retired from the center and the university in 1981, and died in Charlottesville, VA, on 18 February 1998.

Greville had a quiet, pleasant, and unassuming personality, and a wide breadth of interests. He published more than 80 papers. He mostly was a statistician, but among numerical analysts, he is known for his book on generalized inverses coauthored with Adi Ben-Israel [50]. There are also knots for B-splines named after him.

8.5 John Bryce McLeod

John Bryce McLeod was born on 23 December 1929 in Aberdeen, Scotland. After his primary and grammar schools, during the war, he went to the University of Aberdeen and was awarded a scholarship to study at Oxford University in 1950. After 1 year in Canada and his military service, he went back to Oxford, and in 1958, he obtained a doctorate for his dissertation *Some Problems in the Theory of Eigenfunction Expansions* advised by Edward Titchmarsh (1899–1963). McLeod was appointed as a lecturer in mathematics at the University of Edinburgh. He spent 2 years there before returning to Wadham College, Oxford, in 1960 as a fellow. He held the fellowship at Wadham until 1991 and a university lectureship until 1988. He spent several summers and sabbaticals at the University of Wisconsin and the University of Pittsburgh, where he accepted a full professorship in 1987.

Sam Howison, a professor of applied mathematics at Oxford, wrote:

Bryce considered himself a problem-solving mathematician rather than a builder of general theories. He liked to focus on a specific hard problem and to find something new to say about it that was at the same time rigorous, interesting and useful.

At the end of an interview[6], McLeod was asked what advice he would have for mathematicians at the beginning of their career, he said: *Above all, have fun*. He clearly had fun with mathematics, and a positive attitude about the problems he studied, eager to see his way to the crucial steps, and optimistic that he would find them.

McLeod retired from Pittsburgh in 2007 and settled in Abingdon near Oxford, where he died on 20 August 2014.

---

[6] https://www.maths.ox.ac.uk/node/891.

John Bryce McLeod.
Courtesy of Dyrol Lumbard.

8.6 Henri Eugène Padé

Henri Eugène Padé was born in Abbeville, a town in the Picardy region of northern France, on 17 December 1863.[7] He obtained his *Baccalauréat* in 1881, and after two preparation years at the *Lycée Saint Louis* in Paris, he was admitted at the *École normale supérieure* in 1883. Three years later, he obtained the *Agrégation de mathématiques* and began to teach in several secondary schools in France. In 1889–1890, he went to Göttingen, where he translated Klein's Erlangen program into French. In 1892, he defended his *Thèse de Doctorat d'État ès Sciences Mathématiques* under the supervision of Charles Hermite. Its title was *Sur la représentation approchée d'une fonction par des fractions rationnelles*. After teaching for 1 year in a secondary school in Lille, he became *Maître de conférences* (associate professor) at the University of Lille. Then he was nominated as full professor of rational and applied mechanics at the University of Poitiers in 1902, and moved to the University of Bordeaux 1 year later.

---

[7] This biography is reprinted, with permission, from [135].

Henri Eugène Padé in Abbeville.
Courtesy of the Padé family.

The French Academy of Sciences used to propose, from time to time, a problem to be solved. In 1906, the problem was to improve an important point in the study of convergence of algebraic continued fractions. Padé won the *Grand Prix des Sciences Mathématiques*. The second recipient was Robert de Montessus de Ballore, who is well known for a convergence theorem about Padé approximants that bears his name; see Sect. 3.4.

In 1906, Padé, due to this prestigious prize, became dean of the Faculty of Science of the University of Bordeaux. Then the government named him rector of the Academy of Besançon in 1908. At this time, it was a very high position, since there were only 13 academies in France. A rector had to take care of all schools in his academy, from the kindergarten to the university. In 1917, he went to the Academy of Dijon as its rector, and finally, he was rector of the Academy of Aix-Marseille from 1923 to 1934.

Padé died in Aix-en-Provence on 9 July 1953. He is buried there. For his works, a more complete biography, and on his personality, see [483].

8.7 Lewis Fry Richardson

Lewis Fry (the maiden name of his mother) Richardson was born on 11 October 1881 in Newcastle upon Tyne, England. He early showed an independent mind and an empirical approach. In 1898, he entered the Durham College of Science, where he took courses in mathematics, physics, chemistry, botany, and zoology. Then, in 1900, he went to King's College in Cambridge and graduated with a first-class degree in 1903. He spent the next 10 years holding a series of positions in various academic and industrial laboratories. When serving as a chemist with the National Peat Industry Ltd., he had to study the percolation of water. The process was described by the Laplace equation on an irregular domain, and Richardson used finite differences [520]. It was in this paper that he introduced the idea of extrapolation for canceling the first term, in h^2, in the asymptotic expansion of the error. It was in the same paper that he presented his iterative method for solving systems of linear equations together with a strategy for choosing the parameters involved in it. But it was only after much deliberation and correspondence that his paper was accepted for publication. Indeed, as explained in [22, p. 23], the paper was written in two parts: part A was quite general, while part B dealt with the specific case of a dam. Richardson received two referee's reports. As he wrote to Aitken more than 42 years later (31 December 1952)

Lewis Fry Richardson.

[...] for I was appalled to find that the first referee recommended that part A should be omitted and B condensed while the second referee recommended that B should be omitted and A condensed! Perceiving that even referees were not infallible, I decided to persist, and after a lot of bother to myself and to other referees I got both parts published.

Have things changed now?

He submitted this work for a D.Sc. and a fellowship at Cambridge, but it was rejected. The ideas were too new, and the mathematics was considered "approximate mathematics"!

In 1913, Richardson became superintendent of the Eskdalemuir Observatory in southern Scotland. He had no experience in meteorology, but was appointed to bring some theory in its understanding. He again used finite differences. Although he was certainly aware of the difficulty of the problem, since he estimated at 60,000 the number of people who would have to be involved in the computations in order to obtain the prediction of tomorrow's weather [521] before the day actually began, it seems that he did not realize that the problem was ill-conditioned. He also started to write a book on this topic. Over the years, Richardson made important contributions to fluid dynamics, in particular eddy-diffusion in the atmosphere. The so-called Richardson number is a fundamental quantity involving gradients of temperature and wind velocity.

On 16 May 1916, he resigned and joined the Friends' Ambulance Unit (a Quaker organization) in France. He began to think about the causes of wars and how to prevent them. He suggested that the animosity between two countries could be measured, and that some differential equations are involved into the process. He published a book with these ideas, and returned to weather prediction.

In 1920, he became lecturer in mathematics and physics at Westminster Training College, an institution training prospective school teachers up to a bachelor's degree.

In 1926, he again changed his field of research to psychology, to which he wanted to apply the ideas and the methods of mathematics and physics. He established that many sensations are quantifiable, he found methods for measuring them, and modeled them by equations. The same year, he was elected as a Fellow of the Royal Society of London.

Richardson's second paper on extrapolation was published in 1927. The method presented in his 1910 paper [520] was improved in order to cancel the terms in h^2 and in h^4 in the expansion of the error. It consists of two parts. The first one is due to him [523], while the author of the second one is J.A. Gaunt [261]. As told in [22, p. 125], Richardson encouraged students to take part in his scientific work, and when they made a significant contribution, he invited them to be a joint author of the resulting paper. John Arthur Gaunt was the son of missionaries serving in China. He was one of the most brilliant pupils of Richardson's wife, Dorothy, at St Michael School. Then he went to Rugby and Trinity College in Cambridge, and wrote some mathematical papers on quantum mechanics. He was elected as a research fellow at Trinity but left to join the Church Missionary Society. He spent the rest of his life teaching in a school for Chinese boys in Hong Kong. He was captured by the Japanese in 1941 and died in a prisoner of war camp 2 years later.

Richardson left Westminster Training College in 1929 for the position of principal at the Technical College in Paisley, an industrial city near Glasgow. Although he had to teach 16 h a week, he continued his research but returned to the study of the causes of wars and their prevention. He designed a model for the tendencies of nations to prepare for wars, and worked out its applications using historical data from previous conflicts. He also made predictions for 1935, and showed that the situation was unstable, which could be prevented only by a change in the nation's policies.

Richardson wanted to *see whether there is any statistical connection between war, riot, and murder*. He began to accumulate such data, and decided to search for a relation between the probability of two countries going to war and the length of their common border.

To his surprise, the lengths of the borders varied from one source to another. So he investigated how to measure the length of a border, and he realized that it highly depends on the length of the ruler. Using a small ruler allows one to follow more wiggles, more irregularities, than a long that eliminates details. Thus, the smaller the ruler, the larger the result. He noticed that for any natural frontier or coastline, there is a linear relation between the logarithms of the measure of the length and that of the ruler. The slope of the line gives a new mathematical measure of wiggliness. He did not offer any explanation for this property, and noticed it only as a curiosity. At that time, Richardson's results were ignored by the scientific community, and they were only published posthumously. Today, they are considered to be at the origin of fractals later much developed by Benoît Mandelbrot (1924–2010), the nephew of Szolem Mandelbrojt, since 1967 [431].

In 1943, Richardson and his wife moved to their last home, at Kilmun, 25 miles from Glasgow. He returned to his research on differential equations, and solved the associated system of linear equations by the so-called Richardson's method. He mentioned that the idea was suggested to him in 1948 by Arnold Lubin. At home, Richardson was also constructing an analog computer for his meteorological computations. He died on 30 September 1953 in Kilmun.

As mentioned by John Todd [639]:

His work was highly individualistic and his language and symbolism picturesque. For instance he introduced the terms "marching" problem, for initial value problems of the form

$$y'' = ky, \quad y(0), y'(0) \ given,$$

and "jury" problem for a problem of the form

$$y^{vi} - 3y^{iv} + 3y'' - y = \lambda y, \quad y = y'' = y''' = 0 \ for \ x = \pm 1.$$

Another obituary [572] described his personality:

He was a patient and original teacher, delighting in ingenious practical demonstrations with simple apparatus. His writings were utterly individual in style and highly entertaining; but few concessions were made to the reader. A fellow undergraduate at King's College said of him, "Lewis was a rock and flew his colours with superb, audacious gallantry." The audaciousness became with the years an understanding gentleness but the superb gallantry increased.

Richardson was a highly original character, whose contributions to many different fields were prominent but, unfortunately, not appreciated at their real value in his epoch. See [22] for a full-length bibliography, which also contains a detailed biography [339], and [211] for his works.

8.8 Werner Romberg

Werner Romberg in December 1988 in Trondheim.
© Claude Brezinski.

   Werner Romberg was born on 16 May 1909 in Berlin. In 1928, he began studies
in physics and mathematics in Heidelberg, where the Nobel laureate Philipp Eduard
Anton von Lenard (1862–1947)[8] was still quite influential. After 2 years, Romberg
decided to go to the Ludwig-Maximilian University of Munich. He followed mathematics with Constantin Carathéodory (1873–1950) and Oskar Perron (1880–1975),
and physics with Arnold Sommerfeld (1868–1951), his advisor. In 1933, he defended
his thesis *Zur Polarisation des Kanalstrahllichtes* (On the polarization of channel
light beams). The same year, he had to leave Germany because of his left-wing
views, and he went to the USSR. He remained at the Department of Physics and
Technology in Dnepropetrovsk from 1934 to 1937. But his residence permit in the
Soviet Union was not renewed because he had a German passport. Since Poland had

---

[8] His original Hungarian name is Fülöp Lénárd.

not yet been invaded by Nazi Germany, he was able to travel through Warsaw to relatives in Prague, where he briefly stayed in 1938. After obtaining some funding from the Brøgger Committee, he flew to Oslo after the occupation of the Sudetenland but before the occupation of Prague.

Then he obtained a position in Oslo in the autumn of 1938 as the assistant of the physicist Egil Andersen Hylleraas (1898–1965). He also briefly worked at the Technical University of Trondheim with Johan Peter Holtsmark (1894–1975), who was building a Van de Graaff accelerator there. But from 1940 to 1944, Romberg had to escape to Uppsala during the German occupation. In 1941, the Nazi German state stripped him of his German citizenship, and in 1943, recognition of his doctorate was revoked. After the liberation of Oslo, Romberg returned there, and he became a Norwegian citizen in 1947, a citizenship he kept until the end of his life. He worked on a differential analyzer, studied numerical methods, and published a paper on the approximation of a short curved arc by sine functions. In 1949, he joined the Norwegian Institute of Technology in Trondheim (NTH) as an associated professor in physics. In 1955, he published his paper on what is now known as Romberg's method [535]. The method can be found in any textbook on numerical analysis under his name, but without any reference to the original paper, a proof of true fame. In the late 1950s, Romberg made the first attempts to install a digital computer at NTH. However, he lacked the requisite political strength and skills, and he was unable to obtain financial support from the authorities [470]. In 1960, he was appointed head of the Applied Mathematics Department at the NTH. He organized a teaching program in applied mathematics, built a strong research program in numerical analysis, and introduced the first course on mechanical and electronic computers in Trondheim.

In 1968, Romberg returned to Heidelberg, where he accepted a professorship. He built up a group in numerical mathematics, at that time quite underdeveloped in Heidelberg, and was the head of the computing center of the university from 1969 to 1975.

Romberg retired in 1978. In December 1988, the *Romberg Seminar on Quadrature, Interpolation, Extrapolation and Rational Approximations* was held at the University of Trondheim in his honor. The seminar was a small one with only 11 invited speakers, and Romberg gave an informal talk on the introduction of digital computers in Norway. The photo above was taken at this meeting [312].

Werner Romberg died on 5 February 2003.

8.9   John Barkley Rosser

John Barkley Rosser was an American logician born on 6 December 1907. He studied with Alonzo Church (1903–1995). He is known for his part in the Church–Rosser theorem in lambda calculus. He also developed what is now called the Rosser sieve in number theory. In 1936, he proved Rosser's trick, a stronger version of Gödel's first incompleteness theorem that shows that the requirement for ω-consistency may be weakened to consistency. Rather than using the liar paradox sentence equivalent to *I am not provable*, he used a sentence that stated *For every proof of me, there is a shorter proof of my negation.* In prime number theory, he proved Rosser's theorem. The Kleene–Rosser paradox showed that the original lambda calculus is inconsistent.

John Barkley Rosser ca. 1951.
Image courtesy of the UW-Madison Archives, #S07201.

He was the director of the Army Mathematics Research Center at the University of Wisconsin in Madison from 1963 to 1973. He died on 5 September 1989.[9]

8.10 Heinz Rutishauser

Heinz Rutishauser was born on 30 January 1918 in Weinfelden, Switzerland. From 1936 to 1942, he studied mathematics at the *Eidgenössische Technische Hochschule* (ETH) in Zürich. Then, always at ETH, he was an assistant of Walter Saxer (1896–1974), a mathematician specialist in insurance mathematics from 1942 to 1945, and a gymnasium teacher in Glarisegg and Trogen in 1947–1949. In 1948, he married Margrit Wirz. In 1948–1949, he visited Harvard and Princeton to study the state of the art in computing.

From 1949 to 1955, Rutishauser was a research associate at the Institute for Applied Mathematics at ETH Zürich, recently founded by Eduard Stiefel (1909–1978).

---

[9] Information mostly found on http://www.espace-turing.fr/Naissance-de-Barkley-Rosser.html.

He worked together with Ambros Speiser (1922–2003) on the development of the first Swiss computer ERMETH. In 1951, he defended his habilitation *Automatische Rechenplanfertigung* (Automatic construction of computation plans), and became *Privatdozent*. He was a pioneer in *the automatic compilation of a suitably formulated algorithm and thus introduced the concept of what is now known as a compiler* [297]. In particular, he was involved in the definition of the programming language Algol.

Heinz Rutishauser.
© ETH Zurich, Image Archive, Photographer: unknown.

In 1955, he was appointed associate professor, and in 1968, he became full professor and the head of the Group for Computer Science, which later became the Computer Science Institute and ultimately in 1981 the Division of Computer Science at ETH Zürich. Rutishauser died of an heart attack in his office on 10 November 1970.

In numerical analysis, Rutishauser worked on the instability of numerical methods for solving ordinary differential equations. However, he is mostly known for the derivation of the qd-algorithm and the LR-algorithm for computing matrix eigenvalues. The qd-algorithm, based on Hankel determinants, is an extension of the work of the French mathematician Jacques Hadamard (1865–1963) for determining the poles of a rational function given by a power series in inverse powers of the variable. Its recursive rules can be expressed as the product of two triangular matrices, a lower one and an upper one, each of them having only two nonzero diagonals, the main one and the diagonal next to it. Multiplying these matrices together in the reverse order at each step is equivalent to the rules of the qd-algorithm. Thus, Rutishauser had the idea of the LR method. For computing the eigenvalues of a matrix A, he decomposed it into to the product $A = L_0 R_0$, where L_0 is lower triangular and R_0 is upper triangular. Then he constructed the new matrix $A_1 = R_0 L_0$, which is, in turn, decomposed into the product $A_1 = L_1 R_1$, with L_1 lower triangular and R_1 upper triangular, and so on. Under some assumptions on A, if the sequence of matrices (A_k) converges, its limit is an upper triangular matrix whose diagonal contains the eigenvalues of A in decreasing order of magnitude. Rutishauser also contributed to gradient methods for solving systems of linear equations, and he applied Romberg's method to numerical differentiation. For more details, see [299, 300].

8.11 Klaus Samelson

Klaus Samelson was born in Strasbourg when it was the *Imperial Territory of Alsace-Lorraine* on 21 December 1918. Annexation to France had been proclaimed on 5 December, but the process did not gain international recognition until the signing of the Treaty of Versailles in 1919. In his early childhood years, he lived in Breslau, German Empire, the capital of the newly created Prussian Province of Lower Silesia of the Weimar Republic in 1919. Due to these difficult political circumstances, he waited until 1946 to study mathematics and physics at the Ludwig Maximilian University of Munich. After his degree, he briefly worked as a high-school teacher before returning to the university where he obtained a doctorate degree in physics under the guidance of Fritz Bopp (1909–1987), a German theoretical nuclear physicist, with a dissertation on a quantum-mechanical problem.

Samelson became interested in numerical analysis and worked with Robert Sauer (1898–1970), one of the founders of the journal *Numerische Mathematik*. He became involved in early computers as a research associate at the Mathematical Institute of the Technical University of Munich. He was interested in numerical precision in computing eigenvalues of matrices. At that time, computer science was emerging as a new discipline. With Bauer, who was also a doctoral student of Bopp, he studied the structure of programming languages in order to develop efficient algorithms for their translation and implementation. This research led to bracketed structures and stack models. He then played a central role in the conception of Algol. In 1958, he obtained a chair of mathematics at the University of Mainz, and then returned to Munich in 1963, where with Bauer, he developed a curriculum for informatics and computer science. He died on 25 May 1980 after suffering a long and severe illness.

Klaus Samelson around 1968/69.
Credit Repro Uli Benz / TUM. Archiv.

8.12 Daniel Shanks

Daniel (Dan) Shanks was born on 17 January 1917 in Chicago.[10] In 1937, he received a B.Sc. in physics from the University of Chicago. In 1940, he worked at the Aberdeen Proving Ground as a physicist. From 1941 to 1957, he was employed by the Naval Ordnance Laboratory (NOL), located in White Oak, Maryland, first as a physicist and then as a mathematician. There, in 1949, he published a memorandum describing his transformation [567]. He wanted to present this work to the Department of Mathematics of the University of Maryland as a Ph.D. thesis. But without having done any graduate work previously, he had first to complete the degree requirements before his memorandum could be examined as a thesis. Hence, it was only in 1954 that he obtained his Ph.D., and his thesis was published in the *Journal of Mathematical Physics* [568]. Dan considered this paper one of his best two (the second one was his computation of π to 100,000 decimals published with John Wrench [570]).

---

[10] This biography is reprinted, with permission, from [135]. Additional material has been included.

Dan Shanks on his 70th birthday, 17 January 1987.
Courtesy of Phil Eddy.

From 1951 to 1957, he headed the Numerical Analysis Section and then the Applied Mathematics Laboratory. After the NOL, Shanks worked at the David Taylor Model Basin in Bethesda, where an Applied Mathematics Laboratory had been created in 1952 under the leadership of Harry Polachek [500], who, in an oral interview,[11] said: *Some of the key colleagues in the mathematics were Dr. Wrench, Dr. Shanks, Dr. Theilheimer.* Polachek was interested by the calculation of transients [499, 685], which could have motivated Shanks to tackle the problem. See [161] for a history of the David Taylor Model Basin. After that, Shanks spent a year at the National Bureau of Standards, and in 1977, he joined the University of Maryland, where he remained until his death on 6 September 1996.

Dan also served as an editor of *Mathematics of Computation* from 1959 until his death. A special issue of this journal is dedicated to him, 48, Nos. 177–178, 447 pages (1987). He was very influential in this position, which also led him to turn to number theory, a domain in which his book [569] became a classic. In this domain, Shanks is also well known for his SQUFOF algorithm (which stands for SQUare FOrm Factorization) which, as stated in [281], *On a 32-bit computer, SQUFOF is the clear champion factoring algorithm for numbers between 10^{10} and 10^{18}, and will likely remain so.* It is based on continued fractions and quadratic forms (see also

---

[11] http://amhistory.si.edu/archives/AC0196_pola700324.pdf.

[444]). Although Shanks lectured on it, and explained it to a few people in the 1970s, he never published it. Handwritten manuscripts on it were discovered after his death in 1996.

The first author of this book experienced what is written in his obituary [697] when he met him (and Phil Eddy) in December 1976 in Bethesda:

> [...] he was always most solicitous and encouraging of young researchers. He invariably expressed interest in their work and made great efforts to coax from them results of both quality and taste, for Dan's taste in mathematics was very highly developed.

8.13 Johan Frederik Steffensen

Johan Frederik Steffensen.
Courtesy of The Archive of the Department of Mathematical Sciences,
University of Copenhagen.

Johan Frederik Steffensen was born in Copenhagen on 28 February 1873.[12] His father was the supreme judge of the Danish Army, and he himself took a degree in law at the University of Copenhagen. After a short period in Fredericia, in the eastern part of the Jutland peninsula in Denmark, he returned to Copenhagen and began a career in insurance. He was self-taught in mathematics, and in 1912, he earned a Ph.D. for a study in number theory. After 3 years as the managing director of a mutual

---

[12] This biography is reprinted, with permission, from [135] Some modifications have been made.

life assurance society, he turned to teach insurance mathematics at the University of Copenhagen, first as a lecturer and, from 1923 to 1943, as a professor of insurance mathematics. However, he remained interested in the world of business, and was an active member, and even the chairman, of several companies. He published 107 research papers in various fields of mathematics, the first one in 1904 and the last in 1957. He worked on the theory of statistics, interpolation, insurance mathematics, and the calculation of interest. An integral inequality in real analysis, obtained in 1918, is named after him [613]. His book on interpolation [614], published in 1927, can be considered one of the first books in numerical analysis, since its chapters cover interpolation in one and several variables, numerical differentiation, solution of differential equations, and quadrature. His paper on a quadratic method for solving nonlinear equations without having to compute derivatives of the function was published in 1933 [615]. He was president of the Danish Actuarial Society in 1922–1924 and 1930–1933, and of the Danish Mathematical Society in 1930–1936. The University of London invited him to lecture on his research in the theory of statistics and actuarial science in 1930. In the obituary he wrote [473], Maurice Edward Ogborn (1907–2003) reported:

> [...] of his ready help with the derivation of a formula for the rate of interest in an annuity-certain, stated without proof in an eighteenth-century work by William Jones, F.R.S. Steffensen in a few lines, showed how this expression should be expanded in powers of i, when the term involving i^3 disappeared and the higher powers could be neglected, leaving a quadratic which was the basis of the solution.

Steffensen was also invited to give lectures at the Sorbonne in Paris in 1931. He loved English literature, especially Shakespeare. He died on 20 December 1961 from a heart attack.

8.14 Thomas Joannes Stieltjes

Thomas Joannes (or Jan) Stieltjes was born in Zwolle, Netherlands, on 29 December 1856. He was a student at the Polytechnical School of Delft in 1873, but instead of attending lectures, he spent his time reading the works of Gauss and Jacobi. As a consequence, he failed his examinations. Two further failures (in 1875 and 1876) followed. His father, a civil engineer and politician, despaired, but he contacted his friend Hendrikus Gerardus van de Sande Bakhuyzen (1838–1923), who was the director of Leiden University, and Thomas Jan was given a position as an assistant at Leiden Observatory.

Soon afterward, Stieltjes began a correspondence with Charles Hermite, the renowned French mathematician who was the thesis advisor of Padé. Stieltjes originally wrote to Hermite about celestial mechanics, but the subject quickly turned to mathematics. This correspondence lasted for the rest of his life (432 letters totaling 921 pages). On 1 January 1883, the new director of Leiden observatory allowed Stieltjes to put aside his observational work to allow him to work more on mathematics. The following September, he was asked to substitute for a professor at the

Thomas Jan Stieltjes.

University of Delft. From then until December 1883, he lectured on analytical and descriptive geometry. He resigned his post at the observatory at the end of that year.

In 1884, Stieltjes applied for a chair in Groningen. He was initially accepted, but the Department of Education rejected him, since he lacked the required diplomas. In 1884, Hermite and Professor David Bierens de Haan (1822–1895), a Dutch mathematician and historian of science, arranged for an honorary doctorate to be granted to Stieltjes by Leiden University, enabling him to become a professor. In 1885, he was elected a member of the Royal Dutch Academy of Sciences. In 1889, due to Hermite's efforts, he was appointed as a professor of differential and integral calculus at the University of Toulouse, in France, where he died on 31 December 1894 at the age of 38.

Stieltjes is considered the father of the analytic theory of continued fractions. In order to study them with the greatest mathematical rigor, he introduced the notion of what is now called the Riemann–Stieltjes integral. David Hilbert (1862–1943) recognized that Stieltjes's work on the moment problem led him to the spectral theory of operators. Other important contributions to mathematics that he made involved discontinuous functions and divergent series, differential equations, interpolation, the gamma function and elliptic functions, orthogonal and Stieltjes polynomials. He also proved the convergence of Gaussian quadrature methods in the case of a finite interval of integration.

8.15 Thorvald Nicolai Thiele

Thorvald Nicolai Thiele.
Courtesy of Steffen L. Lauritzen.

Thorvald Nicolai Thiele was a brilliant Danish scientist who worked as an actuary, astronomer, mathematician, and statistician. He was born in Copenhagen on 24 December 1838.

Thiele obtained his master's degree in astronomy from the University of Copenhagen in 1860 and his doctoral degree (Sc.D.) in 1866 with a thesis on the orbits of double stars. In 1875 he became professor of astronomy and director of the astronomical observatory at the University of Copenhagen, positions he kept until his retirement in 1907. He was the founder and mathematical director of the Danish insurance company Hafnia from 1872 until his death in Copenhagen on 26 September 1910.

Thiele developed exceptional theoretical skills, not only in his speciality, but also in numerical analysis, mathematical statistics, and actuarial science, combined with a deep involvement in practical applications. In particular, he stressed the importance of empirical model testing by analysis of residuals, and he was an expert in numerical calculation. His interest in statistics arose naturally from his empirical studies in astronomy and mortality investigations, where errors in observations play a significant role. He contributed to the theory of skew distributions. He formulated the canonical form of the linear model with normally distributed errors and reduced the general linear model to canonical form by orthogonal transformations, and he made early fundamental contributions to the analysis of variance and time series. He gave a differential equation for the premium reserve of a life insurance policy that had a tremendous impact on the development of modern life insurance mathematics.

Thiele was the first to propose a mathematical theory of Brownian motion, he introduced the cumulants and the likelihood functions, and was considered one of the greatest statisticians of all times. In the early 1900s, he also developed and proposed a generalization of approval voting to multiple-winner elections called *sequential proportional approval voting*, which was briefly used for party lists in Sweden when proportional representation was introduced in 1909.

In 1906, he published a paper in French about reciprocal differences [636], and his book on interpolation [637] appeared in 1909. He was a man of many talents.

8.16 John Todd

John Todd in 1977.
Photo by Konrad Jacobs.
© Archives of the Mathematisches Forschungsinstitut Oberwolfach.

John (Jack) Todd was born in Carnacally (Ireland) on 16 May 1911. He had an important impact on various aspects of numerical analysis and on the profession. In 1931, he received a B.Sc. from Queen's University in Belfast. Then he went to Cambridge and undertook research with John Edensor Littlewood (1885–1977). After 2 years of research on transfinite superpositions of absolutely continuous functions, Todd was appointed to Queen's University Belfast, where John Semple (1904–1985) was a professor. When Semple moved to King's College in London, he invited Todd to join him. At the beginning of WWII, Todd returned to Belfast to teach at the Methodist College of Belfast in 1940–1941. As part of the war effort, he worked for the British Admiralty from 1941 to 1945. He rapidly persuaded his superiors to establish what became the Admiralty Computing Service, centralizing much of numerical computations for the naval service.

In 1945, Todd was part of a group that visited Germany for investigating mathematics and computers that could be of interest to the Navy. During this mission, they discovered an old hunting lodge, *Lorenzenhof*, in Oberwolfach, in the Black Forest, which had been used as a research center for mathematics since the fall of 1944. But Moroccan soldiers wanted to occupy the building. Todd persuaded them, as he wrote in [640], to *leave the mathematicians and* même les poules *undisturbed*. Thus, the rescue of this beautiful center for organizing conferences is due to him.

Even before setting up the Admiralty Computing Service, Todd had been interested in computing. In 1946, he had to teach a numerical analysis course at King's College in London. He and his wife, Olga Taussky (1906–1995), had little experience in the numerical solution of systems of linear equations. They looked into *Mathematical Reviews*, and they found the review of a paper by Henry Jensen, [356], in which it was stated that *Cholesky's method seems to possess all advantages*. Thus, Todd decided to teach this method, which was thus revived, since it was unknown to numerical analysts because it had been published in a geodetic bulletin [141]. After the war, Todd emigrated to the United States with Olga Taussky (who died on 7 October 1995), and he passed away at his home in Pasadena, California, on 21 June 2007.

8.17 Herman van Rossum

Herman van Rossum, emeritus professor of the University of Amsterdam, died on 25 February 2006. He was 88 years old. During a period of 20 years, he was director of the Institute for Propaedeutical Mathematics at the University of Amsterdam. This institute, emphasizing good teaching of basic mathematics to students in other sciences, was founded by him. Later the institute changed its name to the Institute for Interdisciplinary Mathematics. Herman van Rossum defended his Ph.D. thesis, *A theory of orthogonal polynomials based on the Padé table*, at the University of Utrecht in 1953 under the guidance of Jan Popken (1905–1970). Between 1955 and 1995 he published 26 research papers, mostly on orthogonal polynomials and Padé approximation. Seven of these papers were joint papers with Erik Hendriksen, who

Herman van Rossum.
Courtesy of Tom H. Koornwinder.

was inspired by Herman van Rossum to enter research in orthogonal polynomials and approximation theory.[13]

8.18 Adriaan van Wijngaarden

Adriaan van Wijngaarden was a Dutch mathematician and computer scientist considered the founding father of computer science in the Netherlands. He was born on 2 November 1916, and received a degree in mechanical engineering from Delft University of Technology in 1939. Then he went to England to learn the new technologies that had been developed during the Second World War. In December 1945, he received his Ph.D. for calculations on ship propellers in the Applied Mechanics Department of Delft University under the supervision of the Dutch mathematician Cornelis Benjamin Biezeno (1888–1975). Interested in computers, he became the head of the Computing Department of the new *Mathematisch Centrum* in Amsterdam on 1 January 1947. He made several visits to England and the United States, gathering ideas for the construction of the first Dutch computer, the ARRA, in 1952. He hired Edsger Wybe Dijkstra (1930–2002),[14] and they worked on software for their new computer.

---

[13] From https://staff.fnwi.uva.nl/t.h.koornwinder/links/mathnews.html.

[14] Wynn mentioned Dijkstra in several of his papers for helping him for his Algol programs.

Van Wijngaarden was seriously injured, and his wife killed, in a car accident in 1958 in Edinburgh. After recovering, he focused more on programming languages, and was one of the designers of the original Algol, and later Algol 68. He developed a two-level type of grammar that became known as "van Wijngaarden grammars." In 1961, he became the director of the *Mathematisch Centrum* and remained in that post for the next 20 years. He passed away on 7 February 1987.

Adriaan van Wijngaarden in 1952.
With permission of "Centrum Wiskunde & Informatica (CWI), Amsterdam".

As a Conclusion

This book proposed a *tour d'horizon*, a fresco, on linear and nonlinear extrapolation methods, Padé approximation, formal orthogonal polynomials, continued fractions, and related topics. It also has the unique feature of mixing mathematical concepts and results, their history, a detailed analysis of important publications, personal testimonies by researchers involved in research on these domains, and information, anecdotes, and exclusive remembrances on some actors.

After a mathematical introduction to these themes, we gave an account of their status of knowledge around 1950, that is, at the time most scholars, Aitken, Richardson, Shanks, Wynn, and others, began their research. In the course of the analysis of these works, and in particular of Peter Wynn's papers and reports (all written alone), some more mathematical concepts were introduced and explained. Then the recent developments of the domains were presented.

In published papers and books, research is most of the time introduced in an impersonal way, although developed by people who have existed or are still alive. The human side of research is missing. This is why we asked researchers who personally played (and are still playing) a role in these fields, and whose works are related to those described herein, to send us their testimonies. They show how research in a domain starts and evolves. For the same reason, biographies of other scholars encountered have been included. The human side of research plays a fundamental part in its development. This is why congresses form an important feature of our profession, since they allow researchers to meet other specialists and to build relationships that often last a lifetime. In the domains concerned here or related to them, many congresses, or special sessions in more general ones, or even summer schools, were organized. We cannot list them here, because they were too numerous. The proceedings of many of them were published and can easily be found.

We are aware of not having cited all the interesting extensions of the works mentioned above, an almost impossible task. We also apologize to those who are not quoted here. It does not mean that their achievements were less interesting than those mentioned, but a choice had to be made. We are responsible for this choice, for the unintentional omissions, and for the misprints and even the mistakes contained in this book.

© Springer Nature Switzerland AG 2020
C. Brezinski, M. Redivo-Zaglia, *Extrapolation and Rational Approximation*,
https://doi.org/10.1007/978-3-030-58418-4

If we look back to the *General Introduction* of this book, we put forward the names of five pioneering mathematicians, who, in chronological order of their works, are Henri Eugène Padé, Lewis Fry Richardson, Alexander Craig Aitken, Daniel Shanks, and Peter Wynn. Let us now try to draw a conclusion about the accomplishments of each of them, although it is quite a presumptuous task.

Continued fractions and Padé approximants were known and used long before Henri Padé defended his thesis in 1892. Almost any mathematical book of the nineteenth century, in English, French, German, Italian, etc., contains a chapter on continued fractions. However, it was his thesis, although written in French (but at that time French was one of the most widely used languages for science), that brought light to the topic, since it was the first systematic study of these approximants. Padé arranged them into a double entry table, and he gave a closed formula for the approximants of the exponential function. He also related them to continued fractions. Let us mention that Charles Hermite, Padé's thesis advisor, was a little bit skeptical about the results obtained, since in his report at this occasion, he wrote (see [483]):[1]

> While thinking that the theory of continued fractions could be generalized under other points of view, opening a more interesting and more fruitful path, and that the importance of the results which he has achieved does not correspond as much as one would like to his long and conscientious work, we judge that the thesis of Mr. Padé is worthy of the approval of the Faculty.

Padé published several papers after his thesis, but overwhelmed by his administrative duties, he quite rapidly abandoned research. His papers are gathered in [483] together with their contemporary analyses. However, his contribution certainly opened the way to further work, and it was rapidly known outside France, as illustrated, for example, by van Vleck's talk at the 1903 colloquium of the American Mathematical Society [665] (see Sect. 3.4). Thus, the work of Padé was quite influential in the development, in approximation theory, of continued fractions and of the approximants named after him.

It was for improving the accuracy of a finite difference approximation to a partial differential equation that Lewis Fry Richardson introduced extrapolation in 1910 [520]. He noticed that the error had an expansion in even powers of the step size h, and that by combining two approximations with h and $2h$, the term in h^2, and some years later (in 1927 [523]), also the term in h^4 can be eliminated. Although the idea could be traced back to the improvement of old approximations of π by inscribed and circumscribed polygons, it was the first time it was used for solving a numerical analysis problem. Then extrapolation rapidly found its place in the toolbox of numerical analysts, and it became a favorite method for treating the case of a polynomial expansion of the error. It seems that Richardson, interested in many other topics, was not fully aware of the potentialities of the method he found (nor was he of his discovery of fractals!).

---

[1] *Tout en pensant que la théorie des fractions continues pourrait se généraliser sous d'autres points de vue ouvrant une voie plus intéressante et plus féconde, et que l'importance des résultats auxquels il est parvenu ne répond pas autant qu'on le désirerait à son long et consciencieux travail, nous jugeons que la thèse de Mr. Padé est digne de l'approbation de la Faculté.*

Although it can also be traced back to at least the seventeenth century (see [476]), the Δ^2 process introduced in 1926 by Alexander Craig Aitken was the first nonlinear convergence acceleration method used in numerical analysis [5]. It rapidly became known and used, but was not immediately studied from the theoretical point of view, in particular for its interpretation as an extrapolation method. However, despite the fact that Aitken did not fully exploit its possibilities, the process finally gained recognition, and it had an important impact on the opening of the new field of convergence acceleration and extrapolation methods in numerical analysis.

The purpose of Daniel Shanks when he built his transformation in 1949 [567] was to accelerate the convergence of slowly convergent sequences and series, and to assign a value, the *antilimit* as he called it, to divergent ones. His sequence transformation was based on the analogy between transients (a sum of exponentials) and mathematical sequences, and it was expressed as a ratio of two Hankel determinants. Shanks knew that it was a generalization of the Aitken process, but he was not aware that his transformation had a larger kernel containing other sequences than pure sums of exponentials. He studied the connections of his transformation with continued fractions and Padé approximants, and gave several numerical examples. Shanks produced a more detailed study of his transformation in 1955 [568]. As he explained to C.B., he was implementing it by computing separately the Hankel determinants of their numerators and their denominators by their recurrence relation given in Sect. 3.2.4.

Although it is a quite difficult exercise, let us now try to express our feelings about the work of Peter Wynn. He obviously had an extended mathematical culture running from computer languages and the skills required for the implementation of algorithms, to classical numerical analysis, complex analysis, and abstract algebra. He also had a vast knowledge of the mathematical literature in English, French, German, Italian, and Russian. At that time, it was difficult to be aware even of the existence of several references he cited, not to mention that they had to be found in libraries at a period before the internet existed. Even if his papers are not quoted as much as they should be, they are quite influential. Recall that the Shanks's paper appeared in an applied mathematics and mathematical physics journal. Wynn's paper on the scalar ε-algorithm was published in 1956 in the only journal devoted, at that time, to numerical methods [713]. Without Wynn's paper, the Shanks transformation could have remained unknown, and if not, at least useless in practice. Once known, the name of Wynn sparked interest, and researchers began to look at his other papers, even if they were not directly working on the topic treated. Thus, due to his work, some themes that were, if not completely forgotten, at least neglected, were revived. Padé approximants are a particular case, since they were used by theoretical physicists for some time, but not so much studied by numerical and complex analysts. The same is true for continued fractions, which were mostly used by number theorists. The influence of Wynn on the further development of these domains cannot be denied. In particular, the field of convergence acceleration and extrapolation methods could have remained in an embryonic state without him. Moreover, many of his discoveries have not yet been fully exploited, and we hope that this book will inspire some readers to pursue them and to study the manuscripts found after his death and analyzed in

[122]. Moreover, many numerical analysis methods originated in fact from the ideas of Shanks and Wynn, in particular those concerning the solution of systems of linear and nonlinear equations, and the acceleration of some methods used for that. To conclude, Peter Wynn was a quite eminent computer scientist, numerical analyst, and mathematician whose influence on applied mathematics is still important.

In this book, we tried to cover, as much as possible, the development of summation and extrapolation methods, continued fractions, rational (Padé) approximation, and related topics from their dissemination to the present. We hope that this book, simultaneously of a historical and a mathematical nature, with personal touches provided by the testimonies of the renowned researchers we collected, and presenting Wynn's legacy, will serve as an introduction to the domains concerned, and also as a basis for their further developments. To conclude, we tried to give this book the unique feature of combining mathematics, history, and personal recollections. We hope we succeeded.

Acknowledgments

The authors would like to express their sincere thanks to the contributors who sent us their testimonies, which form an invaluable part of this book: Adhemar Bultheel (Katholieke Universiteit Leuven, Leuven, Belgium), Annie Cuyt (University of Antwerp, Antwerp, Belgium), André Draux (INSA of Rouen, Rouen, France), Walter Gander (ETH Zürich, Zürich, Switzerland), Xing-Biao Hu (Institute of Computational Mathematics, Chinese Academy of Sciences, Beijing, China), William Branham Jones (Department of Mathematics, University of Colorado, Boulder, Colorado, USA), Pierre-Jean Laurent (University of Grenoble-Alpes, Grenoble, France), David Levin (Tel Aviv University, Tel Aviv, Israel), Naoki Osada (Tokyo Woman's Christian University, Tokyo, Japan), Avram Sidi (Technion—Israel Institute of Technology, Haifa, Israel), Ernst Joachim Weniger (Regensburg University, Regensburg, Germany), Jean Zinn-Justin (French Academy of Sciences, Paris, France).

We also acknowledge the contributions of Adrianne Ali (Digital Marketing and Communications manager, SIAM, Philadelphia, USA), Rodolfo Ambrosetti (former director of IBM Cloud Software Services for Latin America), Bernhard Beckermann (University of Lille, Lille, France), Adi Ben-Israel (Center for Operations Research, Rutgers University, Piscataway, USA), Manfred Broy (Technical University of Munich, Munich, Germany), Urban Cegrell (University of Umeå, Umeå, Sweden), Centrum Wiskunde and Informatica (CWI), (Amsterdam, The Netherlands), Stefano Cipolla (University of Padua, Padua, Italy), Jennifer Cong Yan Zhao (liaison librarian for computer science, electrical and computer engineering, and physical geography, Schulich Library of Physical Sciences, Life Sciences, and Engineering, McGill University, Montreal, Canada), Carl de Boor (University of Wisconsin-Madison, Madison, USA), Vladimir Dobrushkin (Center for Fluid Mechanics, Brown University, Providence, USA), Manvendra Krishna Dubey (Los Alamos National Laboratory, Los Alamos, USA), Ellen Embleton (picture curator, The Royal Society, London, UK), Jacob Fontenot (head, Interlibrary Loan Services, LSU Libraries,

Louisiana State University, Baton Rouge, USA), Walter Gautschi (Department of Computer Sciences, Purdue University, West Lafayette, USA), Amy Hackett (director's assistant, School of Computer Science, McGill University, Montreal, Canada), Heike Hartmann (Sammlungen und Archive, ETH-Bibliothek, Zürich, Switzerland), Sandra Herkle (head of Communication and Marketing, Department of Computer Science, ETH, Zürich, Switzerland), Doris Herrmann (Fakultät für Informatik, Technische Universität München, Munich, Germany), Jennifer Hinneburg (Archives of the Mathematisches Forschungsinstitut Oberwolfach, Oberwolfach, Germany), Bettina Kemme (School of Computer Science, McGill University, Montreal, Canada), Andrea Klee (Corporate Communications Center, Technische Universität München, Munich, Germany), Tom H. Koornwinder (University of Amsterdam, Amsterdam, The Netherlands), Steffen L. Lauritzen (Department of Mathematical Sciences, University of Copenhagen, Copenhagen, Denmark), Hervé Le Ferrand (University of Bourgogne, Dijon, France), Tom Louis Lindstrøm (University of Oslo, Oslo, Norway), Doron Shaul Lubinsky (School of Mathematics, Georgia Institute of Technology, Atlanta, Georgia, USA), Dyrol Lumbard (external relations manager, Mathematical Institute, University of Oxford, Oxford, United Kingdom), Jette Lunding Sandqvist (head of Financial Services Actuarial at PwC, Danmark), Jesper Lützen (Department of Mathematical Sciences, University of Copenhagen, Copenhagen, Denmark), Pascal Maroni (University of Paris, Paris, France), Gérard Meurant (former deputy director at CEA, Villeneuve-Saint-Georges, France), André Montpetit (University of Montreal, Montreal, Canada), Katie Nash (University Archives and Records Management, University of Wisconsin-Madison, Madison, USA), Juan Antonio Pérez (Unidad Académica de Matemáticas, Universidad Autónoma de Zacatecas, Zacatecas, Mexico), Rob van Rooijen (Library CWI, Amsterdam, The Netherlands), Hanna Rutishauser, The Los Alamos Monitor (Los Alamos, USA), Héctor René Vega-Carrillo (Unidad Académica de Matemáticas, Universidad Autónoma de Zacatecas, Zacatecas, Mexico).

We would like to express our gratitude to Francis Alexander Norman, professor of mathematics and associated chair at the University of Texas at San Antonio, Department of Mathematics, San Antonio, USA. He informed us about the boxes of documents left by Peter Wynn in the house of his friend Manuel Berriozábal, of the University of Texas at San Antonio, Department of Mathematics, San Antonio, USA, and informed us about the manuscripts they contain. Without them, this material would certainly have been lost. We, and the scientific community, owe them a lot.

David Kramer, our copyeditor, had the heavy job of improving, and sometimes correcting our *insecure grasp of the English grammar* as Peter Wynn once wrote to C.B.. David did an extraordinary and incredibly careful work. We have been very lucky to have him as our copyeditor. We are grateful for his many suggestions, improvements and even corrections of our original text.

We acknowledge the work of Ms. Lavanya Venkatesan, Project Manager, Content Solutions, SPi Global, and her team for producing the final version of the book.

Special thanks are due to our editor and friend Elisabeth Loew. She was always open to our proposals, and she also suggested pertinent recommendations for improving the project. It is always a pleasure to work with her. We acknowledge the work of Ms. Lavanya Venkatesan, Project Manager, Content Solutions, SPi Global, and her team for producing the final version of the book.

The work of C.B. was supported by the Labex CEMPI (ANR-11-LABX-0007-01), University of Lille, France. The work of M.R.-Z. was partially supported by the University of Padua, Italy.

Bibliographies

We preferred to give, in two separate lists, the general bibliography and the works of Peter Wynn. After each reference, the page(s) where it is quoted in this book are written in italics.

Let us mention that in 1960, the journal *Mathematical Tables and Other Aids to Computation* (Math. Tables Aids Comput.) became *Mathematics of Computation* (Math. Comp.). From its Volume 5, the journal *Revue française de traitement de l'information, Chiffres* became simply *Chiffres*.

Concerning the second list, some references in [135] have been corrected. When known, the DOI, the MR number, and the submission date of each paper are given. References are listed according to the publication date. In various of his papers, Wynn mentioned papers of his that were never published.

General Bibliography

1. M. Abramowitz, I.A. Stegun eds., *Handbook of Mathematical Functions with Formulas, Graphs, and Mathematical Tables*, Applied Mathematics Series, vol. 55, United States Department of Commerce, National Bureau of Standards, Washington D.C.; Dover Publications, New York, 1964.
   (Cited on page 149.)
2. N. Achyeser, Über eine Eigenschaft der elliptischen Polynome, Comm. Soc. Math. Kharkof (Zapiski Inst. Mat. Mech.) (4) 9 (1934) 3–8.
   (Cited on page 73.)
3. H.H. Aiken ed., *International Conference on Information Processing, Unesco House, Paris, 15–20 June 1959*, United Nations Educational, Scientific and Cultural Organization, Place de Fontenoy, Paris-7e, Printed by Imprimerie Union, Paris, Unesco 1959.
   (Cited on page 115.)
4. J.R. Airey, The "converging factor" in asymptotic series and the calculation of Bessel, Laguerre and other functions, Phil. Mag., (7) 24 (1937) 521–552.
   (Cited on pages 140 and 141.)
5. A.C. Aitken, On Bernoulli's numerical solution of algebraic equations, Proc. R. Soc. Edinb., 46 (1925–1926) 289–305.
   (Cited on pages 16, 51, 57, and 341.)

© Springer Nature Switzerland AG 2020
C. Brezinski, M. Redivo-Zaglia, *Extrapolation and Rational Approximation*,
https://doi.org/10.1007/978-3-030-58418-4

6. A.C. Aitken, Studies in practical mathematics. II. The evaluation of latent roots and latent vectors of matrix, Proc. R. Soc. Edinb., 57 (1936–1937) 269–304.
    (Cited on pages 51 and 58.)
7. A.C. Aitken, *Determinants and Matrices*, Oliver and Boyd, Edinburgh, 1939.
    (Cited on page 86.)
8. A.C. Aitken, *Gallipoli to the Somme: Recollections of a New Zealand Infantryman*, Oxford University Press, Oxford, 1963.
    (Cited on page 312.)
9. A.C. Aitken, *To Catch the Spirit. The Memoir of A.C. Aitken with a Biograohical Introduction by P.C. Fenton*, University of Otago Press, Dunedin, New Zealand, 1995.
    (Cited on pages 57 and 313.)
10. S.V. Aksenov, M.A. Savageau, U.D. Jentschura, J. Becher, G. Soff, P.J. Mohr, Application of the combined nonlinear-condensation transformation to problems in statistical analysis and theoretical physics, Comput. Phys. Comm., 150 (2003) 1–20.
    (Cited on page 111.)
11. A.A. Albert, Quadratic forms permitting composition, Ann. of Math., 43 (1942) 161–177.
    (Cited on page 155.)
12. A.A. Albert, Non-associative algebras, I - Fundamental concepts and isotopy, Ann. of Math., 43 (1942) 685–707; II - New simple algebras, id., 708–723.
    (Cited on page 155.)
13. G.D. Allen, C.K. Chui, W.R. Madych, F.J. Narcowich, P.W. Smith, Padé approximation and orthogonal polynomials, Bull. Austral. Math. Soc. 10 (1974) 263–270.
    (Cited on page 35.)
14. E.L. Allgower, K. Georg, *Computational Solution of Nonlinear Systems of Equations*, Amererican Mathematical Society, Providence, 1990.
    (Cited on page 195.)
15. D.G.M. Anderson, Iterative procedures for nonlinear integral equations, J. Assoc. Comput. Mach., 12 (1965) 547–560.
    (Cited on page 196.)
16. D.G.M. Anderson, Comments on "Anderson Acceleration, Mixing and Extrapolation", Numer. Algorithms, 80 (2019) 135–234.
    (Cited on page 196.)
17. H. Andoyer, Interpolation, in *Encyclopédie des Sciences Mathématiques Pures et Appliquées*, J. Molk ed., Tome I, vol.4, Fasc. 1, I-21, Gauthier-Villars, Paris, 1904–1912, pp.127–160; reprint by Éditions Gabay, Paris, 1993.
    (Cited on page 175.)
18. G.E. Andrews, R. Askey, Classical orthogonal polynomials, in *Polynômes Orthogonaux et Applications, Proceedings of the Laguerre Symposium held at Bar-le-Duc, October 15–18, 1984*, C. Brezinski, A. Draux, A.P. Magnus, P. Maroni, A. Ronveaux eds., Lecture Notes in Mathematics, vol. 1171, Springer-Verlag, Berlin, 1984, pp. 36–62.
    (Cited on page 70.)
19. I.V. Andrianov, L.I. Manevitch, with help from M. Hazewinkel, *Asymptotology. Ideas, Methods, and Applications*, Kluwer Academic Publishers, Dordrecht, 2002.
    (Cited on page 137.)
20. R. Apéry, Irrationalité de $\zeta(2)$ et de $\zeta(3)$, Astérisque, 61 (1979) 11–13.
    (Cited on page 206.)
21. R.J. Arms, A. Edrei, The Padé tables and continued fractions generated by totally positive sequences, in *Mathematical Essays dedicated to A.J. Macintyre*, Ohio University Press, Athens, Ohio, 1970, pp. 1–21.
    (Cited on pages 31 and 164.)
22. O.M. Ashford, *Prophet or Professor? The Life and Work of Lewis Fry Richardson*, Adam Hilger Ltd., Bristol and London, 1985.
    (Cited on pages 321, 322, and 323.)

23. R. Bacher, B. Lass, Développements limités et réversion des séries, Enseign. Math., 52 (2006) 267–293.
    (Cited on page 211.)
24. Z. Bai, R.W. Freund, A partial Padé-via-Lanczos method for reduced-order modeling, Linear Algebra Appl., 332–334 (2001) 139–164.
    (Cited on page 208.)
25. G.A. Baker Jr., Recursive calculation of Padé approximants, in *Padé Approximants and Their Applications*, P.R. Graves-Morris ed., Academic Press, London and New York, 1973, pp. 83–91.
    (Cited on page 32.)
26. G.A. Baker Jr., *Essentials of Padé Approximants*, Academic Press, New York, 1975.
    (Cited on pages 7, 29, and 314.)
27. G.A. Baker Jr., Counter-examples to the Baker-Gammel-Wills conjecture and patchwork convergence, J. Comput. Appl. Math., 179 (2005) 1–14.
    (Cited on page 74.)
28. G.A. Baker Jr., J.L. Gammel, The Padé approximant, J. Math. Anal. Appl., 2 (1961) 21–30.
    (Cited on page 204.)
29. G.A. Baker Jr., J.L. Gammel, *The Padé Approximant in Theoretical Physics*, Academic Press, New York, 1970.
    (Cited on page 314.)
30. G.A. Baker Jr., J.L. Gammel, J.G. Wills, An investigation of the applicability of the Padé approximant method, J. Math. Anal. Appl., 2 (1961) 405–418.
    (Cited on page 73.)
31. G.A. Baker Jr., P.R. Graves-Morris, *Padé Approximants*, 2nd Edition, Cambridge University Press, Cambridge, 1996.
    (Cited on pages 7, 25, 29, and 314.)
32. J. Baranger, Approximation optimale de la somme d'une série, C.R. Acad. Sci. Paris, 271 A (1970) 149–152.
    (Cited on page 169.)
33. J. Baranger, *Quelques résultats en optimisation non convexe. I. Formules optimales de sommation d'une série. II. Théorèmes d'existence en densité et application au contrôle*, Thèse de Doctorat d'État ès Sciences Mathématiques, Université Scientifique et Médicale de Grenoble, 23 mars1973.
    (Cited on page 169.)
34. M.N. Barber, C.J. Hamer, Extrapolation of sequences using a generalized epsilon-algorithm, J. Austral. Math. Soc., B 23 (1982) 229–240.
    (Cited on page 181.)
35. J. Barkley Rosser, Transformations to speed the convergence of series, J. Res. Natl. Bur. Stand., 46 (1951) 56–64.
    (Cited on page 110.)
36. J.L. Basdevant, Padé approximants, in *Methods in Subnuclear Physics*, vol. IV, M. Nikolic ed., Gordon and Breach, London, 1970, pp. 129–168.
    (Cited on page 204.)
37. J.L. Basdevant, The Padé approximantion and its physical applications, Fortschritte der Physik/Progress of Physics, 20 (1972) 283–331.
    (Cited on pages 29 and 204.)
38. F.L. Bauer, The Quotient-Difference and the epsilon algorithms, in *On Numerical Approximation*, R.E. Langer ed., University of Wisconsin Press, Madison, 1959, pp. 361–370.
    (Cited on pages 91, 95, and 145.)
39. F.L. Bauer, The g-algorithm, SIAM J., 8 (1960) 1–17.
    (Cited on pages 91, 95, 127, and 145.)
40. F.L. Bauer, La méthode de l'intégration numérique de Romberg, in *Colloque sur l'Analyse Numérique*, Librairie Universitaire, Louvain, Belgium, 1961, pp. 119–129.
    (Cited on page 57.)

41. F.L. Bauer, Algorithm 60: Romberg integration, Comm. ACM, 4 (1961) 255.
    (Cited on page 57.)
42. F.L. Bauer, Nonlinear sequence transformations, in *Approximation of Functions*, P. Garabedian ed., Elsevier, New York, 1965, pp. 134–151.
    (Cited on pages 91 and 145.)
43. F.L. Bauer, H. Rutishauser, E. Stiefel, New aspects in numerical quadrature, in *Proc. Symposium on Applied Mathematics*, vol. 15, American Mathematical Society, Providence, 1963, pp. 199–218.
    (Cited on page 57.)
44. H. Baumann, *Generalized Continued Fractions: Definitions, Convergence and Applications to Markov Chains*, Habilitatiosschrift, Universität Hamburg, 2017.
    (Cited on page 202.)
45. B. Beckermann, A connection between the E-algorithm and the epsilon-algorithm, in *Numerical and Applied Mathematics*, C. Brezinski ed., Baltzer, Basel, 1989, pp. 443–446.
    (Cited on page 177.)
46. B. Beckermann, V. Kalyagine, A.C. Matos, F. Wielonsky, How well does the Hermite-Padé approximation smooth the Gibbs phenomenon?, Math. Comp., 80 (2011) 931–958.
    (Cited on page 118.)
47. B. Beckermann, A.C. Matos, Algebraic properties of robust Padé approximants, J. Approx. Theory, 190 (2015) 91–114.
    (Cited on page 197.)
48. B. Beckermann, A.C. Matos, F. Wielonsky, Reduction of the Gibbs phenomenon for smooth functions with jumps by the epsilon-algorithm, J. Comput. Appl. Math, 219 (2008) 329–349.
    (Cited on page 118.)
49. R. Bellman, K.L. Cooke, *Differential-Difference Equations*, Academic Press, New York, 1963.
    (Cited on page 122.)
50. A. Ben-Israel, T.N.E. Greville, *Generalized Inverses: Theory and Applications*, Wiley, New York, 1974.
    (Cited on page 318.)
51. C.M. Bender, S.A. Orszag, *Advanced Mathematical Methods for Scientists and Engineers*, McGraw-Hill, Kogokusha, 1978.
    (Cited on page 29.)
52. A. Berlinet, *Sur quelques Problèmes d'Estimation Fonctionnelle et de Statistique des Processus*, Thèse de Doctorat d'État ès Sciences Mathématiques, Université des Sciences et Techniques de Lille I, 1984.
    (Cited on page 212.)
53. A. Berlinet, Geometric approach to the parallel sum of vectors and application to the vector ε-algorithm, Numer Algorithms, 65 (2014) 783–807.
    (Cited on page 181.)
54. A.F. Berlinet, Ch. Roland, Acceleration schemes with application to the EM algorithm, Comput. Statist. Data Anal., 51 (2007) 3689–3702.
    (Cited on page 195.)
55. A. Berlinet, Ch. Roland, Parabolic acceleration of the EM algorithm, Stat. Comput., 19 (2009) 35–47.
    (Cited on page 195.)
56. A.F. Berlinet, Ch. Roland, Acceleration of the EM algorithm: P-EM versus epsilon algorithm, 56 (2012) 4122–4137.
    (Cited on page 195.)
57. S. Bernstein, Sur les fonctions absolument monotones, Acta Math., 52 (1928) 1–66.
    (Cited on page 162.)
58. J.P. Berrut, Rational functions for guaranteed and experimentally well-conditioned global interpolation, Comput. Math. Appl., 15 (1988) 1–16.
    (Cited on page 197.)

59. D. Bessis, M. Pusterla, Unitary Padé approximants in strong coupling field theory and application to the calculation of the ρ- and f_0-meson Regge trajectories, Il Nuovo Cimento A, 54 (1968) 243–294.
   (Cited on page 204.)

60. W.G. Bickley, L.J. Comrie, D.H. Sadler, J.C.P. Miller, A.J. Thompson, *Bessel functions, part II, Functions of Positive Integer Order*, British Association for the Advancement of Science, Mathematical Tables, vol. X, Cambridge University Press, Cambridge, 1952.
   (Cited on page 66.)

61. W.G. Bickley, J.C.P. Miller, The numerical summation of slowly convergent series of positive terms, Phil. Mag., (7) 22 (1936) 754–767.
   (Cited on pages 66, 109, 110, and 138.)

62. P. Bjørstad, G. Dahlquist, E. Grosse, Extrapolation of asymptotic expansions by a modified Aitken δ^2-formula, BIT, 21 (1981) 56–65.
   (Cited on page 179.)

63. M. Blakemore, G. Evans, J. Hyslop, Comparison of some methods for evaluating infinite range oscillatory integrals, J. Comput. Phys., 22 (1976) 352–376.
   (Cited on page 180.)

64. G. Blanch, On the numerical solution of parabolic partial differential equations, J. Res. Natl. Bur. Stand., 50 (1953) 343–356.
   (Cited on page 170.)

65. D.I. Bodnar, Kh. Yo. Kuchmins'ka, Development of the theory of branched continued fractions in 1996–2016, J. Math. Sci., 231 (2018) 481–494.
   (Cited on page 201.)

66. P.I. Bodnarcuk, W.Ja. Skorobogatko, *Branched Continued Fractions and Applications* (in Ukrainian), Naukowaja Dumka, Kiev, 1974.
   (Cited on page 201.)

67. N. Bogolyubov, N. Krylov, On Rayleigh's principle in the theory of differential equations of mathematical physics and upon Euler's method in the calculus of variation (in Russian), Acad. Sci. Ukraine (Phys. Math.), 3 (1926) 3–22.
   (Cited on page 55.)

68. H.C. Bolton, H.I. Scoins, G.S. Rushbrooke, Eigenvalues of differential equations by finite-difference methods, Math. Proc. Cambridge Philos. Soc., 52 (1956) 215–229.
   (Cited on page 170.)

69. J. Boos, *Classical and Modern Methods in Summability*, Oxford University Press, Oxford, 2000.
   (Cited on page 54.)

70. É. Borel, Fondements de la théorie des séries divergentes sommables, J. Math. Pures Appl., (5), 2 (1896) 103–122.
   (Cited on page 50.)

71. É. Borel, Mémoire sur les séries divergentes, Ann. Sci. Éc. Norm. Supér., (3) 16 (1899) 9–136.
   (Cited on pages 76 and 204.)

72. É. Borel, *Leçons sur les Séries Divergentes*, Gauthier-Villars, Paris, 1901.
   (Cited on page 50.)

73. É. Borel, Sur l'approximation des nombres par des nombres rationnels, C. R. Acad. Sci. Paris, 136 (1903) 1054–1055.
   (Cited on page 74.)

74. F. Bornemann, D. Laurie, S. Wagon, J. Waldvogel, *The SIAM 100-Digit Challenge: A Study in High-Accuracy Numerical Computing*, SIAM, Philadelphia, 2004.
   (Cited on pages 7 and 369.)

75. J.W. Bradshaw, Continued fractions and modified continued fractions for certain series, Amer. Math. Monthly, 45 (1938) 352–362.
   (Cited on page 109.)

76. J.W. Bradshaw, Modified series, Amer. Math. Monthly, 46 (1939) 486–492.
   (Cited on page 109.)

77. J.W. Bradshaw, More modified series, Amer. Math. Monthly, 51 (1944) 389–391.
    (Cited on page 109.)

78. L.C. Breaux, *A Numerical Study of the Application of Acceleration Techniques and Prediction Algorithms to Numerical Integration*, M.Sc. Thesis, Louisina State University in New Orleans, 1971.
    (Cited on pages 82 and 165.)

79. C. Brezinski, Application de l'ε-algorithme à la résolution des systèmes non linéaires, C. R. Acad. Sci. Paris, 271 A (1970) 1174–1177.
    (Cited on pages 118 and 196.)

80. C. Brezinski, Sur un algorithme de résolution des systèmes non linéaires, C. R. Acad. Sci. Paris, 272 A (1971) 145–148.
    (Cited on page 118.)

81. C. Brezinski, *Méthodes d'Accélération de la Convergence en Analyse Numérique*, Thèse de Doctorat d'État ès Sciences Mathématiques, Université Scientifique et Médicale de Grenoble, 26 avril 1971.
    (Cited on pages 109 and 165.)

82. C. Brezinski, Convergence d'une forme confluente de l'ε-algorithme, C.R. Acad. Sci. Paris, 273 A (1971) 582–585.
    (Cited on page 181.)

83. C. Brezinski, Études sur les ε et ϱ-algorithmes, Numer. Math., 17 (1971) 153–162.
    (Cited on pages 94 and 180.)

84. C. Brezinski, L'ε-algorithme et les suites totalement monotones et oscillantes, C. R. Acad. Sci. Paris, 276 A (1973) 305–308.
    (Cited on pages 89, 94, and 180.)

85. C. Brezinski, Some results in the theory of the vector ε-algorithm, Linear Alg. Appl., 8 (1974) 77–86.
    (Cited on page 187.)

86. C. Brezinski, Généralisation de la transformation de Shanks, de la table de Padé et de l'ε-algorithme, Calcolo, 12 (1975) 317–360.
    (Cited on pages 182, 184, and 185.)

87. C. Brezinski, Forme confluente de l'ε-algorithme topologique, Numer. Math., 23 (1975) 363–370.
    (Cited on page 181.)

88. C. Brezinski, Génération de suites totalement monotones et oscillantes, C. R. Acad. Sci. Paris, 280 A (1975) 729–731.
    (Cited on pages 94 and 180.)

89. C. Brezinski, Padé approximants and orthogonal polynomials, in *Padé and Rational Approximation*, E. B. Saff and R. S. Varga eds., Academic Press, New York, 1977, pp. 3–14.
    (Cited on pages 35 and 187.)

90. C. Brezinski, *Accélération de la Convergence en Analyse Numérique*, Lecture Notes in Mathematics, vol. 584, Springer-Verlag, Berlin-Heidelberg, 1977.
    (Cited on pages 7 and 48.)

91. C. Brezinski, Convergence acceleration of some sequences by the ε-algorithm, Numer. Math., 29 (1978) 173–177.
    (Cited on pages 89, 94, and 180.)

92. C. Brezinski, *Algorithmes d'Accélération de la Convergence. Étude Numérique*, Technip, Paris, 1978.
    (Cited on page 7.)

93. C. Brezinski, Rational approximation to formal power series, J. Approx. Theory, 25 (1979) 295–317.
    (Cited on pages 34 and 197.)

94. C. Brezinski, *Padé-Type Approximation and General Orthogonal Polynomials*, ISNM, vol. 50, Birkhäuser-Verlag, Basel, 1980.
    (Cited on pages 7, 25, 33, 34, 37, 41, 45, 71, 96, 116, 187, 188, 192, and 194.)

95. C. Brezinski, A general extrapolation algorithm, Numer. Math., 35 (1980) 175–187.
(Cited on pages 173, 176, and 177.)

96. C. Brezinski, Recursive interpolation, extrapolation and projection, J. Comput. Appl. Math., 9 (1983) 369–376.
(Cited on pages 185 and 199.)

97. C. Brezinski, Prediction properties of some extrapolation methods, Appl. Numer. Math., 1 (1985) 457–462.
(Cited on page 177.)

98. C. Brezinski, How to accelerate continued fractions, in *Informatique and Calcul*, P. Chenin et al. eds., Masson, Paris, 1986, pp. 3–39.
(Cited on page 201.)

99. C. Brezinski, Error estimate in Padé approximation, in *Orthogonal Polynomials and Their Applications*, M. Alfaro et al. eds., Lecture Notes in Mathematics, vol. 1328, Springer-Verlag, Heidelberg, 1988, pp. 1–19.
(Cited on page 35.)

100. C. Brezinski, A survey of iterative extrapolation by the E-algorithm, Det Kong. Norske Vid. Selsk. Skr., 2 (1989) 1–26.
(Cited on page 177.)

101. C. Brezinski, A direct proof of the Christoffel-Darboux identity and its equivalence to the recurrence relationship, J. Comput. Appl. Math., 32 (1990) 17–25.
(Cited on pages 39 and 188.)

102. C. Brezinski, *History of Continued Fractions and Padé Approximants*, Springer-Verlag, Berlin, 1990.
(Cited on pages 2, 7, 26, 31, 74, 75, 89, and 131.)

103. C. Brezinski, *A Bibliography on Continued Fractions, Padé Approximation, Sequence Transformation, and Related Subjects*, Prensas Universitarias, Universidad Zaragoza, Zaragoza, 1991.
(Cited on page 6.)

104. C. Brezinski, Generalizations of the Christoffel-Darboux identity for adjacent families of orthogonal polynomials, Appl. Numer. Math., 8 (1991) 193–199.
(Cited on page 190.)

105. C. Brezinski, *Biorthogonality and Its Applications to Numerical Analysis*, Marcel Dekker, New York, 1992.
(Cited on page 191.)

106. C. Brezinski, The generalizations of Newton's interpolation formula due to Mühlbach and Andoyer, Elect. Trans. Numer. Anal., 2 (1994) 130–137.
(Cited on page 175.)

107. C. Brezinski, Formal orthogonality on an algebraic curve, Annals Numer. Math., 2 (1995) 21–33.
(Cited on page 71.)

108. C. Brezinski, *Projection Methods for Systems of Equations*, North-Holland, Amsterdam, 1997.
(Cited on page 2.)

109. C. Brezinski, Convergence acceleration during the 20th century, J. Comput. Appl. Math., 122 (2000) 1–21.
(Cited on pages 57 and 173.)

110. C. Brezinski, *Computational Aspects of Linear Control*, Kluwer, Dordrecht, 2002.
(Cited on pages 158, 196, and 206.)

111. C. Brezinski, Extrapolation algorithms for filtering series of functions, and treating the Gibbs phenomenon, Numer. Algorithms, 36 (2004) 309–329.
(Cited on page 117.)

112. C. Brezinski, Some pioneers of extrapolation methods, in *The Birth of Numerical Analysis*, A. Bultheel and R. Cools eds., World Scientific Publ. Co., Singapore, 2009, pp. 1–22.
(Cited on page 57.)

113. C. Brezinski, Cross rules and non-Abelian lattice equations for the discrete and confluent non-scalar ε-algorithms, J. Phys. A: Math. Theor., 43 (2010) 205201.
(Cited on page 214.)

114. C. Brezinski, From numerical quadrature to Padé approximation, Appl. Numer. Math., 60 (2010) 1209–1220.
(Cited on pages 33 and 35.)

115. C. Brezinski, Reminiscences of Peter Wynn, Numer. Algorithms, 80 (2019) 5–11.
(Cited on pages 167 and 304.)

116. C. Brezinski, J.P. Chehab, Nonlinear hybrid procedures and fixed point iterations, Numer. Funct. Anal. Optimization, 19 (1998) 465–487.
(Cited on page 195.)

117. C. Brezinski, J.P. Chehab, Multiparameter iterative schemes for the solution of systems of linear and nonlinear equations, SIAM J. Sci. Comput., 20 (1999) 2140–2159.
(Cited on page 195.)

118. C. Brezinski, M. Crouzeix, Remarques sur le procédé Δ^2 d'Aitken, C. R. Acad. Sci. Paris, 270 A (1970) 896–898.
(Cited on page 22.)

119. C. Brezinski, Y. He, X.-B. Hu, M. Redivo-Zaglia, J.-Q. Sun, Multistep ε-algorithm, Shanks transformation, and the Lotka-Volterra system by Hirota's method, Math. Comp., 81 (2012) 1527–1549.
(Cited on page 214.)

120. C. Brezinski, Y. He, X.-B. Hu, J.-Q. Sun, H.-W. Tam, Confluent form of the multistep ε-algorithm, and the relevant integrable system, Stud. Appl. Math., 127 (2011) 191–209.
(Cited on page 214.)

121. C. Brezinski, A. Lembarki, The linear convergence of limit periodic continued fractions, J. Comput. Appl. Math., 19 (1987) 75–77.
(Cited on page 201.)

122. C. Brezinski, F.A. Norman, M. Redivo-Zaglia, The legacy of Peter Wynn, in preparation.
(Cited on pages 5, 64, 157, 167, 168, and 342.)

123. C. Brezinski, M. Redivo-Zaglia, *Extrapolation Methods. Theory and Practice*, North-Holland, Amsterdam, 1991.
(Cited on pages 7, 180, 184, 185, and 354.)

124. C. Brezinski, M. Redivo-Zaglia, A general extrapolation procedure revisited, Advances Comput. Math., 2 (1994) 461–477.
(Cited on page 177.)

125. C. Brezinski, M. Redivo-Zaglia, Generalizations of Aitken's process for accelerating the convergence of sequences, Mat. Apl. Comput., 26 (2007) 171–189.
(Cited on page 179.)

126. C. Brezinski, M. Redivo-Zaglia, Extensions of Drummond's process for convergence acceleration, Appl. Numer. Math., 60 (2010) 1231–1241.
(Cited on page 179.)

127. C. Brezinski, M. Redivo-Zaglia, Software included in the book [123], https://www.math.unipd.it/~michela/extracode/Extrapolation_Library.html.
(Cited on pages 7, 92, and 309.)

128. C. Brezinski, M. Redivo-Zaglia, Padé-type rational and barycentric interpolation, Numer. Math., 125 (2013) 89–113.
(Cited on page 197.)

129. C. Brezinski, M. Redivo-Zaglia, The simplified topological ε-algorithms for accelerating sequences in a vector space, SIAM J. Sci. Comput., 36 (2014) A2227–A2247.
(Cited on pages 184 and 185.)

130. C. Brezinski, M. Redivo-Zaglia, New representations of Padé, Padé-type, and partial Padé approximants, J. Comput. Appl. Math., 284 (2015) 69–77.
(Cited on page 197.)

131. C. Brezinski, M. Redivo-Zaglia, Shanks function transformations in a vector space, Appl. Numer. Math., 116 (2017) 57–63.
(Cited on page 181.)

132. C. Brezinski, M. Redivo-Zaglia, The simplified topological ε-algorithms: software and applications, Numer. Algorithms, 74 (2017) 1237–1260.
(Cited on pages 7, 92, 115, 184, and 185.)

133. C. Brezinski, M. Redivo-Zaglia, EPSfun: a Matlab toolbox for the simplified topological ε-algorithm, Netlib (2017), http://www.netlib.org/numeralgo/, na44 package.
(Cited on pages 7, 92, 184, 185, and 308.)

134. C. Brezinski, M. Redivo-Zaglia, Hirota's bilinear method, Shanks transformation, and the ε-algorithms, Rev. Roumaine Math. Pures Appl., 63 (2018) 361–375.
(Cited on page 214.)

135. C. Brezinski, M. Redivo-Zaglia, The genesis and early developments of Aitken's process, Shanks transformation, the ε-algorithm, and related fixed point methods, Numer. Algorithms, 80 (2019) 11–133.
(Cited on pages 16, 20, 22, 23, 57, 59, 65, 80, 83, 86, 187, 196, 311, 319, 329, 331, and 347.)

136. C. Brezinski, M. Redivo-Zaglia, Extrapolation methods for the numerical solution of nonlinear Fredholm integral equations, J. Integral Equations Appl., 31 (2019) 29–57.
(Cited on page 115.)

137. C. Brezinski, M. Redivo-Zaglia, Y. Saad, Shanks sequence transformations and Anderson acceleration, SIAM Rev., 60 (2018) 646–669.
(Cited on pages 187 and 196.)

138. C. Brezinski, A.C. Rieu, The solution of systems of equations using the vector ε-algorithm, and an application to boundary value problems, Math. Comp., 28 (1974) 731–741.
(Cited on page 119.)

139. C. Brezinski, H. Sadok, Vector sequence transformations and fixed point methods, in *Numerical Methods in Laminar and Turbulent Flows*, C. Taylor et al. eds., Pineridge Press, Swansea, 1987, pp. 3–11.
(Cited on page 196.)

140. C. Brezinski, H. Sadok, Lanczos type algorithms for solving systems of linear equations, Appl. Numer. Math., 11 (1993) 443–473.
(Cited on pages 191 and 193.)

141. C. Brezinski, D. Tournès, *André-Louis Cholesky (1875–1918), Mathematician, Topographer and Army Officer*, Birkhäuser, Basel, 2014.
(Cited on page 336.)

142. C. Brezinski, J. van Iseghem, Padé approximations, in *Handbook of Numerical Analysis*, vol. III, P.G. Ciarlet and J.L. Lions eds., North-Holland, Amsterdam, 1994, pp. 47–222.
(Cited on pages 29, 197, and 199.)

143. C. Brezinski, J. van Iseghem, A taste of Padé approximation, in *Acta Numerica 1995*, A. Iserles ed., Cambridge University Press, Cambridge, 1995, pp. 53–103.
(Cited on pages 29 and 199.)

144. C. Brezinski, G. Walz, Sequences of transformations and triangular recursion schemes, with applications in numerical analysis, J. Comput. Appl. Math., 34 (1991) 361–383.
(Cited on page 177.)

145. M.G. de Bruin, *Generalized C-Fractions and a Multidimensional Padé Table*, Thesis, Amsterdam, 1964.
(Cited on page 143.)

146. M.G. de Bruin, Simultaneous Padé approximation and orthogonality, in *Polynômes Orthogonaux et Applications, Proceedings of the Laguerre Symposium held at Bar-le-Duc, October 15–18, 1984*, C. Brezinski, A. Draux, A.P. Magnus, P. Maroni, A. Ronveaux eds., Lecture Notes in Mathematics, vol. 1171, Springer-Verlag, Berlin, 1984, pp. 74–83.
(Cited on page 198.)

147. D. Bubenik, A practical method for the numerical evaluation of Sommerfeld integrals, IEEE Trans. Antennas Propag., 25 (1977) 904–906.
(Cited on page 180.)

148. R.A. Buckingham, *Numerical Methods*, Pitman and sons, London, 1957
     (Cited on page 51.)
149. R. Bulirsch, J. Stoer, Fehlerabschätzungen und Extrapolation mit rationalen Funktionen bei
     Verfahren vom Richardson-Typus, Numer. Math., 6 (1964) 413–427.
     (Cited on page 120.)
150. R. Bulirsch, J. Stoer, Numerical treatment of ordinary differential equations by extrapolation
     methods, Numer. Math., 8 (1966) 1–13.
     (Cited on page 120.)
151. A. Bultheel, *Laurent Series and their Padé Approximations*, Birkhäuser, Basel, 1987.
     (Cited on page 197.)
152. A. Bultheel, P. González-Vera, E. Hendriksen, O. Njåstad, *Orthogonal Rational Functions*,
     Cambridge University Press, Cambridge, 1999.
     (Cited on page 191.)
153. A. Bultheel, M. van Barel, Padé techniques for model reduction in linear system theory: a
     survey, J. Comput. Appl. Math., 14 (1986) 401–438.
     (Cited on pages 161 and 196.)
154. A. Bultheel, M. van Barel, *Linear Algebra, Rational Approximation and Orthogonal Polyno-
     mials*, North-Holland, Amsterdam, 1997.
     (Cited on page 206.)
155. D. Buoso, A. Karapiperi, S. Pozza, Generalizations of Aitken's process for a certain class of
     sequences, Appl. Numer. Math., 90 (2015) 38–54.
     (Cited on page 180.)
156. H. Cabannes ed., *Padé Approximants Method and Its Applications to Mechanics*, Lecture
     Notes in Physics, vol. 47, Springer, Berlin, Heidelberg, 1976.
     (Cited on page 196.)
157. S. Cabay, L.W. Jackson, A polynomial extrapolation method for finding limits and antilimits
     of vector sequences, SIAM J. Numer. Anal., 13 (1976) 734–752.
     (Cited on page 187.)
158. F. Cala Rodriguez, H. Wallin, Padé-type approximants and a summability theorem by Eier-
     mann, J. Comput. Appl. Math., 39 (1992) 15–21.
     (Cited on page 197.)
159. E. Caliceti, M. Meyer-Hermann, P. Ribeca, A. Surzhykov, U.D Jentschura, From useful
     algorithms for slowly convergent series to physical predictions based on divergent perturbative
     expansions, Phys. Rep., 446 (2007) 1–96.
     (Cited on page 203.)
160. T. Carleman, Sur le problème des moments, C.R. Acad. Sci. Paris, 174 (1922) 1680–1682.
     (Cited on page 77.)
161. R.P. Carlisle, *Where the Fleet Begins. A History of the David Taylor Research Center, 1898–
     1998*, Naval Historical Center, Department of the Navy, Washington, 1998.
     (Cited on page 330.)
162. R. D. Carmichael, General aspects of the theory of summable series, Bull. Amer. Math. Soc.,
     25 (1918) 97–131.
     (Cited on page 126.)
163. C. Carstensen, On a general epsilon algorithm, in *Numerical and Applied Mathematics*, C.
     Brezinski ed., Baltzer, Basel, 1989, pp.437–441.
     (Cited on page 177.)
164. A.L. Cauchy, *Analyse Algébrique*, Imprimerie Royale, Paris, 1821; reprint Éditions Jacques
     Gabay, Sceaux, 1989.
     (Cited on page 26.)
165. J. Chandra, S.M. Robinson, *An Uneasy Alliance. The Mathematics Research Center at the
     University of Wisconsin, 1956–1987*, SIAM, Philadelphia, 2005.
     (Cited on page 318.)
166. T.S. Chihara, *An Introduction to Orthogonal Polynomials*, Gordon and Breach, New York,
     1978.
     (Cited on page 146.)

167. J.S.R. Chisholm, Applications of Padé approximation to numerical integration, Rocky Mountain J. Math., 4 (1974) 159–168.
    (Cited on page 180.)

168. J.S.R. Chisholm, A.K. Common, Generalisations of Padé approximation for Chebyshev and Fourier series, in *E. B. Christoffel: The Influence of His Work on Mathematics and the Physical Sciences*, P.L. Butzer and F. Fehèr eds., Birkhäuser, Basel, 1981, pp. 212–231.
    (Cited on page 197.)

169. E.B. Christoffel, *De motu permanenti electricitatis in corporibus homogeneis*, Inaugural Dissertation, Berlin, 1856.
    (Cited on page 136.)

170. E.B. Christoffel, Über die Gaussische Quadratur und eine Verallgemeinerung derselben, J. Reine Angew. Math., 55 (1858) 61–82.
    (Cited on page 136.)

171. E.B. Christoffel, Observatio arithmetica, Ann. Math. Pura Appl., II, 6 (1875) 148–153.
    (Cited on page 75.)

172. E.B. Christoffel, Sur une classe particulière de fonctions entières et de fractions continues, Ann. Mat. Pura Appl., Serie II, 8 (1877) 1–10.
    (Cited on page 136.)

173. E.B. Christoffel, Lehrsätze über arithmetische Eigenschaften der Irrationalzahlen, Ann. Mat. Pura Appl., II, 15 (1888) 253–276.
    (Cited on page 75.)

174. G. Claessens, On the Newton-Padé approximation problem, J. Approx. Theory, 22 (1978) 150–160.
    (Cited on page 197.)

175. G. Claessens, A useful identity for the rational Hermite interpolation table, Numer. Math., 29 (1978) 227–231.
    (Cited on page 198.)

176. A.M. Cohen, *Numerical Methods for Laplace Transform Inversion*, Springer, New York, 2007.
    (Cited on page 196.)

177. L.J. Comrie, On the construction of tables by interpolation, Mon. Not. Roy. Astron. Soc., 88 (1928) 506–523.
    (Cited on page 65.)

178. A. Connes, *Noncommutative Geometry*, Academic Press, San Diego, 1994.
    (Cited on page 215.)

179. F. Cordellier, Particular rules for the vector ε-algorithm, Numer. Math., 27 (1977) 203–207.
    (Cited on page 101.)

180. F. Cordellier, Démonstration algébrique de l'extension de l'identité de Wynn aux tables de Padé non normales, in *Padé Approximation and Its Applications*, L. Wuytack ed., Lecture Notes in Mathematics, vol. 765, Springer-Verlag, Berlin, 1979, pp. 36–60.
    (Cited on page 92.)

181. F. Cordellier, *Interpolation Rationnelle et autres Questions : Aspects Algorithmiques et Numériques*, Thèse de Doctorat d'État ès Sciences Mathématiques, Université des Sciences et Techniques de Lille, 1989.
    (Cited on page 92.)

182. M. Crouzeix, F. Ruamps, On rational approximations to the exponential, RAIRO Numer. Anal., 11 (1977) 241–243.
    (Cited on page 198.)

183. A.A.M. Cuyt, The epsilon-algorithm and multivariate Padé approximants, Numer. Math., 40 (1982) 39–46.
    (Cited on page 181.)

184. A.A.M. Cuyt. The epsilon-algorithm and Padé approximants in operator theory, SIAM J. Math. Anal., 14 (1983) 1009–1014.
    (Cited on page 181.)

185. A. Cuyt, *Padé Approximants for Operators: Theory and Applications*, Lecture Notes in Mathematics, vol. 1065, Springer-Verlag, Berlin, 1984.
(Cited on page 146.)

186. A.A.M. Cuyt, V. Petersen, B.M. Verdonk, H. Waadeland, W.B. Jones, *Handbook of Continued Fractions for Special Functions*, Springer, Heidelberg, 2008.
(Cited on page 201.)

187. A.A.M. Cuyt, B.M. Verdonk, A review of branched continued fraction theory for the construction of multivariate rational approximants, J. Comput. Appl. Math., 4 (1988) 263–271.
(Cited on page 201.)

188. A. Cuyt, L. Wuytack, *Nonlinear Methods in Numerical Analysis*, North-Holland, Amsterdam, 1987.
(Cited on pages 47, 121, and 180.)

189. G. Dahlquist, A special stability problem for linear multistep methods, BIT, 3 (1963) 27–43.
(Cited on page 198.)

190. J.W. Daniel, Summation of series of positive terms by condensation transformations, Math. Comp., 23 (1969) 91–96.
(Cited on page 111.)

191. J.W. Daniel, Extrapolation with spline-collocation methods for two-point boundary-value problems I: Proposals and justifications, Aequationes Math., 16 (1977) 107–122.
(Cited on page 172.)

192. J.W. Daniel, V. Pereyra, L. Schumaker, Iterated deferred corrections for initial value problems, Acta Cient. Venezolana, 19 (1968) 128–135.
(Cited on page 172.)

193. J.W. Daniel, B.K. Swartz, Extrapolated collocation for two-point boundary-value problems using cubic splines, J. Inst. Maths Applics, 16 (1975) 161–174.
(Cited on page 172.)

194. G. Darboux, Sur l'approximation des fonctions de très-grands nombres et sur une classe étendue de développements en série, J. Math. Pures Appl., 3e série, 4 (1878) 5–56; 377–416.
(Cited on page 137.)

195. J.P. Delahaye, Liens entre la suite du rapport des erreurs et celle du rapport des différences, C. R. Acad. Sci. Paris, 290 A (1980) 343–346.
(Cited on page 19.)

196. J.P. Delahaye, *Sequence Transformations*, Springer-Verlag, Berlin, 1988.
(Cited on page 7.)

197. J.P. Delahaye, B. Germain-Bonne, Résultats négatifs en accélération de la convergence, Numer. Math., 35 (1980) 443–457.
(Cited on page 1.)

198. A.P. Dempster, N.M. Laird, D.B. Rubin, Maximum Likelihood from Incomplete Data via the EM Algorithm (with discussion), J. Roy. Statist. Soc., ser. B, 39 (1977) 1–38.
(Cited on page 195.)

199. H. Denk, M. Riederle, A generalization of a theorem of Pringsheim, J. Approx. Theory, 35 (1982) 355–363.
(Cited on page 202.)

200. H. Dette, W.J. Studden, *The Theory of Canonical Moments with Applications in Statistics, Probability, and Analysis*, John Wiley, New York, 1997.
(Cited on page 42.)

201. F. Diacu, The solution of the n-body problem, Math. Intelligencer, 18 (1996) 66–70.
(Cited on page 203.)

202. J. Dieudonné, *Calcul Infinitésimal*, Hermann, Paris, 1968.
(Cited on page 165.)

203. A. Doliwa, Non-commutative double-sided continued fractions, arXiv:1905.10429, 2019.
(Cited on page 202.)

204. E. de Doncker, An adaptive extrapolation algorithm for automatic integration, ACM SIGNUM Newsletter, 13 (1978) 12–18.
(Cited on page 180.)

205. W.F. Donoghue, *Monotone Matrix Functions and Analytic Continuation*, Springer, New York-Heidelberg-Berlin, 1974.
    (Cited on page 205.)

206. A. Draux, *Polynômes Orthogonaux Formels. Applications*, Lecture Notes in Mathematics, vol. 974, Springer-Verlag, Berlin, 1983.
    (Cited on pages 7, 93, 190, and 193.)

207. A. Draux, The Padé approximants in a non-commutative algebra and their applications, in *Padé Approximation and Its Applications. Bad Honnef 1983*, H. Werner and H.J. Bünger eds., Lecture Notes in Mathematics, vol. 1071, Springer, Berlin-Heidelberg, 1984, pp. 117–131.
    (Cited on page 200.)

208. A. Draux, The epsilon algorithm in a non-commutative algebra, J. Comput. Appl. Math., 19 (1987) 9–21.
    (Cited on pages 181 and 201.)

209. A. Draux, Convergence of Padé approximants in a non-commutative algebra, in *Approximation and Optimization*, J.A. Gómez-Fernandez et. al. eds., Lecture Notes in Mathematics, vol. 1354, Springer, Berlin-Heidelberg, 1988, pp. 118–130.
    (Cited on page 181.)

210. A. Draux, P. Van Ingelandt, *Polynômes Orthogonaux et Approximants de Padé, Logiciels*, Éditions Technip, Paris, 1987.
    (Cited on page 190.)

211. P.G. Drazin ed., *Collected Papers of Lewis Fry Richardson, vol. 1, parts 1 and 2, Meteorology and Numerical Analysis*, Cambridge University Press, Cambridge, 1993.
    (Cited on pages 56 and 323.)

212. T.A. Driscoll, B. Fornberg, A Padé-based algorithm for overcoming the Gibbs phenomenon, Numer. Algorithms, 26 (2001) 77–92.
    (Cited on page 118.)

213. J.E. Drummond, A formula for accelerating the convergence of a general series, Bull., Aust. Math. Soc., 6 (1972) 69–74.
    (Cited on page 179.)

214. S. Dumas, *Sur le Développement des Fonctions Elliptiques en Fractions Continues*, Thèse, Université de Zürich, 1908.
    (Cited on page 73.)

215. S. Duminil, H. Sadok, D.B. Szyld, Nonlinear Schwarz iterations with reduced rank extrapolation, Appl. Numer. Math., 94 (2015) 209–221.
    (Cited on page 187.)

216. J. Dutka, Richardson extrapolation and Romberg integration, Historia Math., 11 (1984) 3–21.
    (Cited on page 57.)

217. F.J. Dyson, Divergence of perturbation theory in quantum electrodynamics, Phys. Rev., 85 (1952) 631–632.
    (Cited on page 203.)

218. R.P. Eddy, Extrapolation to the limit of a vector sequence, in *Information Linkage between Applied Mathematics and Industry*, P.C.C. Wang ed., Academic Press, New York, 1979, pp. 387–396.
    (Cited on page 187.)

219. B.L. Ehle, A-stable methods and Padé approximation to the exponential, SIAM J. Math. Anal., 4 (1973) 671–680.
    (Cited on page 198.)

220. M. Eiermann, On the convergence of Padé-type approximants to analytic functions, J. Comput. Appl. Math., 10 (1984) 219–227.
    (Cited on page 197.)

221. H. Engels, Zur Geschichte der Richardson-Extrapolation, Historia Math., 6 (1979) 280–293.
    (Cited on page 57.)

222. T.O. Espelid, On integrating vertex singularities using extrapolation, BIT 34 (1994) 62–79.
    (Cited on page 180.)

223. T.O. Espelid, K.J. Overholt, DQAINF: An algorithm for automatic integration of infinite oscillating tails, Numer. Algorithms, 8 (1994) 83–101.
     (Cited on page 180.)

224. L. Euler, De seriebus divergentibus, Novi Commentarii academiae scientiarum Petropolitanae, 5 (1760) 205–237; reprinted in *Opera Omnia*, Series 1, vol. 14, pp. 585–617, Eneström-Number E247, Birkhäuser, Basel; translation in English by A. Aycock, arXiv:1808.02841.
     (Cited on pages 49 and 50.)

225. C. Evans, S.N. Pollock, L.G. Rebholz, M. Xiao, A proof that Anderson acceleration improves the convergence rate in linearly converging fixed-point methods (but not in those converging quadratically), SIAM J. Numer. Anal., 58 (2020) 788–810.
     (Cited on page 196.)

226. V. Eyert, A comparative study on methods for convergence acceleration of iterative vector sequence, J. Comput. Phys., 124 (1996) 271–285.
     (Cited on page 196.)

227. W. Fair, *Formal Continued Fractions and Applications*, Ph.D. Thesis, University of Kansas, 1969.
     (Cited on page 201.)

228. W. Fair, Noncommutative continued fractions, SIAM J. Math. Anal., 2 (1971) 226–232.
     (Cited on page 201.)

229. W. Fair, A convergence theorem for noncommutative continued fractions, J. Approx. Theory, 5 (1972) 74–76.
     (Cited on page 201.)

230. W. Fair, Y.L. Luke, Generalized Rational Approximations with Applications to Problems in Control Theory, Interim Technical Report, Contract No. NAS8-20403, National Aeronautics and Space Administration, George C. Marshall Space Flight Center, Huntsville, Alabama 35812, 14 August 1967.
     (Cited on page 201.)

231. H. Fang, Y. Saad, Two classes of multisecant methods for nonlinear acceleration, Numer. Linear Algebra Appl., 16 (2009) 197–221.
     (Cited on page 196.)

232. P. Feldmann, R.W. Freund, Efficient linear circuit analysis by Padé approximation via the Lanczos process, IEEE Trans. Comput.-Aided Design, 14 (1995) 639–649.
     (Cited on page 208.)

233. T. Fessler, W.F. Ford, D.A. Smith, HURRY: An acceleration algorithm for scalar sequences and series, ACM Trans. Math. Software, 9 (1983) 346–354.
     (Cited on page 179.)

234. T. Fessler, W.F. Ford, D.A. Smith, Algorithm 602, HURRY: An acceleration algorithm for scalar sequences and series, ACM Trans. Math. Software, 9 (1983) 355–357.
     (Cited on page 179.)

235. D.A. Field, Convergence theorems for matrix continued fractions, SIAM J. Math. Anal., 15 (1984) 1220–1227.
     (Cited on page 202.)

236. S.-I. Filip, Y. Nakatsukaza, L.N. Trefethen, B. Beckermann, Rational minimax approximation via adaptive barycentric representations, SIAM J; Sci. Comput., 40 (2018) A2427–A2455.
     (Cited on page 197.)

237. S. Filippi, Das Verfahren von Romberg-Stiefel-Bauer als Spezialfall des allgemeinen Prinzips von Richardson, Teil I und II, Mathematik, Technik, Wirtschaft, Zeitschrift für moderne Rechentechnik und Automation, 11 (1964) 49–54; 98–100.
     (Cited on page 172.)

238. S. Filippi, Die Berechnung einiger elementarer transzendenter Funktionen mit Hilfe des Richardson-Algorithmus, Computing, 1 (1966) 127–132.
     (Cited on page 172.)

239. P. Flajolet. Combinatorial aspects of continued fractions, Discrete Math., 32 (1980) 125–161.
     (Cited on page 211.)

240. P. Flajolet, R. Sedgewick, *Analytic Combinatorics*, Cambridge University Press, Cambridge, 2009.
(Cited on page 212.)

241. P. Flajolet, B. Vallée, Continued fraction algorithms, functional operators, and structure constants, Theoret. Comput. Sci., 194 (1998) 1–34.
(Cited on page 211.)

242. J. Fleischer, Nonlinear Padé approximants for Legendre series, J. Math. Phys., 14 (1973) 246–248.
(Cited on page 197.)

243. R. Fletcher, Conjugate gradient methods for indefinite systems, in *Numerical Analysis*, G.A. Watson ed., Lecture Notes in Mathematics, vol. 506, Springer-Verlag, Berlin, 1976, pp. 73–89.
(Cited on page 192.)

244. M. Fliess, Matrices de Hankel, J. Math. Pures Appl., 53 (1974) 197–222.
(Cited on page 206.)

245. D. Foata, Combinatoire des identités sur les polynômes orthogonaux, in *Proceedings of the International Congress of Mathematicians*, August 16–24, 1983, Warsaw, pp. 1541–1553.
(Cited on page 210.)

246. W.F. Ford, A. Sidi, An algorithm for a generalization of the Richardson extrapolation process, SIAM J. Numer. Anal., 24 (1987) 1212–1232.
(Cited on page 177.)

247. G.E. Forsythe, Solving linear algebraic equations can be interesting, Bull. Amer. Math. Soc., 59 (1963) 299–329.
(Cited on page 178.)

248. L. Fox, Romberg integration for a class of singular integrands, Computer J., 10 (1967) 87–93.
(Cited on page 175.)

249. R. Frank, C.W. Ueberhuber, Iterated defect correction for the efficient solution of stiff systems of ordinary differential equations, BIT 17 (1977) 146–159.
(Cited on page 172.)

250. S.P. Frankel, Convergence rates of iterative treatments of partial differential equations, Math. Tables Aids Comput., 4 (1950), 65–75.
(Cited on page 55.)

251. G. Frobenius, Ueber Relationen zwischen den Näherungsbruchen von Potenzreihen, J. Reine Angew. Math., 90 (1881) 1–17.
(Cited on pages 26 and 72.)

252. M. Froissart, Asymptotic behavior and subtractions in the Mandelstam representation, Phys. Rev., 123 (1961) 1053–1057.
(Cited on page 30.)

253. M. Froissart, Approximation de Padé. Application à la physique des particules élémentaires, in *Les Rencontres Physiciens-Mathématiciens de Strasbourg—RCP 25*, Tome 9, 1969, exp. no 2, pp. 1–13.
(Cited on page 204.)

254. B. Gabutti, On two methods fo accelerating convergence of series, Numer. Math., 43 (1984) 439–461.
(Cited on page 170.)

255. B. Gabutti, An algorithm for computing generalized Euler's transformations of series, Computing, 34 (1985) 107–116.
(Cited on page 170.)

256. B. Gabutti, J.N. Lyness, Some generalizations of the Euler-Knopp transformation, Numer. Math., 48 (1986) 199–220.
(Cited on page 170.)

257. W. Gander, G.H. Golub, D. Gruntz, Solving linear equations by extrapolation, in *Supercomputing*, J.S. Kowalik ed., NATO ASI Series, vol. F 62, Springer-Verlag, Berlin, Heidelberg, 1990, pp. 279–293.
(Cited on page 187.)

258. C.R. Garibotti, F.F. Grinstein, A summation procedure for expansions in orthogonal polyno-
      mials, Rev. Brasileira Fís., 7 (1977) 557–567.
      (Cited on page 197.)
259. D. Garreau, B. Georgel, La méthode de Prony en analyse des vibrations, Trait. Signal, 3
      (1986) 235–240.
      (Cited on page 161.)
260. N. Gastinel ed., *Procédures Algol en Analyse Numérique*, Tome II, Éditions du Centre National
      de la Recherche Scientifique, Paris, 1970.
      (Cited on page 7.)
261. J.A. Gaunt, The deferred approach to the limit. Part II: Interpenetrating lattices, Philos. Trans.
      Roy. Soc. London, ser. A, 226 (1927) 350–361.
      (Cited on pages 50, 55, and 322.)
262. W. Gautschi, Computational aspects of three-term recurrence relations, SIAM Rev., 9 (1967)
      24–82.
      (Cited on page 66.)
263. E. Gekeler, Über den ε-Algorithmus von Wynn, ZAMM, 51 (1971) 53–54.
      (Cited on page 118.)
264. E. Gekeler, On the solution of systems of equations by the epsilon algorithm of Wynn, Math.
      Comp., 26 (1972) 427–436.
      (Cited on pages 118 and 196.)
265. I. Gelfand, D. Krob, A. Lascoux, B. Leclerc, V.S. Retakh, J.-Y. Thibon, Noncommuta-
      tive symmetric functions, Adv. Math., 112 (1995) 218–348; https://hal.archives-ouvertes.fr/
      hal-00017721.
      (Cited on page 210.)
266. A. Genz, *The Approximate Calculation of Multidimensional Integrals Using Extrapolation
      Methods*, Ph.D. thesis, University of Kent, Canterbury, 1975.
      (Cited on page 180.)
267. Ja. L. Geronimus, On polynomials orthogonal with respect to a given numerical sequence (in
      Russian; English summary), Zap. Naučno-Isssled. Inst. Mat. Meh. i Har'kov. Mat. Obšč., (4)
      17 (1940) 3–18.
      (Cited on page 70.)
268. Ja. L. Geronimus, On polynomials orthogonal with respect to a given numerical sequence,
      and on a theorem of W. Hahn (in Russian; French summary), Izv. Akad. Nauk SSSR, 4 (1940)
      215–228.
      (Cited on page 70.)
269. Ja. L. Geronimus, Orthogonal polynomials, Amer. Math. Soc. Transl., Ser. 2, 108 (1977)
      37–130.
      (Cited on page 70.)
270. J. Gilewicz, Numerical detection of the best Padé approximant and determination of the
      Fourier coefficients of insufficiently sampled function, in *Padé Approximants and Their
      Applications*, P.R. Graves-Morris ed., Academic Press, New York, 1973, pp. 99–103.
      (Cited on pages 122 and 177.)
271. J. Gilewicz, *Approximants de Padé*, Lecture Notes in Mathematics, vol. 667, Springer-Verlag,
      Berlin-Heidelberg, 1978.
      (Cited on page 7.)
272. J. Gilewicz, Y. Kryakin, Froissart doublets in Padé approximation in the case of polynomial
      noise, J. Comput. Appl. Math., 153 (2003) 235–242.
      (Cited on page 29.)
273. J. Gilewicz, E. Leopold, Padé approximant inequalities for the functions of the class S, in *Padé
      Approximation and Its Applications. Amsterdam 1980*, M.G. de Bruin and H. van Rossum
      eds., Lecture Notes in Mathematics, vol. 888. Springer, Berlin, 1981, pp. 208–219.
      (Cited on page 122.)
274. J. Gilewicz, M. Pindor, Padé-type approximants and errors of Padé approximants, J. Comput.
      Appl. Math., 99 (1998) 155–165.
      (Cited on page 197.)

275. J.W.L. Glaisher, On the transformation of continued products into continued fractions, Proc. London Math. Soc., 5 (1873/1874) 78–89.
   (Cited on page 138.)

276. G.H. Golub, G. Meurant, *Matrices, Moments and Quadrature with Applications*, Princeton University Press, Princeton, 2010.
   (Cited on page 191.)

277. G.H. Golub, D.P. O'Leary, Some history of the conjugate and Lanczos algorithms, SIAM Rev., 31 (1989) 50–102.
   (Cited on page 192.)

278. P. Gonnet, S. Güttel, L.N. Trefethen, Robust Padé approximation via SVD, SIAM Rev., 55 (2013) 101–117.
   (Cited on page 197.)

279. C. González-Concepción, M.C. Gil-Fariña, Padé approximation in economics, Numer. Algorithms, 33 (2003) 277–292.
   (Cited on page 212.)

280. E.T. Goodwin ed., *Modern Computing Methods*, Notes on Applied Science no. 16, National Physical Laboratory, Her Majesty's Stationery, Office, London, 1960.
   (Cited on page 51.)

281. J.E. Gower, S.S. Wagstaff, Jr., Square form factorization, Math. Comp., 77 (2008) 551–588.
   (Cited on page 330.)

282. W.B. Gragg, *Repeated Extrapolation to the Limit in the Numerical Solution of Ordinary Differential Equations*, Ph.D. Thesis, UCLA, Los Angeles, 1963.
   (Cited on page 120.)

283. W.B. Gragg, On extrapolation algorithms for ordinary initial value problems, SIAM J. Numer. Anal., 2 (1965) 384–403.
   (Cited on page 120.)

284. W.B. Gragg, The Padé table and its relation to certain algorithms of numerical analysis, SIAM Rev., 14 (1972) 1–62.
   (Cited on pages 7 and 148.)

285. L.G. Grandi, *Quadratura circula et hyperbolae per infinitas hyperbolas geometrice exhibita*, Francesco Bindi, Pisis, 1710.
   (Cited on page 49.)

286. P.R. Graves-Morris ed., *Padé Approximants and Their Applications*, Academic Press, London and New York, 1973.
   (Cited on page 196.)

287. P.R. Graves-Morris, C.D. Jenkins, Generalised inverse vector-valued rational interpolation, in *Padé Approximation and Its Applications*, H. Werner and H.J. Bünger eds., Lecture Notes in Mathematics, vol. 1071, Springer, Berlin, 1984, pp. 144–156.
   (Cited on page 102.)

288. P.R. Graves-Morris, D.E. Roberts, A. Salam, The epsilon algorithm and related topics, J. Comput. Appl. Math., 122 (2000) 51–80.
   (Cited on page 102.)

289. H.L. Gray, T.A. Atchison, G.V. McWilliams, Higher order G-transformations, SIAM J. Numer. Anal. 8 (1971) 365–381.
   (Cited on page 172.)

290. T.N.E. Greville, On some conjectures of P. Wynn concerning the ε-algorithm, MRC Technical Summary Report # 877, Madison, Wisconsin, May 1968.
   (Cited on page 116.)

291. E. Grimme, D. Sorensen, P. van Dooren, Model reduction of state space systems via an implicitly restarted Lanczos method, Numer. Algorithms, 12 (1996) 1–31.
   (Cited on page 208.)

292. C. Gu, J. Shen, Function-valued Padé-type approximant via the formal orthogonal polynomials and its applications in solving integral equations, J. Comput. Appl. Math., 221 (2008) 114–131.
   (Cited on page 197.)

293. C. Gudermann, Umformung einer Reihe von sehr allgemeiner Form, J. Reine Angew. Math.,
     7 (1831) 306–308.
     (Cited on page 105.)
294. C. Guilpin, J. Gacougnolle, Y. Simon, The ε-algorithm allows to detect Dirac delta functions,
     Appl. Numer. Math., 48 (2004) 27–40.
     (Cited on page 117.)
295. P. Gustafson, Biography of William Branham Jones, Rocky Mountain J. Math., 33 (2003)
     381–393.
     (Cited on page 238.)
296. B. Gustafsson, W. Kress, Deferred correction methods for initial value problems, BIT Numer.
     Math., 41 (20001) 986–995.
     (Cited on page 172.)
297. M.H. Gutknecht, The pionner days of scientific computing in Switzerland, in *A History of
     Scientific Computiong*, Stephen G. Nash ed., ACM Press, 1990, pp. 301–313.
     (Cited on page 327.)
298. M.H. Gutknecht, The Lanczos algorithms and their relations to formal orthogonal polyno-
     mials, Padé approximation, continued fractions, and the qd algorithm, in *Proceedings of the
     Copper Mountain Conference on Iterative Methods*, Copper Mountain, Colorado, 1–5 April
     1990, pp. 1–46.
     (Cited on page 193.)
299. M.H. Gutknecht, Numerical Analysis in Zürich—50 years ago, Zürich Intelligencer, Springer-
     Verlag, July 2007, pp. 10–15, (www.sam.math.ethz.ch/~mhg/pub/ICIAMintell.pdf).
     (Cited on page 328.)
300. M.H. Gutknecht, B.N. Parlett, From qd to LR, or, how were the qd and LR algorithms
     discovered?, IMA J. Numer. Anal., 31 (2011) 741–754.
     (Cited on page 328.)
301. J. Hadamard, Éssai sur l'étude des fonctions données par leur developpement de Taylor, J.
     Math. Pures Appl., 8 (1892) 101–186.
     (Cited on page 134.)
302. J. Hadamard, S. Mandelbrojt, *La Série de Taylor et son Prolongement Analytique*, No. 41,
     Scientia, Paris, 1926.
     (Cited on page 50.)
303. W. Hahn, Über die Jacobischen Polynome und zwei verwandte Polynomklassen, Math.
     Zeitschr., 39 (1935) 634–38.
     (Cited on page 70.)
304. W.R. Hamilton, On continued fractions in quaternions, Phil. Mag., ser. 4, 3 (1852) 371–373;
     4 (1852) 303; 5 (1853) 117–118, 236–238, 321–326.
     (Cited on page 77.)
305. H. Hankel, *Über eine besondere Classe der symmetrischen Determinanten*, Dissertation,
     Leipzig Universität, Göttingen, 1861.
     (Cited on pages 21 and 27.)
306. J.F. Hauer, C.J. Demeure, L.L. Scharf, Initial results in Prony analysis of power system
     response signals, IEEE Trans. Power Syst., 5 (1990) 80–89.
     (Cited on page 161.)
307. F. Hausdorff, Summationsmethoden und Momentfolgen. I, Math. Z., 9 (1921) 74–109.
     (Cited on page 93.)
308. F. Hausdorff, Summationsmethoden und Momentfolgen. II, Math. Z., 9 (1921) 280–299.
     (Cited on page 93.)
309. T. Håvie, Generalized Neville type extrapolation schemes, BIT, 19 (1979) 204–213.
     (Cited on page 175.)
310. T. Håvie, Remarks on a unified theory for classical and generalized interpolation and extrap-
     olation, BIT, 21 (1981) 465–474.
     (Cited on page 177.)

311. T. Håvie, Romberg integration and classical analysis, in *Romberg Seminar on Quadrature, Interpolation, Extrapolation and Rational Approximations*, T. Håvie ed., Det Kongelige Norske Videnskabers Selskab, Skrifter 2, Tapir Forlag, Trondheim, 1989, pp. 60–72.
(Cited on page 57.)

312. T. Håvie ed., *Romberg Seminar on Quadrature, Interpolation, Extrapolation and Rational Approximations*, Det Kongelige Norske Videnskabers Selskab, Skrifter 2, Tapir Forlag, Trondheim, 1989.
(Cited on page 325.)

313. D.R. Hartree, *Numerical Analysis*, Clarendon Press, Oxford, 1952.
(Cited on page 51.)

314. T.L. Hayden, Continued fractions in Banach space, Rocky Mountain J. Math., 4 (1974) 367–369.
(Cited on page 201.)

315. Y. He, X.-B. Hu, J.-Q. Sun, E.J. Weniger, Convergence acceleration algorithm via the lattice Boussinesq equation. SIAM J. Sci. Comput., 33 (2011) 1234–1245.
(Cited on page 213.)

316. Y. He, X.-B. Hu, H.-W. Tam, A q-difference version of the ε-algorithm, J. Phys. A, 42 (2009) 5202–5210.
(Cited on page 214.)

317. E. Hellinger, Zur Stieltjesschen Kettenbruchtheorie, Math. Ann., 86 (1922) 18–29.
(Cited on page 77.)

318. P. Henrici, The quotient difference algorithm, NBS Appl. Math. Series, 49 (1958) 23–46.
(Cited on page 148.)

319. P. Henrici, *Elements of Numerical Analysis*, Wiley, New York, 1964.
(Cited on page 195.)

320. P. Henrici, An algorithm for analytic continuation, SIAM J. Numer. Anal., 3 (1966) 67–78.
(Cited on page 50.)

321. P. Henrici, *Applied and Computational Complex Analysis*, vol. 1, Wiley, New York, 1974.
(Cited on page 46.)

322. P. Henrici, P. Pfluger, Truncation error estimates for Stieltjes fractions, Numer. Math. 9 (1966) 120–158.
(Cited on page 201.)

323. C. Hermite, Sur la fonction exponentielle, C. R. Acad. Sci. Paris, 77 (1873) 18–24, 74–79, 226–233, 285–293.
(Cited on page 28.)

324. C. Hermite, Sur la généralisation des fractions continues algébriques, Ann. Mat. Pura Appl., Ser. 11, 21 (1893) 289–308.
(Cited on page 28.)

325. C. Hespel, Approximation de séries formelles par des séries rationnelles, RAIRO Inform. Théor., 18 (1984) 241–258.
(Cited on page 208.)

326. C. Hespel, G. Jacob, Approximation of nonlinear dynamical systems by rational series, Theoret. Comput. Sci., 79 (1991) 151–162.
(Cited on page 208.)

327. M.R. Hestenes, E. Stiefel, Methods of conjugate gradients for solving linear systems, J. Res. Natl. Bur. Stand., 49 (1952) 409–436.
(Cited on pages 116 and 192.)

328. A. Heyting, Die Theorie der linearen Gleichungen in einer Zahlenspezies mit nichtkommutativer Multiplikation, Math. Ann., 98 (1927) 465–490.
(Cited on pages 10 and 12.)

329. D. Hilbert, Grundzüge einer allgemeinen Theorie der linearen Integralgleichungen, Nach. König. Ges. Wiss. Göttingen, Math. Phys. Klasse, Gött. Nachr., I–IV (1904), 49–91; (1904), 213–259; (1905), 307–338; (1906), 157–227; (1906), 439–480; (1910), 355–419.
(Cited on page 76.)

330. F.B. Hildebrand, *Introduction to Numerical Analysis*, McGraw-Hill, New York, 1956.
     (Cited on page 51.)

331. R. Hirota, *The Direct Method in Soliton Theory*, Cambridge Univ. Press, Cambridge, 1992.
     (Cited on page 214.)

332. J.T. Holdeman Jr., A method for the approximation of functions defined by formal series expansions in orthogonal polynomials, Math. Comp., 23 (1969) 275–287.
     (Cited on page 197.)

333. H.H.H. Homeier, A Levin-type algorithm for accelerating the convergence of Fourier series, Numer. Algorithms, 3 (1992) 245–254.
     (Cited on page 179.)

334. H.H.H. Homeier, Some applications of nonlinear convergence accelerators, Int. J. Quantum Chem., 45 (1993) 545–562.
     (Cited on pages 177 and 179.)

335. H.H.H. Homeier, A hierarchically consistent, iterative sequence transformation, Numer. Algorithms, 8 (1994) 47–81.
     (Cited on page 177.)

336. A.S. Householder, *Principles of Numerical Analysis*, McGraw-Hill, New York, 1953.
     (Cited on page 51.)

337. A.S. Householder, *The numerical treatment of a single nonlinear equation*, McGraw-Hill, New York, 1970.
     (Cited on page 51.)

338. A.S. Householder, The Padé table, the Frobenius identities, and the qd algorithm, Linear Algebra Appl., 4 (1971) 161–174.
     (Cited on page 190.)

339. J.C.R. Hunt, A general introduction to the life and work of L.F. Richardson, in *Collected Papers of Lewis Fry Richardson, Vol. 1, Parts 1 and 2, Meteorology and Numerical Analysis*, P.G. Drazin ed., Cambridge University Press, Cambridge, 1993, pp. 1–27.
     (Cited on page 323.)

340. C. Huygens, *De Circuli Magnitudine Inventa. Accedunt eiusdem problematum quorundam illustrium constructiones*, Lugdunum Batavorum, Johann and Daniel Elzevir, Leiden, 1654.
     (Cited on page 57.)

341. E. Isaacson, Mathematics of Computation: a brief history, Proc. Sympos. Appl. Math., Amer. Math. Soc. Providence, RI, vol. 48, 1994, pp. xvii–xx.
     (Cited on page 87.)

342. A. Iserles, On the A-acceptatbility of Padé approximations, SIAM J. Math. Anal., 10 (1979) 1002–1007.
     (Cited on page 198.)

343. A. Iserles, Complex dynamics of convergence acceleration, IMA J. Numer. Anal., 11 (1991) 205–240.
     (Cited on page 195.)

344. A. Iserles, Convergence acceleration as dynamical system, Appl. Numer. Math., 15 (1994) 101–121.
     (Cited on page 195.)

345. M. Ivan, Some forms of the Lagrange-Waring interpolation polynomial, Automat. Comput. Appl. Math., 13 (2004) 107–122.
     (Cited on page 33.)

346. M. Ivan, *Elements of Interpolation Theory*, Mediamira, Cluj-Napoca, 2004.
     (Cited on page 33.)

347. M. Ivan, A note on the Hermite interpolation polynomial for rational functions, Appl. Numer. Math., 57 (2007) 230–233.
     (Cited on page 33.)

348. C.G.J. Jacobi, Über die Darstellung einer Reihe gegebener Werthe durch einer gebrochnen rationale Funktion, J. Reine Angew. Math.,30 (1845) 127–156.
     (Cited on pages 26 and 123.)

349. C.G.J. Jacobi, Allgemeine Theorie der Kettenbruchähnlichen Algorithmen, in welchen jede Zahl aus drei vorhergehenden gebildet wird, J. Reine Angew. Math., 69 (1868) 29–64.
(Cited on page 28.)

350. H. Jager, A simultaneous generalization of the Padé table, I–VI, Indag. Math., 67 (1964) 193–249.
(Cited on page 28.)

351. K. Jbilou, *Méthodes d'Extrapolation et de Projection. Applications aux Suites de Vecteurs*, Thèse de 3ème cycle, Université des Sciences et Techniques de Lille, 1988.
(Cited on page 185.)

352. K. Jbilou, A general projection algorithm for solving systems of linear equations, Numer. Algorithms, 4 (1993) 361–397.
(Cited on page 187.)

353. K. Jbilou, H. Sadok, Some results about vector extrapolation methods and related fixed point iterations, J. Comp. Appl. Math., 36 (1991) 385–398.
(Cited on pages 185 and 196.)

354. K. Jbilou, H. Sadok, LU implementation of the modified minimal polynomial extrapolation method for solving linear and nonlinear systems, IMA J. Numer. Anal., 19 (1999) 549–561.
(Cited on page 187.)

355. K. Jbilou, H. Sadok, Vector extrapolation methods. Applications and numerical comparison, J. Comput. Appl. Math., 122 (2000) 149–165.
(Cited on page 187.)

356. H. Jensen, An attempt at a systematic classification of some methods for the solution of normal equations, Meddelelse No. 18, Geodætisk Insitut, Kobenhavn, 1944, 45 pp.
(Cited on page 336.)

357. U.D. Jentschura, P.J. Mohr, G. Soff, E.J. Weniger, Convergence acceleration via combined nonlinear-condensation transformations, Comput. Phys. Comm., 116 (1999) 28–54.
(Cited on page 111.)

358. W.B. Jones, E.E. Reed, F.W. Stevenson, Biography of Wolfgang Joseph Thron (1918–2001), Rocky Mountain J. Math., 33 (2003) 395–403.
(Cited on page 305.)

359. W.B. Jones, W.J. Thron, *Continued Fractions. Analytic Theory and Applications*, Addison-Wesley, Reading, 1980.
(Cited on pages 7, 30, and 238.)

360. W.B. Jones, W.J. Thron, Continued fractions in numerical analysis, Appl. Numer. Math., 4 (1988) 143–230.
(Cited on pages 7, 42, and 238.)

361. D.C. Joyce, Survey of extrapolation processes in numerical analysis, SIAM Rev., 13 (1971) 435–490.
(Cited on page 57.)

362. D. Kahaner, Numerical quadrature by the ε-algorithm, Math. Comp. 26 (1972), 689–693.
(Cited on page 180.)

363. S. Kaniel, J. Stein, Least-square acceleration of iterative methods for linear equations, J. Optim. Theory Appl., 14 (1974) 431–437.
(Cited on page 187.)

364. I. Kaplansky, Abraham Adrian Albert, November 9, 1905–June 6, 1972, in *A Century of Mathematics in America*, P. Duren ed., American Mathematical Society, Providence, RI, vol. 1, pp. 245–264.
(Cited on page 155.)

365. L. Karlberg, H. Wallin, Padé-type approximants for functions of Markov-Stieltjes type, Rocky Mountain J. Math., 21 (1991) 437–449.
(Cited on page 197.)

366. H.B. Keller, V. Pereyra, Difference methods and deferred corrections for ordinary boundary value problems, SIAM J. Numer. Anal., 2 (1979) 241–259.
(Cited on page 172.)

367. C.T. Kelley, *Iterative Methods for Linear and Nonlinear Equations*, SIAM, Philadephia, 1995.
(Cited on page 195.)

368. J.E. Kiefer, G.H. Weiss, A comparison of two methods for accelerating the convergence of Fourier series, Comput. Math. Appl., 7 (1981) 527–535.
(Cited on page 117.)

369. R.F. King, An efficient one-point extrapolation method for linear convergence, Math. Comp., 152 (1980) 1285–1290.
(Cited on page 194.)

370. M. Kline, Euler and infinite series, Math. Mag., 56 (1983) 307–314.
(Cited on page 49.)

371. M. Koecher, R. Remmert, Cayley numbers or alternative division algebras, in H.-D. Ebbinghaus et al., *Numbers*, Springer-Verlag, New York, 1991.
(Cited on page 156.)

372. E.G. Kogbetliantz, Report No. 1 on "Maehly's method, improved and applied to elementary functions" subroutines, April 1957, Service Bureau Corp., New York.
(Cited on page 54.)

373. K. Kommerell, *Das Grenzgebiet der elementaren und höheren Mathematik. In ausgewählten Kapiteln dargestellt*, K.F. Köhler, Leipzig, 1936.
(Cited on page 57.)

374. Z. Kopal, Operational methods in numerical analysis based on rational approximations, in *On Numerical Approximation*, R.E. Langer ed., University of Wisconsin Press, Madison, 1959, pp. 25–42.
(Cited on page 87.)

375. A. Korganoff ed., *Méthodes de Calcul Numérique*, Dunod, Paris, 1961.
(Cited on page 51.)

376. M.G. Krein, Infinite J-matrices and a matrix moment problem, Dokl. Akad. Nauk SSSR, 69 (1949) 125–128 (in Russian); translation into English by W. van Assche, arXiv preprint arXiv:1606.07754, 2016.
(Cited on page 77.)

377. D. Krob, D. Leclerc, Minor identities for quasi-determinants and quantum determinants, Commun. Math. Phys., 169 (1995) 1–23.
(Cited on page 210.)

378. A.R. Krommer, C.W. Überhuber, *Computational Integration*, SIAM, Philadelphia, 1998.
(Cited on page 180.)

379. A.S. Kronrod, *Nodes and Weights of Quadrature Formulas*, Nauka, Moscow, 1964; Authorized translation from Russian, Consultants Bureau, New York, 1965.
(Cited on page 35.)

380. E.E. Kummer, Eine neue Methode, die numerischen Summen langsam convergirender Reihen zu berechnen, J. Reine. Angew. Math., 16 (1837) 206–214.
(Cited on page 72.)

381. J.L. Lagrange, Nouvelle méthode pour résoudre les équations littérales par le moyen des séries, Mém. Acad. Roy. Sci. Berlin, 24 (1770) 251–326.
(Cited on page 211.)

382. J.L. Lagrange, Sur l'usage des fractions continues dans le calcul intégral, Nouv. Mém. Acad. Roy. Sci. Berlin, 7 (1776) 236–264.
(Cited on page 26.)

383. J.H. Lambert, Observationes variæ in mathesin puram, Acta Helvetica, 3 (1758) 128–158.
(Cited on page 26.)

384. J.H. Lambert, Mémoire sur quelques propriétés remarquables des quantités transcendantes circulaires et logarithmiques, Mém. Acad. R. Sci. Berlin, 17 (1761/1768) 265–322.
(Cited on page 131.)

385. C. Lanczos, Trigonometric interpolation of empirical and analytical functions, J. Math. Phys., 17 (1938) 123–199.
(Cited on page 53.)

386. C. Lanczos, An iteration method for the solution of the eigenvalue problem of linear differential and integral operators, J. Res. Natl. Bur. Stand., 45 (1950) 255–282.
(Cited on pages 192 and 208.)

387. C. Lanczos, Solution of systems of linear equations by minimized iterations, J. Res. Natl. Bur. Stand., 49 (1952) 33–53.
(Cited on page 192.)

388. C. Lanczos, *Applied Analysis*, Prentice Hall, Englewoods Cliffs, N.J., 1956.
(Cited on page 53.)

389. F.M. Larkin, Some techniques for rational interpolation, Comput. J., 10 (1967) 178–187.
(Cited on page 121.)

390. A. Lascoux, Inversion des matrices de Hankel, Linear Algebra Appl., 129 (1990) 77–102.
(Cited on page 210.)

391. A. Lascoux, Motzkin paths and powers of continued fractions, Séminaire Lotharingien de Combinatoire, 44 (2000) Article B44e.
(Cited on page 210.)

392. A. Lascoux, *Symmetric Functions and Combinatorial Operators on Polynomials*, CBMS Regional Conference Series in Mathematics, vol. 99, American Mathematical Society, Providence, 2003.
(Cited on page 209.)

393. A. Lascoux, P. Pragacz, Bezoutians, Euclidean algorithm, and orthogonal polynomials, Ann. Comb., 9 (2005) 301–319.
(Cited on page 210.)

394. P.-J. Laurent, Un théorème de convergence pour le procédé d'extrapolation de Richardson, C.R. Acad. Sci. Paris, 256 (1963) 1435–1437.
(Cited on pages 108 and 171.)

395. P.-J. Laurent, *Étude de Procédés d'Extrapolation en Analyse Numérique*, Thèse de Doctorat d'État ès Sciences Mathématiques, Université de Grenoble, 15 juin 1964.
(Cited on pages 14, 108, and 170.)

396. P.-J. Laurent, *Approximation et Optimisation*, Hermann, Paris, 1972.
(Cited on page 171.)

397. D. Laurie, Appendix A of [74], Convergence acceleration accompanying code, https://www-m3.ma.tum.de/m3old/bornemann/challengebook/AppendixA/.
(Cited on page 7.)

398. H.T. Laux, *A Numerical Library in C for Scientists and Engineers*, CRC Press, Boca Raton, 1994.
(Cited on page 111.)

399. H. Le Ferrand, The quadratic convergence of the topological epsilon algorithm for systems of nonlinear equations, Numer. Algorithms, 3 (1992) 273–284.
(Cited on pages 119 and 196.)

400. H. Le Ferrand, Robert de Montessus de Ballore's 1902 theorem on algebraic continued fractions : genesis and circulation, arXiv:1307.3669.
(Cited on page 30.)

401. H. Le Ferrand, The rational iteration method by Georges Lemaître, Numer. Algorithms, 80 (2019) 235–251.
(Cited on page 53.)

402. É. Le Roy, Sur les séries divergentes et les fonctions définies par un développement de Taylor, Ann. Fac. Sci. Toulouse, Série 2, 2 (1900) 317–430.
(Cited on page 126.)

403. W. Ledermann, Obituary, A.C. Aitken, DSc, FRS, Proc. Edinburgh Math. Soc., 16 (1968) 151–176.
(Cited on page 313.)

404. G. Lemaître, L'itération rationnelle, Bull. Acad. Roy. Sci. Belgique, Cl. Sci., (5) 28 (1942) 347–364.
(Cited on page 52.)

405. G. Lemaître, Intégration d'une équation différentielle par itération rationnelle, Bull. Acad. Roy. Sci. Belgique, Cl. Sci., (5) 28 (1942) 815–825.
   (Cited on page 53.)

406. C. Lemaréchal, Une méthode de résolution de certains systèmes non linéaires bien posés, C.R. Acad. Sci. Paris, 272 A (1971) 605–607.
   (Cited on page 195.)

407. P. Lepora, B. Gabutti, An algorithm for the summation of series, Appl. Numer. Math., 3 (1987) 523–528.
   (Cited on page 170.)

408. D. Levin, Development of non-linear transformations for improving convergence of sequences, Int. J. Comput. Math., B3 (1973) 371–388.
   (Cited on page 178.)

409. D. Levin, A. Sidi, Two new classes of nonlinear transformations for accelerating the convergence of infinite integrals and series, Appl. Math. Comput., 9 (1981) 175–215.
   (Cited on page 179.)

410. F. Lindemann, Über die Zahl π, Math. Ann., 20 (1882) 213–225.
   (Cited on page 28.)

411. J. Liouville, Sur une classe très étendue de quantités dont la valeur n'est ni algébrique, ni même réductible à des irrationnelles algébriques, J. Math. Pures Appl., (1) 16 (1851) 133–142.
   (Cited on page 74.)

412. L.N. Lipatov, Divergence of the perturbation theory series and the quasiclassical theory, Sov. Phys. JETP, 45 (1977) 216–223, Zh. Eksp. Teor. Fi., 72 (1977) 411–427.
   (Cited on page 204.)

413. J.-J. Loeffel, Saclay Report, DPh-T/76/20, unpublished.
   (Cited on page 204.)

414. K. Löwner, Über monotone Matrixfunktionen, Math. Z., 38 (1934) 177–216.
   (Cited on page 205.)

415. I. Longman, On the generation of rational function approximations for Laplace transform inversion with an application to viscoelasticity, SIAM J. Appl. Math., 24 (1973) 429–440.
   (Cited on page 196.)

416. I.M. Longman, M. Sharir, Laplace transform inversion of rational functions, Geophys. J. R. Astrom. Soc., 25 (1971) 299–305.
   (Cited on page 196.)

417. L. Lorentzen, Computation of limit periodic continued fractions. A survey, Numer. Algorithms, 10 (1995) 69–111.
   (Cited on page 201.)

418. L. Lorentzen, H. Waadeland, *Continued Fractions with Applications*, North-Holland, Amsterdam, 1992.
   (Cited on pages 7 and 42.)

419. L. Lovitch, M.F. Marziani, Borel summability of divergent Born series, Il Nuovo Cimento, A76 (1983) 615–626.
   (Cited on pages 204 and 205.)

420. D.S. Lubinsky, Diagonal Padé approximants and capacity, J. Math. Anal. Appl., 78 (1980), 58–67.
   (Cited on page 197.)

421. D.S. Lubinsky, Rogers-Ramanujan and the Baker-Gammel-Wills (Padé) conjecture, Ann. of Math, 157 (2003) 847–889.
   (Cited on page 73.)

422. D.S. Lubinsky, Reflections on the Baker–Gammel–Wills (Padé) conjecture, in *Analytic Number Theory, Approximation Theory, and Special Functions*, G. Milovanović and M. Rassias eds., Springer, New York, 2014, pp. 561–571.
   (Cited on pages 73 and 74.)

423. S. Lubkin, A method of summing infinite series, J. Res. Natl. Bur. Stand., 48 (1952) 228–254.
   (Cited on pages 18, 63, 65, and 179.)

424. R. Ludwig, Verbesserung einer Iterationsfolge bei Gleichungssystemen, Z. Angew. Math. Mech., 32 (1952) 232–234.
(Cited on page 195.)

425. J.R. Macdonald, Accelerated convergence, divergence, iteration, extrapolation, and curve fitting, J. Appl. Phys., 35 (1964) 3034–3041.
(Cited on page 180.)

426. H.J. Maehly, Methods for fitting rational approximations, Part I: Telescoping procedures for continued fractions, J. ACM, 7 (1960)150–162.
(Cited on page 53.)

427. H.J. Maehly, Methods for fitting rational approximations, Parts II and III, J. ACM, 10 (1963) 257–277.
(Cited on page 53.)

428. A.P. Magnus, Rate of convergence of sequences of Padé-type approximants and pole detection in the complex plane, in *Padé Approximation and Its Applications, Amsterdam 1980*, H. van Rossum and M.G. de Bruin eds., Lecture Notes in Mathematics, vol. 888, Springer Verlag, Berlin, 1981, pp. 300–308.
(Cited on page 197.)

429. A.P. Magnus, On optimal Padé-type cuts, Ann. Numer. Math., 4 (1997) 435–450.
(Cited on page 197.)

430. E. Maillet, *Introduction à la Théorie des Nombres Transcendants et des Propriétés Arithmétiques des Fonctions*, Gauthier-Villars, Paris, 1906.
(Cited on page 28.)

431. B. Mandelbrot, How long os the cost on Britain? Statistical self-similarity and fractional dimension, Science, New Series, 156 (1967) 636–638.
(Cited on page 323.)

432. Y.I. Manin, M. Marcolli, Continued fractions, modular symbols, and noncommutative geometry, Sel. Math., New Ser., 8 (2002) 475521.
(Cited on page 215.)

433. G.I. Marchuk, V.V. Shaidurov, *Difference Methods and Their Extrapolations*, Springer-Verlag, New York, 1983.
(Cited on page 55.)

434. H. Marder, B. Weitzner, A bifurcation problem in E-layer equilibria, Plasma Phys. 12 (1970) 435–445.
(Cited on page 195.)

435. P. Maroni, Une généralisation du théorème de Favard-Shohat sur les polynômes orthogonaux, C. R. Acad. Sci. Paris, Série I, 293 (1981) 19–22.
(Cited on pages 191 and 199.)

436. P. Maroni, Sur quelques espaces de distributions qui sont des formes linéaires sur l'espace vectoriel des polynômes, in *Polynômes Orthogonaux et Applications, Proceedings of the Laguerre Symposium held at Bar-le-Duc, October 15–18, 1984*, C. Brezinski, A. Draux, A.P. Magnus, P. Maroni, A. Ronveaux eds., Lecture Notes in Mathematics, vol. 1171, Springer-Verlag, Berlin, 1985, pp. 184–194.
(Cited on page 191.)

437. P. Maroni, Une théorie algébrique des polynômes orthogonaux. Application aux polynômes orthogonaux semi-classiques, in *Proceedings of the Third International Symposium on Orthogonal Polynomials and Their Applications, Erice (Trapani), June 1–8, 1990*, C. Brezinski, L. Gori and A. Ronveaux eds., IMACS Annals on Computing and Applied Mathematics, Volume 9: Orthogonal Polynomials and Their Applications, J.C. Baltzer AG, Scientific Publishing Company, Basel, Switzerland, 1991, pp. 95–130.
(Cited on page 191.)

438. I. Marx, Remark concerning a non-linear sequence-to-sequence transform, J. Math. and Phys., 42 (1963) 334–335.
(Cited on page 18.)

439. M.F. Marziani, A connection between Borel and Padé summation techniques, Lett. Nuovo Cimento, 37 (1983) 124–128.
   (Cited on page 205.)

440. M.F. Marziani, Convergence of a class of Borel-Padé type approximants, Nuovo Cimento, 99B (1987) 145–154.
   (Cited on page 205.)

441. A.C. Matos, M. Prévost, Acceleration property for the columns of the E-algorithm, Numerical Algor., 2 (1992) 393–408.
   (Cited on page 177.)

442. J. Mawhin, Édouard Le Roy : un père oublié de la méthode de continuation, in *Liber Amicorum Jean Dhombres*, P. Radelet-de Grave ed., Brepols, Turnhout, 2008, pp. 330–363.
   (Cited on page 126.)

443. J.B. McLeod, A note on the ε-algorithm, Computing, 7 (1971) 17–24.
   (Cited on pages 102, 104, 116, 143, and 145.)

444. S. McMath, F. Crabbe, D. Joyner, Continued fractions and parallel SQUFOF, Int. J. Pure Appl. Math., 34 (2007) 17–36.
   (Cited on page 331.)

445. G. Meinardus, G.D. Taylor, Lower estimates for the error of the best uniform approximation, J. Approx. Theory, 16 (1976) 150–161.
   (Cited on page 174.)

446. M. Mešina, Convergence acceleration for the iterative solution of $x = Ax + f$, Comput. Methods Appl. Mech. Eng., 10 (1977) 165–173.
   (Cited on page 187.)

447. G. Meurant ed., *Extrapolation and Fixed Points in Memoriam Peter Wynn (1931–2017)*, Numer. Algorithms, vol. 80, no. 1.
   (Cited on pages 4 and 187.)

448. G. Meurant, J. Duintjer Tebbens, *Krylov Methods for Nonsymmetric Linear Systems, from Theory to Computations*, Springer Nature Switzerland AG, 2020.
   (Cited on page 192.)

449. P. Midy, Scaling trnasformations and extrapolation algorithms for vector sequences, Comput. Phys. Comm., 70 (1992) 285–291.
   (Cited on page 187.)

450. J.C.P. Miller, A method for the determination of converging factors, applied to the asymptotic expansions for the parabolic cylinder functions, Proc. Camb. Phil. Soc., 48 (1952) 243–254.
   (Cited on page 140.)

451. W.E. Milne, *Numerical Calculus*, Princeton University Press, Princeton, 1949.
   (Cited on page 51.)

452. L.M. Milne-Thompson, *Calculus of Finite Differences*, Macmillan and Co., London, 1933.
   (Cited on page 51.)

453. Y. Minesaki, Y. Nakamura, The discrete relativistic Toda molecule equation and a Padé approximation algorithm, Numer. Algorithms, 27 (2001) 219–235.
   (Cited on page 214.)

454. C. Moler, C. van Loan, Nineteen dubious ways to compute the exponential of a matrix, SIAM Review, 20 (1978) 801–836.
   (Cited on page 199.)

455. C. Moler, C. van Loan, Nineteen dubious ways to compute the exponential of a matrix, twenty-five years later, SIAM Review, 45 (2003) 3–49.
   (Cited on page 199.)

456. R. de Montessus de Ballore, Sur les fractions continues algébriques. Bull. Soc. Math. France 30 (1902) 28–36.
   (Cited on pages 30, 73, and 134.)

457. P. Mortreux, M. Prévost, An acceleration property for the E-algorithm for alternate sequences, Adv. Comput. Math., 5 (1996) 443–482.
   (Cited on page 177.)

458. G. Mühlbach, Neville-Aitken algorithms for interpolating by functions of Čebyšev-systems in the sense of Newton and in a generalized sense of Hermite, in *Theory of Approximation with Applications*, A.G. Law and B.N. Sahney eds., Academic Press, New York, 1976, pp. 200–212.
(Cited on page 175.)

459. T. Muir, An overlooked discoverer in the theory of determinants, Philos. Magazine, (5) 18 (1884) 416–427.
(Cited on page 8.)

460. T. Muir, *The Theory of Determinants in the Historical Order of Their Development. Part I. General Determinants up to 1841*, Second Edition, Macmillan and Co., London, 1906.
(Cited on page 8.)

461. A. Nagai, J. Satsuma, Discrete soliton equations and convergence acceleration algorithms, Physics Lett., A 209 (1995) 305–312.
(Cited on page 214.)

462. A. Nagai, T. Tokihiro, J. Satsuma, The Toda molecule equation and the ε-algorithm, Math. Comp., 67 (1998) 1565–1575.
(Cited on page 214.)

463. N. Negoescu, Convergence theorems on non-commutative continued fractions, Mathematica Rev. Anal. Numér. Théor. Approx., 5 (1976) 165–180.
(Cited on page 202.)

464. R. Nevanlinna, Asymptotische Entwicklungen beschränkter Funktionen und das Stieltjessche Momentumproblem, Ann. Acad. Sci. Fenn., ser. A, 18 (5) (1922) 1–53.
(Cited on page 77.)

465. P. Ni, *Anderson Acceleration of Fixed-Point Iteration with Applications to Electronic Structure Computations*, Ph.D. thesis, Worcester Polytechnic Institute, Worcester, MA, 2009.
(Cited on page 196.)

466. W. Niethammer, Numerical application of Euler's series transformation and its generalizations, Numer. Math., 34 (1980) 271–283.
(Cited on page 170.)

467. W. Niethammer, U. Schweitzer, On the numerical analytic continuation of power series with application to the two-body and three-body problems, Comput. Methods Appl. Mech. Engrg., 5 (1975) 239–249.
(Cited on page 170.)

468. Y. Nievergelt, Aitken's and Steffensen's accelerations in several variables, Numer. math., 59 (1991) 295–310.
(Cited on page 196.)

469. T. Noda, The Aitken-Steffensen formula for systems of nonlinear equations. III, Proc. Japan Acad., 62A, (1986) 174–177.
(Cited on page 196.)

470. O. Nordal, Tool or science? The history of computing at the Norwegian University of Science and Technology, in *History of Nordic Computing 2*, J. Impagliazzo, T. Järvi and P. Paju eds., Springer-Verlag, New York, 2009, pp. 121–129.
(Cited on page 325.)

471. N.E. Nörlund, Fraction continues et différences réciproques, Acta Math., 34 (1911) 1–108.
(Cited on pages 47 and 61.)

472. N.E. Nörlund, *Differenzenrechnung*, Springer, Berlin, 1924.
(Cited on page 51.)

473. M.E. Ogborn, Johan Frederik Steffensen, J. Inst. Actuar., 88 (1962) 251–253.
(Cited on page 332.)

474. J.M. Ortega, W.C. Rheinboldt, *Iterative Solution of Nonlinear Equations in Several Variables*, Academic Press, San Diego, 1970.
(Cited on pages 195 and 196.)

475. N. Osada, An acceleration theorem for the ρ-algorithm, Numer. Math., 73 (1996) 521–531.
(Cited on page 109.)

476. N. Osada, The early history of convergence acceleration methods, Numer. Algorithms, 60 (2012) 205–221.
   (Cited on pages 54 and 341.)
477. M.R. Osborne, h^2-extrapolation in eigenvalue problems, Quart. J. Mech., 13 (1960) 156–168.
   (Cited on page 170.)
478. A.M. Ostrowski, A method of speeding up iterations with super-linear convergence, J. Math. Mech., 7 (1958) 117–120.
   (Cited on page 194.)
479. A.M. Ostrowski, *Solution of Equations and Systems of Equations*, Academic Press, New York, 1960.
   (Cited on page 51.)
480. K.J. Overholt, Extended Aitken accelaration, BIT, 5 (1965) 122–132.
   (Cited on page 178.)
481. H. Padé, Sur la représentation approchée d'une fonction par des fractions rationnelles, Ann. Sci. Éc. Norm. Supér. (3), 9 (1892) 3–93.
   (Cited on pages 28 and 72.)
482. H. Padé, Sur la généralisation des fractions continues algébriques, J. Math. Pures Appl., 4e série, 10, (1894) 291–329.
   (Cited on page 28.)
483. H. Padé, *Œuvres, Rassemblées et Présentées par Claude Brezinski*, Librairie Scientifique et Technique Albert Blanchard, Paris, 1984.
   (Cited on pages 7, 77, 320, and 340.)
484. V. Papageorgiu, B. Grammaticos, A. Ramani, Integrable lattices and convergence acceleration algorithms, Physics Lett., A 179 (1993) 111–115.
   (Cited on page 213.)
485. S. Paszkowski, Approximation uniforme des fonctions continues par les fonctions rationnelles, Zastos. Mat., 6 (1963) 441–458.
   (Cited on page 197.)
486. S. Paszkowski, *Zastosowania Numeryczne Wielomianów i Szeregów Czebyszewa* (in Polish), Państwowe Wydawn. Naukowe, Warsaw, 1975.
   (Cited on page 197.)
487. R. Pennacchi, Le trasformazioni razionali di una successione, Calcolo, 5 (1968) 37–50.
   (Cited on page 172.)
488. R. Pennacchi, Somma di serie numeriche mediante la trasformazione quadratica $T_{2,2}$, Calcolo, 5 (1968) 51–61.
   (Cited on page 172.)
489. V. Pereyra, On improving an approximate solution of a functional equation by deferred corrections, Numer. Math., 8 (1966) 376–391.
   (Cited on page 171.)
490. V. Pereyra, Accelerating the convergence of discretization algorithms, SIAM J. Numer. Anal., 4 (1967) 508–533.
   (Cited on page 171.)
491. V. Pereyra, Iterated deferred corrections for nonlinear operator equations, Numer. Math., 10 (1967) 316–323.
   (Cited on page 171.)
492. V. Pereyra, Iterated deferred corrections for nonlinear boundary value problems, Numer. Math., 11 (1968) 111–125.
   (Cited on page 171.)
493. O. Perron, *Die Lehre von den Kettenbrüchen*, Teubner, Stuttgart, 1957.
   (Cited on pages 7, 76, and 142.)
494. A. Peyerimhoff, *Lectures on Summability*, Lecture Notes in Mathematics, vol. 107, Springer-Verlag, Berlin, 1969.
   (Cited on page 54.)

495. P. Pflüger, *Matrizenkettenbrüche*, Dissertation, ETH Zürich, Juris Druck + Verlag, Zürich, 1966.
(Cited on page 201.)

496. R. Piessens, E. de Doncker-Kapenga, C.W. Überhuber, D.K. Kahaner, *Quadpack, A Subroutine Package for Automatic Integration*, Springer, Berlin, 1983.
(Cited on page 180.)

497. H. Poincaré, Sur les intégrales irrégulières des équations linéaires, Acta Math., 8 (1886) 295–344.
(Cited on page 50.)

498. H. Polachek, On the solution of systems of linear equations of high order, Naval Ordnance Laboratory, Memorandum 9522, 1948.
(Cited on page 178.)

499. H. Polachek, Calculation of transient excitation of ship hulls by finite difference methods, Math. Comp,. 13 (1959) 109–116.
(Cited on page 330.)

500. S. Polachek, Harry Polachek (1913–2002), IEEE Ann. Hist. Comput., 25 (2003) 90–92.
(Cited on page 330.)

501. G. Pólya, Algebraische Untersuchungen über ganze Funktionen vom Geschlechte Null und Eins, J. Reine Angew. Math., 145 (1915) 224–249.
(Cited on page 162.)

502. L.S. Pontryagin, On the zeros of some transcendental functions, Amer. Math. Soc. Transl., Ser. 2, 1 (1955) 95–110.
(Cited on page 122.)

503. F.A. Potra, H. Engler, A characterization of the behavior of the Anderson acceleration on linear problems, Linear Algebra Appl., 438 (2013) 1002–1011.
(Cited on page 196.)

504. P. Pragacz, Architectonique des formules préférées d'Alain Lascoux (in English), Séminaire Lotharingien de Combinatoire, 52 (2005) Article B52d.
(Cited on page 210.)

505. M. Prévost, Padé-type approximants with orthogonal generating polynomials, J. Comput. Appl. Math., 9 (1983) 333–346.
(Cited on page 197.)

506. M. Prévost, Acceleration property for the E-algorithm and an application to the summation of series, Adv. Comput. Math., 2 (1994) 319–341.
(Cited on page 177.)

507. M. Prévost, A new proof of the irrationality of $\zeta(2)$ and $\zeta(3)$ using Padé approximants, J. Comput. Appl. Math., 67 (1996) 219–235.
(Cited on page 206.)

508. M. Prévost, T. Rivoal, Application of Padé approximation to Euler's constant and Stirling's formula, Ramanujan J., (2020), https://doi.org/10.1007/s11139-019-00201-9.
(Cited on page 206.)

509. G.B. Price, Some identities in the theory of determinants, Amer. Math. Monthly, 54 (1947) 75–90.
(Cited on page 8.)

510. A. Pringsheim, Irrationalzahlen und Konvergenz unendlicher Prozesse, in *Encyklopädie der mathematischen Wissenschaften mit Einschluß ihrer Anwendungen*, H. Burkhardt and W. Fr. Meyer eds., B.G. Teubner, Leipzig, 1898, Bd. 1–1.
(Cited on page 74.)

511. A. Pringsheim, Über ein Konvergenzkriterium für Kettenbrüche mit positiven Gliedern, Sitzungsber. Bayer. Akad. Wiss., Math-Naturwiss. Kl., 29 (1899) 261–271.
(Cited on page 75.)

512. A. Pringsheim, Ëinige Konvergenz-Kriterien fur Kettenbrüche mit komplexen Gliedern, Sitzungsber. Bayer. Akad. Wiss., Math-Naturwiss. Kl., 35 (1905) 359–380.
(Cited on page 202.)

513. R. Prony, Éssai experimental et analytique sur les lois de la dilatabilité des fluides élastiques et sur celles de la force expansive de la vapeur de l'eau et de la vapeur de l'alkool, à différentes températures, Journal de l'École Polytechnique, Volume 1, Cahier 22, Floréal et Plairial, An III (1795) 24–76.
   (Cited on pages 158 and 160.)

514. W.C. Pye, T.A. Atchison, An algorithm for the computation of the higher order G-transformation, SIAM J. Numer. Anal.,10 (1973) 1–7.
   (Cited on page 172.)

515. L.D. Pyle, A generalized inverse ε-algorithm for constructing intersection projection matrices, with applications, Numer. Math., 10 (1967) 86–102.
   (Cited on page 181.)

516. P. Rabinowitz, Extrapolation methods in numerical integration, Numer. Algorithms, 3 (1992) 17–28.
   (Cited on page 180.)

517. M Raissouli, A. Kacha, Convergence of matrix continued fractions, Linear Algebra Appl., 320 (2000) 115–129.
   (Cited on page 202.)

518. A. Ralston, On economization of rational functions, J. ACM, 10 (1963) 278–282.
   (Cited on page 54.)

519. C. Reutenauer, Michel Fliess and non-commutative formal power series, Intern. J. Control, 81 (2008) 338–343.
   (Cited on page 206.)

520. L.F. Richardson, The approximate arithmetical solution by finite difference of physical problems involving differential equations, with an application to the stress in a masonry dam, Philos. Trans. Roy. Soc. London, ser. A, 210 (1910) 307–357.
   (Cited on pages 13, 14, 50, 54, 108, 119, 170, 321, 322, and 340.)

521. L.F. Richardson, *Weather Prediction by Numerical Process*, Cambridge University Press, Cambridge, 1922.
   (Cited on page 322.)

522. L.F. Richardson, How to solve differential equations approximately by arithmetic, Math. Gazette, 12 (1925) 415–421.
   (Cited on page 55.)

523. L.F. Richardson, The deferred approach to the limit. Part I: Single lattice, Philos. Trans. Roy. Soc. London, ser. A, 226 (1927) 299–349.
   (Cited on pages 13, 14, 50, 54, 55, 108, 170, 322, and 340.)

524. H.W. Richmond, On certain formulae for numerical approximation, J. London Math. Soc., 19 (1944) 31–38.
   (Cited on page 158.)

525. B. Riemann, Sullo svolgimento del quoziente di due serie ipergeometriche in frazione continua infinita, in *Gesammelte Mathematische Werke und Wissenschaftlicher Nachlass*, B.G. Teubner, Leipzig, 1876, pp. 400–406.
   (Cited on page 75.)

526. M. Riesz, Sur le problème des moments, Ark. Mat. Astr. Fys., 16 (12) (1921) 1–23; 16 (19) (1921) 1–21; 17 (16) (1923) 1–52.
   (Cited on page 77.)

527. J.D. Riley, Solving systems of linear equations with a positive definite, symmetric, but possibly ill-conditioned matrix, Math. Tables Aids Comput., 9 (1955) 96–101.
   (Cited on page 178.)

528. J. Rissanen, Recursive evaluation of Padé approximants for matrix sequences, IBM J. Research and Development, 16 (1972) 401–406.
   (Cited on page 199.)

529. D.E. Roberts, On the algebraic foundations of the vector ε-algorithm, in *Clifford Algebras and Spinor Structures*, R. Abłamowicz and P. Lounesto eds., Mathematics and Its Applications, vol. 321, Kluwer, Dordrecht, 1995, pp. 343–361.
   (Cited on page 102.)

530. E. Roblet, *Une Interprétation Combinatoire des Approximants de Padé*, Thèse de Mathématiques Pures, Université de Bordeaux I, 1994.
 (Cited on page 211.)
531. T. Rohwedder, R. Schneider, An analysis for the DIIS acceleration method used in quantum chemistry calculations, J. Math. Chem., 49 (2011) 1889.
 (Cited on page 196.)
532. Ch. Roland, R. Varadhan, New iterative schemes for nonlinear fixed point problems, with applications to problems with bifurcations and incomplete-data problems, Appl. Numer. Math., 55 (2005) 215–226.
 (Cited on page 195.)
533. Ch. Roland, R. Varadhan, C.E. Frangakis, Squared polynomials extrapolation methods with cycling: an application to the positron emission tomography problem, Numer. Algorithms, 44 (2007) 159–172.
 (Cited on page 195.)
534. S. Roman, *The Umbral Calculus*, Academic Press, Orlando, 1984.
 (Cited on page 191.)
535. W. Romberg, Vereinfachte numerische Integration, Kgl. Norske Vid. Selsk. Forsh., 28 (1955) 30–36.
 (Cited on pages 15, 51, 171, and 325.)
536. H. Rutishauser, Anwendungen des Quotienten-Differenzen-Algorithmus, Z. Angew. Math. Phys., 5 (1954) 496–508.
 (Cited on pages 45, 46, and 148.)
537. H. Rutishauser, Ein kontinuierliches Analogon zum Quotienten-Differenzen Algorithmus, Arch. Math. 5 (1954) 132–137.
 (Cited on page 95.)
538. H. Rutishauser, *Der Quotienten-Differenzen-Algorithmus*, Birkhäuser Verlag, Basel, 1957.
 (Cited on pages 45 and 148.)
539. H. Rutishauser, Ausdehnung des Rombergschen Prinzips, Numer. Math., 5 (1963) 48–54.
 (Cited on page 57.)
540. H. Rutishauser, *Description of ALGOL 60*, Springer-Verlag, Berlin Heidelberg, 1967.
 (Cited on page 152.)
541. H. Rutishauser, E. Stiefel, Remarques concernant l'intégration numérique, C.R. Acad. Sci. Paris, sér. A–B, 252 (1961) 1899–1900.
 (Cited on page 57.)
542. D.H. Sadler, Obituary-Miller, Jeffrey-Charles, Quart. J. Roy. Astron. Soc., 23 (1982) 311–313.
 (Cited on page 66.)
543. H. Sadok, Analysis of the convergence of the minimal and the orthogonal residual methods, Numer. Algorithms, 40 (2005) 201–216.
 (Cited on page 187.)
544. E.B. Saff, An extension of Montessus de Ballore's theorem on the convergence of interpolation rational functions, J. Approx. Theory, 6 (1972) 63–67.
 (Cited on page 197.)
545. E.B. Saff, P.R. Graves-Morris, A de Montessus theorem for vector valued rational interpolants, in *Rational Approximation and Interpolation*, P.R. Graves-Morris, E.B. Saff, and R.S. Varga eds., Lecture Notes in Mathematics, vol. 1105, Springer-Verlag, Berlin, 1984, pp. 227–242.
 (Cited on page 197.)
546. A. Salam, Non-commutative extrapolation algorithms, Numer. Algorithms, 7 (1994) 225–251.
 (Cited on pages 12, 102, 103, and 113.)
547. A. Salam, On the vector-valued Padé approximants and the vector ε-algorithm, in *Nonlinear Numerical Methods and Rational Approximation II*, A. Cuyt ed., Kluwer Academic Publishers, Dordrecht, 1994, pp. 291–301.
 (Cited on pages 103 and 113.)

548. A. Salam, An algebraic approach to the vector ε-algorithm, Numer. Algorithms, 11 (1996) 327–337.
     (Cited on pages 102, 103, and 113.)
549. M.G. Salvadori, Extrapolation formulas in linear difference operators, in *Proceedings of the first U.S. National Congress of Applied Mechanics held at Illinois Institute of Technology, Chicago, Illinois, June 11–16, 1951*, American Society of Mechanical Engineers, New York, 1952, pp. 15–18.
     (Cited on page 170.)
550. H.E. Salzer, LXXVII. An alternative definition of reciprocal differences, The London, Edinburgh, and Dublin Philosophical Magazine and Journal of Science (Philos. Mag.), 39: 295 (1948) 649–656.
     (Cited on page 47.)
551. H.E. Salzer, A simple method for summing slowly convergent series, J. Math. and Phys., 33 (1955) 356–359.
     (Cited on page 108.)
552. H.E. Salzer, Formulas for the partial summation of series, Math. Tables Aids Comput., 10 (1956) 149–156.
     (Cited on page 108.)
553. H.E. Salzer, G.M. Kimbro, Improved formulas for complete and partial summation of certain series, Math. Comp., 15 (1961) 23–39.
     (Cited on page 108.)
554. R.J. Schmidt, On the numerical solution of linear simultaneous equations by an iterative method, Phil. Mag., 7 (1941) 369–383.
     (Cited on pages 65 and 86.)
555. C. Schneider, Vereinfachte Rekursionen zur Richardson-Extrapolation in Spezialfällen, Numer. Math., 24 (1975) 177–184.
     (Cited on page 173.)
556. I.J. Schoenberg, Some analytic aspects of the problem of smoothing, in *Studies and Essays Presented to R. Courant on His 60th Birthday, Jan. 8, 1948*, Interscience, New York, 1948, pp. 351–370.
     (Cited on pages 163 and 164.)
557. I.J. Schoenberg, On smooting operations and their generating functions, Bull. Amer. Math. Soc., 59 (1953) 199–230.
     (Cited on pages 163 and 164.)
558. M.-P. Schützenberger. On the definition of a family of automata, Information and Control, 4 (1961) 245–270.
     (Cited on page 206.)
559. I. Schur, *Über eine Klasse von Matrizen, die sich einer gegebenen Matrix zuordnen lassen*, Doctoral Dissertation, Universität Berlin, 1901.
     (Cited on page 209.)
560. I. Schur, Über Potenzreihen, die im Einheitskreises beschränkt sind, J. Reine Angew. Math., 147 (1917) 205–232.
     (Cited on page 10.)
561. F. Schweins, *Theorie der Differenzen und Differentiale, der gedoppelten Verbindungen, der Producte mit Versetzungen, der Reihen, der wiederholenden Functionen, der allgemeinsten Facultäten und der fortlaufenden Brüche*, Heidelberg 1825; full view at https://babel.hathitrust.org/cgi/pt?id=mdp.39015068512063&view=1up&seq=31.
     (Cited on page 7.)
562. D. Scieur, A. d'Aspremont, F. Bach, Regularized nonlinear acceleration, Math. Program., 179 (2020) 47–83.
     (Cited on pages 187 and 196.)
563. W.T. Scott, H.S. Wall, A convergence theorem for continued fractions, Trans. Amer. Math. Soc., 47 (1940) 155–172.
     (Cited on page 44.)

564. R. Sedgewick, P. Flajolet, *An Introduction to the Analysis of Algorithms*, Addison-Wesley, Boston, 1995.
     (Cited on page 212.)

565. L. Seidel, *Untersuchungen über die Konvergenz und Divergenz der Kettenbrüche*, Habilitation, München Universität, 1846.
     (Cited on page 75.)

566. R.E. Shafer, On quadratic approximation, SIAM J. Numer. Anal., 11 (1974) 447–460.
     (Cited on page 198.)

567. D. Shanks, An Analogy between Transient and Mathematical Sequences and Some Nonlinear Sequence-to-Sequence Transforms Suggested by It. Part I, Memorandum 9994, Naval Ordnance Laboratory, White Oak, July 1949.
     (Cited on pages 2, 4, 20, 21, 53, 58, 59, 60, 63, 65, 86, 329, and 341.)

568. D. Shanks, Non-linear transformations of divergent and slowly convergent sequences, J. Math. and Phys., 34 (1955) 1–42.
     (Cited on pages 20, 21, 58, 59, 61, 65, 86, 329, and 341.)

569. D. Shanks, *Solved and Unsolved Problems in Number Theory*, Chelsea Publishing Co., New York, 1985.
     (Cited on page 330.)

570. D. Shanks, J.W. Wrench, Jr., Calculation of π to 100,000 decimals, Math. Comp., 16 (1962) 76–79.
     (Cited on pages 58 and 329.)

571. B. Shawyer, B. Watson, *Borel's Methods of Summability*, Clarendon Press, Oxford, 1994.
     (Cited on page 204.)

572. P.A. Sheppard, Dr. L.F. Richardson, F.R.S., Nature, 172 (1953) 1127–1128.
     (Cited on page 323.)

573. W.F. Sheppard, Central difference formulæ, Proc. London Math. Soc., 31 (1899) 449–488.
     (Cited on page 54.)

574. J. Sherman, On the numerators of the convergents of the Stieltjes continued fractions, Trans. Amer. Math. Soc., 35 (1933) 64–87.
     (Cited on page 76.)

575. J. Shohat, On Stieltjes continued fractions, Amer. J. Math., 54 (1932) 79–84.
     (Cited on page 76.)

576. J. Shohat (alias Jacques Chokhate), Sur les polynômes orthogonaux généralisés, C.R. Acad. Sci. Paris, 207 (1938) 556–558.
     (Cited on page 69.)

577. J.A. Shohat, J.D. Tamarkin, *The Problem of Moments*, American Mathematical Society, Providence, Rhode Island, 1943.
     (Cited on pages 7, 70, and 77.)

578. A. Sidi, A new method for deriving Padé approximants for some hypergeometric functions, J. Comput. Appl. Math., 7 (1981) 37–40.
     (Cited on page 111.)

579. A. Sidi, An algorithm for a special case of a generalization of the Richardson extrapolation process, Numer. Math. 38 (1982) 299–307.
     (Cited on page 179.)

580. A. Sidi, Convergence and stability properties of minimal polynomial and reduced rank extrapolation algorithms, SIAM J. Numer. Anal., 23 (1986) 197–209.
     (Cited on page 187.)

581. A. Sidi, Application of vector extrapolation methods to consistent singular linear systems, Appl. Numer. Math., 6 (1989/90) 487–500.
     (Cited on page 187.)

582. A. Sidi, On a generalization of the Richardson extrapolation process, Numer. Math., 57 (1990) 365–377.
     (Cited on pages 177 and 179.)

583. A. Sidi, Convergence of intermediate rows of minimal polynomial and reduced rank extrapolation tables, Numer. Algorithms, 6 (1994) 229–244.
(Cited on page 187.)

584. A. Sidi, Further results on convergence and stability of a generalization of the Richardson extrapolation process, BIT, 36 (1996) 143–157.
(Cited on page 177.)

585. A. Sidi, Extension and completion of Wynn's theory on convergence of columns of the epsilon table, J. Approx. Theory, 86 (1996) 21–40.
(Cited on page 180.)

586. A. Sidi, A complete convergence and stability theory for a generalized Richardson extrapolation process, SIAM J. Numer. Anal., 34 (1997) 1761–1778.
(Cited on page 177.)

587. A. Sidi, *Practical Extrapolation Methods. Theory and Applications*, Cambridge University Press, Cambridge, 2003.
(Cited on page 7.)

588. A. Sidi, Acceleration of convergence of general linear sequences by the Shanks transformation, Numer. Math., 119 (2011) 725–764.
(Cited on page 180.)

589. A. Sidi, Minimal polynomial and reduced rank extrapolation methods are related, Adv. Comput. Math., 43 (2016) 151–170.
(Cited on page 187.)

590. A. Sidi, *Vector Extrapolation Methods with Applications*, SIAM, Philadelphia, 2017.
(Cited on pages 185 and 196.)

591. A. Sidi, A convergence study for reduced rank extrapolation on nonlinear systems, Numer. Algorithms, 84 (2020) 957–982.
(Cited on pages 187 and 196.)

592. A. Sidi, D. Levin, Prediction properties of the t-transformation, SIAM J. Numer. Anal., 20 (1983) 589–598.
(Cited on page 177.)

593. W. Siemasko, Thiele-type branched continued fractions for two-variable functions, J. Comput. Appl. Math., 9 (1983) 137–153.
(Cited on page 201.)

594. B. Simon, *Loewner's Theorem on Monotone Matrix Functions*, Springer Nature Switzerland AG, 2019.
(Cited on page 205.)

595. S. Skelboe, Extrapolation methods for computation of the periodic steady-state response of nonlinear circuits, Report IT 7, Institute of Circuit Theory and Telecommunication, Technical University of Denmark, October 1976.
(Cited on pages 119 and 196.)

596. S. Skelboe, Extrapolation methods for computation of the periodic steady-state response of nonlinear circuits, in *Proc. 1977 IEEE Int. Symp. on Circuits and Systems*, pp. 64–67.
(Cited on page 119.)

597. S. Skelboe, Computation of the periodic steady-state response of nonlinear networks by extrapolation methods, IEEE Trans. Circuits Syst., 27 (1980) 161–175.
(Cited on page 119.)

598. I.V. Śleszyński, On the convergence of continued fractions (in Russian), Mém. Soc. Naturalistes Nouv. Russie, 10 (1889) 201–256.
(Cited on page 75.)

599. D.A. Smith, W.F. Ford, Acceleration of linear and logarithmic convergence, SIAM J. Numer. Anal., 16 (1979) 223–240.
(Cited on page 179.)

600. D.A. Smith, W.F. Ford, Numerical comparison of nonlinear convergence accelerators, Math. Comp., 38 (1982) 481–499.
(Cited on page 179.)

601. D.A. Smith, W.F. Ford, A. Sidi, Extrapolation methods for vector sequences, SIAM Rev., 29 (1987) 199–233; Erratum, SIAM Rev., 30 (1988) 623–624.
   (Cited on page 187.)

602. H.J.S. Smith, Note on continued fractions, Mess. math., (2) 6 (1877) 1–14.
   (Cited on page 74.)

603. S.B. Smith, *The Great Mental Calculators*, Columbia University Press, New York, 1983.
   (Cited on page 311.)

604. P. Sonneveld, A fast Lanczos-type solver for nonsymmetric linear systems, SIAM J. Sci. Stat. Comput., 10 (1989) 35–52.
   (Cited on page 193.)

605. V.N. Sorokin, Hermite-Padé approximations for Nikishin systems and the irrationality of $\zeta(3)$, Uspekhi Mat. Nauk, 49 (1994) 167–168 (in Russian); Russian Math. Surveys, 49 (1994) 176–177.
   (Cited on page 206.)

606. V.N. Sorokin, A transcendence measure for π^2, Mat. Sbornik, 187 (1996) 87–120 (in Russian); English translation in Sb. Math., 187 (1996) 1819–1852.
   (Cited on page 28.)

607. J.-M. Souriau, Une méthode pour la décomposition spectrale et l'inversion des matrices. C.R. Acad. Sci. Paris, 227 (1948) 1010–1011.
   (Cited on page 207.)

608. P.E. Spicer, F.W. Nijhoff, P.H. van der Kamp, Higher analogues of the discrete-time Toda equation and the quotient-difference algorithm, Nonlinearity, 24 (2011) 2229–2263.
   (Cited on page 214.)

609. K. Spielberg, Representation of power series in terms of polynomials, rational approximations and continued fractions, J. ACM, 8 (1961) 613–627.
   (Cited on page 54.)

610. K. Spielberg, Efficient continued fraction approximations to elementary functions, Math. Comp., 15 (1961) 409–417.
   (Cited on page 54.)

611. H. Stahl, The convergence of diagonal Padé approximants and the Padé conjecture, J. Comput. Appl. Math., 86 (1997) 287–296.
   (Cited on page 73.)

612. H. Stahl, V. Totik, *General Orthogonal Polynomials*, Cambridge University Press, Cambridge, 1992.
   (Cited on page 31.)

613. J.F. Steffensen, On certain inequalities between mean values, and their application to actuarial problems, Skand. Aktuarietidskr. (Scand. Actuar. J.), 1 (1918) 82–97.
   (Cited on page 332.)

614. J.F. Steffensen, *Interpolation*, The Williams and Wilkins Company, Baltimore, 1927.
   (Cited on page 332.)

615. J.F. Steffensen, Remarks on iteration, Skand. Aktuarietidskr. (Scand. Actuar. J.), 16 (1933) 64–72.
   (Cited on pages 51, 194, and 332.)

616. M.A. Stern, Theorie der Kettenbrüche und ihre Anwendung, J. Reine Angew. Math., 10 (1833) 1–22, 154–166, 241–274, 364–376; 11 (1834) 33–66, 142–168, 277–306, 311–350.
   (Cited on page 74.)

617. M.A. Stern, Über die Kennzeichen der Convergenz eines Kettenbruchs, J. Reine Angew. Math., 37 (1848) 255–272.
   (Cited on page 75.)

618. H.J. Stetter, Asymptotic expansions for the error of discretization algorithms for non-linear functional equations, Numer. Math. 7 (1965) 18–31.
   (Cited on page 171.)

619. H.J. Stetter, The defect correction principle and discretization methods, Numer. Math. 29 (1978) 425–443.
   (Cited on page 171.)

620. E. Stiefel, Altes und Neues über numerische Quadratur, Z. Angew. Math. Mech., 41 (1961) 408–413.
     (Cited on page 57.)
621. E. Stiefel, Some examples of numerical methods and the philosophy behind them, in *Information Processing 1962*, C.M. Popplewell ed., North-Holland, Amsterdam, 1963, pp. 17–20.
     (Cited on page 57.)
622. T.J. Stieltjes, Recherches sur quelques séries semi-convergentes, Ann. Sci. Éc. Norm., Paris, sér. 3, 3 (1886) 201–258.
     (Cited on page 50.)
623. T.J. Stieltjes, Sur la réduction en fraction continue d'une série procédant suivant les puissances descendantes d'une variable, Ann. Fac. Sci. Toulouse, 1ère série, 3 (1889) H1–H17.
     (Cited on pages 148 and 211.)
624. T.J. Stieltjes, Recherches sur les fractions continues, Ann. Fac. Sci. Toulouse, 8 (1894), J1–J122; 9 (1895), A1–A47.
     (Cited on pages 75 and 76.)
625. J. Stoer, Über zwei Algorithmen zur Interpolation mit rationalen Funktionen, Numer. Math., 3 (1961) 285–304.
     (Cited on pages 120 and 121.)
626. T. Ström, Strict error bounds in Romberg quadratur, BIT, 7 (1967) 314–321.
     (Cited on page 57.)
627. J.-Q. Sun, X.-K. Chang, Y. He, X.-B. Hu, An extended multistep Shanks transformation and convergence acceleration algorithm with their convergence and stability analysis, Numer. Math., 125 (2013) 785–809.
     (Cited on page 214.)
628. J.-Q. Sun, Y. He, X.-B. Hu, H.-W. Tam, Q-difference and confluent forms of the lattice Boussinesq equation and the relevant convergence acceleration algorithms, J. Math. Phys., 52 (2011) 023522.
     (Cited on page 213.)
629. B. Swirles Jeffreys, John Arthur Gaunt (1904–1944), Notes Rec. R. Soc. Lond., 44 (1990) 73–79.
     (Cited on page 55.)
630. J.J. Sylvester, On the relation between the minor determinants of linearly equivalent quadratic functions, Philos. Magazine (4) 1 (1851) 295–365 (1851); and Collected Math. Papers, vol. 1, Cambridge University Press, Cambridge, 1904, pp. 241–256.
     (Cited on page 7.)
631. J.J. Sylvester, On a fundamental rule in the algorithm of continued fractions, Phil. Mag., Series 4, 6:39 (1853) 297–299.
     (Cited on page 74.)
632. G. Szegö, *Orthogonal Polynomials*, Colloquium Publications, vol. XXIII, American Mathematical Society, Providence, Rhode Island, USA, 1939.
     (Cited on pages 70 and 146.)
633. R.C.E. Tan, Computing derivatives of eigensystems by the vector ε-algorithm, IMA J. Numer. Anal., 7 (1987) 485–494.
     (Cited on page 187.)
634. R.C.E. Tan, Computing derivatives of eigensystems by the topological ε-algorithm, Appl. Numar. Math., 3 (1987) 539–550.
     (Cited on page 187.)
635. R.C.E. Tan, Implementation of the topological ε-algorithm, SIAM J. Sci. Stat. Comput., 9 (1988) 839–848.
     (Cited on page 187.)
636. T.N. Thiele, Différences réciproques, Overs. Danske Vids. Selsk. Forhandl. (Sitzber. Akad. Kopenhagen) (1906) 153–171.
     (Cited on pages 121 and 335.)
637. T.N. Thiele, *Interpolationsrechnung*, Teubner, Leipzig, 1909.
     (Cited on pages 46, 61, and 335.)

638. A. Thue, Über Annäherungswerte algebraischer Zahlen, J. Reine Angew. Math., 135 (1909) 284–305.
(*Cited on page 206.*)

639. J. Todd, Obituary: L.F. Richardson (1881–1953), Math. Tables Aids Comput., 8 (1954), 242–245.
(*Cited on pages 55 and 323.*)

640. J. Todd, Oberwolfach—1945, in *General Inequalities 3, 3rd International Conference on General Inequalities, Oberwolfach, April 26–May 2, 1981*, E.F. Beckenbach and W. Walter eds., Springer Basel, 1983, pp. 19–22.
(*Cited on page 336.*)

641. J. Todd, D.H. Sadler, Admiralty Computing Service, Math. Tables Aids Comput., 2 (1947) 289–297.
(*Cited on page 80.*)

642. O. Toeplitz, Über die lineare Mittelbildungen, Prace Mat.-Fiz., 22 (1911) 113–118.
(*Cited on page 54.*)

643. A. Toth, C.T. Kelley, Convergence analysis for Anderson acceleration, SIAM Numer. Anal., 53 (2015) 805–819.
(*Cited on page 196.*)

644. R.R. Tucker, *Error Analysis, Convergence, Divergence, and the Acceleration of Convergence*, Ph.D. Thesis, Oregon State University, Corvallis, Oregon, 1963.
(*Cited on page 18.*)

645. R.R. Tucker, Remark concerning a paper by Imanuel Marx, J. Math. and Phys., 45 (1966) 233–234.
(*Cited on page 18.*)

646. R.R. Tucker, The δ^2-process and related topics, I, Pacific J. Math., 22 (1967) 349–359.
(*Cited on page 18.*)

647. R.R. Tucker, The δ^2-process and related topics, II, Pacific J. Math., 28 (1969) 455–463.
(*Cited on page 18.*)

648. R.R. Tucker, A geometric derivation of Daniel Shanks e_k transform, The Faculty Review, Bulletin of the Carolina A&T State University, 65 (1973) 60–63.
(*Cited on page 180.*)

649. H.W. Turnbull, Matrices and continued fractions, I, Proc. Roy. Soc. Edin., 53 (1933) 151–163.
(*Cited on page 77.*)

650. H.W. Turnbull, Matrices and continued fractions, II, Proc. Roy. Soc. Edin., 53 (1933) 208–219.
(*Cited on page 77.*)

651. K.T. Vahlen, Über Näherungswerte und Kettenbrüche, J. Reine Angew. Math., 115 (1895) 221–233.
(*Cited on page 74.*)

652. G. Valent, W. van Assche, The impact of Stieltjes' work on continued fractions and orthogonal polynomials: additional material, J. Comput. Appl. Math., 65 (1995) 419–447.
(*Cited on page 76.*)

653. W. van Assche, The impact of Stieltjes work on continued fractions and orthogonal polynomials, in *Stieltjes Collected Papers*, vol. I, G. van Dijk ed., Springer, Berlin, 1993, pp. 5–37.
(*Cited on page 76.*)

654. W. van Assche, Hermite-Padé rational approximation to irrational numbers, Comput. Methods Funct. Theory, 10 (2011) 585–602.
(*Cited on page 206.*)

655. A. van der Sluis, *General Orthogonal Polynomials*, Thesis, Wolters, Groningen, 1956.
(*Cited on page 71.*)

656. H.A. van der Vorst, Bi-CGSTAB: a fast and smoothly converging variant of Bi-CG for the solution of nonsymmetric linear systems, SIAM J. Sci. Stat. Comput., 13 (1992) 631–644.
(*Cited on page 193.*)

657. M. van Dyke, Extension of Goldstein's series for the Oseen drag of a sphere, J. Fluid Mech., 44 (1970) 365–372.
(Cited on page 180.)

658. J. van Iseghem, Vector Padé approximants, in *Numerical Mathematics and Applications*, R. Vichnevetsky and J. Vignes eds., North-Holland, Amsterdam, 1986, pp. 73–77.
(Cited on pages 143, 191, and 198.)

659. J. van Iseghem, An extended cross rule for vector Padé approximants, Appl. Numer. Math., 2 (1986) 143–155.
(Cited on page 199.)

660. J. van Iseghem, *Approximants de Padé Vectoriels*, Thèse de Doctorat d'État ès Sciences Mathématiques, Université des Sciences et Technologies de Lille, 1987.
(Cited on page 198.)

661. J. van Iseghem, Convergence of the vector QD-algorithm. Zeros of vector orthogonal polynomials, J. Comput. Appl. Math., 25 (1989) 33–46.
(Cited on page 199.)

662. H. van Rossum, *A Theory of Orthogonal Polynomials Based on the Padé Table*, Thesis, University of Utrecht, Van Gorcum, Assen, 1953.
(Cited on page 71.)

663. E.B. van Vleck, On the convergence of continued fractions with complex elements, Trans. Amer. Math. Soc., 2 (1901) 215–233.
(Cited on page 75.)

664. E.B. van Vleck, On the convergence of the continued fraction of Gauss and other continued fractions, Ann. Math., (2) 3 (1901–1902) 1–18.
(Cited on page 76.)

665. E.B. van Vleck, Selected topics in the theory of divergent series and continued fractions, Am. Math. Soc. Colloquium Publ., vol. I, Boston Colloquium, 1903, pp. 75–187.
(Cited on pages 72 and 340.)

666. E.B. van Vleck, On an extension of the 1894 memoir of Stieltjes, Trans. Amer. Math. Soc., 4 (1903) 297–332.
(Cited on page 76.)

667. A. van Wijngaarten, A transformation of formal series. I, Indag. Math., 56 (1953) 522–533; II, id., 534–543.
(Cited on page 50.)

668. A. van Wijngaarten, Process analyse, in *Cursus: Wetenschappelijk Rekenen B*, Stichting Mathematisch Centrum, Amsterdam, 1965, pp. 51–60.
(Cited on page 111.)

669. J.-M. vanden Broeck, L.W. Schwartz, A one-parameter family of sequence transformations, SIAM J. Math. Anal., 10 (1979) 658–666.
(Cited on page 181.)

670. V.S. Varadarajan, *Euler through Time: A New Look at Old Themes*, Amer. Math. Soc., Providence, 2006.
(Cited on page 50.)

671. V.S. Varadarajan, Euler and his work on infinite series, Bull. Amar. Math. Soc., 44 (2007) 515–539.
(Cited on page 50.)

672. R. Varadhan, Ch. Roland, Squared extrapolation methods (SQUAREM): A new class of simple and efficient numerical schemes for accelerating the convergence of the EM algorithm, Department of Biostatistics, Working Paper, Johns Hopkins University, 63 (2004) 1–70.
(Cited on page 195.)

673. R.S. Varga, On higher order stable imlicit methods for solving parabolic partial differential equations, J. Math. Phys. 40 (1961) 220–231.
(Cited on page 199.)

674. D. Vekemans, Algorithm for the E-prediction, J. Comput. Appl. Math., 85 (1997) 181–202.
(Cited on page 177.)

675. G. Viennot, A combinatorial theory for general orthogonal polynomials with extensions and applications, in *Polynômes Orthogonaux et Applications, Proceedings of the Laguerre Symposium held at Bar-le-Duc, October 15–18, 1984*, C. Brezinski, A. Draux, A.P. Magnus, P. Maroni, A. Ronveaux eds., Lecture Notes in Mathematics, vol. 1171, Springer, Berlin, Heidelberg, 1985, pp. 139–157.
(Cited on page 210.)

676. G. Viennot, A combinatorial interpretation of the quotient-difference algorithm, in *Formal Power Series and Algebraic Combinatorics*, D. Krob, A.A. Mikhalev, A.V. Mikhalev eds., Springer, Berlin, Heidelberg, 2000, pp. 379–390.
(Cited on page 210.)

677. H. von Koch, Sur un théorème de Stieltjes et sur les fractions continues, Bull. Soc. Math. France, 23 (1895) 33–40.
(Cited on page 76.)

678. H.F. Walker, P. Ni, Anderson acceleration for fixed-point iterations, SIAM Numer. Anal., 49 (2011) 1715–1735.
(Cited on page 196.)

679. H.S. Wall, *On the Padé Approximants Associated with the Continued Fraction and Series of Stieltjes*, PhD Thesis, University of Wisconsin, Madison, 1927.
(Cited on page 73.)

680. H.S. Wall, On the Padé approximants associated with the continued fraction and series of Stieltjes, Trans. Amer. Math. Soc., 31 (1929) 91–115.
(Cited on pages 73 and 76.)

681. H.S. Wall, On the Padé approximants associated with a positive definite power series, Trans. Am. Math. Soc., 33 (1931) 511–532.
(Cited on page 73.)

682. H.S. Wall, On the Padé table for a power series having a corresponding continued fraction in which the coefficients have limiting values, Bull. Am. Math. Soc., 38 (1932) 181.
(Cited on page 73.)

683. H.S. Wall, *Analytic Theory of Continued Fractions*, Van Nostrand, New York, 1948.
(Cited on pages 7 and 76.)

684. H. Wallin, The convergence of Padé approximants and the size of the power series coefficients, Appl. Anal., 4 (1974) 235–251.
(Cited on page 30.)

685. T.S. Walton, H. Polachek, Calculation of transient motion of submerged cables, Math. Comp., 14 (1960) 27–46.
(Cited on page 330.)

686. G. Walz, *Asymptotics and Extrapolation*, Akademie Verlag, Berlin, 1996.
(Cited on pages 57, 137, and 177.)

687. D.D. Warner, *Hermite Interpolation with Rational Functions*, Ph.D Thesis, University of California, San Diego, 1974.
(Cited on page 198.)

688. J.H.M. Wedderburn, On continued fractions in non-commutative quantities, Ann. Math., ser. 2, 15 (1913–1914) 101–105.
(Cited on pages 77 and 142.)

689. J.A.C. Weideman, Computing the dynamics of complex singularities of nonlinear PDEs, SIAM J. Appl. Dyn. Sys., 2 (2003) 171–186.
(Cited on page 197.)

690. K. Weierstrass, Neuer Beweis des Satzes, daß jede ganze rationale Funktion einer Veränderlichen dargestellt werden kann als ein Produkt aus linearen Funktionen derselben Veränderlichen, Sitzungsber. d. Königl.-Preuß Akad. der Wissensch. zu Berlin, Jahrg. 1891, 1085–1101; also in *Ges. Werke*, vol. 3, 1903, pp. 251–269.
(Cited on page 50.)

691. L. Weiss, R.N. McDonough, Prony's method, z-transform, and Padé approximation, SIAM Rev., 5 (1963) 145–149.
(Cited on pages 158 and 160.)

692. E.J. Weniger, Nonlinear sequence transformations for the acceleration of convergence and the summation of divergent series, Comput. Phys. Rep., 10 (1989) 189–371.
     (Cited on pages 7, 68, and 180.)
693. E.J. Weniger, On the derivation of iterated sequence transformations for the acceleration of convergence and the summation of divergent series, Comput. Phys. Comm., 64 (1991) 19–45.
     (Cited on page 179.)
694. J.M. Whittaker, M.S. Bartlett, Alexander Craig Aitken, 1895–1967, Biographical Memoirs of Fellows of the Royal Society, 14 (1968) 1–14.
     (Cited on page 313.)
695. D.V. Widder, Necessary and sufficient conditions for the representation of a function as a Laplace integral, Trans. Amer. Math. Soc., 33 (1931) 851–892.
     (Cited on page 162.)
696. E.P. Wigner, J. von Neumann, Significance of Loewner's theorem in the quantum theory of collisions, Ann. of Math., (2) 59 (1954) 418–433.
     (Cited on page 205.)
697. H.C. Williams, Daniel Shanks (1917–1996), Notices Amer. Math. Soc., 44 (1997) 813–816.
     (Cited on page 331.)
698. R. Wilson, Divergent continued fractions and polar singularities, Proc. Lond. Math. Soc., 26 (1927) 159–168; 27 (1928) 497–512; 28 (1928) 128–144.
     (Cited on page 73.)
699. J. Wimp, *Sequence Transformations and Their Applications*, Academic Press, New York, 1981.
     (Cited on pages 7 and 173.)
700. J. Wimp, *Computation with Recurrence Relations*, Pitman, Boston, 1984.
     (Cited on page 66.)
701. J. Worpitzky, Untersuchungen über die Entwickelung der monodromen und monogenen Funktionen durch Kettenbrüche, Friedrichs-Gymnasium und Realschule Jahresbericht, Berlin, 1865, pp. 3–39.
     (Cited on pages 75 and 201.)
702. L. Wuytack, A new technique for rational extrapolation to the limit, Numer. Math., 17 (1971) 215–221.
     (Cited on page 120.)
703. P.E. Zadunaisky, A method for the estimation of errors propagated in the numerical solution of a system of ordinary differential equations, in *Proc. Astron. Union, Symposium 25*, Academic Press, New York, 1966, pp. 281–287.
     (Cited on page 172.)
704. P.E. Zadunaisky, On the estimation of errors propagated in the numerical integration of ordinary differential equations, Numer. Math., 27 (1976) 21–39.
     (Cited on page 172.)
705. V.L. Zaguskin, *Handbook of Numerical Methods for the Solution of Algebraic and Transcendental Equations*, Pergamon Press, Oxford, 1961.
     (Cited on page 51.)
706. F.-Z. Zhang ed., *The Schur Complement and Its Applications*, Springer, New York, 2005.
     (Cited on pages 9 and 14.)
707. J. Zinn-Justin, Strong interactions dynamics with Padé approximants, Phys. Rept., 1 (1971) 55–102.
     (Cited on page 205.)
708. J. Zinn-Justin, *Quantum Field Theory and Critical Phenomena*, Clarendon Press, Oxford, 1989.
     (Cited on page 204.)
709. Z. Zlatev, I. Dimov, I. Faragó, I., A. Havasi, *Richardson Extrapolation. Practical Aspects and Applications*, De Gruyter, Berlin, Boston, 2017.
     (Cited on page 55.)
710. J. Zygarlicki, J. Mroczka, Prony's method used for testing harmonics and interharmonics in electrical power systems, Metrol. Meas. Syst., 19 (2012) 659–672.
     (Cited on page 161.)

References of Peter Wynn

711. P. Wynn, A note on Salzer's method for summing certain convergent series, J. Math. and Phys., 35 (1956) 318–320. https://doi.org/10.1002/sapm1956351318. MR 0086910. Submitted 19 July 1955.
(Cited on pages 108 and 239.)

712. P. Wynn, On a procrustean technique for the numerical transformation of slowly convergent sequences and series, Math. Proc. Cambridge Philos. Soc., 52 (1956) 663–671. https://doi.org/10.1017/S030500410003173X. MR 0081979. Submitted 31 October 1955.
(Cited on pages 47, 86, 95, 108, 111, and 115.)

713. P. Wynn, On a device for computing the $e_m(S_n)$ transformation, Math. Tables Aids Comput., 10 (1956) 91–96. https://doi.org/10.2307/2002183. MR 0084056.
(Cited on pages 8, 22, 62, 65, 86, and 341.)

714. P. Wynn, On a cubically convergent process for determining the zeros of certain functions, Math. Tables Aids Comput., 10 (1956) 97–100. https://doi.org/10.1090/s0025-5718-1956-0081547-9. MR 0081547.
(Cited on page 158.)

715. P. Wynn, Central difference and other forms of the Euler transformation, Quart. J. Mech. Appl. Math., 9 (1956) 249–256. https://doi.org/10.1093/qjmam/9.2.249. MR 0080782. Submitted 28 July 1955.
(Cited on pages 54 and 104.)

716. P. Wynn, On the propagation of error in certain non-linear algorithms, Numer. Math., 1 (1959) 142–149. https://doi.org/10.1007/BF01386380. MR 0107988. Submitted 26 February 1959.
(Cited on pages 90, 91, 92, and 213.)

717. P. Wynn, A sufficient condition for the instability of the q-d algorithm, Numer. Math., 1 (1959) 203–207. https://doi.org/10.1007/BF01386385. MR 0109426. Submitted 25 March 1959.
(Cited on page 91.)

718. P. Wynn, Converging factors for continued fractions, I, II, Numer. Math., 1 (1959) 272–307; 308–320. https://doi.org/10.1007/BF01386391, https://doi.org/10.1007/BF01386392. MR 0116158. Submitted 25 March 1959.
(Cited on pages 66, 138, and 139.)

719. P. Wynn, Über einen Interpolations-algorithmus und gewisse andere Formeln, die in der Theorie der Interpolation durch rationale Funktionen bestehen, Numer. Math., 2 (1960) 151–182. https://doi.org/10.1007/BF01386220. MR 0128597. Submitted 31 July 1959.
(Cited on pages 81, 119, and 121.)

720. P. Wynn, The rational approximation of functions which are formally defined by a power series expansion, Math. Comp., 14 (1960) 147–186. https://doi.org/10.2307/2003209. MR 0116457. Submitted 5 November 1959.
(Cited on pages 54, 69, 76, 146, 147, 149, and 200.)

721. P. Wynn, Confluent forms of certain non-linear algorithms, Arch. Math., 11 (1960) 223–236. https://doi.org/10.1007/BF01236936. MR 0128068. Submitted 5 October 1959.
(Cited on pages 25, 47, and 94.)

722. P. Wynn, A note on a confluent form of the ε-algorithm, Arch. Math., 11 (1960) 237–240. https://doi.org/10.1007/BF01236937. MR 0128069. Submitted 27 March 1959.
(Cited on page 96.)

723. P. Wynn, On the tabulation of indefinite integrals, BIT, 1 (1961) 286–290. https://doi.org/10.1007/BF01933245.
(Cited on pages 63 and 149.)

724. P. Wynn, L'ε-algoritmo e la tavola di Padé, Rend. Mat. Roma, (V) 20 (1961) 403–408. MR 0158206.
(Cited on pages 69, 76, 146, 147, and 148.)

725. P. Wynn, The epsilon algorithm and operational formulas of numerical analysis, Math. Comp., 15 (1961) 151–158. https://doi.org/10.2307/2004221. MR 0158513.
(Cited on page 87.)

726. P. Wynn, On repeated application of the epsilon algorithm, Revue française de traitement de l'information, Chiffres, 4 (1961) 19–22. MR 0149145.
(Cited on pages 62, 87, and 153.)

727. P. Wynn, A comparison between the numerical performances of the Euler transformation and the ε-algorithm, Revue française de traitement de l'information, Chiffres, 4 (1961) 23–29.
(Cited on pages 54 and 104.)

728. P. Wynn, The numerical transformation of slowly convergent series by methods of comparison, Part I, Revue française de traitement de l'information, Chiffres, 4 (1961) 177–210. MR 0162350.
(Cited on pages 105, 128, and 130.)

729. P. Wynn, A sufficient condition for the instability of the ε-algorithm, Nieuw Arch. Wiskd., 9 (1961) 117–119. MR 0139252.
(Cited on pages 91 and 92.)

730. P. Wynn, A note on a method of Bradshaw for transforming slowly convergent series and continued fractions, Amer. Math. Monthly, 69 (1962) 883–889. https://doi.org/10.2307/2311237. MR 0146559.
(Cited on pages 66 and 110.)

731. P. Wynn, Upon a second confluent form the ε-algorithm, Proc. Glasgow Math. Assoc., 5 (1962) 160–165. https://doi.org/10.1017/S2040618500034535. MR 0139253. Submitted 21 July 1961.
(Cited on pages 91 and 97.)

732. P. Wynn, Acceleration techniques for iterated vector and matrix problems, Math. Comp., 16 (1962) 301–322. https://doi.org/10.2307/2004051. MR 0145647.
(Cited on pages 100, 103, 182, and 196.)

733. P. Wynn, A comparison technique for the numerical transformation of slowly convergent series based on the use of rational functions, Numer. Math., 4 (1962) 8–14. https://doi.org/10.1007/BF01386291. MR 0136500. Submitted 17 July 1961.
(Cited on page 128.)

734. P. Wynn, Numerical efficiency profile functions, Koninkl. Nederl. Akad. Wet., 65A (1962) 118–126. https://doi.org/10.1016/S1385-7258(62)50011-5. Submitted 30 September 1961.
(Cited on pages 150 and 151.)

735. P. Wynn, The numerical efficiency of certain continued fraction expansions, IA, IB, Koninkl. Nederl. Akad. Wet., 65A (1962) 127–137; 138–148. https://doi.org/10.1016/S1385-7258(62)50012-7, https://doi.org/10.1016/S1385-7258(62)50013-9. Submitted 30 September 1961.
(Cited on pages 54, 150, and 151.)

736. P. Wynn, On a connection between two techniques for the numerical transformation of slowly convergent series, Koninkl. Nederl. Akad. Weten., 65A (1962) 149–154. Submitted 30 September 1961.
(Cited on pages 54, 66, and 110.)

737. P. Wynn, Una nota su un analogo infinitesimale del q-d algoritmo, Rend. Mat. Roma, 21 (1962) 77–85. Submitted 11 September 1961.
(Cited on page 96.)

738. P. Wynn, A note on fitting certain types of experimental data, Stat. Neerl., 16 (1962) 145–150. https://doi.org/10.1111/j.1467-9574.1962.tb01061.x
(Cited on page 158.)

739. P. Wynn, Note on the solution of a certain boundary-value problem, BIT, 2 (1962) 61–64. https://doi.org/10.1007/BF02024783. MR 0155445.
(Cited on page 161.)

740. P. Wynn, An arsenal of Algol procedures for complex arithmetic, BIT, 2 (1962) 232–255. https://doi.org/10.1007/BF01940171. MR 0166945.
(Cited on page 152.)

741. P. Wynn, The numerical transformation of slowly convergent series by methods of comparison. Part II, Revue française de traitement de l'information, Chiffres, 5 (1962) 65–88. MR 0149146.
(Cited on pages 105 and 106.)

742. P. Wynn, Acceleration technique in numerical analysis with particular reference to problems in one independent variable, in *Information Processing 1962, Proc. IFIP Congress 62, Munich, 27 August–1 September 1962*, C.M. Popplewell ed., North-Holland, Amsterdam, 1963, pp. 149–156.
(Cited on pages 114 and 152.)

743. P. Wynn, Singular rules for certain non-linear algorithms, BIT, 3 (1963) 175–195; also Report MR 59, Mathematisch Centrum, Amsterdam, 1963. https://doi.org/10.1007/BF01939985. MR 0166946.
(Cited on pages 90, 92, 154, 184, and 213.)

744. P. Wynn, Note on a converging factor for a certain continued fraction, Numer. Math., 5 (1963) 332–352. https://doi.org/10.1007/BF01385901. Submitted 5 February 1963.
(Cited on page 139.)

745. P. Wynn, On a connection between the first and the second confluent forms of the ε-algorithm, Niew. Arch. Wisk., 11 (1963) 19–21. Submitted 29 October 1962.
(Cited on page 97.)

746. P. Wynn, Continued fractions whose coefficients obey a non-commutative law of multiplication, Arch. Rat. Mech. Anal., 12 (1963) 273–312. https://doi.org/10.1007/BF00281229. Submitted 13 August 1962.
(Cited on pages 103, 116, 142, 143, 144, 181, 201, and 202.)

747. P. Wynn, A numerical study of a result of Stieltjes, Revue française de traitement de l'information, Chiffres, 6 (1963) 175–196.
(Cited on pages 76 and 162.)

748. P. Wynn, Converging factors for the Weber parabolic cylinder function of complex argument, Proc. Kon. Nederl. Akad. Weten., IA, IB, 66 (1963) 721–736; 737–754. Submitted 29 June 1963.
(Cited on page 140.)

749. P. Wynn, Partial differential equations associated with certain non-linear algorithms, Z. Angew. Math. Phys., 15 (1964) 273–289. https://doi.org/10.1007/BF01607018. MR 0166944. Submitted 1 September 1963.
(Cited on pages 63, 126, and 213.)

750. P. Wynn, General purpose vector epsilon-algorithm Algol procedures, Numer. Math., 6 (1964) 22–36. https://doi.org/10.1007/BF01386050. MR 0166947. Submitted 12 July 1963.
(Cited on pages 152 and 153.)

751. P. Wynn, On some recent developments in the theory and application of continued fractions, SIAM J. Numer. Anal., Ser. B, 1 (1964) 177–197. https://doi.org/10.1137/0701015. Submitted 4 November 1963.
(Cited on pages 130, 144, 201, and 202.)

752. P. Wynn, Four lectures on the numerical application of continued fractions, in *Alcune Questioni di Analisi Numerica*, A. Ghizzetti ed., Series: C.I.M.E. Summer Schools, vol. 35, Springer, Heidelberg, 1965, pp. 111–251. https://doi.org/10.1007/978-3-642-11027-6_2
(Cited on page 132.)

753. P. Wynn, A note on programming repeated application of the epsilon-algorithm, Revue française de traitement de l'information, Chiffres, 8 (1965) 23–62; Errata, 156. MR 0181081.
(Cited on pages 62, 111, 153, and 170.)

754. P. Wynn, Upon systems of recursions which obtain among the quotients of the Padé table, Numer. Math., 8 (1966) 264–269. https://doi.org/10.1007/BF02162562. Submitted 5 May 1965.
(Cited on pages 89, 181, and 209.)

755. P. Wynn, On the convergence and stability of the epsilon algorithm, SIAM J. Numer. Anal., 3 (1966) 91–122. https://doi.org/10.1137/0703007. Submitted 16 September 1965.
(Cited on pages 93 and 111.)

756. P. Wynn, Upon a conjecture concerning a method for solving linear equations, and certain other matters, MRC Technical Summary Report 626, University of Wisconsin, Madison, April 1966.
(Cited on pages 116 and 194.)

757. P. Wynn, Complex numbers and other extensions to the Clifford algebra with an application to the theory of continued fractions, MRC Technical Summary Report 646, University of Wisconsin, Madison, May 1966.
(Cited on pages 143, 144, and 201.)

758. P. Wynn, Upon the diagonal sequences of the Padé table, MRC Technical Summary Report 660, University of Wisconsin, Madison, May 1966.
(Cited on page 121.)

759. P. Wynn, Upon an invariant associated with the epsilon algorithm, MRC Technical Summary Report 675, University of Wisconsin, Madison, July 1966.
(Cited on pages 90 and 97.)

760. P. Wynn, On the computation of certain functions of large argument and parameter, BIT, 6 (1966) 228–259. https://doi.org/10.1007/BF01934356.
(Cited on page 141.)

761. P. Wynn, An arsenal of Algol procedures for the evaluation of continued fractions and for effecting the epsilon algorithm, Revue française de traitement de l'information, Chiffres, 9 (1966) 327–362.
(Cited on page 154.)

762. P. Wynn, Accelerating the convergence of a monotonic sequence by a method of intercalation, MRC Technical Summary Report 674, University of Wisconsin, Madison, January 1967.
(Cited on pages 111 and 114.)

763. P. Wynn, A general system of orthogonal polynomials, Quart. J. Math. Oxford (2), 18 (1967) 81–96. https://doi.org/10.1093/qmath/18.1.81. Submitted 8 September 1966.
(Cited on pages 69 and 148.)

764. P. Wynn, Transformations to accelerate the convergence of Fourier series, in *Gertrude Blanch Anniversary Volume*, B. Mond, G. Blanch eds., Wright Patterson Air Force Base, 1967, pp. 339–379; also as MRC Technical Summary Report 673, University of Wisconsin, Madison, July 1966. MR 0215553.
(Cited on pages 116 and 170.)

765. P. Wynn, A note on the convergence of certain noncommutative continued fractions, MRC Technical Summary Report 750, University of Wisconsin, Madison, May 1967.
(Cited on pages 144, 201, and 202.)

766. P. Wynn, Upon the Padé table derived from a Stieltjes series, SIAM J. Numer. Anal., 5 (1968) 805–834. https://doi.org/10.1137/0705060. MR 0239734. Submitted 22 March 1968, revised 5 July 1968.
(Cited on pages 76 and 122.)

767. P. Wynn, Vector continued fractions, Linear Algebra Appl., 1 (1968) 357–395. https://doi.org/10.1016/0024-3795(68)90015-3. Submitted 5 March 1968.
(Cited on pages 144, 201, and 202.)

768. P. Wynn, Upon the definition of an integral as the limit of a continued fraction, Arch. Rat. Mech. Anal., 28 (1968) 83–148. https://doi.org/10.1007/BF00283861. Submitted 24 May 1967.
(Cited on pages 63, 76, 125, 132, and 165.)

769. P. Wynn, Zur Theorie der mit gewissen speziellen Funktionen verknüpften Padéschen Tafeln, Math. Z., 109 (1969) 66–70. https://doi.org/10.1007/BF01135574. Submitted 17 April 1968.
(Cited on page 122.)

770. P. Wynn, Upon a recursive system of flexible rings permitting involution, Report CRM-50, Centre de Recherches Mathématiques, Université de Montréal, Montréal, November 1970.
(Cited on page 155.)

771. P. Wynn, Upon the inverse of formal power series over certain algebra, Report CRM-53, Centre de Recherches Mathématiques, Université de Montréal, Montréal, November 1970.
(Cited on page 154.)

772. P. Wynn, Upon a hierarchy of epsilon arrays, Technical Report 46, Louisiana State University, New Orleans, October 1970.
(Cited on pages 112 and 113.)

773. P. Wynn, A note on the generalised Euler transformation, The Computer Journal, 14 (1971) 437–441; Errata 15 (1972) 175. https://doi.org/10.1093/comjnl/14.4.437.
     (Cited on pages 54, 106, and 170.)

774. P. Wynn, The abstract theory of the epsilon algorithm, Report CRM-74, Centre de Recherches Mathématiques, Université de Montréal, Montréal, February 1971.
     (Cited on pages 103, 113, and 181.)

775. P. Wynn, Upon a class of functions connected with the approximate solution of operator equations, Report CRM-103, Centre de Recherches Mathématiques, Université de Montréal, Montréal, June 1971, presented at the *Workshop on Padé approximants, Marseille, 1975.*
     (Cited on pages 163 and 164.)

776. P. Wynn, A note upon totally monotone sequences, Report CRM-139, Centre de Recherches Mathématiques, Université de Montréal, Montréal, November 1971.
     (Cited on pages 93 and 94.)

777. P. Wynn, A transformation of series, Calcolo, 8 (1971) 255–272. https://doi.org/10.1007/BF02575517. Submitted 1 September 1971.
     (Cited on pages 54 and 107.)

778. P. Wynn, Difference-differential recursions for Padé quotients, Proc. London Math. Soc., S. 3, 23 (1971) 283–300. https://doi.org/10.1112/plms/s3-23.2.283. Submitted 4 May 1970.
     (Cited on page 121.)

779. P. Wynn, Upon the generalized inverse of a formal power series with vector valued coefficients, Compo. Math. 23 (1971) 453–460. http://www.numdam.org/item/CM_1971__23_4_453_0. MR 306224. Submitted 13 January 1971.
     (Cited on page 156.)

780. P. Wynn, Über orthonormale Polynome und ein assoziiertes Momentproblem, Math. Scand., 29 (1971) 104–112. Submitted 27 April 1971.
     (Cited on pages 76 and 149.)

781. P. Wynn, On an extension of a result due to Pólya, J. Reine Angew. Math., 248 (1971) 127–132. https://doi.org/10.1515/crll.1971.248.127. Submitted 22 November 1969.
     (Cited on page 162.)

782. P. Wynn, Convergence acceleration by a method of intercalation, Computing, 9 (1972) 267–273. https://doi.org/10.1007/BF02241602. Submitted 6 August 1971.
     (Cited on pages 62, 87, and 114.)

783. P. Wynn, Invariants associated with the epsilon algorithm and its first confluent form, Rend. Circ. Mat. Palermo, 21 (1972) 31–41. https://doi.org/10.1007/BF02844229. Submitted January 1972.
     (Cited on page 97.)

784. P. Wynn, Hierarchies of arrays and function sequences associated with the epsilon algorithm and its first confluent form, Rend. Mat. Roma, 5, Serie VI (1972) 819–852. Submitted 15 May 1972.
     (Cited on page 113.)

785. P. Wynn, A note on a partial differential equation, Report CRM-22, Centre de Recherches Mathématiques, Université de Montréal, Montréal, 1972.
     (Cited on pages 63 and 127.)

786. P. Wynn, Sur les suites totalement monotones, C.R. Acad. Sci. Paris, 275 A (1972) 1065–1068. MR 0310480. Accepted 6 November 1972.
     (Cited on pages 93 and 94.)

787. P. Wynn, Transformation de séries à l'aide de l'ε-algorithm, C.R. Acad. Sci. Paris, 275 A (1972) 1351–1353. MR 0311068. Accepted 18 December 1972.
     (Cited on pages 93, 94, 114, and 170.)

788. P. Wynn, Upon a convergence result in the theory of the Padé table, Trans. Amer. Math. Soc., 165 (1972) 239–249. https://doi.org/10.2307/1995884. Received 26 October 1970, revised 21 May 1971.
     (Cited on pages 76 and 123.)

789. P. Wynn, A convergence theory of some methods of integration, Report CRM-193, Centre de Recherches Mathématiques, Université de Montréal, Montréal, May 1972.
(Cited on pages 63 and 99.)

790. P. Wynn, The partial differential equation of the Padé surface, Report CRM-197, Centre de Recherches Mathématiques, Université de Montréal, Montréal, June 1972.
(Cited on pages 63, 127, and 128.)

791. P. Wynn, The algebra of certain formal power series, Report CRM-216, Centre de Recherches Mathématiques, Université de Montréal, Montréal, September 1972.
(Cited on page 156.)

792. P. Wynn, On some extensions of Euclid's algorithm, and some consequences thereof, Report CRM, Centre de Recherches Mathématiques, Université de Montréal, Montréal, 1972.
(Cited on page 85.)

793. P. Wynn, Upon some continuous prediction algorithms. I, II, Calcolo, 9 (1973) 197–234; 235–278. https://doi.org/10.1007/BF02576490, https://doi.org/10.1007/BF02575582. Submitted 20 June 1972.
(Cited on pages 98, 99, 100, 125, 165, and 239.)

794. P. Wynn, Distributive rings permitting involution, Report CRM-281, Centre de Recherches Mathématiques, Université de Montréal, Montréal, 1973.
(Cited on page 157.)

795. P. Wynn, On the zeros of certain confluent hypergeometric functions, Proc. Amer. Math. Soc., 40 (1973) 173–183. https://doi.org/10.2307/2038658. Submitted 7 July 1972, revised 26 October 1972.
(Cited on page 151.)

796. P. Wynn, Accélération de la convergence de séries d'opérateurs en analyse numérique, C.R. Acad. Sci. Paris, 276 A (1973) 803–806. MR 0317519. Accepted 12 March 1973.
(Cited on pages 93, 94, and 114.)

797. P. Wynn, On the intersection of two classes of functions, Rev. Roumaine Math. Pures Appl., 19 (1974) 949–959.
(Cited on page 164.)

798. P. Wynn, Extremal properties of Padé quotients, Acta Math. Hungar., 25 (1974) 291–298. https://doi.org/10.1007/BF01886088. Submitted 14 July 1972.
(Cited on page 124.)

799. P. Wynn, Sur l'équation aux dérivées partielles de la surface de Padé, C.R. Acad. Sci. Paris, 278 A (1974) 847–850. MR 0341910.
(Cited on pages 63, 127, and 128.)

800. P. Wynn, A numerical method for estimating parameters in mathematical models, Report CRM-443, Centre de Recherches Mathématiques, Université de Montréal, Montréal, August 1974.
(Cited on pages 118 and 196.)

801. P. Wynn, Some recent developments in the theories of continued fractions and the Padé table, Rocky Mountain J. Math., 4 (1974) 297–324. https://doi.org/10.1216/RMJ-1974-4-2-297. Submitted 8 February 1973.
(Cited on pages 63 and 123.)

802. P. Wynn, How to integrate without integrating, Colloque Euromech 58, Toulon, 1975, unpublished.
(Cited on page 63.)

803. P. Wynn, Upon a class of functions connected with the approximate solution of operator equations, Ann. Mat. Pura Appl., 104 (1975) 1–29. https://doi.org/10.1007/BF02417008. Submitted 10 October 1972.
(Cited on pages 163 and 164.)

804. P. Wynn, Five lectures on the numerical application of continued fractions, Orientation Lecture Series 5, MRC, University of Wisconsin, Madison.
(Cited on page 132.)

805. P. Wynn, The algebra of certain formal power series, Riv. Mat. Uni. Parma, (4) 2 (1976) 155–176. http://www.rivmat.unipr.it/vols/1976-2/indice76.html. MR 0447220. Submitted 28 August 1974.
(Cited on pages 156 and 157.)

806. P. Wynn, An array of functions, Report, School of Computer Science, McGill University, Montreal, 1976.
(Cited on pages 85 and 125.)

807. P. Wynn, A continued fraction transformation of the Euler-MacLaurin series, Report, School of Computer Science, McGill University, Montreal, 1976.
(Cited on page 85.)

808. P. Wynn, A convergence theory of some methods of integration, J. Reine Angew. Math., 285 (1976) 181–208. https://doi.org/10.1515/crll.1976.285.181. MR 0415119. Submitted 22 March 1974.
(Cited on pages 63, 99, 125, and 165.)

809. P. Wynn, The calculus of finite differences over certain systems of numbers, Calcolo, 14 (1977) 303–341. https://doi.org/10.1007/BF02575990. MR 0503568. Submitted 30 August 1976.
(Cited on page 165.)

810. P. Wynn, The transformation of series by the use of Padé quotients and more general approximants, in *Padé and Rational Approximation. Theory and Applications*, E.B. Saff and R.S. Varga eds., Academic Press, New York, 1977, pp. 121–144. https://doi.org/10.1016/B978-0-12-614150-4.50015-0
(Cited on pages 76 and 124.)

811. P. Wynn, The evaluation of singular and highly oscillatory integrals by use of the anti-derivative, Calcolo, 15, Fasc. IV bis, (1978) 1–103; also as Report, School of Computer Science, McGill University, Montreal, 1976. Submitted 17 July 1977.
(Cited on pages 63, 125, and 164.)

812. P. Wynn, The work of E.B. Christoffel on the theory of continued fractions, in *E.B. Christoffel: The Influence of His Work on Mathematics and the Physical Sciences*, P.L. Butzer, F. Fehér eds., Birkhäuser Verlag, Basel, 1981, pp. 190–202. https://doi.org/10.1007/978-3-0348-5452-8_11. MR 0661065. Submitted 9 October 1979.
(Cited on pages 125 and 135.)

813. P. Wynn, Remark upon developments in the theories of the moment problem and of quadrature, subsequent to the work of Christoffel, in *E.B. Christoffel: The Influence of His Work on Mathematics and the Physical Sciences*, P.L. Butzer, F. Fehér eds., Birkhäuser Verlag, Basel, 1981, pp. 731–734. https://doi.org/10.1007/978-3-0348-5452-8_60. Submitted 28 April 1980.
(Cited on pages 125, 135, and 137.)

814. P. Wynn, The convergence of approximating fractions, Bol. Soc. Mat. Mexicana, 26 (1981) 57–71. MR 0742016.
(Cited on page 125.)

Translations

815. A.Ya. Khintchin, *Continued Fractions*, Translated from Russian by Peter Wynn, P. Noordhoff N.V., Groningen, 1963.
(Cited on page 166.)

816. A.N. Khovanskii, *The Application of Continued Fractions and Their Generalizations to Problems in Approximation Theory*, Translated from Russian by Peter Wynn, P. Noordhoff N.V., Groningen, 1963.
(Cited on pages 129 and 166.)

Index

In this index, all the occurrences of some keywords and authors' names that appear too often in the text have not been included. We have inserted only the most important ones. Keywords are in roman characters, names are in italics.

Symbols

$1/d$-orthogonality, 191, 199
A-acceptability, 198
E-algorithm, 176
G-transformation, 172
H-algorithm, 196
$S\beta$-algorithm, 185
Δ^2 process, 16
η-algorithm, 90, 95
ω-algorithm, 98
ε-algorithm, 22, 85, 180
  abstract theory, 103
  confluent, 181
  confluent form, 25, 94–98, 112, 133, 134
  matrix, 100, 116, 143
  repeated application, 87, 153
  simplified topological, 184
  topological, 183
  vector, 100, 143, 145, 194
ε-array, 23, 62, 86
  hierarchy, 112, 113
ϱ-algorithm, 47, 90, 95, 108
  confluent form, 47, 95, 99

d-orthogonality, 199
g-algorithm, 90, 95
h^2-extrapolation, 54, 170
n-body problem, 203
qd table, 45
qd-algorithm, 42, 45, 46, 58, 91, 148, 149, 214, 328
  confluent form, 94, 96
z-transform, 158

A

Abramowitz, Milton, (1915–1958), 149
abstract algebra, 154
accuracy control, 150
accuracy–through–order conditions, 26, 198
Achyeser, Naoum, (1901–1980), 73
Airey, John Robinson, (1868–1937), 140, 141
Aitken
  process, 4, 16, 57, 194
  scheme, 120

© Springer Nature Switzerland AG 2020
C. Brezinski, M. Redivo-Zaglia, *Extrapolation and Rational Approximation,*
https://doi.org/10.1007/978-3-030-58418-4

Printed in the United States
by Baker & Taylor Publisher Services